Materials Science of Novel Oxide-Based Electronics

MATERIALS RESEARCH SOCIETY
SYMPOSIUM PROCEEDINGS VOLUME 623

Materials Science of Novel Oxide-Based Electronics

Symposium held April 24–27, 2000, San Francisco, California, U.S.A.

EDITORS:

David S. Ginley
National Renewable Energy Laboratory
Golden, Colorado, U.S.A.

John D. Perkins
National Renewable Energy Laboratory
Golden, Colorado, U.S.A.

Hiroshi Kawazoe
Tokyo Institute of Technology
Yokohama, Japan

Dennis M. Newns
IBM T.J. Watson Research Center
Yorktown Heights, New York, U.S.A.

Andrey B. Kozyrev
Electrotechnical University
St. Petersburg, Russia

Materials Research Society
Warrendale, Pennsylvania

CAMBRIDGE UNIVERSITY PRESS
Cambridge, New York, Melbourne, Madrid, Cape Town,
Singapore, São Paulo, Delhi, Mexico City

Cambridge University Press
32 Avenue of the Americas, New York NY 10013-2473, USA

Published in the United States of America by Cambridge University Press, New York

www.cambridge.org
Information on this title: www.cambridge.org/9781107413047

Materials Research Society
506 Keystone Drive, Warrendale, PA 15086
http://www.mrs.org

First published 2000
First paperback edition 2013

Single article reprints from this publication are available through
University Microfilms Inc., 300 North Zeeb Road, Ann Arbor, MI 48106

CODEN: MRSPDH

ISBN 978-1-107-41304-7 Paperback

This work was supported in part by the Office of Naval Research under Grant Number ONR:
N00014-99-1-1048. The United States Government has a royalty-free license throughout the
world in all copyrightable material contained herein.

CONTENTS

*Invited Paper

NEW IDEAS AND MAGNETISM

FERROELECTRICS AND RELATED MATERIALS

*Invited Paper

TRANSPARENT CONDUCTORS

*Invited Paper

FILM DEPOSITION METHODS

*Invited Paper

PREFACE

Oxide materials are emerging as potential candidates for a variety of existing and new opto-electronic and microwave applications. Critical to enabling this development is the need to have high-quality materials and to be able to construct viable heterostructures. Considerable progress has been made in the last few years on the fabrication and processing of a diverse variety of high-quality, metal oxide films for application as active components in a new generation of electronic devices. The existence of improved films has subsequently enabled a deeper understanding of both the oxides' intrinsic properties and their potential application in real world electronics. Because of their special properties or combinations thereof, oxide films are leading to new classes of devices as well as potentially replacing some common semiconductor devices. This symposium focused on the materials growth, characterization, processing, and application of oxide films employed as the active elements in devices.

The symposium "Materials Science of Novel Oxide-Based Electronics" was held April 24–27, 2000 at the 2000 MRS Spring Meeting in San Francisco, California. There was a diverse collection of just over 100 papers, which dealt with a broad spectrum from new device designs and theory to materials deposition. The symposium and proceedings volume are organized into topical areas:

▼ Applications
▼ New Ideas and Magnetism
▼ Ferroelectrics and Related Materials
▼ Transparent Conductors
▼ Film Deposition Methods

There were significant new developments reported in each of these areas. This has lead to the repetition of this general symposium topic at the 20001 MRS Spring Meeting. It is an exciting time for the emerging field of oxide-based opto-electronics and microwaves. A new understanding of basic materials properties is being combined with the development of novel heterostructures to lead to whole new classes of devices potentially affecting everything from energy conversion to computer memory.

David S. Ginley
John D. Perkins
Hiroshi Kawazoe
Dennis M. Newns
Andrey B. Kozyrev

August 2000

ACKNOWLEDGMENTS

We wish to thank all of the speakers, authors, and referees for their contributions to the success of the symposium and of these proceedings.

It is our pleasure to acknowledge with gratitude the support provided for the symposium by: National Renewable Energy Laboratory, IBM T.J. Watson Research Center, Office of Naval Research, and the Materials Research Society. Without this help, we could not have successfully organized the symposium or produced the proceedings volume.

We gratefully acknowledge the invaluable editorial and organizational assistance from Mrs. Carole Allman of the National Renewable Energy Laboratory in Golden, Colorado, for her indispensable assistance in the organization of the symposium, and in her diligence and care in assembling the proceedings volume.

MATERIALS RESEARCH SOCIETY SYMPOSIUM PROCEEDINGS

Volume 578— Multiscale Phenomena in Materials—Experiments and Modeling, I.M. Robertson, D.H. Lassila, R. Phillips, B. Devincre, 2000, ISBN: 1-55899-486-6

Volume 579— The Optical Properties of Materials, J.R. Chelikowsky, S.G. Louie, G. Martinez, E.L. Shirley, 2000, ISBN: 1-55899-487-4

Volume 580— Nucleation and Growth Processes in Materials, A. Gonis, P.E.A. Turchi, A.J. Ardell, 2000, ISBN: 1-55899-488-2

Volume 581— Nanophase and Nanocomposite Materials III, S. Komarneni, J.C. Parker, H. Hahn, 2000, ISBN: 1-55899-489-0

Volume 582— Molecular Electronics, S.T. Pantelides, M.A. Reed, J. Murday, A. Aviram, 2000, ISBN: 1-55899-490-4

Volume 583— Self-Organized Processes in Semiconductor Alloys, A. Mascarenhas, D. Follstaedt, T. Suzuki, B. Joyce, 2000, ISBN: 1-55899-491-2

Volume 584— Materials Issues and Modeling for Device Nanofabrication, L. Merhari, L.T. Wille, K.E. Gonsalves, M.F. Gyure, S. Matsui, L.J. Whitman, 2000, ISBN: 1-55899-492-0

Volume 585— Fundamental Mechanisms of Low-Energy-Beam-Modified Surface Growth and Processing, S. Moss, E.H. Chason, B.H. Cooper, T. Diaz de la Rubia, J.M.E. Harper, R. Murti, 2000, ISBN: 1-55899-493-9

Volume 586— Interfacial Engineering for Optimized Properties II, C.B. Carter, E.L. Hall, S.R. Nutt, C.L. Briant, 2000, ISBN: 1-55899-494-7

Volume 587— Substrate Engineering—Paving the Way to Epitaxy, D. Norton, D. Schlom, N. Newman, D. Matthiesen, 2000, ISBN: 1-55899-495-5

Volume 588— Optical Microstructural Characterization of Semiconductors, M.S. Unlu, J. Piqueras, N.M. Kalkhoran, T. Sekiguchi, 2000, ISBN: 1-55899-496-3

Volume 589— Advances in Materials Problem Solving with the Electron Microscope, J. Bentley, U. Dahmen, C. Allen, I. Petrov, 2000, ISBN: 1-55899-497-1

Volume 590— Applications of Synchrotron Radiation Techniques to Materials Science V, S.R. Stock, S.M. Mini, D.L. Perry, 2000, ISBN: 1-55899-498-X

Volume 591— Nondestructive Methods for Materials Characterization, G.Y. Baaklini, N. Meyendorf, T.E. Matikas, R.S. Gilmore, 2000, ISBN: 1-55899-499-8

Volume 592— Structure and Electronic Properties of Ultrathin Dielectric Films on Silicon and Related Structures, D.A. Buchanan, A.H. Edwards, H.J. von Bardeleben, T. Hattori, 2000, ISBN: 1-55899-500-5

Volume 593— Amorphous and Nanostructured Carbon, J.P. Sullivan, J. Robertson, O. Zhou, T.B. Allen, B.F. Coll, 2000, ISBN: 1-55899-501-3

Volume 594— Thin Films—Stresses and Mechanical Properties VIII, R. Vinci, O. Kraft, N. Moody, P. Besser, E. Shaffer II, 2000, ISBN: 1-55899-502-1

Volume 595— GaN and Related Alloys—1999, T.H. Myers, R.M. Feenstra, M.S. Shur, H. Amano, 2000, ISBN: 1-55899-503-X

Volume 596— Ferroelectric Thin Films VIII, R.W. Schwartz, P.C. McIntyre, Y. Miyasaka, S.R. Summerfelt, D. Wouters, 2000, ISBN: 1-55899-504-8

Volume 597— Thin Films for Optical Waveguide Devices and Materials for Optical Limiting, K. Nashimoto, R. Pachter, B.W. Wessels, J. Shmulovich, A.K.-Y. Jen, K. Lewis, R. Sutherland, J.W. Perry, 2000, ISBN: 1-55899-505-6

Volume 598— Electrical, Optical, and Magnetic Properties of Organic Solid-State Materials V, S. Ermer, J.R. Reynolds, J.W. Perry, A.K.-Y. Jen, Z. Bao, 2000, ISBN: 1-55899-506-4

Volume 599— Mineralization in Natural and Synthetic Biomaterials, P. Li, P. Calvert, T. Kokubo, R.J. Levy, C. Scheid, 2000, ISBN: 1-55899-507-2

Volume 600— Electroactive Polymers (EAP), Q.M. Zhang, T. Furukawa, Y. Bar-Cohen, J. Scheinbeim, 2000, ISBN: 1-55899-508-0

Volume 601— Superplasticity—Current Status and Future Potential, P.B. Berbon, M.Z. Berbon, T. Sakuma, T.G. Langdon, 2000, ISBN: 1-55899-509-9

Volume 602— Magnetoresistive Oxides and Related Materials, M. Rzchowski, M. Kawasaki, A.J. Millis, M. Rajeswari, S. von Molnár, 2000, ISBN: 1-55899-510-2

Volume 603— Materials Issues for Tunable RF and Microwave Devices, Q. Jia, F.A. Miranda, D.E. Oates, X. Xi, 2000, ISBN: 1-55899-511-0

Volume 604— Materials for Smart Systems III, M. Wun-Fogle, K. Uchino, Y. Ito, R. Gotthardt, 2000, ISBN: 1-55899-512-9

MATERIALS RESEARCH SOCIETY SYMPOSIUM PROCEEDINGS

Prior Materials Research Society Symposium Proceedings available by contacting Materials Research Society

Applications

BURIED OXIDE CHANNEL FIELD EFFECT TRANSISTOR

J.A. MISEWICH and A.G. SCHROTT
IBM Research, T.J. Watson Research Center, Yorktown Heights, NY 10598

ABSTRACT

A room temperature oxide channel field effect transistor with the channel on the surface was recently demonstrated at IBM which showed switching characteristics similar to conventional silicon FETs. In this paper we introduce a new architecture for the oxide channel transistor where the oxide channel material is buried below the gate oxide layer. This buried channel architecture has several significant advantages over the surface channel design in coupling capacitance, channel mobility, and channel stability. We will discuss the design and fabrication of the buried channel oxide FET and we will present results from these devices which demonstrate a higher transconductance.

INTRODUCTION

The drive toward higher density semiconductor integrated circuits has been fueled for decades by the scalability of silicon MOSFET technology. Unfortunately there are fundamental physical effects which might limit the scaling of this technology beyond 30 nm.[1] Driven by the search for a potentially scalable technology, a novel field effect transistor (FET) has recently been proposed [2-5] which utilizes a material capable of undergoing the Mott metal-insulator transition [6-8] as the channel material. Such an FET is similar to a conventional silicon MOSFET in that there are source and drain electrodes on either end of a channel and a gate oxide and gate electrode which produce a field terminating in the channel. The channel, however, consists of a material capable of undergoing the Mott metal-insulator transition rather than a semiconductor. A Mott insulator is a material in which the electrons are localized as a result of the Coulomb interaction between electrons rather than due to the ionic potential as in most insulators. In our devices the transition between insulating and metallic states of the channel is induced by the gate field leading to strong switching characteristics. One promising class of oxide materials for the channel is the cuprate family of perovskite structure materials related to high temperature superconductors. Due to the widespread interest in high temperature superconductivity the cuprates have been extensively characterized and a substantial materials knowledge base exists for these materials. One of the challenges of such devices is that the surface charge density the gate field must produce is estimated [2,3] to be ~10^{14} carriers/cm^2, requiring the use of gate oxides with high dielectric constant and high breakdown field. Again the cuprates are attractive since they offer epitaxial compatibility with gate materials such as strontium titatate (STO) and barium strontium titanate (BST) which have been demonstrated to have a high dielectric constant and high dielectric breakdown field [9-11] capable of producing the required surface charge density. Despite the challenges, a buried oxide-channel field effect transistor (OxFET) is of interest because of the potential to scale such a device beyond the silicon scaling limits due to the absence of impurity doping in these materials and since the charge separation layer at the source and drain contacts is expected to be on the ~1 Angstrom scale rather than ~100 Angstrom.[3]

We would like to make an important distinction. At first glance, the structures we fabricate resemble those of the superconducting FET structures proposed and built by several

Mat. Res. Soc. Symp. Proc. Vol. 623 © 2000 Materials Research Society

groups [12-15]. The structures are almost identical in most cases, however, the details differ as do the experimental conditions and the goals. The most important contrast is that all the data shown in this paper and in references [2,4,5] are obtained in *room temperature* operation. In the superconducting FET proposals, the goal was to induce (or inhibit) superconductivity in a channel by modulating the carrier concentration with a gate potential. Our work concentrates exclusively on the normal state properties of the cuprate films.

DEVICE ARCHITECTURE

An OxFET with the channel material on the surface of the device was recently demonstrated at IBM in which the channel material was $Y_{1-x}Pr_xBa_2Cu_3O_{7-\delta}$ (YPBCO).[2] The devices which were demonstrated had a simple architecture illustrated in figure 1 which was chosen for ease of fabrication. In this structure all the materials fabrication is done via pulsed laser deposition (PLD) directly on an unpatterned substrate and the source and drain contacts are directly deposited on the films through a stencil mask. The substrate in this case is niobium doped strontium titanate (Nb:STO) which is a conducting material. The substrate therefore also acted as the gate electrode. The gate dielectric (STO) is deposited on the substrate first, and then the channel material (YPBCO) was deposited. The surface channel FET had characteristic curves similar to conventional silicon MOSFET curves with a transconductance of ~ 2 μS at a drain voltage of 1 volt and a gate voltage of 10 volts. ON/OFF ratios of up to 10^4 were observed. However, despite the ease of fabrication of the surface channel devices there are several disadvantages to the surface channel architecture.

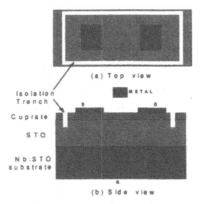

Figure 1: Surface channel architecture (a) top view (b) side view.

In the surface channel devices, the substrate also serves as the gate electrode. Therefore, there is a very large capacitance associated with the gate and there is a strong coupling capacitance to other devices. This makes operation at high frequencies impossible. Another significant disadvantage of the surface channel architecture is a result of moisture sensitivity of the cuprate materials as a class. Since the cuprate is an unshielded top layer, it slowly deteriorates due to water contamination, making necessary the deposition of thick cuprate films to protect the switching layer near the gate oxide/channel interface. However, the source and drain electrodes are only in direct contact with the upper (unswitched) part of the cuprate film. Thus the surface

4

channel architecture has two limiting problems: First, there is a potential resistance (in series with the channel) which could limit the "ON" state performance of the device due to conduction through unswitched cuprate layers from the source and drain electrodes to the switched part of the cuprate channel near the gate oxide/channel interface. Second, there is a potential resistance (in parallel with the channel) which could limit the "OFF" state performance of the device due to conduction from the source to the drain through the unswitched part of the cuprate channel. Due to the short electric field screening length in cuprate channels in the conducting state, OxFET devices are strongly dependent upon the quality of the cuprate near the gate oxide/channel interface. This is because only the cuprate within an electric field screening length of the gate oxide/channel interface is switched. Of course, long-range order is paramount to the best transport in any material. This is particularly true in our devices due to the interface nature of the switching. Any grain boundaries in the interfacial film will degrade the mobility of the devices. Therefore it is important to make the highest quality interface. Although the substrates we start with are atomically smooth, in the surface channel devices, the cuprate is deposited only after the gate oxide is deposited. After depositing a ~1000 Angstrom gate oxide film on the atomically smooth substrate, the cuprate interface might not be as smooth as would be the case if the cuprate were deposited first.

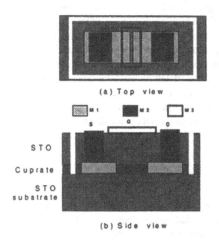

Figure 2 : Buried channel device architecture (a) top view (b) side view.

The buried channel OxFET architecture provides a solution to several of these limitations although the fabrication is now more complicated. The buried channel OxFET architecture is shown in Fig. 2a (top view) and 2b (side view). The substrate is nonconducting undoped strontium titanate (STO) and the gates are now independent for each device since they are deposited on the top of the structure through a lithographic process. The cuprate channel is now directly on the atomically smooth substrate ensuring the highest quality interface. The cuprate channel is also protected from moisture by the gate oxide layer. The source and drain electrodes make direct contact with the cuprate material at the gate oxide/channel interface eliminating the potential series resistance. Since the cuprate is protected from moisture by the gate oxide layer, the cuprate channel layer can eventually be made thinner thereby reducing the potential parallel resistance. Also, since the substrate is insulating the large gate capacitance and coupling capacitance to other devices is eliminated.

FABRICATION

Fabrication of the devices starts by putting the first metallization layer (M1) on the undoped STO substrate. This is accomplished using standard lithographic techniques. The M1 metallization is critical since it defines the channel geometry. Since the M1 process involves a lithographic step and a liftoff step, we were concerned about the quality of the STO surface after M1 processing. AFM measurements showed that the STO surface in the channel between the source and drain M1 electrodes remained atomically flat. Further, ion scattering studies [16] have shown that the initially Ti terminated Kawasaki [17] processed STO substrate is left with a Sr rich surface after liftoff, oxygen ashing, and cleaning. This is fortuitous since a Sr terminated surface is better in preventing copper oxide precipitation [18] during the pulsed laser deposition of the cuprate channel. After M1 metallization, liftoff, oxygen ashing, and cleaning, the substrate with M1 is placed in a PLD chamber for thin film deposition. The cuprate film used for these experiments was La_2CuO_4 (LCO). The LCO was grown at a temperature of 700 C in an oxygen background pressure of 10 mTorr. The LCO film thickness was approximately 100 Angstrom. Immediately after the LCO growth, a 1000 Angstrom film of STO was grown on top the cuprate. The oxygen pressure was 250 mTorr and the temperature was 760 C for the STO film growth. After deposition, the films were slowly cooled to room temperature in 1 atm of oxygen. After film deposition, processing continued with lithographic definition of M2 vias to make contact with the buried M1 metallization. A combination of chemical etching and ion milling was used to make vias down to the M1 metallization. The vias were then metallized and M2 was defined in the liftoff. Next, the M3 metallization, which defines the gate electrodes, was accomplished with another lithographic step, metallization, and liftoff. Finally, the devices were isolated from each other by making a trench around each device down to the STO substrate. This was done using lithographic techniques to define the trenches and ion milling and chemical etching to make the trenches.

RESULTS

The completed devices were tested using a Hewlett-Packard 4145 semiconductor parameter analyzer with a grounded source electrode. In Fig. 3 we plot the drain current (I_d) versus gate voltage (V_g) for a constant drain voltage of 1 volt. In Fig. 4 we plot the transconductance (dI_d/dV_g) versus gate voltage. The devices show a transconductance of up to 45 μS at 1 volt drain voltage and a gate voltage of 2 volts for a channel length of 1 μm and width = 150 μm. This is an improvement over the transconductances seen in the surface channel devices. We expect that the better interface quality of the buried channel devices leads to a higher mobility and therefore a higher transconductance. One limitation of the buried channel devices is that the "OFF" state conductance is still high in these devices. The maximum $\Delta R/R$ we observed was 240%. However, we have not yet optimized the process for "OFF" state conductance. This is most likely the result of a parallel conduction channel associated with the unswitched layers in the films that we have made so far.

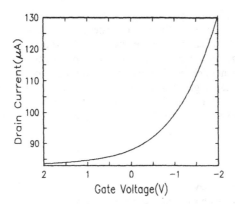

Figure 3: Drain current versus gate voltage for a buried channel architecture device.

An interesting observation is the gate field dependent mobility. This is illustrated in Fig. 4. Since transconductance is proportional to mobility, this curve suggests a linear increase in mobility with gate field above threshold. This is consistent with a picture of a mobility transition in these materials in going from the insulating state to the metallic state.

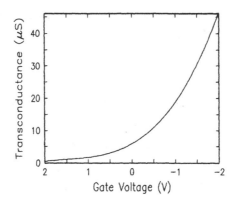

Figure 4: Transconductance versus gate voltage for a buried channel architecture device.

The Mott metal-insulator transition induced by electric field is expected to occur at a surface charge density of 10^{14} carriers/cm^2 [3]. In order to determine whether the present gate oxide films are achieving the theoretical transition charge density we have measured the capacitance of our devices as a function of voltage. The surface charge density is limited by two factors, the dielectric constant for the film and the breakdown potential for the film. Although the bulk dielectric constant for STO is ~300 [17], the value usually decreases in thin films. The dielectric constant we determine for our film (thickness = 1000 Angstrom) is 180. Further, the dielectric constant decreases with voltage. The breakdown potential for the tested film was 8 volts. Thus we find that the maximum surface charge density we attain in the present films is < 0.5×10^{14} carriers/cm^2. This indicates that more work needs to be done with the gate dielectric to optimize the surface charge density. Processing improvements might be made to improve the quality of films of the current thickness. For example, in RF sputtered STO thin films it was found that a dielectric constant higher than the bulk value can be attained under proper conditions [19]. In addition, detailed studies of the dielectric constant and the breakdown potential versus film thickness are necessary to find the optimal film thickness.

SUMMARY

We have demonstrated devices of a buried channel architecture for room temperature oxide field effect transistors with a channel capable of undergoing the Mott metal-insulator transition. This architecture utilizes a buried interface for the gate oxide/cuprate channel to improve the quality of the interface, protect the cuprate from moisture, reduce contact resistance to the channel, and reduce gate capacitance and coupling capacitance to other devices. An improvement in transconductance of these devices expected from the better mobility of a higher quality gate oxide/channel interface was observed.

ACKNOWLEDGEMENTS

The authors gratefully wish to acknowledge many stimulating discussions with B.A. Scott, D.M. Newns, C.C. Tsuei, and A. Gupta.

REFERENCES

1. H.-S. P. Wong, D.J. Frank, P.M. Solomon, C.H.J. Wann, and J.J. Welser, Proc. IEEE **87**, 537 (1999).

2. D.M. Newns, J.A. Misewich, C.C. Tsuei, A. Gupta, B.A. Scott, and A. Schrott, Appl. Phys. Lett. **73**, 780 (1998).

3. C.Zhou, D.M. Newns, J.A. Misewich, and P. Pattnaik, Appl. Phys. Lett. **70**,598 (1997).

4. T. Doderer, C.C. Tsuei, W. Hwang, and D.M. Newns, submitted for publication in Phys. Rev. Lett..

5. J.A. Misewich and A.G. Schrott, submitted for publication.

6. N. Mott, "Metal-Insulator Transitions", (Taylor and Francis, London) 1990.

7. Y. Tokura, Physica C **235-240**, 138 (1994).

8. T.V. Ramakrishnan, J. Solid State Chem.**111**, 4 (1994).

9. H.-M.Christen, J. Mannhart, E.J. Williams, Ch. Gerber, Phys. Rev. B **49**, 12095 (1994).

10. K. Abe and S. Komatsu, Jpn. J. Appl. Phys., Part 2 **32**, L1157 (1993).

11. T. Hirano, M. Ueda, K. Matsui, T. Fujii, K. Sakuta, and T. Kobayashi, Jpn. J. Appl. Phys., Part 2 **31**, L1346 (1992).

12. J. Mannhart, Supercond. Sci. Technol. (UK) **9**, (1996) 49-67.

13. A. Levy, J.P. Falck, M.A. Kastner, R.J. Birgeneau, and A. T. Fiory, Phys. Rev. B **51**, 648 (1995).

14. A.T. Fiory, A.F. Hebard, R.H. Eick, P.M. Mankievich, R.E. Howard, and M.L. O'Malley, Phys. Rev. Lett. **65**, 3441 (1990).

15. V. Talyansky, S.B. Ogale, I. Takeuchi, C. Doughty, and T. Venketesan, Phys. Rev. B **53**, 14575 (1996).

16. M. Copel, A.G. Schrott, and J.A. Misewich, to be published.

17. M. Kawaski, K. Takahashi, T. Maeda, R. Tsuchiya, M. Shinohara, O. Ishiyama, T. Yonezawa, M. Yoshimoto, and H. Koinuma, Science **266**, 1540 (1994).

18. P. Tsuchiya, M. Kawasaki, H. Kubota, J. Nishino, H. Sato, H. Akoh, H. Koinuma, Appl. Phys. Lett. **71**, 1570 (1997).

19. Jpn. J. Appl. Phys. **37**, 5651 (1998).

Simulation of a simplified design for a nanoscale metal-oxide field effect transistor

D. M. Newns, W. M. Donath, and P.C. Pattnaik

IBM T. J. Watson Research Center, P.O.B. 218 Yorktown Heights, NY 10598

Abstract

We describe simulations on a simplified design for a metal-oxide nanoscale Field Effect Transistor (FET). The device features an oxide channel with a high dielectric constant ferroelectric as the gate insulator. In the present model, the gate and source/drain electrodes are unconventionally placed on opposite sides of the channel. Simulations are quantum mechanical and are based on a simplified transport model. Results on a 10 nm. channel device show adequate conductance and ON/OFF ratio, while simulation of a ring oscillator yields an estimated device switching time of 300 fs..

I. INTRODUCTION

The imminent breakdown of scaling of device size (Moore's Law) in the conventional Si/SiO_2 MOSFET at the 50-60 nm. channel length scale has been well documented in technical publications [1]. Hence in the present situation there is a need for intensive study of new technology pathways leading to the possibility of continuing size shrinkage , especially those offering the possibility of considerable dynamic range of scaling and leading to increased device switching speed.

In the present paper we shall present a simplified simulation study of a MOSFET design quite radically modified both as regards materials and physical design, which is predicted to operate down to a relatively aggressive scale of 10 nm. channel length, with a very fast switching time of 300 fs. Key points in the design distinguishing it from the conventional

MOSFET enable it to function at such a small scale. The device operates via an oxide channel material with only majority carriers, enabling the bulky *pn* junctions of the conventional MOSFET to be dispensed with. The concept of high dielectric constant gate oxide is extended to use of a ferroelectric material with a very high dielectric constant, thus enabling (i) reduction of the channel length into the nanoscopic regime, without intervention of unacceptable short channel effects, and (ii) a thick enough gate oxide to avoid tunneling.

Following an early theoretical suggestion regarding the Mott Transition Field Effect Transistor (MTFET) [2], and early experimental work [3], the experimental approach to studying all-oxide transistors with ferroelectric gate oxide and cuprate channel materials has already made considerable progess [4]. Currents of 700 μA have been switched, ON/OFF ratios of 10^4 observed, and progress made in enhancing mobility [5] [6], though bulk mobilities have not yet been reached. In concert with the first pass theoretical analysis derived in the present paper, the outlook for an all-oxide FET technology seems promising.

II. THE MODEL

A sketch of the 2D device model as simulated is illustrated in Fig. 1. The device, with a nominal 10 *nm*. channel length, has several features which, apart from scale, are quite different from a conventional FET. First of all, the geometry involves a back gate rather than the conventional front gate. The gate oxide is assumed to consist of a ferroelectric dielectric material with a dielectric constant (permittivity) ε_{ox} of several hundred, and to have a thickness $d = 15$ nm.. Experimental work supports operation of very high dielectric constant films at this thickness [7]. The aspect ratio of the gate oxide structure is quite different from a conventional SiO_2 one in having a thickness of the same order as its length. The channel material shares a planar interface with the gate oxide (this will from the fabrication point of view be an epitaxial interface). The total length of channel material in the model is taken to be 16 *nm*., and the height to be 5 atomic layers (2 *nm*. assuming a perovskite structure material). The channel properties are assumed to be isotropic in the plane of the figure; if an anisotropic material such as a cuprate is used for the channel material, the orientation

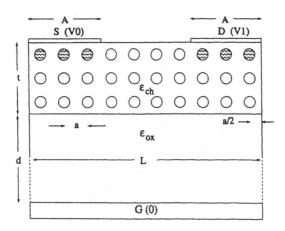

FIG. 1. Sketch of channel region (not to scale), showing atom rods (circles), dielectric constants (ϵ_{ch} for channel, ϵ_{ox} for gate oxide) and dimensions. Shaded atom rods denote doping.

needs to be with the c-axis, which is the low-conductance one, normal to the direction of current flow. The dielectric constant ϵ_{ch} of the channel material is assumed to be quite large (it is taken as 25). A thin layer of channel material under each source and drain electrode is heavily doped to reduce contact resistance at the electrodes.

As regards transport, in which the metal atoms of the channel material undoubtedly play a critical role, we adopt a discrete model in which 'atoms' are modelled as thin rods transverse to the direction of current flow. Since in contrast the dielectric properties of oxides stem from the polarizabilities of both of anionic and cationic species, the approximation taken for the dielectric constant is to model it as a continuum.

The electrostatic boundary conditions which we have taken are that the gate is grounded, with the source at V_0 and the drain at V_1. This is quite general. On all edges where there is an exposed oxide surface, the normal derivative of the electric field has been taken as zero, based on the assumption that the dielectric constant of all exterior material will be much lower than that of the ferroelectric gate oxide or of the channel material.

We use a Green's function approach to the electrostatic calculation, wherein (Appendix A and Ref. [8]) the electrostatic potential on atom rod i is written as

$$V(i) = f_{BS} \sum_{i \neq j} G(i,j)Q(j) + V_0 \Gamma_S(i) + V_1 \Gamma_D(i). \tag{1}$$

Here i, j are 2D discrete vectors describing an atom rod in the Fig. 2 array. $Q(j)$ is the charge per unit length on atom rod j, which can be related to the expectation value q_j of charge on an actual atom at 2D coordinates j by $Q_j = q_j/a$, where a is Lattice constant. The definition of the Green's function $G(i,j)$ in (1) is that $f_{BS}G(i,j)$ is the potential at atom rod i due to unit charge per unit length on rod j, with S, D and G grounded and zero normal derivative of potential on the other device faces. The Green's function G is symmetric, $G(i,j) = G(j,i)$. We choose to work in atomic units $e = \hbar = 1$, when within our definition of the Green's functions the electrostatic prefactor f_{BS} is $f_{BS} = 4\pi/\varepsilon_{ch}$, where ε_{ch} is the permittivity of the channel material.

In (1) the Green's function $\Gamma_S(i)$ is the potential on site i when unit potential is on the Source, other electrodes being grounded, and with zero normal derivative of potential on the exposed oxide surfaces. The Green's function $\Gamma_D(i)$ is defined in an anologous manner. All the Green's functions, G, Γ_S, and Γ_D, are calculated numerically once and for all at the beginning of the program, for details see Ref. [8] and Appendix A.

Our treatment of the channel electronic properties takes into account two key requirements: (i) the model must be fully quantum, since oxide materials are highly degenrate carrier (electron/hole) gases in their conducting states (ii) it must be possible to allow for abrupt changes in potential over distances of order 1-2 lattice spacings, such as are seen in Fig. 2.

These requirements for a highly inhomogeneous quantum model are nontrivial to fulfil. The solution adopted in the present first pass treatment is to use a real space based approach. Each site carries a fully quantum density of states $\rho(\epsilon)$, expressing the oxide band width of order 1 eV. At equilibrium, the system's fully quantum spatial distribution of density of states, charge etc. will be correctly reproduced in this model. Intersite carrier transfer under dynamic conditions, which has a typical range of 3-4 lattice constants in cuprates at 300K, is truncated to one lattice constant in our first pass treatment, with a corresponding adjustment

14

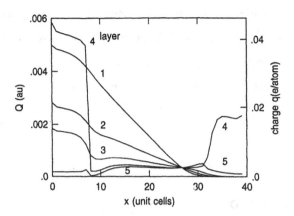

FIG. 2. charge Q per unit rod length (RH scale) as a function of number of rods along channel in pinch-off regime. $V_0 = 0.7$ V, $V_1 = -0.3$V. Channel layer 1 lies closest to the gate oxide. Notation and parameters as previous figure

of the intersite tunneling rate to reproduce the experimental value by enhancement of the prefactor C_0 (see below) by a factor of 3-4. The error involved in this approximation to the carrier dynamics lies in neglect of ballistic effects in the carrier trajectory over distances of order 1-1.5 nm. - effects which we do not believe will significantly impact the conclusions of this paper.

It is assumed that conduction is p−type, as in cuprates, though this is non-critical. As in semiconductor modelling, the valence band position in energy space is allowed to vary according to the local electrostatic potential $V(i)$, so that the top of the valence band is defined to lie at $E_{B0} - V(i)$, E_{B0} being a reference level for the valence band relative to the source/drain electrode material Fermi level. The valence band chemical potential $\phi_p(i)$ is also varying in space.

As regards equilibrium properties, the charge per unit length on the atom rod i is given, assuming Fermi statistics, by (in atomic units)

$$Q(i) = \frac{1}{a} \sum_k \frac{1}{e^{(\phi_p(i) - \epsilon_k)/k_B T} + 1} = \frac{1}{a} \int^{E_{B0} - V(i)} \frac{\rho(\epsilon) d\epsilon}{e^{(\phi_p(i) - \epsilon_k)/k_B T} + 1}, \tag{2}$$

where $\rho(\epsilon)$ is the density of states of the band eigenstates ϵ_k per Cu atom. A constant density of states may be a reasonable approximation for effectively 2D electron systems such as the cuprates, in which case we take $\rho(\epsilon) = \rho_0$. This approximation is attractive both in terms of appropriateness for the cuprates and in terms of convenience since it leads to analytic results. Then doing the integral under the assumption that the band width $>> k_B T$, as is entirely reasonable since band widths are of the order of $1\ eV$, we get a closed form expression for the charge

$$Q(i) = \rho_D k_B T \log(1 + e^{-(\phi_p(i) - E_{B0} + V(i))/k_B T}), \qquad (3)$$

where $\rho_D = \rho_0/a$ is the rod DOS.

The current flow per unit rod length between two adjacent sites i, j is written as

$$J_{i\leftarrow j} = C_0 k_B T \log\left[\frac{1 + e^{-(\phi_p(j) - E_{B0} + \widehat{V}(i,j))/k_B T}}{1 + e^{-(\phi_p(i) - E_{B0} + \widehat{V}(i,j))/k_B T}}\right], \qquad (4)$$

where $\widehat{V}(i, j) = \max(V(i), V(j))$, and C_0 is a constant to be obtained from experimental transport data. The expression (9), like (8), is based on Fermi statistics. In the classical limit, and for low chemical potential gradients (ohmic regime) we retrieve the conventional form in which conductivity is proportional to carrier concentration

$$J_{i\leftarrow j} = \frac{C_0}{\rho_0 k_B T}(\phi_p(i) - \phi_p(j))q(i). \qquad (5)$$

The constant C_0 is seen to be the sheet conductance per square in the channel material in the metallic regime (Fermi function $= 1$), and amounts to several quanta of conductance. The value of C_0 we use is that of crystalline LSCO at doping 0.15 [9].

The source and drain are treated as sites linked to their neighboring atomic sites, and with

$$\phi_{pS} = -V_0, \phi_{pD} = -V_1. \qquad (6)$$

Doped regions are defined under the source and drain electrodes (see Fig. 1), for the top 2 atomic layers only. This is implemented by locating a background charge $Q_0 = -0.15|e|/a$ on the doped atomic layers.

16

III. NUMERICAL APPROACH

The first step in the numerical approach is to precompute the Green's Functions G, Γ_S, Γ_D (Appendix A).

The sequence of program steps to update from time step n to time step $n+1$ is as follows.

1. Assuming a charge distribution $Q^{(n)}(i)$, the electrostatic potentials on the rods are calculated from the Green's functions using equation (1).

2. Inverting (3) to obtain

$$\phi_p^{(n)}(i) = E_{B0} - V(i) - k_B T \log \left[e^{Q^{(n)}(i)/\rho_D k_B T} - 1 \right], \tag{7}$$

the chemical potential on all sites is calculated.

3. We now have the electrostatic and chemical potentials on all sites, enabling the calculation of all link currents $J_{i \leftarrow j}^{(n)}$ from (4).

4. We then update the charges by means of charge continuity

$$Q^{(n+1)}(i) = Q^{(n)}(i) + \sum_{j \subset NN_i} \Delta t J_{i \leftarrow j}^{(n)}, \tag{8}$$

where the sum is over the nearest neighbor atom rods (including source or drain if applicable) to atom rod i, and Δt is the time step (taken as 1 au or $2.4x10^{-17}$ sec.).

The sequence can be repeated by returning to step 1. We found this approach at all times stable and problem-free. The initial conditions chosen were normally that the channel be empty of charge. Computation of an entire set of current characteristics (Fig. 3) typically took 20 minutes on a single 604-chip RS6000 machine.

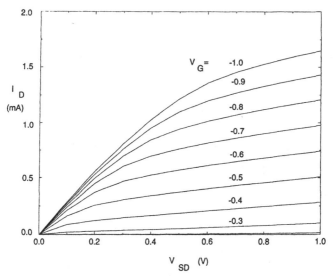

FIG. 3. Drain current for device of channel width 40 nm. vs. drain to source voltage for various gate to source voltages (labels on curves). Parameters other than voltages as previous figure.

IV. RESULTS

Some of the parameter choices have been mentioned in the foregoing. For completeness: Source region overlies atom rods 0-7, drain region overlies rods 33-40. Rods are spaced 0.4 nm. apart, $A = 3.2$ nm., channel length$= 9.6$ nm., $t = 2.0$ nm. (5 layers in channel; layers 4 and 5 are doped to $q = 0.15$ under source and drain electrodes), length of device in unit cells $M_x = 40$, number of grid points (see Appendix A) $N = 400$, $\rho_0 = 0.25$ eV^{-1}, $E_{B0} = -0.35$ eV, $\varepsilon_{ox} = 500$, $\varepsilon_{ch} = 25$, $d = 15$ nm.. Time step Δt is 1 au or $2.4x10^{-17}$ sec..

The choice $\rho_0 = 0.25$ eV^{-1} for the DOS (we use a spinless model since the nature of the carriers in materials such as the HiTc cuprates is controversial) implies a mass of ≈ 0.4 electron masses.

We have run simulations in a wide range of parameter regimes; in Fig. 2 we illustrate the behavior in the 'pinch-off' regime where the gate voltage is smaller than the source-drain voltage drop, so that the current is in the saturation regime. In Fig. 2, there is seen to be very little charge in the undoped layers near the drain, illstrating the pinch-off effect.

18

There is also found to be a pronounced jump in both potential and chemical potential at the drain end of the channel; the potential in the atomic layer closest to the gate (layer 1) is nevertheless close to an equipotential, due to the high dielectric constant of the gate oxide, a feature consistent with short channel effects being under control.

In Fig 3 we display the current-voltage characteristics for the device, assuming a width to channel length ratio of 4:1. Two key points are worth noting. First, the currents switched by the device are large, of order 1 mA in its 'ON' state; the current is somewhat enhanced because *all* the layers 1-5 in the channel are in the degenerate quantum state or close to it, a result which might not hold for different choices of carrier mass. Secondly, the characteristics are similar to those of a conventional FET except that the saturation in the pinch-off regime is not absolute (the curves retain small but finite slope).

The lack of total drain current saturation is due to the penetration of the drain field into the channel ('short channel effect'), despite precautions, such as the introduction of a gap between the doped region and the gate oxide interface, to reduce field penetration through the ferroelectric.

The logarithm of the drain current vs. the gate voltage for several sorce-drain voltages is plotted in Fig. 4. Two useful measures of device performance can be read off from this

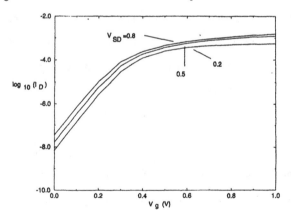

FIG. 4. Logarithm of drain current (mA) for device of channel width 40 nm. vs. gate to source voltage for various source to drain voltages (labels on curves). Parameters other than voltages as previous figure.

type of plot. The ON/OFF ratio is determined as in the range 4-8x10^4 for these curves; it is adequate if the criterion is avoidance of excessive OFF-state heating. The subthreshold slope, the gate voltage swing to produce one order of magnitude change in drain current, is 77 meV per decade, of similar magnitude to that in conventional FET devices operating at low voltages.

The static characteristics of the device show that in terms of (i) 'ON' current, (ii) gain (related to slope of the saturated characteristics) (ii) ON/OFF ratio, and (iv) subthreshold slope its performance is of appropriate type for computer switching applications.

In order to investigate the dynamic behavior of the device, we modelled an 11-stage ring oscillator, consisting of 11 CMOS inverters in series. Both the n-type and p-type device characteristics are derived from the data in Fig. 3. The static gate-source and gate-drain capacitances, which are more nearly voltage-independent when 2 complementary devices are connected together as in a ring oscillator, were calculated from charge-voltage curves and then replaced by approximate constant capacitances. The wiring capacitance in a compact ring oscillator, estimated from the 'rule of thumb' 2 pF per centimeter, is much less than the internal device capacitances and is neglected. The circuit is then solved by standard methods.

The ring oscillator results show that (i) the nanoscale FET device switches adequately (ii) the switching time per transistor is approximately 300 fs, a very fast switching time, which is in some measure due to the high channel conductance arising from the relatively low DOS assumed in the calculation.

V. CONCLUSION

The performance of a nanoscopic switch depends not only on the properties of one or more materials, but on the overall engineering solution to exploit the material properties. The purpose of this paper is to provide a first-pass study of a simplified engineering solution to the problem of the nanoscale, all-oxide FET. We find that from the point of view of both static and dynamic characteristics the proposed model of a 10 nm. channel device performs quite satisfactorily.

Table I: Characteristics of Metal Oxide nanoFET

Device feature	**Consequence**
Carrier density $\simeq 10^{21}$	High 'ON' conductance. Carriers confined at O/C interface.
High permittivity oxide	Supports high carrier density. Reduces short channel effect
No *pn* junction	Majority carrier device. Compact design
Metal oxide channel	Supports high carrier density. Supports epitaxial interface

Finally, we present a summary of the characteristics of the device in the accompanying table.

VI. APPENDIX A

A. Calculation of G

$G(i,j)$ is given by the expression

$$G(i,j) = \Theta(y_i - y_j)\frac{(y_i - y_j)}{L} + \sum_{k>0} \frac{1}{kL}(e^{k|y_i - y_j|} + v_k e^{-k(y_i + y_j)}) \cos kx_i \cos kx_j + X_0^{(j)} \left[1 + \frac{ry_i}{d}\right] \tag{9}$$

$$+\sum_{k>0} X_k^{(j)} \frac{1}{kL}(e^{k(y_i - t)} + v_k e^{-k(y_i + t)}) \cos kx_i, \tag{10}$$

where $X_k^{(j)}$ is the solution to the linear equation set

$$\sum_k R_{\alpha k} X_k^{(j)} = U_\alpha(x_j, y_j), \tag{11}$$

with R and U given by

$$R_{\alpha k} = (1 + \frac{rt}{d})\delta_{k0} + (1 + v_k e^{-2kt}) \cos kx_\alpha (1 - \delta_{k0});\ \ \alpha \subset S, D; \tag{12}$$

$$R_{\alpha k} = \gamma \frac{r}{d}\delta_{k0} + \gamma k(1 - v_k e^{-2kt}) \cos kx_\alpha (1 - \delta_{k0});\ \ \ \alpha \subset C, \tag{13}$$

and

$$U_\alpha(x,y) = \frac{t-y_0}{L} - \sum_{k>0} \frac{1}{kL}(e^{k(y-t)} + v_k e^{-k(y+t)}) \cos kx \cos kx_\alpha; \quad \alpha \subset S, D; \tag{14}$$

$$U_\alpha(x,y) = \frac{\gamma}{L} + \gamma \sum_{k>0} \frac{1}{L}(e^{k(y-t)} + v_k e^{-k(y+t)}) \cos kx \cos kx_\alpha; \quad \alpha \subset C. \tag{15}$$

Here the x_α constitute a supporting grid of dimension N along the top edge of the device (see Fig. 1), with overlap in the Source (S), Drain (D), and Channel (C) regions. $v_k = (1 - r\coth(kd))/(1 + r\coth(kd))$, where $r = \varepsilon_{ox}/\varepsilon_{ch}$. (x_i, y_i) are the coordinates of charges i, $k = n\pi/L$, n integer, and γ is a nonzero constant upon which the results have no dependence.

This expression for G is not manifestly symmetric, but the numerical results were symmetric to a high order of accuracy, which supplies a valuable check on the overall calculation.

Γ_S is given by

$$\Gamma_i^S = X_0^S \left[1 + \frac{ry_i}{d}\right] + \sum_{k>0} X_k^S \frac{1}{kL}(e^{k(y_i-t)} + v_k e^{-k(y_i+t)}) \cos kx_i,$$

where analogously to the G calculation

$$\sum_k R_{\alpha k} X_k^S = U_\alpha^S, \tag{16}$$

where R is as given, while U^S is given by

$$U_\alpha^S = 1; \ \alpha \subset S; \ U_\alpha^S = 0; \ \alpha \subset C, D. \tag{17}$$

The expression for Γ_D is obtained by a straightforward extension, or by symmetry.

REFERENCES

[1] E.J. Lerner, IBM Research (3), p.10 (1998); S. Thompson, P. Packan, and M. Bohr, Intel Technology Journal, Q3 1998, p.1.

[2] C. Zhou, D.M. Newns, J.A. Misewich, and P.C. Pattnaik, Appl. Phys. Lett. 70, 598 (1997); D.M. Newns, J. Misewich and C. Zhou, USA Patent disclosure #: YO895-0318, 1996.

[3] A. Levy, J.P. Falck, M.A. Kastner, W.J. Gallagher, A. Gupta, and A.W. Kleinsasser, J. Appl. Phys. **69**, 4439 (1991); A. Levy, J.P. Falck, M.A. Kastner, and R.J. Birgeneau, Phys. Rev. **B51**, 648 (1995); S. Hontsu, H. Tabata, N. Nakamori, J. Ishii, and T. Kawai, Jpn. J. Appl. Phys. **35**, L774 (1996); S.B. Ogale, V. Talyansky, C.H. Chen, R. Ramesh, R.L. Greene, and T. Venkatesan, Phys. Rev. Lett. **77**, 1159 (1996).

[4] D.M. Newns et al., Appl. Phys. Lett. **73**, 780 (1998); D.M. Newns et al., 'The Mott Transition Field Effect Transistor: a Nanodevice?', Proceedings of the 5th International Workshop on Oxide Electronics, U. Maryland (1998), J. Electroceramics (Kluwer), in press; Jum & Alex MRS 1998.

[5] T. Doderer *et al.*, 'Charge transport in the normal state of electron or hole doped $YBa_2Cu_3O_{7-x}$', preprint.

[6] New Jim & Alex paper

[7] M. Izuha, K. Abe, and N. Fukushima, Jpn. J. Appl. Phys. **36**, 5866 (1997).

[8] D.M. Newns, W. Donath and P.C. Pattnaik, submitted to Applied Physics Letters.

[9] H. Takagi et al., Phys. Rev. Lett. **69**, 2975 (1992).

[10] T. Ito, K. Takenaka, and S. Uchida, Phys. Rev. Lett. **25**, 3995 (1993).

OPTIMIZING FABRICATION OF BURIED OXIDE CHANNEL FIELD EFFECT TRANSISTORS

A.G. SCHROTT, J.A. MISEWICH, M. COPEL, D.W. ABRAHAM, and D.A. NEUMAYER,
IBM Research, T.J. Watson Research Center, Yorktown Heights, NY 10598

ABSTRACT

In this paper we describe improved methods of fabrication for an oxide channel field effect transistor (OxFET) similar in architecture to a conventional FET. We demonstrate that a substrate treatment consisting of a low power oxygen ashing followed by annealing yields a strontium (A-site) terminated surface in single-crystal strontium titanate (STO) . This surface termination of the substrate results in pulsed laser deposited cuprate-channel films of improved quality.

INTRODUCTION

The second generation of oxide channel field effect transistor (OxFET) architecture with a buried oxide channel, recently demonstrated at IBM [1], provides a solution to several of the limitations present in the first generation of OxFets[2], and offers the potential to scale beyond the limits of silicon technology. The second generation OxFET architecture is shown schematically in Fig. 1. The substrate is nonconducting undoped strontium titanate (STO), and the cuprate channel is now directly on the atomically smooth substrate ensuring a higher quality channel film. The latter is also protected from moisture by the gate oxide layer. The source and drain electrodes make contact with the cuprate material at the gate oxide/channel interface reducing the potential series resistance. Since the cuprate is protected from moisture by the gate oxide layer, the cuprate channel layer can eventually be made thinner, thereby reducing the potential parallel resistance. Also, since the substrate is insulating the large gate capacitance and coupling capacitance to other devices is eliminated. In these devices the gate oxide/channel interface quality and gate oxide dielectric strength are critical to the device performance.

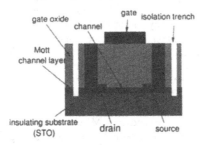

Fig. 1. Schematic representation of the OxFET structure with the buried channel layer

Kawasaki and coworkers [3] have demonstrated that surface termination plays a substantial role in addition to atomic smoothness of the perovskite substrates in obtaining perfect two dimensional epitaxy of the heterostructures used in the current architecture. It is well known that substrates with the highest degree of smoothness are obtained through an etching process which

25

leaves a STO substrate terminated in Ti [4]. Unfortunately, titanium termination leads to precipitates in the epitaxially grown cuprates[3]. These precipitates are very detrimental to the performance of our second architecture devices because they could lead to grain boundaries in the channel and in addition, they could protrude through the gate oxide film and become the locus for short circuits between the source/drain and the gate.

In this paper we assess the role of the STO substrate surface termination on the electronic conductivity of the channel films, and we report on the methods utilized in order to produce strontium terminated surfaces for fabrication of OxFET's with high quality channels. Furthermore, we report on different possible paths to achieve a gate oxide with high dielectric constant and high breakdown voltage.

EXPERIMENT AND RESULTS

Processing steps and measurements

As it has been already described in more detail in Ref. [1], the fabrication of our OxFET devices starts with the first metallization layer (M1), deposited through a lithographic mask on a STO substrate which had undergone the Kawasaki process [4] by the vendor. In order to produce a Sr terminated surface for the subsequent growth of the channel film, we used to deposit about a monolayer of SrO [3] by ablating from a Sr peroxide target. However, this method was not convenient for our La_2CuO_4 (LCO) channel because it produced Sr doping at the substrate/channel interface, and the conductive $La_{2-x}Sr_xCuO_4$ interfacial layer was detrimental to the switching of the channel. Serendipitously, we discovered a facile and benign method of preparing Sr terminated surfaces in STO by a mild O_2 ash, which we will discuss in the next section.

Typically, following the ultrasound cleaning after the M1 lift-off, the sample undergoes a mild reactive ion etching (ashing) in O_2. This is accomplished in a parallel plate, reactive ion etching tool, model Jupiter III plasma system. Its chamber dimensions are 25 cm diameter x 3.25 cm high. We use flowing O_2 at 500 mtorr, and 50 watts of power at 13.56 MHz. The substrate is then introduced into the deposition chamber where the channel layer and the gate insulator are sequentially deposited by pulsed layer deposition (PLD), following the interval deposition method [6], and with the substrate held at 750 $^\circ$C. Prior to deposition, the sample is annealed at 700 $^\circ$C for 3 hr. in vacuum followed by 1 hr in 1 atm. of O_2. The vacuum annealing step plays the role of alloying the Ti adhesion layer with the Pt metallization used for M1. However, this step reduces the near-surface region of the STO substrate making it conductive. The O_2 annealing restores the lost oxygen and consequently the insulating character of the substrate.

Since it defines the channel geometry, the M1 metallization is critical. The M1 process involves a lithographic step and a lift-off step, which could impair the quality of the STO surface after M1 processing. Atomic force microscope (AFM) measurements showed that the STO surface in the channel between the source and drain M1 electrodes remained atomically flat after lift-off. Furthermore, the STO surface becomes even flatter after ashing. As for the surface termination of the substrate, ion scattering experiments show only a slight compositional change with respect to the surface obtained by the Kawasaki process, see Figs 2a and 2b. Figure 2 shows ion scattering spectra, taken with an incident beam of 200KeV He ions, of STO substrates after various treatments. Fig 2a is a spectrum of a ultrasonically cleaned STO (Kawasaki process) substrate. For comparison, Fig. 2b shows the spectrum corresponding to a sample that had photoresist spun-on, and was later ultrasonically stripped with solvents, to reproduce the process related to the M1 lift-off step.

Channel film epitaxy

Remarkably, ashing produces distinct changes in the STO surface. Figure 2c shows the ion scattering spectrum corresponding to a substrate after it was ashed in O_2 for 1 min., the peaks corresponding to Sr and Ti exhibit a noticeable broadening indicating a high degree of disorder. In addition, this spectrum shows that the signal from surface Sr (signal closest to the energy position indicated by the corresponding arrow, app. 193 KeV) has increased with respect to those of Fig. 2 a, b, while that from surface Ti (signal closest to the energy position indicated by the corresponding arrow, app. 187 KeV) has decreased, indicating a more strontium rich surface termination. Furthermore, the spectrum of Fig 2d, which corresponds to the sample of Fig. 2c after undergoing the annealing described above, shows a high degree of order (evidenced by the narrow peaks), as well as an inversion of the surface termination. The latter is evidenced by the peak intensities and positions, the Sr peak increased and moved to a higher energy position, whereas the Ti peak decreased and shifted to a lower energy, compared to those of Fig. 2a, respectively. Additional experiments indicate that a 3 min ashing step leaves an almost purely Sr terminated STO surface. This is fortunate since A-site terminated surfaces are very difficult to attain in perovskites [5].

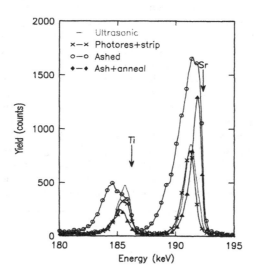

Fig. 2 . IonScattering spectra for different sample treatments. a (——), ultrasound cleaning; b (x— x), after photoresist stripping; c (o—o), after ashing; d (Δ—Δ) after annealing sample in c.

In order to establish the affects of ashing on the subsequent properties of the channel film, we performed the following experiment. A substrate with M1 metallization was subjected to different ashing times, keeping the power constant. Since the design of M1 defines 4 equivalent quadrants, we could selectively mask each quadrant with small squares of STO in order to determine a different ashing time for each of the four quadrants of a substrate. Therefore, we could compare in the same sample the impact of different substrate treatments on the smoothness and conductivity of the channel film. After deposition of the channel film, the surface topography was investigated by AFM, and the frequency dependent film conductivity was measured using a HP 4275 LCR meter. The contact to the source and drain electrodes was achieved by pushing the tungsten probes through the film onto the underlying M1 metallization.

AFM scans were performed on a 10 nm thick LCO channel film, deposited on a sample that was ashed in the following way. Quadrant 1 (Q1), no ashing; quadrant 2 (Q2), 25 sec.; quadrant 3 (Q3), 50 sec.; quadrant 4 (Q4), 75 sec. The flatness of the film improves with the STO ashing in going from quadrant 1 (control), to quadrant 4. As for the effect of Sr termination on the PLD growth of $YBa_2Cu_3O_{7-\delta}$ (YBCO), a similar experiment was carried out on a 10 nm YBCO film, except that in this case, the ashing times were 0, 50, 100, 160 sec for Q1 through Q4, respectively. Table 1 shows the mean roughness and root mean square (RMS) values in nm for each quadrant of both the LCO and YBCO samples.

	Q1	Q2	Q3	Q4
LCO	0.9 (1.2)	0.95 (1.3)	0.4 (0.64)	0.33 (0.46)
YBCO	0.43 (0.59)	0.35 (0.45)	0.38 (0.46)	0.16 (0.22)

Table 1. Mean roughness (RMS) for each quadrant

Figure 3 shows the drain current vs. drain voltage (IV) curves for the LCO sample. There is a clear correlation between better conductivity and ashing time. Furthermore, the impedance vs. frequency ($Z\Omega$) curves for Q1 and Q4 shown in Fig.4 indicate an interesting capacitive trend. Both the impedance and the phase decrease with frequency. In general, these curves are typical of granular conduction, and from Fig.4 we infer that Q4 is less granular than Q1. Thus, our result seem to indicate that Sr surface termination in STO contributes to a better long range order in the channel film. The formation of grains may be partly attributed to the large lattice mismatch between LCO and STO, and it seems that the Sr surface termination tends to off-set that effect through an enhanced chemical interaction. For YBCO, we observed a similar trend. However, contact problems between the film and M1 have so far prevented us from obtaining quantitative conductivity data. We are currently engaged in solving this problem through the development of improved M1 electrodes. Nevertheless, the results from table 1 suggest that the enhanced chemical interaction provided by the Sr termination is very beneficial for the growth of YBCO which has a small lattice mismatch and exhibits a very flat surface for the film on Q4.

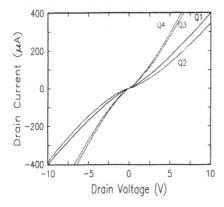

Fig. 3. IV curves for each quadrant

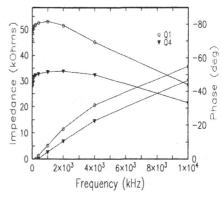

Fig. 4. $Z\Omega$ curves for Q1 and Q4

Gate oxide quality

Another important element that contributes to the performance of the device is the gate oxide. It has been estimated that a surface charge density of about 10^{14} /cm^2 would fully metallize the channel [7]. Several factors conspire to prevent attaining the optimal surface charge in a

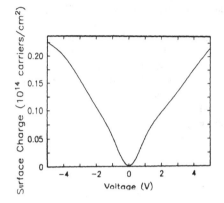

Fig. 5 . Surface charge vs. Voltage.
Sample with a 160 nm thick STO gate oxide.

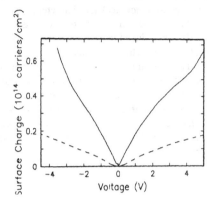

Fig. 6. Surface charge vs. Voltage
120 nm thick BST capacitors; (- - -),
Ti/Pt electrode; (——) , Pt electrode

consistent fashion; stress induced defects, dielectric constant and dielectric homogeneity. We noticed that growing the gate oxide film at the same temperature as the channel film (700-750 $^\circ$C) usually leads to film stress, which probably constitutes the source of the electrical breakdown that often occurs at the boundaries of the M1 pads in enhancement type devices.

In order to improve the gate oxide, we tried different approaches. One approach consists of depositing first a thin STO film at lower temperature to induce a polycrystalline growth, and subsequently continue the growth at higher temperature. Figure 5 shows a plot of surface charge density vs. voltage for a device with a 160 nm PLD STO film grown with the temperature scheme just described. The surface charge density is less than what is needed for full metallization of the film, and smaller than what is obtained by growing the film at constant temperature, indicating that this growth method has a detrimental effect on the dielectric constant. In a second approach, a barium strontium titanate (BST) film, prepared by chemical solution deposition (CSD), is deposited over a thin PLD STO film grown at the same temperature as the channel. This approach would heal stress related cracks in the PLD STO film while still maintaining a high dielectric constant. The spin solution was prepared under nitrogen with stirring. 25.46 gm (0.185 mole) barium, 6.95 g (0.0793 mole) strontium, and 150 ml butoxyethanol were refluxed until all the barium and strontium metal had reacted. In a separate flask 75.13 g (0.264 mole) titanium isopropoxide and 150 ml butoxyethanol was refluxed for 3h and isopropanol distilled away. Both solutions were allowed to cool to room temperature before mixing. The combined solution was filtered in vacuo and 77.1 g (0.53 mol) of ethylhexanoic acid was added. The solution volume was adjusted with additional butoxyethanol to make a 4.9 weight % Ba, 1.45 weight % Sr and 2.36 weight % Ti to prepare the $Ba_{.70}Sr_{.33}Ti_{.97}$ stock solution as determined by ICP. The spin solution was prepared by diluting the stock solution 1:1 with an equal volume of butoxyethanol, and was loaded into a syringe and a 0.45 μm and 0.1 μm filter was attached. The solution was syringed on the substrate until completely wetted. The substrate was then spun for 30-60 sec at 2500 rpm. for 1-2 min in an AG-Associates Model 610 rapid thermal annealer. Layer thickness was ~ 20 nm / layer for a 0.1M spin solution concentration. Thicker films were fabricated by depositing multiple layers.

The spin coat method of making a gate oxide is promising. However, to optimize the dielectric constant and leakage characteristics, an annealing step in O_2 is needed. The latter step may introduce electrode related problems as it is shown next. Figure 6 shows the surface charge density induced in capacitors made with 100 nm thick CSD BST on Nb doped STO substrates. The dashed curve corresponds to a sample where the Pt electrodes were evaporated over a Ti adhesion layer through a photolithographic mask, whereas the solid curve corresponds to a sample with Pt electrodes deposited through a stencil mask without adhesion layer. Figure 6 suggests that although CSD BST is potentially a good gate oxide, there is a strong effect on the total capacitance due to the dielectric heterogeneity induced by the oxidized adhesion layer. Avoiding the latter requires a new method to define the gate electrodes. It consists of evaporating a blanket Pt film, following subtraction of the unwanted metal by ion milling through a negative lithographic mask.

SUMMARY

In this paper we investigated several critical issues in the fabrication of oxide based FETs in order to improve performance. We have demonstrated a method to invert the B-site terminated perovskite surface into an A-site terminated surface. Fabrication of devices on A-site terminated substrates leads to improved conductivity. We also underscored the importance of the gate electrode in achieving sufficient surface charge values.

REFERENCES

1. A.G. Schrott, J.A. Misewich, B.A. Scott, A. Gupta, D. M. Newns, D.W. Abraham in *Multicomponent Oxide Films for Electronics*, edited by M.E. Hawley, D.H. Blank, C-B. Eom, and S.K. Streiffer (Mat. Res. Soc. Proc. **574**, Pittsburgh, PA, 1999) pp. 243-248; J.A. Misewich, A.G. Schrott, to be published.

2. D.M. Newns, J.A. Misewich, C.C. Tsuei, A. Gupta, B.A. Scott, and A.G. Schrott, Appl. Phys Lett. **73**, p.780 (1998).

3. P. Tsuchiya, M. Kawasaki, H. Kubota, J. Nishino, H. Sato, H. Akoh, H. Koinuma, Appl. Phys. Lett. **71**, 1570 (1997).

4. M. Kawaski, K. Takahashi, T. Maeda, R. Tsuchiya, M. Shinohara, O. Ishiyama, T. Yonezawa, M. Yoshimoto, and H. Koinuma, Science **266**, 1540 (1994).

5. T. Ohnishi, K. Takahashi, M. Nakamura, M. Kawasaki, M. Yoshimoto, and H. Koinuma, Appl. Phys. Lett. **74**, 2531 (1999).

6. G. Koster, G.J.H.M . Rijnders, D.H.A. Blank, and H. Rogalla, Apl. Phys. Lett. **74**, 3729 (1999).

7. C. Zhou, D.M. Newns, J.A. Misewich, and P.C. Pattnaik, Appl. Phys. Lett. **70**, p. 598 (1997).

Electrical Characteristics of LSMCD-derived SrBi$_{2.4}$Ta$_2$O$_9$ Thin Films Using TiO$_2$ Buffer Layer for Metal/Ferroelectric/Insulator/Semiconductor Field Effect Transistor Devices

Joo Dong Park and Tae Sung Oh
Department of Metallurgical Engineering and Materials Science,
Hong Ik University, Seoul 121-791, Korea

ABSTRACT

Pt/SBT/TiO$_2$/Si structure was proposed for metal/ferroelectric/insulator/semiconductor field effect transistor (MFIS-FET) applications. SrBi$_{2.4}$Ta$_2$O$_9$ (SBT) thin films of 400 nm thickness were prepared using liquid source misted chemical deposition (LSMCD) on Si(100) substrates with TiO$_2$ buffer layers deposited by DC reactive sputtering with the thickness ranging from 5 nm to 200 nm and electrical properties of MFIS structures were investigated. Memory window and maximum capacitance of the Pt/SBT/TiO$_2$/Si structure increased with decreasing the thickness of TiO$_2$ buffer layer. The Pt/SBT(400 nm)/TiO$_2$(10 nm)/Si structure exhibited C-V hysteresis loop with the memory window of 1.6 V at ±5 V, and could be applicable for MFIS-FET applications.

INTRODUCTION

Recently, SrBi$_2$Ta$_2$O$_9$-based ferroelectric thin films of Bi-layered perovskite structure have been extensively investigated for non-volatile ferroelectric random access memory (FRAM) applications [1-2]. One of the most promising FRAM devices is the metal-ferroelectric-semiconductor field effect transistor (MFS-FET), in which the ferroelectric thin film is used as a gate material [3]. Since the current between the source and drain is controlled by the spontaneous polarization of the ferroelectric thin film and can be sensed without reversal of the ferroelectric polarization, nondestructive read out (NDRO) operation is possible in the MFS-FET configurations. MFS-FET devices also provide high integration density due to their structures with a single transistor per unit cell. However, interfacial reactions between ferroelectric thin film and Si substrate make it difficult to obtain the desired electrical characteristics for MFS-FET devices [4]. As an alternative solution, metal-ferroelectric-insulator-semiconductor (MFIS) structure has been proposed to improve the interfacial properties using buffer layers such as CeO$_2$, Y$_2$O$_3$, and SrTiO$_3$ [5-7].

In this work, we have prepared the Pt/SrBi$_2$Ta$_2$O$_9$/TiO$_2$/Si structure using TiO$_2$ as a buffer layer between SrBi$_2$Ta$_2$O$_9$ (SBT) ferroelectric thin film and Si substrate, and investigated the electrical properties of the Pt/SBT/TiO$_2$/Si structure with variation of the TiO$_2$ thickness. Although SBT thin film has attracted much attention for FRAM applications due to its fatigue-free characteristics and a small coercive field, high annealing temperature such as 800 ℃ is needed to obtain the optimum ferroelectric properties of SBT film [1]. Thus, the buffer material for MFIS structure with SBT thin film should be stable and thickness of the buffer layer should be thick enough to prevent interdiffusion between SBT film and Si during such annealing process. Since TiO$_2$ maintains its metastable phase, anatase, up to near 900 ℃ [8] and exhibits high dielectric constant [9], it is expected that TiO$_2$ acts as a good diffusion barrier and exhibits a low gate voltage even with sufficient thickness.

Mat. Res. Soc. Symp. Proc. Vol. 623 © 2000 Materials Research Society

EXPERIMENTAL DETAILS

To fabricate the Pt/SBT/TiO$_2$/Si structure, TiO$_2$ buffer layers of the thickness ranging from 5 nm to 200 nm were deposited on RCA-cleaned p-type Si(100) substrate using DC reactive sputtering at room temperature. Thickness of the TiO$_2$ films was measured by ellipsometry. Deposition of the SBT film on TiO$_2$/Si was conducted using liquid source misted chemical deposition (LSMCD) which has been considered as an attractive technique to fabricate multicomponent oxide films such as SBT [10]. As bismuth volatility occurs during annealing process of the LSMCD-derived SBT thin film, addition of excess bismuth oxide into the coating solution is required to prevent the bismuth deficiency of the SBT film. In this study, Sr, Bi, and Ta-2-ethylhexanoate of 0.5 M concentration were mixed as Sr : Bi : Ta mole ratio of 1 : 2.4 : 2 and diluted with n-butyl acetate to prepare the SrBi$_{2.4}$Ta$_2$O$_9$ (SBT) coating solution of 0.05 M concentration. TiO$_2$/Si substrates were charged in the LSMCD deposition chamber. After evacuating the chamber below 3×10^{-5} torr, nitrogen gas was introduced into the chamber to keep it at 600 ~ 700 torr. Then, the coating mist was generated from the SBT coating solution using ultrasonic transducer of 1.65 MHz and carried into the chamber using argon gas. The substrates were rotated at 1 ~ 3 rpm to spread the mist uniformly on top of the substrates. After depositing for 5 minutes, films were dried using halogen lamp at 150 ℃ and 400 ℃ for 5 minutes, respectively. This procedure was repeated several times to obtain the desired film thickness of 400 nm. Annealing of the LSMCD-derived SBT films was conducted at 800 ℃ for 1 hour in oxygen ambient. Sputter deposition of 200 nm-thick Pt was followed to form the gate electrodes of 200 μm diameter using the shadow mask and post-metallization annealing was performed at 600 ℃ for 10 minutes in oxygen atmosphere for stabilization of the Pt/SBT interface. Native oxide on the backside of the Si substrate was etched using buffered hydrofluoric acid and aluminum was sputtered to make ohmic contact. Consequently, Pt(200 nm)/SBT(400 nm)/TiO$_2$(5~200 nm)/Si structure was fabricated.

Crystalline phases of the SBT/TiO$_2$/Si structure were characterized using X-ray diffractometry (XRD) with an incident angle of 5°. Remanent polarization and coercive field of the Pt/SBT/Pt film capacitor were characterized using RT66A ferroelectric tester. C-V characteristics of the Pt/SBT/TiO$_2$/Si structures were measured by HP 4194A impedance/gain phase analyzer.

RESULTS AND DISCUSSION

XRD patterns of the annealed SBT films prepared on the as-deposited TiO$_2$(200 nm)/Si and Pt/Ti/SiO$_2$/Si substrates were shown in Fig. 1. After annealing at 800 ℃ for 1 hour in oxygen atmosphere, SBT films on the TiO$_2$/Si and platinized Si substrates were fully crystallized to bismuth layered perovskite structure without preferred orientation. As illustrated in Fig. 1, XRD patterns of the SBT film prepared on the Pt/Ti/SiO$_2$/Si substrate were identical to those prepared on the TiO$_2$/Si substrate, indicating that the crystallization behavior of the LSMCD-derived SBT film was not much affected with the underlayer.

Fig. 2 illustrates the ferroelectric hysteresis curves of the Pt/SBT/Pt capacitors. LSMCD-derived SBT film with thickness of 400 nm exhibited well-developed hysteresis curves. At applied voltage of ±5 V, the remanent polarization (2P$_r$) and coercive field (E$_c$) of the SBT film

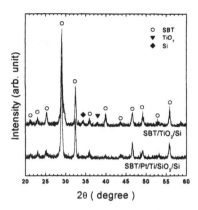

Figure 1. XRD patterns of the SBT films prepared on TiO$_2$/Si and Pt/Ti/SiO$_2$/Si substrates.

were 8.3 μC/cm^2 and 31 kV/cm, respectively. Well-symmetric C-V behavior of the Pt/SBT/Pt capacitor was obtained at 1 MHz with applied voltage of ±5 V and also shown in Fig. 2.

Fig. 3 shows C-V curves of the Pt/SBT/TiO$_2$/Si structures, which were measured at 1 MHz with oscillation level of 30 mV. The applied bias voltage was swept from negative to positive and vice versa with the sweep rate of 0.05 V/sec. With the bias voltage ranging from ±2 V to ±7 V, the Pt/SBT/TiO$_2$/Si structure with 10 nm-thick TiO$_2$ buffer layer exhibited clockwise directional hysteresis, indicating well-defined ferroelectric switching behavior of the SBT film [5,6]. For the Pt/SBT/TiO$_2$/Si structure with 5 nm-thick TiO$_2$ buffer layer, however, counterclockwise directional hysteresis was observed at ±7 V, implying charge injection into the Pt/SBT/TiO$_2$/Si structure.

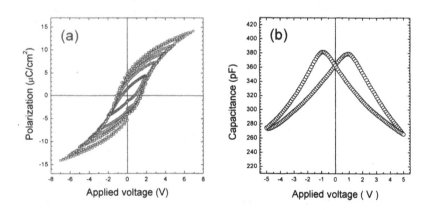

Figure 2. (a) Ferroelectric hysteresis loops and (b) C-V curves of the Pt/SBT/Pt capacitors.

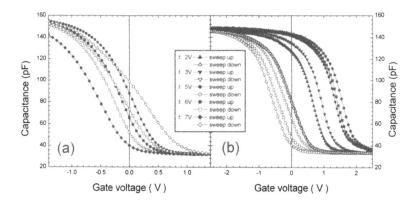

Figure 3. C-V curves of the Pt/SBT/TiO₂/Si structures with TiO₂ thickness of (a) 5 nm and (b) 10 nm.

Auger depth profiles of Pt/SBT/TiO₂/Si structures were shown in Fig. 4. Although interdiffusion between SBT and Si was observed for the Pt/SBT/TiO₂/Si structure with 5 nm-thick TiO₂ buffer layer after annealing at 800 ℃ for 1 hour in oxygen atmosphere, it was suppressed efficiently by TiO₂ buffer layer thicker than 10 nm. From these results, it is considered that TiO₂ acts as a good diffusion barrier in the Pt/SBT/TiO₂/Si structures.

Fig. 5 shows C-V characteristics of the Pt/SBT/TiO₂/Si structures measured at 1 MHz with varying the TiO₂ thickness ranging from 10 nm to 50 nm. At the bias voltage of ±5 V, the regions of accumulation, depletion and inversion were clearly defined in the C-V curves of the Pt/SBT/TiO₂/Si structure regardless of the TiO₂ thickness.

As shown in Fig. 6, the memory window of the Pt/SBT/TiO₂/Si structures, defined as the difference between the applied gate voltages near the flatband capacitance, increased with increasing the applied gate voltage and decreasing the TiO₂ thickness. This was resulted from the

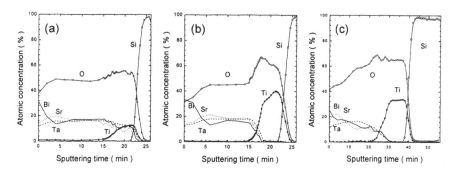

Figure 4. Auger depth profiles of the Pt/SBT/TiO₂/Si structures with TiO₂ thickness of (a) 5 nm, (b) 10 nm, and (c) 30 nm.

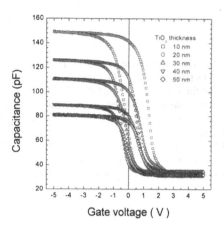

Figure 5. C-V hysteresis curves of the Pt/SBT/TiO$_2$/Si structures at ±5 V.

fact that the electric field across the ferroelectric layer increased with decreasing the TiO$_2$ thickness. In the ideal MFIS structures, the memory window, *i.e.*, width of hysteresis, has been reported to be identical to the coercive field of the ferroelectric layer and proportional to the applied voltage [6,11]. In the real systems, however, SiO$_2$ forms easily at the interface between insulator and Si due to the high annealing temperature [5,6]. Thus, voltage drop can occur due to the formation of SiO$_2$ and MFIS structures exhibit memory window smaller than coercive voltage of the ferroelectric films at the same applied voltage.

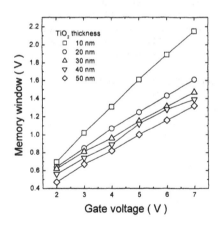

Figure 6. Memory windows of the Pt/SBT/TiO$_2$/Si structures vs. applied gate voltage.

CONCLUSIONS

We verified that TiO_2 thin film was suitable as a buffer layer for deposition of SBT film on Si substrate to prepare the MFIS structure. At applied voltage of ±5 V, the remanent polarization ($2P_r$) and coercive field (E_c) of the LSMCD-derived SBT film were 8.3 $\mu C/cm^2$ and 31 kV/cm, respectively. Using the TiO_2 buffer layer thicker than 10 nm, interdiffusion between SBT and Si was suppressed even after annealing at 800 ℃ for 1 hour in oxygen ambient. The memory window and maximum capacitance of the $Pt/SBT/TiO_2/Si$ structure became larger with decreasing the thickness of TiO_2 buffer layer, and the $Pt/SBT(400$ nm$)/TiO_2(10$ nm$)/Si$ structure exhibited the memory window of 1.6 V at ±5 V. With these results, it could be suggested that the $Pt/SBT/TiO_2/Si$ structure is applicable to NDRO-type MFIS-FET applications.

ACKNOWLEDGMENT

This work was supported by Korea Science and Engineering Foundation (Project No. : 981-0806-037-2).

REFERENCES

1. J. F. Scott and C. A. Paz de Araujo, Science, **246**, 1400 (1989).
2. J. F. Scott, Ferroelectrics Review, **1**, 85 (1998).
3. J. L. Moll and Y. Tarui, IEEE Trans. Electron Devices, **ED-10**, 338 (1963).
4. Y. Shichi, S. Tanimoto, T. Goto, K. Kuroiwa, Y. Tarui, Jpn. J. Appl. Phys., **33**, 5172 (1994).
5. Y. T. Kim and D. S. Shin, Appl. Phys. Lett, **71**, 3507 (1997).
6. H. N. Lee, Y. T. Kim, and S. H. Choh, J. Kor. Phys. Soc., **34**, 454 (1999).
7. E. Tokumitsu, K. Itani, B. Moon, and H. Ishiwara, Jpn. J. Appl. Phys., **34**, 5202 (1995).
8. P. Alexsandrov, J. Koprinarova, and D. Todorov, Vacuum, **47**, 1333 (1996).
9. D. K. Schroder, Semiconductor Materials and Device Characterization, 2nd ed. (John Wiely & Sons, 1998), Chap. 6.
10. L. D. McMillan, M. Huffman, T. L. Roberts, M. C. Scott, and C. A. Paz de Araujo, Integrated Ferroelectrics, **4**, 319 (1994).
11. S. L. Miller and P. J. McWhorter, J. Appl. Phys., **72**, 5999 (1992).

WORK FUNCTION STUDY FOR THE SEARCH OF EFFICIENT TARGET MATERIALS FOR USE IN HYPERTHERMAL SURFACE IONISATION USING A SCANNING KELVIN PROBE

U. Petermann, I.D. Baikie, B. Lägel, K.M. Dirscherl
Department of Applied Physics, Robert Gordon University, Aberdeen, UK

ABSTRACT

In order to search for efficient target materials for use in Hyperthermal Surface Ionisation (HSI), a new mass spectroscopy ionisation technique, we have performed a study of high and low work function (ϕ) surfaces as part of an ongoing project. HSI relies on high and low work function surfaces for the production of positive (pHSI) and negative (nHSI) ions, respectively.
Using a novel UHV Scanning Kelvin Probe we have followed the oxidation kinetics of polycrystalline Re at different temperatures and examined the effects of oxidation, flash annealing and sputter-anneal cleaning cycles via high resolution work function topographies. Our results indicate that oxidised Re is the best candidate for pHSI in terms of ionisation efficiency and ϕ change. The peak work function change of 2.05 eV occurred at 900 K to 950K.
For the nHSI materials Calcium exhibited the best performance with respect to the ionisation efficiency indicating a wf of 2.9 eV. We will present data in terms of mass fragmentation using an HSI-Time-of-Flight (TOF) system and time stability of the work function.

INTRODUCTION

The objective of this study was to examine high (Rhenium Re, Tungsten W, Molybdenum Mo and Platinum Pt) and low (Calcium Ca, Lanthanum Hexaborid LaB$_6$, Gadolinium Gd) work function materials as target surfaces for Hyperthermal Surface Ionisation (HSI) [1]. In this application the sample vapour is ionised by supersonic collision with a suitable target surface and the resulting ion-beam is then mass analysed by a TOF spectrometer. As this technique does not use electron impact (EI) filaments the amount of cracking products is low [2]. Analytical merits of this new HSI-TOF technique include high sensitivity, controllable selectivity, informative mass spectra and an atmospheric pressure inlet. The target work function is a key parameter in determining the HSI efficiency: as the solute molecules approach a high or low work function surface some neutrals will, upon transfer of an electron to the target surface or versus visa, cross the electronic potential curve to form positive or negative ions, respectively.
The newly developed Ultra-High-Vacuum (UHV) compatible Scanning Kelvin Probe (SKP) offers unparalleled sensitivity to changes in surface dipole due to adsorption, contamination and structure. The Kelvin method [3,4] produces the average ϕ of the sample under the tip and not that biased to low ϕ patches as is the case for photo- and thermionic-emission. The SKP has a 1-2 meV work function resolution and employs a sample-to-tip tracking algorithm [5] without which ϕ measurements can be difficult to interpret [6,7]. This technique has previously been applied, together with STM, AES, LEED, TPD and EELS for research into semiconductor surface processing [8,9], co-adsorption [10] and thin metallic films [11,12,13] on metal surfaces as well as surface charge imaging of oxides and operational electronic devices [14]. We have also utilised the SKP to image contamination [15,16,17], sputter induced surface roughness [18] and the temperature/dose dependency of the Höfer precursor in the oxidation of Si(111) 7x7 [19,20,21].

37

Oxidation of the clean metal surfaces produces a strong dipole layer (negative end outwards) which significantly increases ϕ and therefore is used in the study for high work function materials.

The sensitivity of pHSI and nHSI was investigated using Di-Nitro-Toluene (DNT). Hydrogen was used as the carrier gas for the make-up flow for the supersonic molecular beam interface.

EXPERIMENTAL METHOD

Kelvin Method

If two conductors (or semiconductors) A and B with a dissimilar work function ϕ_A and ϕ_B are first brought in contact a flow of electron from the material with the higher work function will continue to the material with the lower work function until an equilibrium is reached and the electrochemical potential of electrons are equal (see Fig. 1b). When the system of A and B reaches the thermal equilibrium, in which the Fermi levels ε are alike, the contact potential difference, V_{cpd}, of the junction is established representing the difference in work function between the materials A and B.

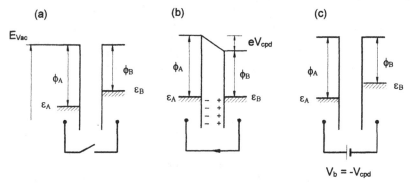

Fig. 1: Electron energy diagram of the traditional Kelvin Probe: a) separated charge-free plates b) both plates electrically connected and c) balanced circuit via an adjustable battery V_b, ε_A and ε_B to the Fermi levels of material A and B, respectively.

The inclusion of the "backing potential" V_b in the external circuit permits biasing of one plate with respect to the other: at a unique point, where $V_b = -V_{cpd}$, the electrical field between the plates vanishes [seeFig. 1c)], resulting in a null output signal.

By vibrating one of the plates a alternating flow of charges Q is induced, which can be described as time-varying current $i(t)$ following:

$$i(t) = \frac{dQ}{dt} = \frac{d}{dt}\left[V_c \cdot C_K(t)\right] \tag{1}$$

where V_c is the overall voltage across the capacitor and $C_K(t)$ denotes the time-altering Kelvin capacitor. V_c can be expressed by

$$V_c = V_{cpd} - V_b. \tag{2}$$

The work function difference between the plates is thus equal and opposite to the DC potential necessary to produce a zero output signal

Hyperthermal Surface Ionisation

The Hyperthermal Surface Ionisation can be described in terms of surface-molecule or molecule-surface electron-transfer process, where the molecule carries kinetic energy. This energy can be directly converted in order to bridge the gap between the molecular ionisation potential and the surface work function (positive ions-pHSI) or the surface work function and the molecular electron affinity (negative ions-nHSI) [22]. If a molecule approaches a surface at a given critical distance R_C, an electron can be spontaneously exchanged between the surface and the molecule. At R_C the stabilisation energy due to the image potential can overcome the energy difference ΔE and an electron transfer from the neutral into an ionic configuration can occur as shown in Fig. 2.

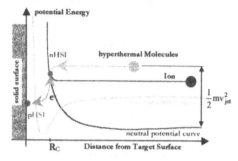

Fig. 2: Mechanism of Hyperthermal Surface Ionisation

EXPERIMENTAL

The polycrystalline sample ($25 \times 12 \times 0.025$ mm^3, 99.99%, GoodFellow) were mounted onto the manipulator using Ta pressure contacts. The UHV chamber base pressure after bakeout is 8×10^{-11} Torr. Clean surfaces were generated by direct current flash annealing to temperatures above 2000 K. Oxygen (99.995%) was inlet from a separately pumped gas inlet system and mass spectra were frequently taken to verify O_2 purity. Auger electron spectra were taken with a VG Rear View LEED optics assembly used in Auger electron mode.

LaB$_6$ was evaporated onto Re substrate using a electron beam evaporator EBN1 (Applied Oxford Research). The absolute work function scale has been generated by initially calibrating the Kelvin probe (KP) tip using Photoelectric (PE) measurements and calibration samples [23].

The absolute work function values of the cleaned samples used in the study are listed below.

Re	5.1 ±0.1 eV
W	4.65±0.1 eV
Mo	4.5 ±0.1 eV
Pt	5.7 ±0.1 eV
LaB$_6$	2.72±0.1 eV
Ca	2.9 ±0.1 eV
Gd	3.1 ±0.1 eV

The HSI experiments were carried out with HD Technology TOF instrument with a supersonic molecular beam (SMB) interface (HYPERJET).

RESULT AND DISCUSSION

High work function materials

Fig. 3 a) shows $\Delta\phi$ data during oxidation of the Re at substrate temperatures of 300-900 K. The peak ϕ value of the oxidised surface is strongly temperature dependant resulting in a $\Delta\phi$ of 1050, 1240, 1560 and 2050 meV for 300, 500, 700 and 900 K, respectively. The peak work functions remains constant after an O_2 exposure of 4×10^4 L ($1L=1$Langmuir$=1\times10^{-6}$ Torr sec) These $\Delta\phi$ results show a good agreement with measurements of ϕ carried out by Fusy et al.[24] and Zehner and Farnsworth [25] using retarding potential technique. At saturation coverage of oxygen they obtained maximum $\Delta\phi$ of 1.8eV at 830 K on recrystallized Re ribbon and 1.7eV on Re(0001) at 1050 K, receptively.

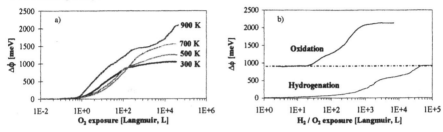

Fig. 3: a) $\Delta\phi$ during Re oxidation at substrate temperatures of 300-900 K b) $\Delta\phi$ upon a hydrogenation and a subsequent oxidation of Re at 900 K.

In our HSI application the sample gas is first mixed with a low molecular weight carrier gas (H_2) and injected through a small diameter nozzle to create a supersonic molecular beam. In the beam the incoming sample gas molecules can achieve kinetic energies in the region of 2-8 eV. The interaction of the H_2 carrier gas on the oxidised Re surface was investigated. Fig. 3 b) shows the change in ϕ during the uptake hydrogen at 900 K followed by a subsequent oxidation at the same temperature. During this experiment the temperature was kept at 900 K at all time. The hydrogenation of Re resulted in $\Delta\phi$ of ~0.910 eV, while the subsequent oxidation additionally contributed ~1.2 eV resulting in total ϕ change of ~2.11 eV for both uptakes. This value is in good agreement with $\Delta\phi$ of pure oxidation of 2.2eV at 900 K, showing no significant reduction in ϕ due to hydrogenated Re surface. This indicates that the adsorbed hydrogen is replaced with the more electronegative oxygen.

The work function values upon oxidation of the metals Re, W, Mo and Pt as a function of the substrate temperature are exhibited in Fig. 4. It clearly indicates that Re produced the highest work function of all 4 metals tested peaking at a ϕ value of 7.15 ± 0.10 eV at ~900K. Mo, W and Pt showed a maximum work function of 6.45eV at 1000K, 6.33eV at 1000K and 6.32eV at 750K, respectively.

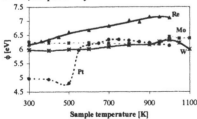

Fig. 4: Stable work function values upon oxidation of the tested metals Re, W, Mo and Pt versus substrate temperature

The oxidation after pre-adsorption of H_2 on Mo, W and Pt did reveal a change more than ±50 meV in the ϕ value of the clean surface oxidation.

Stability of the low work function materials

The work function as a function of time was monitored at a base pressure of ≈ 4 × 10^{-9} Torr for Ca, Gd and LaB_6 as shown in Fig. 5. Gd and LaB_6 were cleaned by flashing up to 1200K and 1700K, respectively. Ca was mechanically cleaned until a shiny surface is exposed and admitted into the vacuum chamber. It can be seen that Ca, Gd and LaB_6 all indicated a increase in work function with time. Gd has shown the best ϕ stability with an increase of 70 meV over 30 hours.

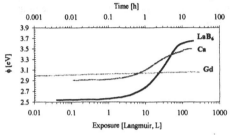

Fig. 5: Change of the work function with time of Ca, Gd and LaB_6 due to adsorption of residual gases at a base pressure of ≈ 4 × 10^{-9} Torr

The target materials were accordingly tested in positive and negative HSI mode with explosive compounds (DNT). Fig. 6 shows the mass spectra of DNT in a) nHSI mode on Ca at ≈ 500 K, b) pHSI mode on Re at 950 K and c) nHSI mode on Gd at 800 K.

HSI study

The sensitivity and mass spectral characteristics were observed to vary according to the temperature of the rhenium surface and therefore the surface work function.
With the calcium and gadolinium surfaces, operating in negative ion mode, both the explosive compounds were detected. DNT (molecular mass M = 182) shows molecular ions at m/z 182, and very little significant fragmentation is observed in the spectrum of Ca, shown in Fig. 6 a).

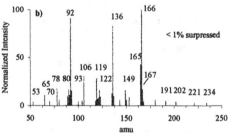

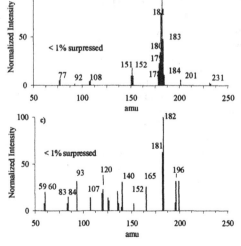

Fig. 6: HSI-TOF mass spectra obtained with a) DNT on Ca surface in negative ion mode at ≈ 500 K b) DNT on Re surface in positive ion mode at 950K and c) DNT on Gd surface in negative ion mode at 800K.

The fragmentation becomes higher in the mass spectrum of DNT in nHSI mode on Gd [Fig. 6 c)] compared to Ca. In pHSI mode using DNT on Re, the spectrum in Fig. 6 b) shows only fragments of the DNT molecules with the most abundant ions at m/z 92, 136 and 166. No molecular ions of DNT at m/z 182 were detected.

CONCLUSION

We have demonstrated that the Scanning Kelvin Probe is a useful tool in a search of efficient target materials for use in Hyperthermal Surface Ionisation. The work function study indicates that oxidised Re is the best candidate for pHSI with $\phi = (7.15 \pm 0.1)$ eV in the temperature range of 900-950 K, while Ca gave the best results for nHSI using the explosives DNT.

ACKNOLEDGEMENTS

This work was supported by EPSRC and DERA.

REFERENCES

1. A. Danon and A. Amirav, Rev. Sci. Instrum., **58**, 1724 (1987).
2. A. Amirav. Org. Mass Spectrom. **26**, 1 (1991).
3. Lord Kelvin, Philos. Mag. **46**, 82 (1898).
4. W.A. Zisman, Rev. Sci. Instrum., **3**, 367 (1932).
5. I.D. Baikie, E. Venderbosch, J.A. Meyer and P.J Estrup, Rev. Sci. Instrum., **62**, 725 (1991).
6. I. D. Baikie, S. Mackenzie, P.J. Z. Estrup and J.A. Meyer, Rev. Sci. Instrum., **62**, 1326 (1991).
7. B. Ritty, F. Wachtel, R. Manquenouille, F.Ott and J.B. Donnet, J. Phys. E. **15**, 310 (1982).
8. I.D. Baikie, Mat.Res. Soc. Proc, **204**, 363 (1991).
9. I.D. Baikie, U. Petermann and B. Lägel, Surf. Sci. **433-435**, 249 (1999).
10. J.A. Meyer, I.D. Baikie, G.P. Lopinski, J.A. Prybyla and P.J. Estrup, J. Vac. Sci. Technol., **8**, 2468 (1990).
11. E. Kopatzhi, H-G Keck, I.D. Baikie, J.A. Meyer and R.J. Behm, Surf. Sci. **345**, L11 (1996).
12. J.A. Meyer, I.D. Baikie, E. Kopatzki and R.J. Behm, Surf. Sci. **365**, L647 (1996).
13. I.D. Baikie, U. Petermann and B. Lägel, Surf. Sci. **433-435**, 770 (1999).
14. I.D. Baikie and G.H. Bruggink, Mat. Res. Soc. Proc., **309**, 35 (1993).
15. I. D. Baikie, E. VenderBosch and B. Hall, Mat. Res. Soc. Proc., **261**, 149 (1992).
16. B. Lägel, I.D. Baikie and U. Petermann, Mat. Res. Soc. Proc., **510**, 619 (1998).
17. B. Lägel, I.D. Baikie and U. Petermann, Surf. Sci. **433-435**, 622 (1999).
18. I.D. Baikie and G.H. Bruggink, Mat. Res. Soc. Proc., **306**, 311 (1993).
19. U. Petermann, I.D. Baikie and B. Lägel, Thin Solid Films **343-344**, 492 (1999).
20. U. Höfer, P. Morgen, W. Wurth, Phy. Rev. **B40**, 1130 (1989).
21. I.D. Baikie, Mat. Res. Soc. Proc., **259**, 149 (1992).
22. A. Amirav and A. Danon, J. Phys. Chem. **93**, 5549 (1989)
23. B. Lägel, I.D. Baikie, K.M. Dirscherl and U. Petermann, to be published in Mat. Res. Soc. Proc.L (2000)
24. J. Fusy, B. Bigeard, A. Cassuto, Surf. Sci. **46**, 177 (1974).
25. D.M. Zehner, H.E. Farnsworth, Surf. Sci. **30**, 335(1972)

SURFACE INVESTIGATIONS ON SINGLE CRYSTAL ANATASE TIO₂

R. HENGERER*, L. KAVAN**, B. BOLLIGER***, M. ERBUDAK***, and M. GR TZEL*
*Laboratory of Photonics and Interfaces, Swiss Federal Institute of Technology, CH-1015 Lausanne.
**J. Heyrovsky Institute of Physical Chemistry, Academy of Sciences of the Czech Republic, CZ-18223 Prague. ***Laboratory of solid state physics, Swiss Federal Institute of Technology, CH-8093 Z rich.

ABSTRACT

Utilizing a chemical transport reaction, we succeeded in growing large and clean anatase TiO₂ single crystals whose surfaces could be characterized by standard physical and electro-chemical techniques. The examination of the structure of the clean (101) and (001) faces by low energy electron diffraction (LEED) and secondary electron imaging (SEI) showed that these surfaces are bulk terminated and thermodynamically stable. Impedance spectroscopy in aqueous solution revealed a slight difference in the flatband potential between the (101) and the (001) faces. This shift is also manifested in a different photocurrent onset potential and can be rationalized by a different water adsorption on the two surface structures. Voltammetry in aprotic solutions showed a different lithium insertion behavior for the two surfaces. This is explained by a different structural transparency of the anatase lattice in the two directions. Both findings favor the (001) over the (101) surface. These orientational dependencies may have some important technological relevance for the mesoscopic TiO₂ films used in solar cells and lithium batteries.

INTRODUCTION

TiO₂ is a wide-bandgap semiconductor that exists mainly in two polymorphic phases, anatase and rutile. So far, only surface properties of rutile are investigated systematically [1]. Due to the lack of suitably large and pure single crystals, investigations on specific surfaces are missing [2]. It is anatase, however, that plays a crucial role in a number of charge-separating devices, like dye-sensitized solar cells [3] or rocking-chair lithium batteries [4]. These devices are based on highly porous films with very large surface areas that consist of nano-crystalline anatase mainly exhibiting the (101) and (001) faces. Although they show already a remarkable performance, fundamental understanding of surface processes, like dye adsorption or charge transfer, is still deficient. Moreover, charge-transfer processes should generally depend on the crystallographic orientation, so separate characterizations of different surfaces are desirable.

Here we report the first successful imaging of the surface lattice of TiO₂ single crystals in the anatase structure by secondary-electron imaging (SEI) and low-energy electron diffraction (LEED), respectively. On both faces, which are originally bulk-terminated, we have observed reversible structural transitions, induced by sputtering and subsequent annealing. In addition, electrochemical and photoelectrochemical studies in aqueous media, as well as Li^+ insertion experiments were performed. We point at salient electrochemical differences between the two surfaces and relate them to the different surface structures.

EXPERIMENT

Anatase single-crystals were grown from the gas phase by a chemical transport reaction as described elsewhere [5]. The transparent and colorless crystals ordinarily have a truncated bipyramidal habitus exhibiting mainly the (101) and (001) surfaces with surface areas between 1 and 4 mm^2. The surface orientation of the crystals was proved by X-ray diffraction (Laue camera). Atomic force microscopy confirmed that the corrugation of the working surfaces was typically below 1 nm. Prior to insertion into vacuum, all crystals were rinsed in water, ethanol, and acetone and contacted with silver glue to a copper plate in direct contact with an oven. For electrochemical measurements the crystals were reductively doped by hydrogen at 500-600°C for 24 h, contacted by a Ga-In alloy to a copper wire and mounted using an epoxy resin.

We used a home-made electron gun in connection with a display-type, retarding grid spherical collector system for LEED. The experiments were performed in an ultrahigh-vacuum chamber with a total pressure in the lower 10^{-10} Torr range.

Electrochemical experiments in aqueous solutions were carried out in conventional three-electrode cells using a potentiostat. The reference electrode was a standard Ag/AgCl electrode in saturated KCl solution, a Pt-wire served as the counter electrode. Illumination of the working electrode was accomplished with a 450 W high-pressure xenon lamp through a quartz window in the cell. The light intensity was about 100 mW/cm^2 (white light); the distance between cell window and working electrode was approximately 1 cm. The studies of Li insertion were carried out in an Ar - filled glove box in a dry solution (10 to 15 ppm H_2O) of 1 M $LiN(CF_3SO_2)_2$ in ethylene carbonate (EC) and 1,2-dimethoxyethane (DME) (1:1 by mass). Li/Li^+ served as a reference, Pt as the counter electrode.

RESULTS

Surface structure

Fig. 1a) and b) show very sharp LEED images from the anatase (101) and (001) surfaces, respectively. The surface lattice constants, calculated from the LEED spots, correspond to the bulk values within an error limit of 1% for both faces. The symmetry of the patterns shows that the surfaces are not reconstructed. This result confirms experimentally a basic assumption underlying e.g. the theoretical calculation of adsorption of water molecules on these two anatase surfaces [9].

During sputtering, we have observed . a structural transition on both surfaces to titanium monoxide [7]. This phase transition is obviously induced by preferential sputtering of oxygen atoms. Upon annealing at 900K in UHV for 30min. the stoichiometry as well as the original anatase structure at the (101) face is restored. On the (001) face, on the other hand, a new pattern came into view (Fig. 1c). The symmetry and the lattice constants did not change, but three additional spots can be detected between the original main spots in the [100] as well as in the [010] direction. This hints toward a (1x4) superstructure with different domains aligned perpendicular to each other. A missing oxygen row model, where each 4th row of bridging oxygen

a)   b)   c)

Figure 1: LEED patterns of a) the (101) and b) the (001) surfaces of anatase. c)LEED pattern of the (001)surface after sputtering and annealing. (The electron energies are a) 102eV, b) 191eV and c) 122eV.)

atoms along the [100] (resp. [010]) direction is missing, might explain this superstructure[7].

Exposing the sample to oxygen or introducing air into the experimental chamber recovers the (1x1) surface order. These observations demonstrate hence the possibility to obtain clean and structurally perfect anatase surfaces.

Electrochemistry in aqueous media: photo-oxidation of water

Fig. 2 displays the photocurrent for water oxidation at anatase electrodes in 0.1 M HCl as a function of applied potential. The electrodes were illuminated by chopped white light of a Xe-lamp. The photocurrent raises with the square root of the applied voltage as predicted by the equation [2]:

$$i_{ph}^2 = 2e\varepsilon_0\varepsilon_r \, (I_0^2\alpha_o^2/N)(U-U_{fb}) \qquad (1)$$

(U is the electrode potential, U_{fb} the flatband potential, N the donor density, I_0 the illumination intensity and α_o the optical absorption coefficient). The onset of photocurrent is at about -0.28 V vs. Ag/AgCl for the (001) orientation and -0.24 V for the (101) orientation. This hints towards a more negative flatband potential (E_{fb}) for the (001). The start of the dark-reduction of water at potentials negative to E_{fb}, too, shows a negative shift for the (001) surface. For a more precise determination of the flatband potential, impedance spectra were also recorded. The observed shift equaled (0.06 – 0.02) V on the average [8].

In aqueous media, the E_{fb} is controlled by the acid-base equilibrium involving surface OH groups [2]. Recent theoretical calculations indicated a significant difference in the adsorption of water for these two surfaces [9]. The clean (101) surface, which contains half-penta-/ half- hexacoordinated Ti atoms, adsorbs water non-dissociatively. On the other hand, the clean (001) surface contains only pentacoordinated Ti atoms. On the (001) surface, water is spontaneously dissociated, which leads to the anchoring of isolated OH groups to Ti atoms. All surface Ti atoms stay pentacoordinated even if they carry terminal hydroxyl groups. Consequently, the more acidic (001) surface attracts less protons, which explains the negative shift of E_{fb}.

The demonstrated orientation dependence of the flatband potential may have technological relevance for the anatase-based solar cells. A more negative flatband potential allows higher open

a)   b)

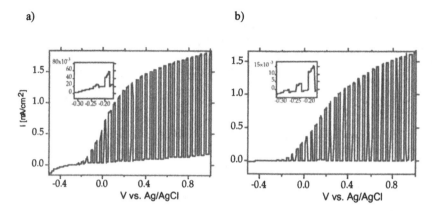

Figure 2: Photocurrent densities for a) the (101) and b) the (001) surface of anatase. The insets show a magnification of the region where the photocurrent starts.

circuit photovoltages (OCV) to be achieved in solar cells. This shift could be, presumably, achieved by the preferential use of (001) surfaces. Such a surface is available, e.g. via self-organization of rod-like particles at controlled conditions [10].

Electrochemistry in aprotic medium: Li⁺ insertion

Fig. 3 displays cyclic voltammograms of anatase (101) and (001) faces in 1 M $LiN(CF_3SO_2)_2$ + EC/DME (the scan rate is 0.1 mV/s). The cathodic/anodic peaks correspond to the insertion/extraction of Li⁺ to/from the anatase lattice:

$$TiO_2 + x\,(Li^+ + e^-) \rightarrow Li_xTiO_2 \qquad (2)$$

where the insertion coefficient, x is usually ≤ 0.5 [11]. The easier Li⁺ insertion through the (001) face is clearly apparent. In contrast to polycrystalline electrodes, which show the ratio of inserted-to-extracted charge close to 100 %, the macroscopic single crystals exhibit considerably lower extraction charges as compared to those for insertion. This is apparently caused by the diffusion of Li⁺ into the bulk crystal, from where the Li⁺ ions are not completely recuperated at the time base of the reverse scan.

Diffusion coefficients for the diffusion of lithium ions in the anatase lattice were obtained from chronoamperometric measurements. They showed noticeably higher values for the (001) surface for both insertion and extraction [8]. For both surfaces they follow an Arrhenius-type temperature dependence (shown for insertion in Fig. 3 b; the extraction case shows the same characteristics, but is omitted here due to the lack of space):

$$D = D_0 \exp(-E_a/kT) \qquad\qquad (3)$$

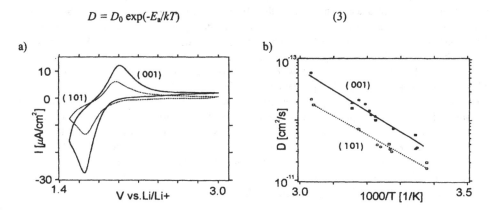

Figure 3: a) Cyclic voltammogram of anatase in 1 M LiN(CF$_3$SO$_2$)$_2$ + EC/DME. b) Chemical diffusion coefficients for Li+ insertion into anatase.

where E_a is the activation energy. Hence we can estimate the activation energies for Li$^+$ insertion to be between 0.6 and 0.7 eV and for extraction about 0.5 eV for both orientations of the anatase crystal. These values are in a remarkable agreement with the activation energy predicted theoretically by ab-initio calculations ($E_a = 0.6$ eV) and also with the values reported for poly-crystalline anatase (0.51 to 0.66) [12].

Both voltammetric and chronoamperometric experiments point at more facile Li$^+$ insertion through the (001) face. This difference in the diffusion coefficients (and the close agreement of the activation energies) fits theoretical predictions that Li$^+$ in anatase moves along zigzag channels connecting the octahedral voids in the lattice [12]. The transport of Li$^+$ to a certain distance perpendicular to the (001) face requires fewer jumps between the equilibrium positions than the same excursion underneath the (101) face.

The conclusion that the (001) face is more permeable for Li$^+$ also rationalizes our previous experimental finding that rod-like self-organized anatase nanocrystals show more facile accommodation of Li$^+$ as compared to ordinary, statistically oriented nanocrystals [13]. The surface of an array of self-organized particles (rods) is composed of (001) faces i.e. it is a mosaic structure, which mimics the (001) face of a single crystal [10]. On the other hand, the ordinary and non-oriented crystals expose prevailingly their (101) faces. This finding may have straightforward practical implications for anatase-based lithium batteries.

CONCLUSIONS

The goal of this work was to investigate the surface geometry of the anatase (101) and (001) surfaces and to relate it to orientation dependences of charge-transfer processes. First, both surfaces were studied by SEI and LEED. We have observed a structural transition to titanium monoxide on the surfaces, induced by preferential sputtering. Upon annealing, the stoichiometry as well as the original bulk-terminated anatase structure at the (101) face is restored. The (001) surface, on the other hand, shows a (1x4) anatase superstructure after sputtering and annealing, which is resolvable by missing oxygen rows. The original bulk-terminated (001) structure, which has a (1x1) symmetry as the (101) face, can be restored by exposure to oxygen.

Electrochemical experiments show clearly an orientation dependence of charge transfer processes on anatase electrodes. In aqueous solutions, the (101) and the (001) surface have different flatband potentials due to different chemisorption of water on the anatase surface. This gives rise to a negative shift of the onset potential for photo-oxidation of water for the (001) surface. In aprotic solvents, differences in the insertion of lithium ions are also due to the anisotropy of the tetragonal bulk anatase lattice, which is less dense in the (001) planes as compared to the (101) planes. Therefore, the propagation of Li^+ into anatase is faster in the c-axis direction. The demonstrated orientational effects on the charge transfer on single crystals may have some technological relevance for the mesoscopic films used in the solar cells and lithium batteries.

ACKNOWLEDGEMENTS

This work was supported by the Swiss National Science Foundation.

REFERENCES

1. V. E. Henrich and P. A. Cox, *The surface science of metal oxides* (Cambridge University Press, Cambridge, 1994), pp.43-49.

2. H. O. Finklea, *Semiconductor Electrodes*, (Elsevier, Amsterdam 1988), pp. 30-145.

3. B. O'Regan and M. Gr tzel, *nature* **353**, 737 (1991).

4. S. Y. Huang, L. Kavan, M. Gr tzel and I. Exnar, *J. Electrochem. Soc.* **142**, 142 (1995).

5. L. Kavan, M. Gr tzel, S. E. Gilbert, C. Klemenz and H. J. Scheel, *J. Am. Chem. Soc.* **118**, 6716 (1996).

6. M. Erbudak, M. Hochstrasser, E. Wetli and M. Zurkirch, *Surf. Rev. Lett.* **4**, 179 (1997).

7. R. Hengerer, B. Bolliger, M. Erbudak and M. Gr tzel, *Surf. Sci.* **460**, 162 (2000).

8. R. Hengerer, L. Kavan, P. Krtil and M. Gr tzel, *J. Electrochem. Soc.* **147**, 4 (2000).

9. A. Vittadini, A. Selloni, F. Rotzinger and M. Gr tzel, Phys. Rev. Lett. **81,** 2954 (1998).

10. S. Burnside, V. Shklover, C. Barbe, P. Comte, F. Arendse, K. Brooks and M. Gr tzel, Chem. Mater. **10,** 2419 (1998).

11. L. Kavan, M. Gr tzel, J. Rathousky and A. Zukal, J. Electrochem. Soc. **143,** 394 (1996).

12. S. Lunell, A. Stashans, H. Lindstr m and A. Hagfeldt, J. Am. Chem. Soc. **119,** 7374 (1997).

13. P. Krtil, D. Fattachova, L. Kavan, S. Burnside and M. Gr tzel, Solid State Ionics, in press.

INTERFACE STABILITY IN HYBRID TRANSITION METAL-OXIDE MAGNETIC JUNCTIONS

J. Z. SUN*[t], K. P. ROCHE** and S. S. P. PARKIN**
*IBM T. J. Watson Research Center, PO Box 218, Yorktown Heights, NY 10598
**IBM Almaden Research Center, 650 Harry Road, San Jose, CA 95120
[t]e-mail: jonsun@us.ibm.com

Abstract

Recent experiments revealed an apparently bias-dependent tunneling magnetoresistance between a transition metal (such as Fe, Co) and an oxide barrier such as $SrTiO_3$. We examine the materials issues involved in this type of hybrid transition metal-oxide junctions. The junction interface is shown to be unstable against thermal treatment or high-bias current stress. We conclude that the junction magnetoresistance is largely determined by the formation of an interface oxide layer different from the barrier or the transition metal electrode themselves.

Introduction

Doped manganites such as $La_{0.67}Sr_{0.33}MnO_3$ (LSMO) have attracted much attention because of their strong conduction band spin-polarization. Recently, for study of spin-polarized tunneling, junctions were fabricated between LSMO and a ferromagnetic transition metal such as Co[1-4]. In some of such structures, a thin layer of eptaxially grown $SrTiO_3$ (STO) was used as tunneling barrier[2-4]. In these junctions it was observed that the sign of the junction's tunneling magnetoresistance is bias-dependent. Similar junction MR sign-reversal upon changing bias condition has also been observed in magnetic tunneling junctions between two Permalloy ($Fe_{20}Ni_{80}$) electrodes with a double-layer barrier material Al_2O_3/Ta_2O_5 sandwiched in between[5]. Since this type of tunneling spectroscopy has the potential of revealing the electronic structure of the electrodes involved[2-5], it is important to understand and control the role materials chemistry play at the junction interface. This is the objective of this study.

Two model systems were chosen. One is a trilayer structure of LSMO-STO-Fe, the other of LSMO-STO-$Co_{80}Fe_{20}$ (LSMO-STO-CoFe). Junctions made from these trilayer films reveal an unstable interface between the transition metal and the STO barrier. The interface property appears to be the controlling factor for the junction's magnetoresistance behavior – both in terms of its field-dependence and its bias dependence, including the sign of the junction magnetoresistance (MR).

Materials synthesis and device fabrication

The trialyers were deposited on (110) cut $NdGaO_3$ substrates, 1 cm × 1 cm in size. The bottom layer was a 600 Å thick LSMO film, epitaxially grown using laser ablation. Then a thin layer of $SrTiO_3$ film, nominally 30 Å thick, was epitaxially deposited under the same condition using laser ablation. For both layers, the deposition was carried out in 300 mTorr of oxygen background pressure at a substrate block temperature of 780 C. The substrates were thermally anchored to the holder with silver paste. A Nd-YAG laser was used for the ablation process, operating in frequency tripled mode (355nm) with a repetition rate of 10 Hz. The energy density on target surface was around 2 J/cm^2 per pulse. This deposition procedure is similar to what has been used before for all-oxide deposition, the films were cooled to ambient temperature in 300 Torr of oxygen. They were then transported to another vacuum system, where the transition metals, 100 Å of either Fe or CoFe, were sputter deposited. A brief oxygen plasma cleaning was given to the films immediately prior to the deposition of the transition metal electrode. The films were capped with 100 Å of titanium before being removed from vacuum for lithographic processing.

The trilayer films were processed using 1× contact optical lithography to form the current perpendicular (CPP) junction geometry. Maximum temperature exposure during the lithography process was 90 C for an accumulated duration of about 5 minutes. Pattern transfer was done using Ar ion milling, with a neutralized beam of 500 eV and 0.3 mA/cm^2 in flux. The film was thermally anchored to a water-cooled rotating table using thermally conducting grease. An SiO$_2$ layer, about 1500 Å thick, was used to isolate the top contact level from the bottom LSMO. Subsequently a lift-off step was applied to remove the junction level photoresist, and to open a self-aligned hole structure on the SiO$_2$ layer for junction contact from the top. A Ti layer of 600 Å was sputtered down as top contact.

Overview of junction behavior

Transport measurement of junction resistance and magnetoresistance was carried out in a close-cycle refrigerator-cooled system. For $R(H)$ measurements, the magnetic field was continuously swept at a frequency of 0.077 Hz. In all measurements, the positive terminal denotes the base LSMO electrode. Thus a positive bias has current flowing from LSMO through the STO barrier into the top transition metal electrode. Electrons flow in the opposite direction. The bottom electrode, LSMO, is a 33% hole-doped metal. The transition metal counter electrodes on top are n-type metals. The barrier is SrTiO$_3$ which when fully oxygenated is a semiconductor with a

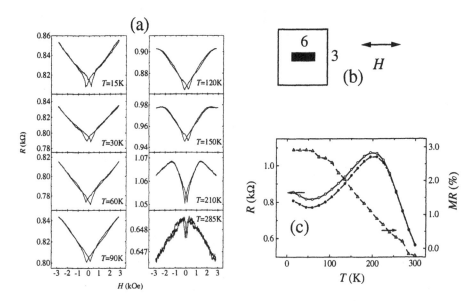

Figure 1: Temperature dependence of junction resistance and junction MR. (a)The evolution of junction $R(H)$ as a function of temperature; (b)Junction geometry and the direction of sweeping field. Numbers indicate feature size in μm.(c)The temperature dependence of the junction resistance-high and resistance-low states as defined from traces obtained in (a). Right: the temperature dependence of junction MR. The junction is made from an LSMO-STO-Fe trilayer. The over-all features and shapes of all junctions appeared similar, although the details of individual junction's $R(H)$ show large variations from device to device, as will be discussed below.

calculated band-gap around 5.1 eV[10].

For this study we define the sign of junction MR to be the same as $[R(H) - R(0)]/R(H)$, where $R(H)$ is the resistance value at ± 3.5 kOe, and $R(0)$ is that in the low-field region. All $R(H)$ curves shown are 10-trace averaged results.

A representative temperature dependence of a junction's resistance and MR is shown in Fig.1. The sheet resistance of the base LSMO film was verified to be about 30 to 60 Ω at low temperature (13 K). Due to the temperature dependence of the resistance of LSMO, the $R_\square$ was about a factor of 20 higher at ambient. This is important to keep in mind, as magnetoresistance measurements in such geometry is only accurate if the junction resistance is at least ten times larger than the sheet resistance $R_\square$ of the base electrode[11, 12]. This condition, while well satisfied at low temperatures, was not satisfied near ambient for this particular junction. This in combination with the diminishing value of junction MR caused an apparent MR reversal in data shown here for $T = 210$ and 285 K. These temperature dependence data are shown here to give a general view of the transport behavior of this type of junctions. For the remainder of this paper, we will only focus on the junction MR behavior at the low temperature end of 13 K.

Junction MR variations

A large variation of junction MR was observed from device to device and from chip to chip. Junction MR could be both positive and negative in value regardless of the choice of top transition metal material, and it showed sensitive dependence on the processing-history of the junction. These observations lead us to conclude that it is the interface chemistry that is determining the junction's MR behavior in these devices.

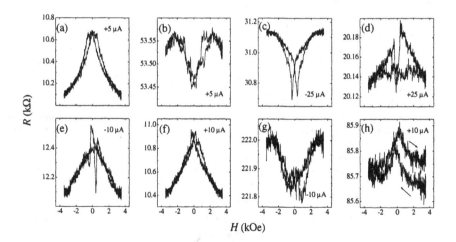

Figure 2: The shape and sign of MR for several junctions. (a)An LSMO-STO-Fe junction showing only negative MR; (b)Another LSMO-STO-Fe junction on the same chip, 200 μm away, showing a positive MR at the same bias current. (c) and (d): The junction in (b) has an asymmetric bias-dependence of MR, shown here as an example. (e) and (f): A LSMO-STO-CoFe junction that shows a negative MR during initial measurement; (g) and (h): the same junction as in (e) and (f), but after a high-current bias up to ± 0.375 mA. The junction resistance increased by a factor of 20, and the junction MR changed sign for negative biases. The drift of $R(H)$ in (h) is related to the junction resistance creep after high-bias stressing, as will be discussed below.

Fig.2 briefly illustrates our observations. For one LSMO-STO-Fe junction, a negative MR was observed at all bias voltages up to 1 V. An example is shown in Fig.2(a). For another junction on the same chip, just 200 μm away, a positive MR was observed with an asymmetric bias-dependence(Fig.Fig.2(b)-(d)). For a LSMO-STO-CoFe junction, a negative MR was initially observed in both bias directions at low voltage. Upon biasing up to about 1 V, junction resistance irreversibly increased by a factor of 20. From then on, the MR changed sign for negative biases (Fig.2(e)-(h)).[15]

Data in Fig.2 indicate that junction MR is sensitive to details of junction interface condition. The choice of transition metal electrode alone is insufficient to determine even its sign.

The shape of junction $R(H)$ as shown in Fig.2 contains large high-field slopes in the kOe field range. This indicates the magnetic spins at the junction interface are loosely coupled to the bulk of the electrodes, or they are under a relatively strong random anisotropy potential. This is obvious because the transition metal electrode films themselves have magnetic coercivities only in the range of several tens of Orsteds. The much higher field-scale involved in these $R(H)$ data, and the gradual slope as opposed to sharp switching in $R(H)$ vs H are both indications that the magnetic moments relevant to spin-polarized tunneling are not rotating in concert with the magnetic domain structures in the bulk of the transition metal electrode. Since spin-dependent tunneling is sensitive only to the magnetic states several monolayers deep into the electrode[13, 14], this anomalous magnetic layer at the junction interface may be very thin (a monolayer or two) and still be sufficient to dominate the MR behavior of the junction. It is conceivable therefore that an oxidized interface transition region between the SrTiO$_3$ barrier and the top transition metal electrode is responsible for these observations.

Shape of $R(H)$ and junction inhomogeneity

Another aspect concerning the complexity of the shape of the $R(H)$ is that its shape appears to depend on junction bias. This is shown in Fig.2 (compare the shape of (b) and (d), for example). A bias-dependent *shape* of $R(H)$ makes sense only if one assumes the conduction across the junction barrier is inhomogeneous and non-linear at the same time – this will cause different current paths to dominate transport at different bias levels. Since different current path samples different local magnetic environments, the over-all shape of $R(H)$ can therefore be bias-dependent.

Junction barrier inhomogeneity is certainly present over a macroscopic length-scale. Of all junctions measured on the same chip, a scatter of junction resistance over two orders of magnitude is seen, both for room temperature and for low temperature (13 K) junction resistance. The microscopic materials origin of such inhomogeneity is however not characterized.

A natural consequence of inhomogeneous junction conduction is locally concentrated current flow. The average current density for measurements presented in this work is of the order of 10^3 A/cm^2. Because of the inhomogeneities, the local current density could be much higher.

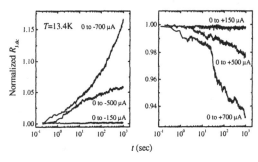

Figure 3: At high bias currents, junction resistance show significant creep. The direction of resistance creep depends on polarity of the bias-current, the rate of creep increases with increasing magnitude of the bias-current. The junction was made with LSMO-STO-CoFe.

Resistance creep under high-bias: oxygen electromigration

For junctions under large current bias, a creep of resistance was observed. Prior to irreversible change at the damage-threshold, junction resistance creeps reversibly to higher or lower value, depending on the direction of the bias. This is summarized in Fig.3. The junction presented here is 6×2 μm^2 in size. The average current density applied, at 700 μA, is about 5.8×10^3 A/cm^2. Application of a magnetic field up to 3 kOe did not change the rate and amplitude of the creep.

This phenomenon of junction resistance creep is qualitatively similar to what was observed in some other oxide junction structures under high current stress – such as in a superconducting tunneling junction with an indium-oxide barrier[16], or as in the grain-boundary junction of a cuperates superconductor $YBa_2Cu_3O_{7-\delta}$ thin film[17]. The *reversible* nature of the creep upon bias reversal suggests the involvement of electromigration. The most mobile element in these structures is oxygen[16, 17]. Hence, oxygen-related electromigration, and the resulting preferential oxidation of the transition metal electrode is the most likely cause of this reversible creep phenomenon.

The instability of these trilayer junctions was further demonstrated by observing the change of junction transport characteristics upon a mild heat-treatment of the junction. Fig.4 shows an example. Here a LSMO-STO-Fe junction underwent a 220 C and 10 min. heat-treatment in 5 kOe of applied magnetic field. The heat-treatment was originally designed for resetting the magnetic orientation of the electrodes. Instead of a simple change in the shape of $R(H)$ as would have been the case if magnetic anisotropy was the only factor, the junction exhibited marked change in its transport behavior. The junction resistance after annealing was a factor of 5 higher, the junction MR changed sign, and the junction developed a marked asymmetry against bias. This clearly indicates an alteration to the nature of transport at the junction interface. It also suggests that a tunneling into an Fe electrode with a more oxidized interface with STO could bring about a negative junction MR, whereas an Fe electrode not as strongly oxidized at the junction interface may give a positive MR with the presence of a lot of loose spins.

Summary

In summary, the junction interface between $SrTiO_3$ and transition metal electrodes such as Fe and CoFe is unstable. Formation of some type of interface oxide with the transition metal is likely. The nature of this interface oxide layer will to a large degree determine the junction MR behavior. This interface oxide layer is process-sensitive, and it can be altered by transport current across the junction. These factors should be carefully taken into account when one attempts to obtain any quantitative understanding of the magnetoresistance in magnetic tunneling junctions.

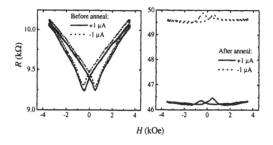

Figure 4: The resistance of an LSMO-STO-Fe junction before and after a 200C, 10 min anneal. After anneal, a 5× increase of junction resistance, as well as a sign change of the junction MR, is observed. The junction after heat-treatment also developed a strong asymmetry against bias.

Acknowledgement

The authors wish to thank Bill Gallagher's group for help during sample preparation.

References

[1] J. Z. Sun, D. W. Abraham, K. Roche, and S. S. P. Parkin, Appl. Phys. Lett. **73**, 1008 (1998).

[2] J. M. De Teresa, A. Barthélémy, J. P. Contour, A. Fert, R. Lyonnet, F. Montaigne, A. Vaurès, and P. Seneor, MRS Symposium, 1999 Spring Meeting, San Francisco (1999).

[3] J. M. De Teresa, A. Barthelemy, A. Fert, J. P. Contour, R. Lyonnet, F. Montaigne, P. Seneor, and A. Vaurès, Phys. Rev. Lett. **82**, 4288 (1999).

[4] Jose Maria De Teresa, Agnès Barthélémy, Albert Fert, Jean Pierre Contour, François Montaigne, and Pierre Seneor, Science **286**, 507 (1999).

[5] Manish Sharma, Shan X. Wang, and Janice H. Nickel, Phys. Rev. Lett. **82**, 616 (1999).

[6] J. Z. Sun, W. J. Gallagher, P. R. Duncombe, L. Krusin-Elbaum, R. A. Altman, A. Gupta, Yu Lu, G. Q. Gong, and Gang Xiao, Appl. Phys. Lett. **69**, 3266 (1996).

[7] J. Z. Sun, L. Krusin-Elbaum, A. Gupta, Gang Xiao, P. R. Duncombe, and S. S. P. Parkin, IBM J. of Res. and Dev. **42**, 89 (1998).

[8] J. Z. Sun, L. Krusin-Elbaum, P. R. Duncombe, A. Gupta, and R. B. Laibowitz, Appl. Phys. Lett. **70**, 1769 (1997).

[9] Yu Lu, X. W. Li, G. Q. Gong, Gang Xiao, A. Gupta, P. Lecoeur, J. Z. Sun, Y. Y. Wang, and V. P. Dravid, Phys. Rev. B **54**, R8357 (1996).

[10] L. F. Mattheiss, Phys. Rev. B **6**, 4718, 4740(1972).

[11] R. J. Pedersen and F. L. Vernon, Appl. Phys. Lett. **10**, 29 (1967).

[12] R. J. M. Van de Veerdonk, J. Nowak, R. Meservey, J. S. Moodera and W. J. M. De Jone, Appl. Phys. Lett. **71**, 2839(1997).

[13] Jagadeesh S. Moodera, Janusz Nowak, and Rene J. M. van de Veerdonk, Phys. Rev. Lett. **80**, 2941 (1998).

[14] Jagadeesh S. Moodera, Janusz Nowak, Lisa R. Kinder, Paul M. Tedrow, René J. M. Van de Veerdonk, Bart A. Smits, Maarten Van Kampen, Henk J. M. Swagten, Wim J. M. De Jonge, Phys. Rev. Lett. **83**, 3029(1999).

[15] For a more complete description of the evolutoin of $R(H)$ at different biases for these junctions, see J. Z. Sun, K. P. Roche and S. S. P. Parkin, Phys. Rev. B **61**, (April 1, 2000).

[16] K. H. Gundlach and H. Konishi, Appl. Phys. Lett. **46**, 441 (1985).

[17] B. H. Moeckly, D. K. Lathrop, and R. A. Buhrman, Phys. Rev. B **47**, 400 (1993).

NEW STRUCTURE MODEL FOR LITHIUM NICKEL BATTERIES

M.A. MONGE*, E. GUTIÉRREZ-PUEBLA*, I. RASINES*, J.A. CAMPA**
*Instituto de Ciencia de Materiales de Madrid, CSIC, Cantoblanco, E-28049 Madrid, Spain.
rasines@icmm.csic.es
** Facultad de Ciencias Geológicas, Universidad Complutense de Madrid, E-28040 Madrid, Spain.

ABSTRACT

After growing black colored single crystals of $Li_xNi_{1-x}O$ (x = 0.27) and solving its crystal structure, a model is proposed as an alternative to the structural type admitted for $Li_xNi_{1-x}O$. The new rhombohedral cell consists of a cubic close packing of oxygens in which the alternation of two kind of mixed layers containing Li and Ni in different ratios can be detected along the c direction. This model implies a Li order incompatible with the alternation of Ni layers, one of them pure and the other containing some Li. The results of magnetization measurements look consistent with the new structural type, and reveal that $Li_xNi_{1-x}O$ behaves as mictomagnetic.

INTRODUCTION

$Li_xNi_{1-x}O$ shows a practical interest since it could replace more expensive lithium cobalt oxide as a cathode material for long-life 4V rechargeable lithium batteries, and for intermediate temperature fuel cells. A recent electrochemical test [1] of a $Li_xNi_{1-x}O$ half cell has shown a high initial discharge capacity, 158 mAhg^{-1}, but very fast fading just after several tens of charge/discharge cycles. When used as a cathode material, $Li_xNi_{1-x}O$ exhibits higher electrical conductivity and better H_2/O_2 fuel cell performance than cobalt lithium oxide [2], although its thermal stability at intermediate temperatures is relatively poorer than that of the later.

In $Li_xNi_{1-x}O$ x can vary [3] between $x=0$ and $x<0.5$, corresponding to Ni oxidation states of 2^+ and 3^+. Samples with $x=0.5$ have never been obtained, since they always contain some Ni^{2+}. For small x values, $0<x\leq 0.2$, $Li_xNi_{1-x}O$ keeps the NaCl structure type of NiO. For $0.2<x<0.5$, it is [4] rhombohedral, space group (S.G.) $R\overline{3}m$, with Li at $3a$ (0,0,0), Ni at $3b$ (0,0,½), O at $6c$ (0,0,z) and $z \cong 0.25$. Its unit cell is a close packing of oxide anions in which ordered layers of Ni and Li parallel to the (111) plane of the parent cubic NiO lattice are present. Many profile refinements of X-ray and neutron diffraction data have confirmed this model, although more recently [5-7] the presence of a fraction of Ni atoms in the Li layers has been detected.

The magnetic properties of $Li_xNi_{1-x}O$ have been interpreted in all the ways imaginable, probably because a variety of compositions of this material occur depending on the synthesis conditions, and because the magnetization measurements were performed on polycrystalline samples of various qualities [8-10]. Since the samples used for all the investigations on $Li_xNi_{1-x}O$ were polycrystalline powders, we proposed to grow single crystals of this material, with a view to obtain more information on its crystal structure and intrinsic magnetic properties.

EXPERIMENT

Crystal Growth

$Li_xNi_{1-x}O$ single crystals were grown using platinum crucibles and reagent grade products.

Octahedral black colored crystals of $Li_{0.27}Ni_{0.73}O$ (Fig 1b) used for crystal structure determinations and magnetization experiments, showed edges of 0.2-0.3 mm and were ground from one mixture of 10 g $LiBO_2$, 1 g R_2O_3 and 0.5 g NiO, that was heated until 1200 °C at 4 °Ch^{-1}, soaked for 2 h, cooled down at 4 °Ch^{-1} to 920 °C and subsequently cooled to room temperature by turning the power off.

X-ray Structure Determinations

Special care was paid to the selection of single crystals. Each of those chosen was tested for perfection and scattering power on a Brucker-Siemens SMART diffractometer equipped with a CDC bidimensional detector, and normal focus 2.4 kW sealed tube X-ray source (graphite monochromated molybdenum radiation, $\lambda = 0.71073$ Å). Two crystals showing well defined faces were finally chosen and fully analyzed, leading to identical positions, site occupancies, and thermal factors. Here the only one that gave a slightly lower R value at the end of the refinement will be considered.

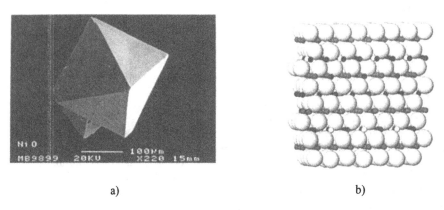

a)   b)

Figure 1a. One $Li_{0.27}Ni_{0.73}O$ crystal showing 0.2 mm edges. Figure 1b. View of the oxygen cubic close packing of $Li_{0.27}Ni_{0.73}O$ along c following the ABC layer sequence. The small spheres lie at positions occupied: only by nickel, Ni(1) and Ni(2), (black); half and half by Ni(3) and Li(1), (grey); or mainly by lithium, Li(1), (white).

Crystal data were collected over a quadrant of the reciprocal space by a combination of two 10 s exposure sets covering 0.3° in ω. Most of the calculations were carried out with SMART [11] software for data collection and data reduction, and SHELXT [11] for structure solution and refinements.

Magnetic Measurements

A SQUID magnetometer (Quantum Design) operating from 300 to 2.0 K at various fields between 1 and 50 kOe was used to perform the dc magnetic measurements in single crystals of $Li_{0.27}Ni_{0.73}O$ under field cooling and zero field cooling conditions. Diamagnetic corrections for magnetic susceptibilities [12] were taken into account. The field dependence of magnetization was measured at 5, 40, 110 and 200 K in magnetic fields ranging from –40 kOe to +40 kOe.

RESULTS

Crystal Structure

From a study of the weak reflections, the new hexagonal cell of parameters $a = 6.020(2)$ and $c = 14.726(5)$ Å became evident. Weak reflections unambiguously indexed in this unit cell, 1 2 -4 or -1 3 -4 for instance, could not have been indexed in the smaller one [3,4] with unit-cell parameters $a = 2.88$, $c = 14.2$ Å. The R3 (No. 146) S.G. [13] was obtained during the course of the structure solution after trying the hexagonal groups in which the only extinction is $-h+k+l \neq 3n$. In this S.G. anisotropic refinement for most of the atoms was performed. The small diference between the R value of 0.07 for observed reflections, $I > 2\sigma(I)$, and the one of 0.08 for all reflections, indicates the adequacy of the model for the weak reflections too.

Three Ni positions have been found in the asymmetric unit, two of them, Ni(1) and Ni(2), fully occupied. The electron density at the Ni(3) position, 15.4(1) e⁻, could be generated by sharing either one half of nickel, Ni(3), and one half of lithium, Li(1), or by 55 % of nickel, Ni(3), and 45 % of vacant sites. The standard deviations denounced some non resolvable disorder involving O(3) due to the fact of being in the polyhedron of a shared position. The high standard deviation found for Li(2) can be attributed to the small electron density at its site. The two possibilities indicated for the Ni(3) site lead to two formulas: I) without vacants, $Li_{0.27}Ni_{0.73}O$, and Ni average charge of 2.37; and II), with Ni vacants, $Li_{0.10}Ni_{0.74}O$, and a Ni charge of 2.57. At no time the total amounts of Ni and Li in the formula I were constrained to the unit. The $Ni_{0.73}Li_{0.27}O$ composition, with no vacancies, was obtained after refining the population factor in two separate shared positions.

The structure of $Li_{0.27}Ni_{0.73}O$ can be envisaged as a cubic close packing (ccp) of oxygens in which the cations occupy the octahedral holes; the anions follow the characteristic ABC layer sequence (Figure 1b); and the unit-cell volume per oxygen atom, 19.2 Å³, is comparable to those [14] of other ccp oxygen arrangements like cubic salt-like NiO, 18.2 Å³, and Li spinels, which vary from 16.2 to 19.6 Å³ for $LiNi_2O_4$ and $Li_{0.3}Mn_{0.7}Ti_2O_4$ respectively.

The cations in $Li_{0.27}Ni_{0.73}O$ give rise along the c direction to the alternation of L1 and L2 mixed layers containing Li and Ni in different proportions. For model I, L1 is formed by Ni(2) and $[Li(2)_{0.76}+Ni(4)_{0.24}]$, and L2 is also a mixed layer constituted by Ni(1) and $[Ni(3)_{0.52}+Li(1)_{0.48}]$, with Ni:Li proportions near to 4:1 and 5:3 respectively. In the case of model II, L2 would be formed by $[Ni(3)_{0.55}+ \; _{0.45}]$, and the concentration of Ni vacants of 45% would likely lead to an ordering of the Ni atoms which is not detected and, hence, this hypothesis looks less probable. From the analysis of the L1 and L2 layers it becomes evident that the way in which the lithium tends to ordinate itself forces the formation of the superstructure with parameters $a = 6.020(2)$ and $c = 14.726(5)$ Å ruling off, for the $Li_{0.27}Ni_{0.73}O$ composition, the small rhombohedral cell [3,4] of hexagonal parameters $a = 2.88$ and $c = 14.2$ Å as well as the monoclinic one defined in some articles [6,15]. In conclusion, the cation distribution in the oxygen ccp leads to the hexagonal unit-cell, and the order of Li atoms in L1 and L2 gives rise to the superstructure determined here.

In a recent study [15] of electron diffraction on polycrystalline $Li_{0.32}Ni_{0.51}O$, it has been presumed the existence of a hexagonal unit cell with parameters twice as long as those [3,4] of small rhombohedral $Li_xNi_{1-x}O$, in order to explain a possible ordering of vacancies among the *lithium layers*. In the present article it has been shown using single crystal X-ray diffraction that this cell exists, although what makes the parameter double is not the ordered Li vacants but the situation of Li in the supposed Ni layer to give L1.

Ni-O and Li-O selected distances are shown in Table I together with a conventional δ parameter which indicates the distortions of NiO_6 and LiO_6 octahedra. The large standard deviations which show the distances involving $O(3)$ or $Li(2)$ agree with the non resolvable $O(3)$ disorder, the high standard deviation obtained after refining $Li(2)$, and the most irregular octahedra of Table I, $Ni(3)O_6$ and $Li(2)O_6$, with δ values of 550 and 447. These are precisely the octahedra which lie at positions shared by Ni and Li. Table I also includes the shortest Ni-Ni distances for $Li_{0.27}Ni_{0.73}O$, which are larger than in Ni metal, 2.49 Å. They vary between 2.8(2) and 3.05(5) Å, being the intralayer Ni-Ni distances slightly larger than those between layers.

Table I. Selected Distances (Å) and δ Distortions, $\delta = 10^4 \Sigma (d-\bar{d})^2$, for $Li_{0.27}Ni_{0.73}O$

Ni(1)-O(2)	3x2.17(7)	Ni(3),Li(1)-O(1)	2.10(8)	Ni(1)-Ni(2)	3x2.94(4)[a]
Ni(1)-O(4)	3x2.12(6)	Ni(3),Li(1)-O(2)	2.05(9)	Ni(1)-Ni(3)	2x3.04(5)[c]
Ni(1)-O	2.14(7)	Ni(3),Li(1)-O(2')	2.14(8)		
$\delta Ni(1)$	14	Ni(3),Li(1)-O(3)	2.23(14)	Ni(2)-Ni(2)	3.02(3)[b]
		Ni(3),Li(1)-O(4)	2.27(7)	Ni(2)-Ni(3)	2.85(5)[a]
Ni(2)-O(1)	2.07(5)	Ni(3),Li(1)-O(4')	2.32(7)	Ni(2)-Ni(3')	2.88(5)[c]
Ni(2)-O(2)	2.03(6)	Ni(3),Li(1)-O	2.18(9)		
Ni(2)-O(2')	2.07(6)	$\delta Ni(3),Li(1)$	550	Ni(3)-Ni(3)	3.03(8)[c]
Ni(2)-O(3)	2.13(12)			Ni(3)-Ni(4)	2.8(2)[a]
Ni(2)-O(4)	2.08(6)	Li(2),Ni(4)-O(2)	3x2.06(14)		
Ni(2)-O(4')	2.07(5)	Li(2),Ni(4)-O(4)	3x2.2(2)	[a]Ni-Ni interlayer distance	
Ni(2)-O	2.07(8)	Li(2),Ni(4)-O	2.13(17)	[b]Ni-Ni intra L1 distance	
$\delta Ni(2)$	52	$\delta Li(2),Ni(4)$	447	[c]Ni-Ni intra L2 distance	

The fully occupied Ni positions, Ni(1) and Ni(2), are disposed along the stacking c direction in such a way that each Ni(1) is surrounded by six Ni(2), three of them in the upper layer and the other three in the lower. Similarly those sites occupied mainly by Li(2) are always located between six positions, three in each of the next layers, shared by Li, Li(1), and Ni, Ni(3). The Ni-Ni connections through oxygen triangles run not exactly along the c axis, but approximately in the (223) direction, because of the distribution of the octahedral sites in the ccp of the anions. Consequently, due to the existence of ordered mixed Li/Ni layers, in this material there are not isolated Ni clusters as it has been recently supossed [17], but a disposition in which the Ni atoms are always connected through oxygens. The same connection through oxygen triangles exists among those positions occupied mainly by Li, Li(2), and those that are shared by Li and Ni. This fact together with the non existence of pure Ni layers seems to indicate that the simple idea of a 2D process for Li diffusion in this material is contested.

Magnetic Properties

The dc magnetic susceptibility, M/H, of $Li_{0.27}Ni_{0.73}O$ crystals and its reciprocal are represented in Figure 2a. Below 28 K the susceptibility after field cooling (FC) is slightly larger than that after zero field cooling (ZFC). Both, ZFC and FC, show a peak, broader for FC, at 14 and 12 K respectively. Below 12 K the ZFC is more temperature dependent. At higher temperatures, for the interval 110-300 K, the reciprocal mass susceptibility, H/M, follows (R = 0.99992) the Curie-Weiss law $X_g^{-1} = 251.00(23) - 16033(49)T$. The molar ($Li_{0.27}Ni_{0.73}O$) susceptibility gives a Curie constant of 0.3314(31) cm^3Kmol^{-1} and a Weiss temperature of 63.88(58) K. In the curve of Figure 2b, which shows the field dependence of magnetization at 5 K, a hysteresis loop can be seen. These results, like those obtained [10] from a polycrystalline sample with x = 0.025, are characteristic of superparamagnetic systems or highly disordered paramagnetic materials.

Moreover, a weak ferromagnetic signal starting at 160 K is clearly observed. The dc susceptibility at low field (10 Oe) indicates a very broad maximum at temperatures different from those observed at strong field (1 kOe). On the other hand, the magnetic behavior of $Li_{0.27}Ni_{0.73}O$ agrees more reliably with the mixed layers structure here determined, than with that including Ni clusters or that based on the alternation of layers fully occupied by either Li or Ni.

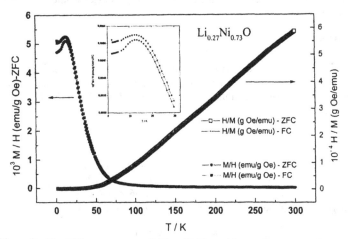

Figure 2a The DC mass susceptibility, M/H, and its reciprocal, as functions of temperature, for $Li_{0.27}Ni_{0.73}O$.

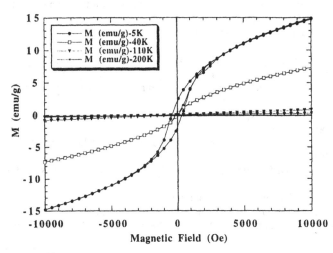

Figure 2b Field dependence of magnetization at different temperatures for $Li_{0.27}Ni_{0.73}O$.

CONCLUSIONS

After describing how to grow $Li_{0.27}Ni_{0.73}O$ single crystals, the present paper has reported the crystal structure solution of this material which has led to the first example of a novel structure type. The more prominent aspect of this model is the mixed quality of the L1 and L2 layers, which alternate along the c direction containing both Li and Ni in different proportions. This shows that the way in which the Li atom tends to order itself forces the appearance of a new cell and is incompatible with the existence of pure Ni layers alternating with more or less doped Li ones, as it was assumed until now. Evidence on the behavior of $Li_{0.27}Ni_{0.73}O$ as mictomagnetic material has also been presented.

ACKNOWLEDGEMENTS

The authors thank Dr. J.L. Martínez who peformed the magnetic measurements, and acknowledge the financial support of the Spanish DGESIC, Projects MAT99-0892 and PB97-0246, the Fundación *Domingo Martínez*, and the CSIC to the Unidad Asociada GBTYS of the University of Santiago de Compostela.

REFERENCES

1. Y.S. Lee, Y.K. Sun, and K.S. Nahm, K.S., Solid State Ionics **118**, p.159 (1999).
2. S. Tao, Q. Wu, Zh. Zhan, and G. Meng, Solid State Ionics **124**, 53 (1999).
3. L.D. Dyer, B.S. Borie, and G.P. Smith, J. Am. Chem. Soc. **76**, p. 1,499 (1954).
4. J.B. Goodenough, D.G. Wickham, and W.J. Croft, J. Phys. Chem. Solids **5**, p. 107 (1958).
5. A. Bajpai, and A. Banerjee, Phys. Rev. B **55**, p. 12,439 (1997).
6. A. Hirano, R. Kanno, Y. Kawamoto, Y. Takeda, K. Yamaura, M. Takano, K. Ohyama, M. Ohashi, and Y. Yamaguchi, Solid State Ionics **78** , p. 12 (1995).
7. V. Massarotti, D. Capsoni, M. Bini, P. Mustarelli, and S. Marini. Ionics **1**, p. 421 (1995).
8. T. Shirakami, M. Takematsu, A. Hirano, R. Kanno, K. Yamaura, M. Takano, and T. Atake, Mater. Sc. Engineer. B **4** p. 70 (1998).
9. K. Yamaura, M. Takano, A. Hirano, and R.J. Kanno, Solid State Chem. **127**, p. 109 (1996).
10. M.Takematsu, T. Shirakami, T. Atake, A. Hirano, and R. Kanno in *Solid State Ionics: New Developments,* edited by B.V.R. Chowdari, World Scientific, Singapore, 1996, p. 330.
11. *SMART* and *SHELXTL*, Siemens Energy and Automation Inc. Analytical Instrumentation, 1996.
12. E.A. Boudreaux, and L.N. Mulay, *Theory and Applications of Molecular Paramagnetism*, Wiley, New York, 1976, p. 494-495.
13. *International Tables for Crystallography*, edited by A.J.C. Wilson, Kluwer, Dordrecht, 1995, Vol. C, p. 219, 484-485.
14. International Centre for Diffraction Data. *Powder Diffraction File.* Card Nos. 47-1049, **41**-890, and **41**-63.
15. J.P. Peres, F. Weill, and C. Delmas, Solid State Ionics **116**, p. 19 (1999).
16. M. Evain, F. Boucher, and O. Gourdon, Chem. Mat. **10** , p. 3,068, 1998.
17. D. Merzt, Y. Ksari, F. Celestini, J.M. Debierre, A. Stepanov, and C. Delmas, Phys. Rev. B **61**, p. 1240 (2000).

New Ideas and Magnetism

MATERIALS DESIGN FOR THE LOW-RESISTIVITY p-TYPE ZnO AND TRANSPARENT FERROMAGNET WITH TRANSITION METAL ATOM DOPED ZnO: PREDICTION vs. EXPERIMENT

K. SATO*, H. KATAYAMA-YOSHIDA* and T. YAMAMOTO**
*Department of Condensed Matter Physics, The Institute of Scientific and Industrial Research (ISIR), Osaka University, 8-1 Mihogaoka, Ibaraki, Osaka 567-0047, Japan.
**Department of Electronic and Photonic System Engineering, Kochi University of Technology, Kochi 782-8502, Japan.

ABSTRACT

We propose a new valence control method of codoping with doping Ga (or In, Al) donor and N acceptor at the same time for the fabrication of a low-resistivity p-type ZnO based upon the *ab initio* calculation. We compare our predicted materials design to fabricate a low resistivity p-type ZnO with the recent successful codoping. Based on the success in the valence control of ZnO, we propose a materials design to fabricate the ferromagnetic Mn-doped p-type ZnO upon codoping. It is shown that the anti-ferromagnetic state is more stable than the ferromagnetic ones due to the anti-ferromagnetic super-exchange interaction, if we have no mobile holes. Upon codoping with the mobile holes, it is shown that the ferromagnetic state becomes more stable than the anti-ferromagnetic ones due to the ferromagnetic double-exchange interaction. However, it is shown that the anti-ferromagnetic state is more stable upon electron doping due to the anti-ferromagnetic super-exchange interaction. We calculate the chemical trends of the magnetic state in V-, Cr-, Fe-, Co-, and Ni-doped (25 at%) in ZnO, and predict that all of these materials show the ferromagnetic ground states without electron and hole doping.

INTRODUCTION

It is well known that the fabrication of the low-resistivity p-type ZnO ($E_g = 3.4$ eV) is difficult because of the compensation [1 – 3]. Blue and ultraviolet laser application and high-power device-application using ZnO has been hampered by their high resistivity in p-type ZnO. The origin of the difficulty to fabricate the low-resistivity p-type ZnO are, (i) the compensation which occurs due to its low solubility, and (ii) the deep energy levels of acceptors, for example ZnO:N (300 meV), with increasing E_g (decreasing the dielectric constant). If the acceptor energy level is 300 meV (corresponds to about 3000 K), we can only activate the carrier density less than 1/100000 of the concentration of the dopant at the room temperature (300 K). In order to fabricate the low-resistivity p-type materials, (i) we should avoid compensation with increasing solubility of the dopant (reducing "the formation energy" of the dopant in the thermal non-equilibrium crystal growth), and (ii) we should reduce the energy level of acceptors upon doping. To do so, we propose an effective new valence control method, which is so called "the codoping method" (using both n- and p-type reactive codopant at the same time, see Fig.1 -(a)), for the fabrication of low-resistivity p-type ZnO based upon ab initio electronic structure calculations [4, 5]. We find that the codoping form an acceptor and donor complexes during the thermal non-equilibrium crystal growth in MBE or MOCVD with making the vapor pressure imbalance between the acceptors and donors. Then we can freeze the acceptor and donor complexes into the crystal which is metastable in the thermal equilibrium. (i) Then, the codoping method contributes

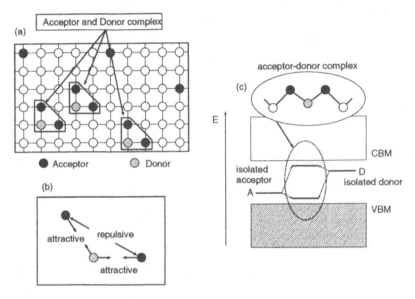

Figure 1: Codoping forms (a) acceptor donor complex with (b) attractive acceptor-donor interactions and repulsive acceptor-acceptor interactions. In order to make p-type semiconductors, we should make imbalance between the acceptor and donor vapor pressure during the crystal growth. (c) Acceptor (A) level is lowered and donor (D) levels is raised with formation of acceptor and donor complex upon codoping.

to reduce "the formation energy" of the dopant due to the reduction of the lattice relaxation energy and Madelung energies in the kinetics of the thermal non-equilibrium crystal growth conditions [see Fig. 1 -(b)]. (ii) The codoping method contributes to increase the carrier mobility due to the short-range dipole scattering or screened short-range scattering (a long-range Coulomb interaction is dominant in the case of simple doping). (iii) The codoping also contributes to reduce the energy level of acceptors due to the formation of donor and acceptor complex (a donor level is raised and an acceptor levels is lowered with forming donor and acceptor complex [see Fig. 1 -(c)].

By the way, if p-type ZnO become available, a ZnO based magnetic semiconductor will be most promising candidate for a ferromagnetic semiconductor. Diluted magnetic semiconductors (DMS) are semiconductors which contain some magnetic atoms as impurities, and they have possibility to bring new ideas to create a functional material with making use of the carrier control techniques in semiconductors, $e.g.$ its magnetic property is controllable by changing the carrier density. Such a trial had already become realistic owing to the first fabrication of (In, Mn)As [6] and the following intensive investigations on III-V DMS [7 – 9], and it was successful for these compounds to control their magnetic behavior by using hetrojunctions [8] or light irradiation [9]. However, the Curie temperature (T_c) of the III-V based DMS is as low as about 100 K, and the low solubility of Mn in them prevents us to realize a large magnetization. In the II-VI compound ZnO it is plausible that the Mn^{2+}

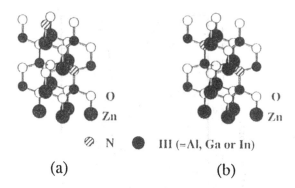

⊘ N ● III (=Al, Ga or In)

(a)   (b)

Figure 2: Crystal structure of supercells for (a) ZnO:2N and (b) codoped p-type ZnO with N acceptors and the reactive donors, Al, Ga or In species.

with d^5 high spin configuration is realized. Moreover, high solubility of Mn in ZnO matrix is expected, because it is in the same transition metal series as Zn. Taking these into account, in this paper, we also propose a ZnO based magnetic semiconductor as a candidate to realize a high T_c and a large magnetization magnetic semiconductor. To search another candidate of a functional magnetic material, the magnetism of $3d$ transition metal impurities in ZnO was also investigated within the same framework.

CALCULATIONS

LOW-RESISTIVITY p-TYPE ZnO:[$Ga_{Zn}+2N_O$, $Al_{Zn}+2N_O$, $In_{Zn}+2N_O$]

In the cases of materials design for the low-resistivity p-type ZnO, our calculations are performed in the framework of local-density-functional theory within the local-density approximation. We have used the augmented-spherical-wave (ASW) method [10] in which we use a parameterized form of the exchange-correlation energy of the homogeneous electron gas given by Hedin and Lundquist and von Barth and Hedin [11]. For valence electrons, we have employed $3d, 4s$, and $4p$ orbitals for Cu, Zn and Ga atoms and outermost s and p orbitals for the other atoms. The Madelung energy which reflects the long-range electrostatic interactions in the system are assumed to be restricted to a sum over monopoles. We studied doped and codoped ZnO with periodic boundary conditions by generating super-cells that contain the acceptor and donor. The super-cell consists of 16 molecules of ZnO as shown in Fig. 2. For p-type ZnO doped with N alone, we replace one of the O atoms by a N atom. For p-type ZnO codoped with the reactive donor of the group III elements and two N atoms, we replace two of the O atoms by the N atoms and one of the Zn atoms by the donor atom. Brillouin zone integration was carried out for $24k$-points in the irreducible wedge of the first Brillouin zone.

Based upon ab initio electronic structure calculation, we propose a codoping method for the fabrication of the low-resistivity p-type ZnO using reactive codopants of donors, such as

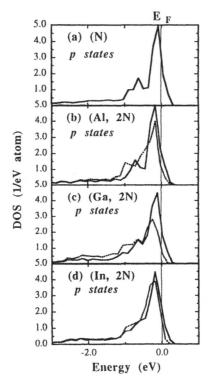

Figure 3: Site-decomposed density of states (DOS) of p-states at the N-sites for (a) ZnO:N, (b) ZnO:[Al+2N], (c) ZnO:[Ga+2N] and (d) ZnO:[In+2N]. Dotted curve indicates the DOS at the N atom sites close to the reactive donor codopnts; the solid curve indicates the DOS at the sites of next-nearest-neighbor N atoms.

Ga, Al or In. First, we show the site-decomposed density of p-states at the N-site in p-type ZnO doped with N alone in Fig. 3-(a). As shown in the figure, a hole band is generated at the top of the valence band, however, these states are largely localized at the impurity site. Applying our codoping method to ZnO using N and group III elements, the acceptor levels in the band gap are lowered due to the strong interaction between the N acceptor and reactive donor codopants and the delocalized states are obtained as shown in Fig. 3-(b) to 3-(d). Next, the calculated energy differences in the Madelung energy among n-type ZnO:Al, ZnO:Ga, ZnO:In and p-type ZnO:N, ZnO:2N and codoped p-type ZnO:[Al+2N], ZnO:[Ga+2N], or ZnO:[In+2N] are given in Table 1. These shows that the codoping with forming the acceptor and donor complexes reduces the long range Coulombic interaction upon doping. Then, during the thermal non-equilibrium crystal growth method such as MBE or MOCVD at the low temperature, we can increase the solubility of the dopants upon kinetics in which the long range Coulomb interaction is dominant. Thus, we find the

Table 1: Calculated differences in the Madelung energy among n-type ZnO:Al, ZnO:Ga, ZnO:In and p-type ZnO:N, ZnO:2N and codoped p-type ZnO:[Al+2N], ZnO:[Ga+2N], or ZnO:[In+2N] in unit of eV.

Doping type	Madelung energy relative to the undoped ZnO (eV)
ZnO:Al	-6.44
ZnO:Ga	-13.72
ZnO:In	-9.73
ZnO:N	+0.79
ZnO:2N	+0.91
ZnO:[Al+2N]	-4.74
ZnO:[Ga+2N]	-12.06
ZnO:[In+2N]	-7.79

delocalized and shallower acceptor states of N for p-type codoped with N and Ga (or In, Al) donors. Our prediction of the materials design to fabricate a low resistivity p-type ZnO [4, 5] is confirmed by the recent successful codoping experiments using Ga and N codoping in the laser MBE crystal growth [12]. The resistivity is about 2Ωcm in the codoped p-type ZnO [12].

FERROMAGNETIC ZnO-BASED DMS

The KKR Green's function method based on the local density approximation [13] with the parameterization by Morruzi, Janak and Williams [14] was employed in the cases of materials design for the ZnO-based magnetic semiconductors. The relativistic effect was taken into account by the scalar relativistic approximation without spin-orbit coupling. The form of the potential is restricted in the muffin-tin type. The wave functions in each muffin-tin sphere were expanded into real harmonics up to $l = 2$, where l is the angular momentum defined at each site. Independent ones of 64 k sampling points in the first Brillouin zone were calculated. ZnO has the Wurtzite structure in which anions and cations respectively form hexagonal close packed lattices separated with each other along the c-axis by the internal coordinate u [15]. To simulate (Zn, Mn)O the supercell illustrated in fig. 4 was employed. The supercell consists of 8 molecules of ZnO. Two of Zn atoms in the supercell were substituted with two Mn atoms. This substitution leads to Mn concentration of 25%. According to the recent experiments by Fukumura et al., Mn content in ZnO can be increased up to about 35% by the pulsed laser deposition technique [16], therefore the present supercell calculation is not so far from the real situation. We adopted the lattice constants of $a_0 = 3.27$ Åand $c_0 = 5.26$ Å, where a_0 and c_0 are the lattice constants of the primitive unit cell of the Wurtzite structure. They were interpolated from the experimental results by Fukumura et al. The internal coordinate u is not available for doped ZnO, therefore we used $u = 0.345$ which is the value for pure ZnO [15]. Muffin-tin radii of $0.1437a$ and $0.1338a$, where a is the lattice constant of the super cell, were used for metal atoms and anions, respectively. In order to obtain a suggestion for a material design of a ferromagnetic semiconductor, first, the total energies (TE) per unit supercell were calculated for both anti-ferromagnetic and ferromagnetic arrangements of Mn magnetic moments. Comparing

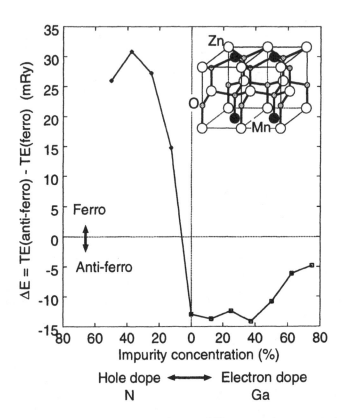

Figure 4: Stability of the ferromagnetic ordering of Mn magnetic moments in ZnO. The simple orthorhombic supercell used in the present calculations is also shown.

calculated TE's, stability of the ferromagnetic arrangement was examined as a function of the concentration of carriers. Next, magnetic behaviors of transition metal atoms such as Sc, Ti, V, Cr, Fe, Co, Ni and Cu in ZnO were studied in addition to Mn.

The calculations were performed for both hole doping and electron doping in the (Zn, Mn)O system. In the cases of hole doping, some of O atoms in the supercell were substituted with N atoms, and for electron doping cases Ga atoms were put instead of Zn atom. Fig. 4 shows calculated energy difference $\Delta E = TE$(anti-ferromagnetic case) $- TE$(ferromagnetic case) as a function of impurity concentration. As shown in the figure, the anti-ferromagnetic ordering was more stable than the ferromagnetic one if no carrier dopants was introduced. The substitution of Zn with Mn does not bring any carriers in the system, therefore, the anti-ferromagnetic ordering was favored owing to the anti-ferromagnetic super-exchange interaction between Mn magnetic moments. With increasing hole concentration, the anti-ferromagnetic states becomes unstable and the ferromagnetic state is the ground state. It is very likely that these behavior can be explained by the

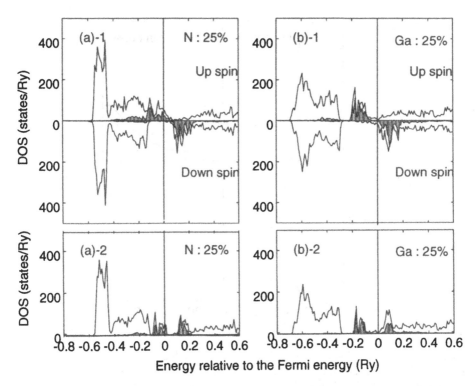

Figure 5: (a) Total DOS (solid line) and local density of Mn-d states (shaded area) in the hole doped (Zn, Mn)O. N atoms are doped up to 25%. (b) Total DOS (solid line) and local density of Mn-d states (shaded area) in the electron doped (Zn, Mn)O. Ga atoms are doped up to 25%. The DOS are shown both for the ferromagnetic states [(a)-1 and (b)-1] and for the anti-ferromagnetic states [(a)-2 and (b)-2].

double exchange mechanism [17]. On the contrary, it seems that the electron doping does not participate in the stabilization of the ferromagnetic ordering. In order to get more detailed knowledge, density of states (DOS) are calculated for typical cases. Fig. 5 shows the total DOS and local density of d states at the Mn site of the hole doped (Zn, Mn)O which contains N atoms up to 25%. Zn $3d$ states and O $2p$ states of the host matrix are located around -0.5 Ry and -0.3 Ry relative to the Fermi energy, respectively. There are holes in the valence up spin band and so called half-metallic situation is realized. The rather shallow $2p$ states of N are well mixed with the Mn $3d$ states as shown in fig. 5-(a)-1, however, the large part of the holes consists of the Mn $3d$ states. A difference between the ferromagnetic case (fig. 5-(a)-1) and the anti-ferromagnetic case (fig. 5-(a)-2) is appeared in the width of each d band. In the ferromagnetic situation, the d band is widened because of the hybridization between Mn $3d$ states. The exchange splitting is larger than the band width in the present cases, therefore the ferromagnetic states can lower its band energy by introducing a sufficient

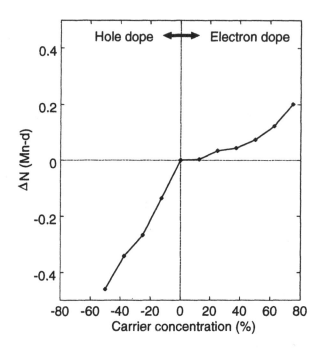

Figure 6: Electron number difference of Mn 3d states as a function of carrier concentration. Δ N = [Mn 3d states population number of carrier doped (Zn, Mn)O] - [Mn 3d states population number of undoped (Zn, Mn)O]. In the case of hole doping (electron doping), the differences of population number in majority (minority) spin states are plotted.

carriers. This is because the ferromagnetic state is more stable than the anti-ferromagnetic state in the hole doping (Zn, Mn)O. On the other hand, in the case of electron doping, the doped electrons are supplied mainly to the Ga 4s and Zn 4s states and does not occupy Mn 3d states as shown in Fig. 5-(b)-1. That is to say we have no mobile d-electron. As a result, an energy gain by the double exchange interaction is not expected. To clarify this feature, the difference of the electron number in the Mn 3d states is plotted as a function of carrier concentration in Fig. 6, where ΔN is population difference of Mn 3d majority spin states (minority spin states) between hole-doped (electron-doped) (Zn, Mn)O and undoped (Zn, Mn)O. As clearly shown in the figure, electrons are removed mainly from the Mn 3d states as holes are introduced, while electrons does not populate in the Mn 3d states in spite of electron doping.

Taking the double exchange mechanism into account, it is likely that the other transition metal atoms show a ferromagnetic ordering without any doping treatment. From the point of practical applications, such feature might be very desirable in some cases and lead to a possibility to tune its magnetic character by alloying between different transition metal atoms in ZnO for examples. Fig. 7 shows a chemical trend of the magnetic states for 3d transition metal atoms in ZnO. As expected, magnetic moments of transition metal

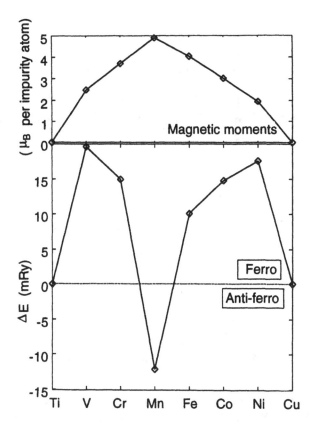

Figure 7: Chemical trend of the magnetic states for 3d transition metal atoms in ZnO. Total magnetic moments per transition metal atom are also shown.

atoms couples with each other ferromagnetically except for Mn. Besides they have almost saturated magnetic moments as shown in the same figure. As changing the valence number, the electron configuration differs from d^5, and resulting mobile carriers participate to the ferromagnetic ordering through the double exchange mechanism. For a typical case, the calculated DOS of (Zn, Fe)O and (Zn, Cr) are shown in fig. 8. For both cases, the high spin states were realized because of the large exchange splitting. In the case of (Zn, Fe)O, residual one valence electron of Fe relative to Mn occupies the Fe-3d states because the Fe 3d states were located at lower energy than the Mn-3d states. In the (Zn, Cr)O case, the Cr-3d states were located at higher energy than the Mn-3d states, so holes were introduced in the Cr-3d states. As a result, there were itinerant carriers in both (Zn, Fe)O and (Zn, Cr)O systems.

CONCLUSIONS

In this paper, we proposed these two materials design for ZnO related substances.

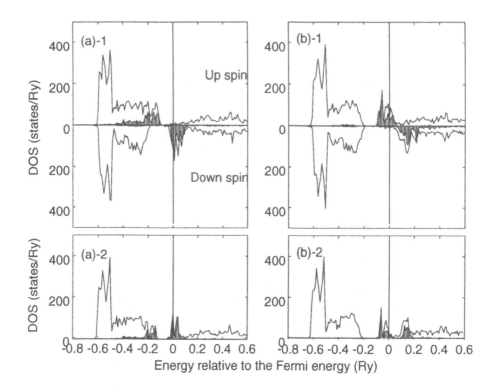

Figure 8: (a) Total DOS (solid line) and local density of Fe-d states (shaded area) of the (Zn, Fe)O. (b) Total DOS (solid line) and local density of Cr-d states (shaded area) of the (Zn, Cr)O. The DOS are shown both for the ferromagnetic states [(a)-1 and (b)-1] and for the anti-ferromagnetic states [(a)-2 and (b)-2].

First, we have studied systematically the electronic structure of p-type ZnO by simultaneous codoping of acceptors and reactive donors using *ab initio* electronic structure calculations. We proposed materials design for various codoping candidate to fabricate a low-resistivity p-type wide band-gap semiconductors; (1) ZnO:[Ga$_{Zn}$+2N$_O$, Al$_{Zn}$+2N$_O$, and In$_{Zn}$+2N$_O$].

Next, we studied the stability of the ferromagnetic state of (Zn, Mn)O from the first principles calculations by using the KKR method based on the LDA. The transition from the anti-ferromagnetic state to the ferromagnetic state took place as the holes were doped, but we found no transition by the electron doping. The ferromagnetism of the $3d$ transition metal atoms in ZnO was also investigated. It was found that the ferromagnetic ordering was the ground state for V, Cr, Fe, Co and Ni. Analyzing the calculated density of states, it was suggested that the ferromagnetism in the present cases were originated from the competition between the ferromagnetic double exchange interaction and the anti-ferromagnetic super-exchange interaction. We propose the hole doped (Zn, Mn)O and the ZnO doped with the 3d transition metal atoms for candidates of a magnet which is transparent in a visible ray. It will have a great impact on an industrial application in the magneto-optical devices.

ACKNOWLEDGMENTS

This work was partially supported by a JSPS Research for the Future Program in the area of Atomic-Scale Surface and Interface Dynamics, JST-ACT, a Grant-In-Aid for Scientific Research on the Priority Area and Sanken-COE from the Ministry of Education, Science, Sports, and Culture. Super-computational time was partially provided by the Institute for Solid State Physics, University of Tokyo and Computational Center, Osaka University. We thank Prof. Akai for fruitful discussions and providing us with his KKR band structure calculation package.

REFERENCES

1. M. Kasuga and S. Ogawa, Jpn. J. Appl. Phys., **22**, 794 (1983).
2. Y. Sato and S. Sato, Thin Solid Films, **281-282**, 445 (1996).
3. K. Minegishi, Y. Koiwai, Y. Kikuchi, K. Yano, M. Kasuga and A. Shimizu, Jpn. J. Appl. Phys., **36** L1453 (1997).
4. T. Yamamoto and H. Katayama-Yoshida, Jpn. J. Appl. Phys. **38**, L166 (1999).
5. H. Katayama-Yoshida and T. Yamamoto, Applied to Japanese Patent (Fabrication method of low-resistivity p-type ZnO: JP H10-287966), and applied to USP and ECP.
6. H. Munekata, H. Ohno, S. von Molnar, Armin Segmüller, L. L. Chang, L. Esaki, Phys. Rev. Lett. **63**, 1849 (1989).
7. H. Ohno, H. Munekata, T. Penney, S. von Molnar, L. L. Chang, Phys. Rev. Lett. **68**, 2664 (1992).
8. H. Munekata, A Zaslavsky, P. Fumagalli, R. J. Gambino, Appl. Phys. Lett. **63**, 2929 (1993).
9. S. Koshihara, A. Oiwa, M. Hirasawa, S. Katsumoto, Y. Iye, C. Urano, H. Takagi, H. Munekata, Phys. Rev. Lett. **78**, 4617 (1997).
10. A. R. Williams, J. Kübler and C. D. Gellatt, Phys. Rev. **B19**, 6094 (1979).
11. L. Hedin and B. I. Lundquist, J. Phys. **C4**, 3107 (1971), U. von Barth and L. Hedin, J. Phys. **C5**, 1629 (1972).
12. M. Joseph, H. Tabata, T. Kawai, Jpn. J. Appl. Phys. **38**, L1205 (1999).
13. H. Akai, M. Akai, S. Blügel, B. Drittler, H. Ebert, K. Terakura, R. Zeller, P. H. Dederichs Prog. Theo. Phys. Supplement **101**, 11 (1990).
14. V. L. Moruzzi, J. F. Janak, A. R. Williams, *Calculated Electronic Properties of Metals* (Pergamon, U.S.A, 1978), p. 15.
15. R. W. G. Wyckoff *Crystal Structures, 2nd. Ed.* (Krieger, 1986), p. 112.
16. T. Fukumura, Z. W. Jin, A. Ohtomo, H. Koinuma, M. Kawasaki, Appl. Phys. Lett. to be published.
17. H. Akai, Phys. Rev. Lett. **81**, 3002 (1998).

THERMAL STABILITY AND SEMICONDUCTING PROPERTIES
OF EPITAXIAL La$_{0.7}$Sr$_{0.3}$MnO$_3$ FILMS

K. S. SO, K. H. WONG, W. B. WU
Department of Applied Physics, The Hong Kong Polytechnic University, Hung Hom, Kowloon, Hong Kong, People's Republic of China.

ABSTRACT

La$_{0.7}$Sr$_{0.3}$MnO$_3$ (LSMO) perovskite oxide films have been grown on (001) LaAlO$_3$ (LAO) by pulsed laser deposition. The films were deposited in an ambient oxygen pressure of 0.1 mTorr to 200 mTorr and under different substrate temperatures. Their structural properties were examined by X-ray diffractometry. Heteroepitaxial growth was confirmed for films deposited at 650°C or above. Electrical measurements suggest that the charge carrier concentration of the films varies with their oxygen content and shows high stability against further thermal treatment. Semiconducting LSMO films at room temperature were obtained for deposition at oxygen pressure ≤ 60 mTorr. The epitaxial LSMO films have been used as the semiconducting channel of a ferroelectric field effect transistor. Heteroepitaxial Pb(Zr$_{0.52}$Ti$_{0.48}$)O$_3$/LSMO/LAO structures have been fabricated and characterized.

INTRODUCTION

In recent years, great attention has been paid to the maganese perovskites due to their unique physical properties such as colossal magnetoresistance (CMR) [1-7] and half metallic with spin polarization [8-10] at the Fermi level. These properties make the maganese perovskites a useful materials for magnetoresistive devices such as magnetic random access memory and magnetic field sensors. By controlling the electrical transport properties of these materials in thin film form, however, the materials can be used as semiconducting layers in electronic devices such as ferroelectric field effect transistor (FeFET) [11,12]. La$_{0.7}$Sr$_{0.3}$MnO$_3$ (LSMO), like other mangenese oxides, is a candidate for fabricating such FeFET. In our present study, where the LSMO films were fabricated by pulsed laser deposition (PLD), we found that the LSMO is a more suitable material than other mangenese oxides due to its easily tuned semiconducting properties, high thermal stability [13] and low processing temperature. Heteroepitaxial Pb(Zr$_{0.52}$Ti$_{0.48}$)O$_3$(PZT)/LSMO/ LaAlO$_3$ (LAO) structures with semiconducting LSMO channel have been fabricated.

EXPERIMENT

A 248 nm KrF excimer laser with a repetition rate 10 Hz was used to ablate stoichoimetric targets of LSMO and PZT. Films were deposited on (001) LAO single-crystal substrate under different conditions. The laser fluences were kept at 6 J/cm^2 and 3 J/cm^2 for irradiating the LSMO and PZT targets respectively. The target-substrate distance was 45 mm. The substrates were cleaned in acetone and rinsed with deionized water. They were then glued to the surface of the substrate heater by silver paste. The position of the heater was aligned to the centre of the plume in order to obtain the best uniformity of a grown film. The typical layer thickness for LSMO and PZT were 200 nm and 150 nm, respectively. LSMO/LAO and PZT/LSMO/LAO heterostructure were characterized by four–circle X-ray diffractometry using CuK$_\alpha$ radiation.

Mat. Res. Soc. Symp. Proc. Vol. 623 © 2000 Materials Research Society

The resistivity-temperature (R-T) curves of the LSMO films were measured by a standard four-point probe technique over a temperature range from 77 K to room temperature.

RESULTS

Different electrical transport properties in LSMO films were observed in the films deposited under oxygen ambient pressure from 0.1 m Torr to 200 m Torr. Fig. 1 shows the X-ray diffraction (XRD) of θ-2θ scans of LSMO films at the deposition pressure of 0.1, 10 and 200 m Torr. It is noted that the diffraction angle for the (002) LSMO has shifted by more than one degree for LSMO films deposited at 0.1 m Torr and 200 m Torr. This suggests that the lattice constant changes with oxygen content of the LSMO films. The R-T measurement shown in Fig. 2 confirmed the change of electrical transport properties of the films with different deposition pressure. The R-T curves of the films grown at above 60 m Torr were metallic like at room temperature. As the deposition pressure decreases, the metal-semiconductor (ferromagnetic-paramagnetic) transition was down shifted to lower temperatures. A semiconducting film at room temperature can be obtained at the deposition oxygen pressure range between 0.1 m Torr and 60 m Torr.

The thermal stability of the LSMO films was investigated by post deposition annealing. The LSMO films were first deposited at substrate temperature of 700°C and under the same oxygen pressure of 200 m Torr. The as-deposite films were in situ annealed at different ambient oxygen pressure from 4×10^{-6} to 10 Torr for 60 minutes. The full width at half maximum (FWHM) of the (002) LSMO reflection rocking curves were around 0.9°. Figure 3 shows the R-T curves of LSMO films under different annealing pressure. It is seen that the resistivity of the films follows

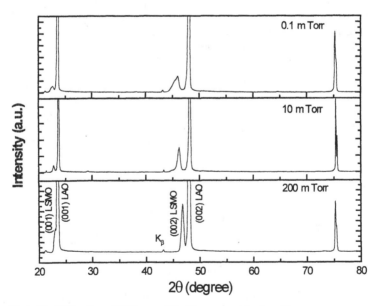

Fig.1. The XRD pattern of θ-2θ scans of the films grown at the deposition oxygen pressure of 0.1, 10 and 200 m Torr, respectively.

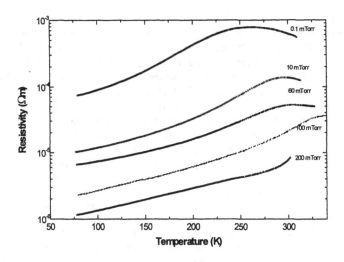

Fig.2. The resistivity vs temperature of the films grown at the deposition
oxygen pressure between 0.1 and 200 m Torr.

the same profile over a wide temperature range. The results shown above imply that the
electrical transport behaviour of the epitaxial LSMO films is insensitive to the post deposition
heat treatments. Apparently, the oxygen content of the films has not changed during the
annealing process.

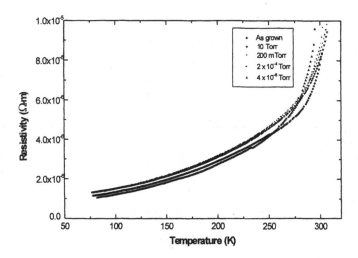

Fig. 3. The R-T curves of a as-grown LSMO film and films in situ annealed
at 700°C for 60 minutes in different oxygen ambient.

Epitaxial LSMO films grown on (001) LAO substrate were obtained over a wide range of substrate temperatures from 500 to 700°C at deposition oxygen pressure of 200 m Torr. Fig. 4 shows the X-ray diffraction of θ-2θ scans of the LSMO films deposited on (001) LAO at 450, 500 and 700°C. There is no reflection for the film deposited at 450°C. For films deposited above 500°C, apart from the (00l) reflections from the LSMO films and the LAO substrates, no trace of other reflections was detected. The FWHM of ω scan rocking curve of the films deposited at 500 and 700°C were 1.15° and 0.85°, respectively. Apparently the orientation of the LSMO films improves with increasing substrate temperature. For both films the φ scans of (022) LSMO and (022) LAO show the characteristic four-fold symmetery. It can be concluded from the above results that cube-on-cube epitaxial LSMO films were fabricated on LAO at 500°C. It is much lower than the previously reported growth temperature of $La_xCa_{1-x}MnO_3$ (LCMO) and LSMO films at 650°C [14] and 700°C [9], respectively. The low growth temperature of LSMO is therefore compatible with the processing technology used in silicon industry.

In order to prepare the FeFET structure, ferroelectric PZT layer was deposited on top of the semiconducting LSMO film at 550°C. The X-ray diffraction θ-2θ scan of the heterostructure PZT/LSMO/LAO is shown in Figure 5. Single phase of PZT and LSMO were observed. The FWHM of ω scan rocking curve of PZT and LSMO are 0.83° and 1.13°, respectively. The inset shows the φ scans of the heterostructure PZT/LSMO/LAO. The four-fold symmetry of these three layers at the same reflection angles indicates that the PZT and LSMO films are cube-on-cube grown on the LAO substrate.

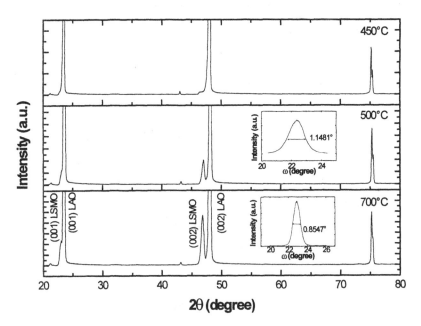

Fig.4. Specular θ-2θ scans for heterostructures LSMO/LAO at 450°C, 500°C and 700°C, respectively. The insets show the ω scans of the corresponding films.

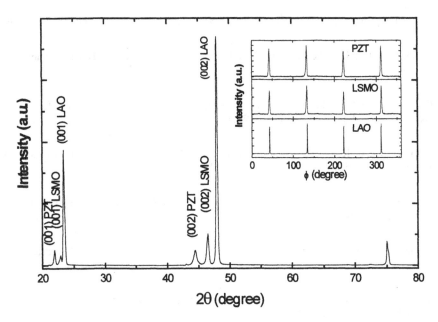

Fig.5. The θ-2θ scans for heterostructures PZT/LSMO/LAO. The inset shows the 360° φ scans of the (022) family reflections of the corresponding films.

CONCLUSIONS

In conclusion, high quality LSMO films grown on LAO substrates were obtained by PLD method at deposition temperature as low as 500°C. The conducting characteristics of the LSMO films can be controlled by tuning the oxygen content during deposition. Post-deposition heat treatment has no effect whatsoever on the electrical transport properties of the films. A ferroelectric PZT layer has been grown on top of the LSMO film. A FeFET structure based on heteroepitaxial PZT/LSMO/LAO has been fabricated and characterized.

ACKNOWLEDGMENTS

The work described in this paper was substantially supported by a grant from the Research Grants Council of the Hong Kong Special Administration Region (Project No. PolyU 5160/98P). One of us (K. S. So) is grateful for the award of research studentship from the Hong Kong Polytechnic University.

REFERENCES

1. M. McCormack, S. Jin, T. H. Tiefel R. M. Fleming, J. M. Philips, and R. Ramesh, Appl. Phys. Lett. **64**, 3045 (1994)

2. H. L. Ju, C. Kwon, Q. Li R. L. Greene, and T. Venkatessan, Appl. Phys. Lett. **65**, 2108 (1994)

3. S. Jin, T. H. Tiefel, M. McCormack, R. A. Fastnacht, R. Ramesh and L. H. Chen, Science **264**, 413 (1994)

4. G. C. Xiong, Q. Li, H. L. Ju, S. N. Mao, L. Senapati, X. X. Xi, R. L. Greene, and T. Venkatesan, Appl. Phys. Lett. **66**, 1427 (1995)

5. Z. Trajanovic, C. Kwon, M. C. Robson, K.-C. Kim, M. Rajeswari, R. Ramesh, T. Venkatesan, S. E. Lofland, S. M. Bhagat, and D. Fork, Appl. Phys. Lett. **69**,1005 (1996).

6. E. Gommert, H. Cerva, A. Rucki, R.v. Helmolt, J. Wecker, C. Kuhrt and K. Samwer, J. Appl. Phys. **81**, 5496 (1997)

7. J. Fontcuberta, M. Bibes, B. Martinez, V. Trik, C. Ferrater, F. Sanchez, and M. Varela, Appl. Phys. Lett. **74**, 1743 (1999)

8. Yu Lu, X. W Li, G. Q. Gong, G. Xiao, A Gupta, P. Lecoeur, J.Z. Sun, Y. Y. Wang, and V. P. Dravid, Phys. Rev. B **54**, R8357 (1996).

9. J. Z. Sun, W. J. Gallagher, P. R. Duncombe, L. Krusin-Elbaum, R. A. Altman, A. Gupta, Yu Lu, G. Q. Gong, and Gang Xiao, Appl. Phys. Lett. **69** 3266 (1996).

10. M Viret, M. Drouet, J. nassar, J. P. Contour, C. Fermon, and A. Fert, Europhys. Lett. **39**, 545 (1997).

11. Yukio Watanable, Appl. Phys. Lett. **66**, 1770 (1995)

12. S. Mathews, R. Ramesh, T. Venkatesan, J. Benedetto, Science **276**, 238 (1997)

13. Wenbin Wu, K. H. Wong, X.-G. Li, and C. L. Choy, J. Appl. Phys. **87**,3006 (2000).

14. Y. S. Leung, and K. H. Wong, Appl. Surf. Sci. **127-129**, 491 (1998).

TRANSPORT ANISOTROPY IN PLD-MADE La$_{0.75}$Sr$_{0.25}$MnO$_3$ FILMS

P. JOHNSSON, S.I. KHARTSEV, A.M. GRISHIN
Condensed Matter Physics, Department of Physics, Royal Institute of Technology,
Stockholm, S-100 44, Sweden.

ABSTRACT

A sequence of epitaxial La$_{0.75}$Sr$_{0.25}$MnO$_3$ (LSMO) films with thickness ranging from 2400 to 50 Å have been prepared by pulsed laser deposition onto (110) SrTiO$_3$ (STO) substrates. Compared with our previous results on LSMO/STO(100) films [1], films on STO(110) substrates exhibit strong anisotropy of electrical resistivity ρ. ρ measured in [1$\bar{1}$0] direction is comparable with the resistivity of LSMO/STO(100) films while ρ in [001] direction is 25 times higher than in STO(100) case. The maximum value of anisotropy parameter $\rho_{[001]} / \rho_{[1\bar{1}0]} = 25$ is reached for thick films at the low temperature of 90 K. Distinct crossover from 3D to 2D case has been observed. For thick films anisotropy monotonously decreases with the temperature increase. Films thinner than 200 Å exhibit a maximum of anisotropy parameter, which shifts to lower temperatures with the thickness decrease. The maximum temperature coefficient of resistivity (TCR) was found to be around 2% if measured along [001] direction and about 50 % higher in [1$\bar{1}$0] in-plane direction. We explain the observed effects in terms of the crystalline properties of fabricated films.

INTRODUCTION

Since the discovery of colossal magnetoresistance (CMR) in 1994 [2] there have been many papers published about various aspects of CMR in film and in bulk samples. Several papers reported effects of anisotropy observed in CMR materials.

Suzuki and Hwang measured magnetoresistance in (110) La$_{0.7}$Sr$_{0.2}$MnO$_3$ films in magnetic field applied perpendicular and parallel to the film plane and current. In weak magnetic fields the anisotropy of magnetoresistance occurred depending on whether current was along the [001], or [1$\bar{1}$0] direction, which are soft and hard magnetic axes correspondingly. The observed effects have been relied on the stress induced by the film-substrate lattice mismatch [3]. O'Donnell *et. al.* have found anisotropy of magnetoresistance in La$_{0.7}$Ca$_{0.3}$MnO$_3$ films regarding the mutual orientation of in-plane magnetic field and electric current. The anisotropy decreases with an increase of the applied field and depends on temperature with a maximum increase of resistivity about 11 % close to the ferromagnetic transition temperature [4]. Mira *et. al* [5] have compared Seebeck coefficients in orthorhombic Pr$_{0.67}$Sr$_{0.33}$MnO$_3$ and rhombohedral La$_{0.67}$Sr$_{0.33}$MnO$_3$ single crystals and found that Seebeck coefficient in the orthorhombic crystal is different in the *ab*-plane and along the *c*-axis at temperatures above the ferromagnetic phase transition temperature, T_c. Rhombohedral La$_{0.67}$Sr$_{0.33}$MnO$_3$ did not show this property and the authors explained this in terms of the difference of Mn-O-Mn bond length and angles. They are different in-*ab*-plane and out-of-plane for Pr$_{0.67}$Sr$_{0.33}$MnO$_3$ but are the same in La$_{0.67}$Sr$_{0.33}$MnO$_3$.

Zeng and Wong grew La$_{0.67}$Ca$_{0.33}$MnO$_3$ (001) films on NdGaO$_3$ substrates by facing target sputtering technique and found resistance anisotropy in-film-plane [6]. Atomic force microscopy images indicated film grown in the step flow mode, with step edges parallel to the *c*-axis.

To investigate the in-plane anisotropy we deposited $La_{0.75}Sr_{0.25}MnO_3$ (LSMO) films on (110) $SrTiO_3$ (STO) substrates, with various thickness of 50, 100, 200, 1200 and 2400 Å. Strong in-plane anisotropy of resistivity has been revealed, with a distinct difference in behavior between the ultrathin (50-200 Å) films and the thicker (1200, 2400 Å) films. The anisotropy found in LSMO/STO(110) thin film structures is also very unlike the above referred effects which are either relatively small and/or disappear below T_c. Since the transport properties of a uniform cubic crystal should be isotropic, we believe that our results can be explained by the granular structure of LSMO/STO(110) films.

EXPERIMENT

A 248 nm KrF excimer laser was used to ablate a stochiometric target of composition $La_{0.75}Sr_{0.25}MnO_3$. Films were grown on (110) STO 5×5 mm^2 single crystal substrates under the following conditions: substrate temperature was 730 °C, laser radiation energy density was 3-4 J/cm^2, pulse repetition rate was 30 Hz and the distance between the target and the substrate was 55 mm. The background pressure did not exceed 10^{-7} Torr. Depositions was carried out in an oxygen pressure of 200 mTorr and finalized by *in situ* annealing in oxygen pressure of 500 Torr at the same temperature for 10 minutes. The deposition rate has been determined by measuring film thickness with atomic force microscope (AFM).

Electrical measurements were performed by standard four-probe dc technique. The four silver contact pads ($\emptyset$ = 0.8 mm) were deposited by thermal evaporation. Magnetoresistance was measured in a field of 7 kOe. The different in-plane directions of the substrate ($[1\bar{1}0]$ and [001]) were determined by x-ray diffraction (XRD). A Siemens D5000 powder diffractometer was used for all XRD measurements. AFM measurements were performed with a Burleigh Aris 3300 Personal SPM system using a high aspect ratio (10:1) silicon tip.

RESULTS

Transport Characteristics

In Fig. 1 the resistivity in the [001] and the $[1\bar{1}0]$ directions are showed. The size of the unit cell in the $[1\bar{1}0]$ direction is $\sqrt{2}$ times longer than in the [001] direction. The resistivity in the [001] direction is about one order of magnitude higher than in the $[1\bar{1}0]$ direction, except for the 50 Å film, where the resistivity in the [001] direction is twice of that in the $[1\bar{1}0]$ direction. All films show a maximum of $d\rho / dT$, which is usually observed close to the ferromagnetic transition temperature, T_c.

Also in Fig. 1, the anisotropy parameter, defined as a ratio between the resistivity in the [001] and $[1\bar{1}0]$ direction $\rho_{[001]} / \rho_{[1\bar{1}0]}$, is plotted versus the temperature. The anisotropy parameter is the largest for the 2400 and 1200 Å films and it decreases with increasing temperature. The 200 and 100 Å films show non-monotonous temperature dependence with maxima around 300-350 K. Even more irregular behavior is exhibited by the 50 Å film.

In the lower panel in Fig. 1 the magnetoresistance (MR), defined as $\Delta\rho/\rho_0 \equiv (\rho_0 - \rho_{7kOe}) / \rho_0$, versus temperature is shown. Similar to T_c it shows a regular behavior with the maximum MR moving to higher temperatures for thicker films. The peak MR-value is higher when measured with current going in the $[1\bar{1}0]$ direction. There is a change in tendency passing to the thicker

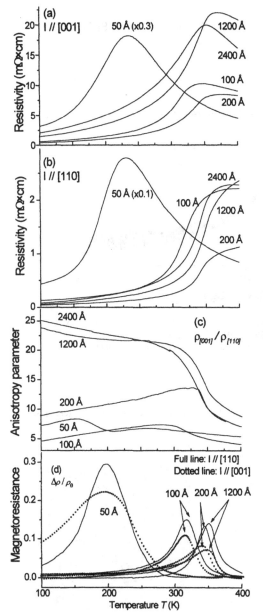

FIG. 1 Temperature dependence of the resistivity with (a) the current along the [001] direction and (b) along the [1 $\bar{1}$ 0] direction. Panel (c) presents the anisotropy parame ter $\rho_{[001]} / \rho_{[1\bar{1}0]}$ versus temperature, and panel (d) shows the magnetoresistance $\Delta\rho / \rho_0$.

1200 Å film. The ultrathin films have a decrease in peak MR value, in both directions, with increasing of film thickness. This trend is broken with the 1200 Å film, where the peak MR-value is lower in the [001] direction, but increasing relative the 200 Å film, for the [1 $\bar{1}$ 0] direction.

The temperature coefficient of resistivity, TCR = $d \ln\rho / dT$ has been calculated for the films in the different directions. TCR in the [001] direction is between 1.1 % and 2.2 % with the peak value for the 100 Å film. In the [1 $\bar{1}$ 0] direction the values are higher, between 2.6 % and 3.3 %. Maximum value is for the 200 Å film, and like in the [001] direction the values decrease monotonically with a thickness changing from the optimal.

X-Ray Diffraction

A θ-2θ scan show only ($hh0$) peaks for both (110) SrTiO$_3$ substrate and deposited La$_{0.75}$Sr$_{0.25}$MnO$_3$ films indicating that fabricated films are single phase and highly (110) oriented. The film-substrate lattice mismatch is quite small ($\sim$ - 0.6 %) which leads to peaks that are hard to resolve clearly. In Fig. 2 the (220) peaks of the five films of different thickness are shown. The film peaks can be seen as shoulders on the high-angle wing of the substrate peak. The vertical arrow indicates where a peak from bulk LSMO would be positioned. It can be seen that the shoulder moves closer to the position of bulk LSMO (220) reflection with increasing thickness of the film, but saturating at $\sim$1200 Å, since there is no difference between the 1200 and 2400 Å peaks.

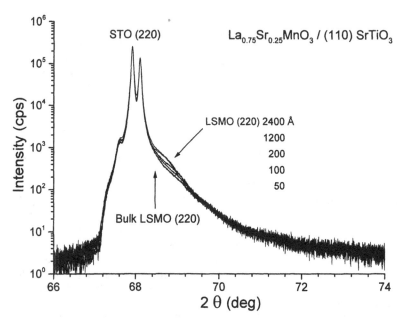

FIG 2. θ-2θ- XRD patterns of clean (110) SrTiO$_3$ substrate and five 50-2400 Å thick La$_{0.75}$Sr$_{0.25}$MnO$_3$ films grown on (110) SrTiO$_3$ single crystal. The (220) LSMO Bragg reflections can be seen at the high angle shoulder on the (220) SrTiO$_3$ substrate reflection. The vertical arrow indicates where a peak for bulk LSMO would be.

Atomic Force Microscopy

AFM measurements were performed in order to find correlation between the transport proper-

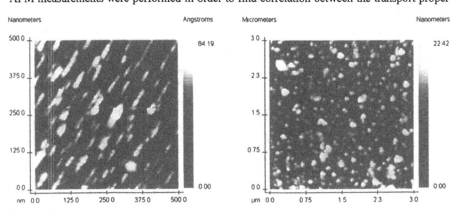

FIG 3. Atomic Force Microscopy (AFM) images of the 100 Å (left) and 1200 Å thick La$_{0.75}$Sr$_{0.25}$MnO$_3$ films. The 100 Å film shows elongated grains while the thicker film shows grains with a circular shape.

ties and films surface morphology. In Fig. 3 AFM-images of the 100 and 1200 Å films are shown. The 100 Å film consists of elongated grains with a typical size of 700×150 Å. The long axis of the grains coincides with the [1$\bar{1}$0] direction. The 200 Å film (not shown) has grains which are slightly longer (~ 900 Å) and 2-3 times wider (~ 400 Å). The second film, shown in Fig. 3, is the 1200 Å film. Here the grains are circular and have typical diameter of about 900 Å. This morphology does not show any anisotropic features. The 2400 Å film (not shown) consists of grains that start to coalesce and show no preferential orientation.

CONCLUSIONS

The presented transport properties can be compared with the results obtained for films of the same composition, fabricated under the same conditions on (001) STO substrates. In Fig. 4 are results from our previous work on thickness dependence of La$_{0.75}$Sr$_{0.25}$MnO$_3$ films on (100) SrTiO$_3$ single crystal substrates. [1] There is a clear dependence of transition temperature and resistivity on film thickness. The films fabricated onto (110) SrTiO$_3$ substrate do not show this regular behavior of resistivity versus thickness, as is evident from Fig. 1. Transition temperature (determined as the position of the maximum $d\rho / dT$ vs. T) shows a behavior similar to Fig. 4.

With the shown morphology of the 100 Å film, the resistivity in the direction along the long [1$\bar{1}$0] axis of the grains will be lower than the resistivity in the perpendicular direction. Since the grain boundaries can be considered as resistances connected in series, the total film resistance should be proportional to the number of grain boundaries in given direction. Therefore, the aspect grain length-to-width ratio of 4-5, obtained from the AFM images, should corresponds to the parameter anisotropy of resistivity. Since it has been found to be about 7 at room temperature, one can infer the anisotropy observed in ultrathin films nicely fits the model of grain boundary resistance.

The next thicker film, 200 Å, has a resistivity in the short direction that is 2 times lower than the

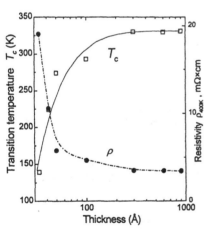

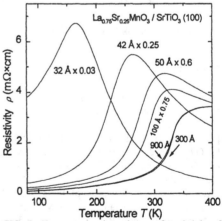

FIG. 4 Dependence of transition temperature an resistivity on film thickness for La$_{0.75}$Sr$_{0.25}$MnO$_3$ film on (100) SrTiO$_3$ substrates. The lines are meant as guide for the eyes [1].

FIG 5. Temperature dependence of resistivity for La$_{0.75}$Sr$_{0.25}$MnO$_3$ films on (100) SrTiO$_3$ substrates. The curves have been scaled by different factors to fit in the same graph [1].

100 Å film, but the width of the grains have doubled. This explains the resistivity change in the short direction. In the long direction the resistivity has decreased 3 times, but the grain length is only slightly longer. This can be explained if we take into account the effect of growth of the resistivity with film thinning observed in our earlier work on films on (100) STO (see Fig. 5). The change in resistivity between a 100 Å and 300 Å LSMO film on (100) STO at 300 K is about 3 times. Thus we may conclude the difference in resistivity in the long direction for the 200 and 100 Å films can be explained by intrinsic LSMO property to increase resistivity with film thickness decrease. We also assume that the grain growth in the short direction eliminates the cause of the thickness dependence of resistivity.

The behavior of 1200 and 2400 Å films presents a bigger mystery, since a film, getting thicker, is expected to become more isotropic as a bulk LSMO. There is a one common feature for all LSMO/STO(110) films: decrease of the parameter of anisotropy at high temperatures. This behavior also supports the model of grain boundary resistance. It is reasonable to assume the grain boundaries to be made up of disorded manganite with T_c lower than the transition temperature T_c inside the grain, higher peak value of resistivity and larger width of para-to-ferromagnetic phase transition [7]. Since the disordered manganite in the grain boundaries becomes more conducting at higher temperatures, the contribution of the grain boundary resistance to total film resistivity relatively decreases. Therefore, the main, as we believe, source of anisotropy is effective only at low temperatures.

To summarize, $La_{0.75}Sr_{0.25}MnO_3$ films have been grown on (110) $SrTiO_3$ single crystal by pulsed laser deposition technique. The films show a strong anisotropy of transport properties that is larger for the thicker films. Atomic force microscopy images show that ultrathin 100 Å and 200 Å films consists of elongated grains, while the thicker films (1200 and 2400 Å) show circular shaped grains. Although details of anisotropy in thick films remain a puzzle, the universal behavior of anisotropy parameter at high temperatures for films with various thickness bring us to the conclusion the granular structure might be responsible for strong transport anisotropy in $La_{0.75}Sr_{0.25}MnO_3/SrTiO_3$ (110) films.

ACKNOWLEDGEMENTS

We would like to thank C. Canalias for AFM measurements.

REFERENCES

[1] S. I. Khartsev, P. Johnsson, A. M. Grishin, J. Appl. Phys. **87**, 2394, 2000.
[2] S. Jin, T. H. Teifel, M. McCormack, R. A. Fastnacht, R. Ramesh, L. H. Chen, Science, **264**, 413, 1994.
[3] Y. Suzuki, H. Y. Hwang, J. Appl. Phys. **85**, 4797, 1999.
[4] J. O'Donnel, J. N. Eckstein, M. S. Rzchowski, Appl. Phys. Lett. **76**, 218, 2000.
[5] J. Mira, A. Fondado, L. E. Hueso, J. Rivas, F. Rivadulla, M. A. López Quintela, Phys. Rev. B, **61**, 5857, 2000.
[6] X. T. Zeng, H. K. Wong, Appl. Phys. Lett. **72**, 740, 1998.
[7] J. E. Evetts, M. G. Blamire, N. D. Mathur, S. P. Isaac. B.-S. Teo, L. F. Cohen, J. L. Macmanus-Driscoll, Phil. Trans. R. Soc. Lond. A, **356**, 1593, 1998.

MAGNETIC AND MAGNETO-OPTIC PROPERTIES OF PULSED LASER DEPOSITED $Ce_xY_{3-x}Fe_5O_{12}$ FILMS

Hyonju Kim, Alex Grishin, R. Sbiaa*, H. Le Gall*, and K. V. Rao
Department of Condensed Matter Physics, Royal Institute of Technology, S-100 44 Stockholm, Sweden
* Laboratoire de Magnetisme de Bretagne, CNRS/UPRESA 6135, Universite, 29285 Brest, France

ABSTRACT

We present experimental studies of crystalline, magnetic, and magneto-optic (MO) properties and their correlation with the processing parameters of Ce-substituted yttrium iron garnet ($Ce_xY_{3-x}Fe_5O_{12}$) thin films grown epitaxially onto single crystal $Gd_3Ga_5O_{12}$ (111) substrates by Nd:YAG pulsed laser deposition technique. Rutherford backscattering (RBS) analysis and microprobe energy dispersive spectrometry (EDS) prove the background oxygen pressure, used for films growth, control films stoichiometry. Oxygen ambient pressure appears to be an important parameter to grow $Ce_xY_{3-x}Fe_5O_{12}$ films with good crystallinity, magnetic properties, as well as enhanced MO effect. It is found that the film fabricated at 50 mTorr oxygen pressure exhibits a maximum Faraday rotation (FR) $\theta_F = 1.78$ deg/μm at $\lambda = 633$ nm, a maximum saturation magnetization $4\pi M_s = 1255$ Gauss with a minimum in-plane coercivity $H_c = 35$ Oe, and the narrowest full width at half maximum FWHM = 0.06° of the (444) Bragg reflection rocking curve. The analog of the Verdet constant $V = \theta_F / 4\pi M_s$, also found to be dependent on the oxygen ambient pressure, reaches the value as high as 1.4 deg/μm·kG at 633 nm, suggesting that this material is useful for MO applications.

INTRODUCTION

It is well known that the rare-earth iron garnets (ReIG) with the substitutions of either the diamagnetic Bi^{3+} and Pb^{2+} ions and/or of lighter rare-earth ions like Ce^{3+}, Pr^{3+} and Nd^{3+} exhibit enhanced magneto-optical (MO) effects in the wavelength range of 0.4-2 μm (Ref. 1 and refs. therein). Recently, cerium substitution to yttrium iron garnet ($Ce_xY_{3-x}Fe_5O_{12}$, Ce:YIG) has been shown to enhance significantly the Faraday effect with paramagnetic dispersion relation not only in the UV region (centered at $h\nu = 3.1$ eV) but also in the IR region at $h\nu = 1.4$ eV. [2] Ce:YIG appears to possess good magnetooptical quality, the figure of merit ψ, defined as a ratio of Faraday rotation (FR) angle θ_F [deg/cm] to the absorption coefficient α [cm^{-1}]. It has been shown, [3] that ψ of Ce:YIG is much larger than for YIGs ($\psi = 250$ deg) in the wavelength region of 1.51-1.57 μm, reaching $\psi = 1200$ deg due to a considerable decrease in the optical absorption. Ce:YIG is thus found to be of considerable interest as universal MO media applicable both in the near-infrared as well as in the visible region.

Following our earlier work, [4] we have carried out comprehensive study on correlation between the crystalline, magnetic, and magnetooptic properties, and composition of the $Ce_xY_{3-x}Fe_5O_{12}$ films. Here we present results on Nd:YAG pulsed laser deposited $Ce_xY_{3-x}Fe_5O_{12}$ films of epitaxial quality exhibiting the maximum FR of 1.78 deg/μm at $\lambda = 633$ nm.

Mat. Res. Soc. Symp. Proc. Vol. 623 © 2000 Materials Research Society

EXPERIMENT

Ce-substituted yttrium iron garnet films ($Ce_xY_{3-x}Fe_5O_{12}$) were prepared by the pulsed laser deposition technique using a Nd:YAG laser ($\lambda = 355$ nm, 10 Hz pulse frequency, and 10 ns pulse width) at an energy density of 90 mJ/pulse. The films were deposited onto (111)-oriented $Gd_3Ga_5O_{12}$ (GGG) single-crystal substrates at 790°C for 30 minutes at various oxygen ambient pressures of 10, 25, 50, 75 and 100 mTorr. Film thickness, determined by the α–step meter, showed 0.4, 0.35, 0.2, 0.2, and 0.2 μm for the films grown at 10, 25, 50, 75, and 100 mTorr, respectively. Thus, deposition rate appears to decrease in the beginning with an increase of the oxygen ambient pressure, but after 50 mTorr it is found to saturate and show 7nm/min under the deposition conditions investigated in this work.

Crystalline properties were investigated using x-ray diffraction (XRD, Cu-$K\alpha$ radiation) θ-2θ scan, rocking curve and φ-scan measurements. Vibrating sample magnetometer (VSM) was used to study the magnetic properties of the films at room temperature. Faraday effect was measured by observing the rotation of the plane of polarization in transmission light in magnetic field (up to 15 kOe) applied perpendicular to the film plane. To determine the stoichiometry of the deposited films, the Rutherford backscattering spectrometry with 2 MeV $^4He^+$ beam and the microprobe Energy Dispersive spectrometry (10 kV accelerating voltage, and 232.5×167 μm^2 beam aperture) have been employed.

RESULTS AND DISCUSSION

From the XRD θ-2θ scan, (111)-oriented epitaxial growth has been confirmed over the examined angle range (5-125 °) for all the $Ce_xY_{3-x}Fe_5O_{12}$ films fabricated at different oxygen ambient pressures. As shown in Fig. 1, which is the expanded sections for (444) Bragg reflections, Ce:YIG (444) diffraction peak appears to move towards higher angles with an increase of oxygen pressure, indicating a decrease in the out-of-the-film lattice constant a. a lattice parameter, determined from the Ce:YIG (444) diffraction peak, is shown in Fig. 2 and the Table. It can be clearly seen that the lattice constant increases as the oxygen ambient pressure

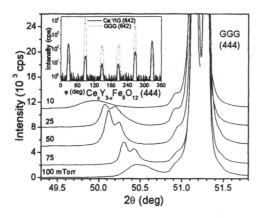

Fig. 1 X-ray diffraction (XRD, CuK_α radiation) θ-2θ scans for $Ce_xY_{3-x}Fe_5O_{12}$ films grown by PLD onto $Gd_3Ga_5O_{12}$ (111) (GGG) single crystal at various oxygen pressures. XRD patterns for 10, 25, 50, and 75 mTorr are offset for clarity. XRD φ-scans for Ce:YIG film (50 mTorr) and GGG substrate (642) reflection are shown in the insert. Detector and sample positions ($2\theta_{det}$, θ_{sample}) have been chosen as follows: (54.474°, 49.514°) and (55.579°, 49.995°) for Ce:YIG (642) and GGG (642) reflections respectively.

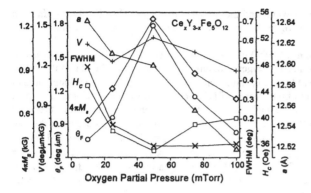

Fig. 2 Oxygen ambient pressure dependencies of saturation magnetization $4\pi M_s$, Faraday rotation θ_F, lattice constant a, Verdet constant V, FWHM of rocking curve for the (444) Bragg reflection, and the in-plane coercivity H_c, measured at room temperature.

used for the growth is lowered.

Ce substitution for Y^{3+} (ionic radius of 0.95 Å) at the dodecahedral sites of garnet structure is known to result in lattice expansion because of its larger ionic radius of 1.13 Å. According to the Vegard rule [5], variation of the lattice constant of substituted yttrium iron garnets depends linearly on the amount of substitution. Thus, the expansion of the lattice constant at lower oxygen pressure seems to indicate a significant increase of Ce incorporation into the YIG lattice. This fact also nicely corresponds to the phenomenon of *discriminated thermalization* of ablated species in multicomponent laser plume, described recently.[6] It is as follows: to get congruent target-to-film material transport, all the ablated constituents should be delivered to the substrate in a right proportion. However, laser ablated particles lose their kinetic energy (*thermalized*) by colliding with the background gas molecules. Since the momentum loss in an elastic collision is more pronounced for light species, the kinetic energy of the ablated particles and the number of species reaching the substrate can vary depending on the pressure regime used for the deposition. In the Ce-substituted YIG, Fe, Y, and Ce have atomic masses of 56, 89, and 140 respectively. At low background pressure, Fe and Y can be easily thermalized, while the transport of the heaviest Ce could be still ballistic. Thus, this can cause a significant loss of Fe with an increase of the background pressure, but an additional incorporation of Ce at lower background pressure, influencing the stoichiometry of the laser ablated $Ce_xY_{3-x}Fe_5O_{12}$ films.

As shown in Fig.1, the films grown at 10 or 25 mTorr O_2 are found to have relatively poor crystallinity. The stoichiometry change and corresponding expansion of the lattice constant are expected to cause the structural disorder, which is responsible for the relatively poor crystallinity of the films grown at lower oxygen background pressures. On the other hand, the film grown at

TABLE Properties of $Ce_xY_{3-x}Fe_5O_{12}$ films prepared by PLD at various oxygen ambient pressures

O_2 pressure (mTorr)	Deposition rate (nm/min)	Ce:Y	Fe:Y	a (Å)	FWHM (deg)	H_c (Oe)	$4\pi M_s$ (G)	FR (deg/µm)	V (deg/µm·kG)
10	13.3	0.21	0.20	12.64	0.47	45	559	0.76	1.36
25	11.7	0.17	0.16	12.61	0.17	38	780	0.96	1.23
50	6.7	0.17	0.12	12.60	0.06	35	1255	1.78	1.41
75	6.7	0.11	-	12.56	0.06	39	882	1.15	1.30
100	6.7	-	-	12.52	0.07	40	710	0.83	1.16

50 mTorr O_2 shows the narrowest width and the highest intensity of Ce:YIG (444) diffraction peak, suggesting the highest homogeneity within film thickness: lowest strain and uniform composition.

The full width at half maximum (FWHM) of the rocking curve of (444) Bragg reflection also exhibits a minimum value at 50 mTorr oxygen pressure as represented in Fig. 2 and the Table. The narrow FWHM of Ce:YIG (444), comparable with FWHM = 0.02° of the rocking curve of GGG (444) reflection, proves the high degree of (111)-orientation of the PLD-made Ce:YIG films. The in-plane texture of the Ce:YIG films has been examined by measuring the x-ray diffraction from off-normal {642} planes. The insert in Fig. 1 presents the φ - scans for the film prepared at 50 mTorr O_2, measured for both Ce:YIG(642) ($2\theta_{det}$ = 54.474°, θ_{sample} = 49.514°) and GGG(642) ($2\theta_{det}$ = 55.579°, θ_{sample} = 49.995°) reflections. Six-fold symmetry of the φ-scans corresponds to three equivalent off-normal {642} planes appearing twice each during 360° in-plane sample rotation. Coincidence of Ce:YIG and GGG peaks position indicates the following film-to-substrate epitaxial relationship:

$$(111) \text{ Ce:YIG } \| \text{ } (111) \text{ GGG},$$
$$[110] \text{ Ce:YIG } \| \text{ } [110] \text{ GGG}.$$

Fig. 3 shows the RBS spectra for $Ce_xY_{3-x}Fe_5O_{12}/Gd_3Ga_5O_{12}$ structures, measured using 2 MeV $^4He^+$ beam. In the figure, the lines with elements indicate the position of the each element surface yield. The RBS spectrum of GGG is also given in the figure for comparison. It is observed that the Gd signal from the $Ce_xY_{3-x}Fe_5O_{12}/Gd_3Ga_5O_{12}$ structure shifts toward lower energy as the oxygen ambient pressure is reduced, indicating a decrease of the film thickness, as also confirmed by α-step meter measurement. The same tendency is also observed for the Ga signal. It appears to be difficult to estimate the stoichiometry of the films using RBS under the

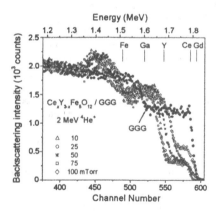

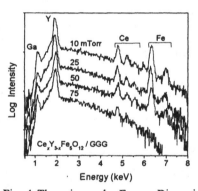

Fig. 3 Random RBS spectra for $Ce_xY_{3-x}Fe_5O_{12}$ films on $Gd_3Ga_5O_{12}$ (GGG) substrates measured using 2 MeV $^4He^+$ beam. GGG spectrum is also given for comparison. The lines with elements indicate the position of the each element surface yield.

Fig. 4 The microprobe Energy Dispersive spectra (EDS, SEM *ZEISS DSM 942*, accelerating voltage of 10 kV, beam aperture of 232.5×167μm^2) recorded for $Ce_xY_{3-x}Fe_5O_{12}$ films grown at different ambient oxygen pressures. The intensity is expressed in log- scale. The spectra have been shifted vertically for clarity.

conditions investigated in this work, due to the complexity of films composition and the overlap between the signals from the film and the substrate. To determine the relative composition, we proceeded our study with the microprobe EDS measurements. Fig. 4 shows the EDS spectra of the films represented in log-scale, where the spectra have been shifted vertically to make the difference between spectra easily discernable. The relative ratio of Ce and Fe to Y, as given in the Table, is found to decrease significantly with an increase of the oxygen ambient pressure. This EDS measurement, thus, nicely confirms our experimental results and their interpretation on enhanced incorporation of Ce ions in YIG lattice at low oxygen pressure.

Magnetic hysteresis measurements in magnetic field perpendicular $H_\perp$ and parallel $H_\parallel$ to the film plane were performed using VSM at room temperature. Fig. 5 shows M-$H_\parallel$ loops of the Ce:YIG films measured in magnetic field $H_\parallel$ parallel to the film surface. In $H_\parallel$ geometry, magnetization saturates at rather low field ($\leq$ 50 Oe) and loops measured at mutually perpendicular in-plane orientations did not show in-plane anisotropy. When the applied field is normal to the film surface, M-$H_\perp$ curves saturate at higher fields and plots in M vs. ($H_\perp$-$4\pi M$) scale coincide with M-$H_\parallel$ hysteresis loops. These testify to the planar magnetization and its dependence on the internal magnetic field. The saturation magnetization $4\pi M_s$ and the coercivity H_c are also found to be dependent on the oxygen partial pressure used for the deposition, showing a maximum value of $4\pi M_s$ = 1255 Gauss and a minimum coercivity of H_c = 35 Oe at 50 mTorr. As represented in Fig. 2 and the Table, the best magnetic properties, achieved for the film grown at 50 mTorr oxygen pressure can be ascribed to the superior crystalline properties, *i.e.*, the highest intensity of Ce:YIG (444) reflection (Fig. 1) and the narrowest FWHM of the rocking curve.

Faraday rotation θ_F - $H_\perp$ loops measured at 633 nm at room temperature in perpendicular magnetic field $H_\perp \leq$ 15 kOe are shown in Fig. 6. The specific FR θ_F [deg/μm] is also found to depend on the oxygen ambient pressure (Fig. 2) and reaches a maximum of 1.78 deg/μm for the film grown at 50 mTorr O_2. This maximum FR at 50 mTorr O_2 also seems to correlate with the obtained superior crystalline and magnetic properties as well as with the optimum Ce incorporation in the film.

The analog of Verdet constant V [deg·μm^{-1}·kG^{-1}] = $\theta_F/4\pi M_s$ of the fabricated Ce:YIG films has been determined at 633 nm and presented in both Fig. 2 and the Table. It also exhibits a maximum as high as 1.41 deg·μm^{-1}·kG^{-1} for the film grown at 50 mTorr oxygen pressure,

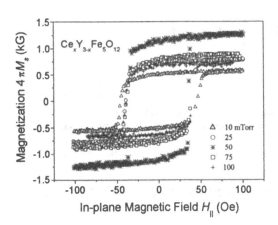

Fig. 5 In-plane magnetization loops of $Ce_xY_{3-x}Fe_5O_{12}$ films grown at various oxygen pressures. Strong paramagnetic contribution from $Gd_3Ga_5O_{12}$ substrate has been subtracted. The $4\pi M_s$ and H_c are found to be dependent on the oxygen ambient pressure (for the value, refer to the Table).

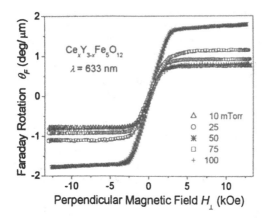

Fig. 6 Faraday rotation (λ = 633 nm) hysteresis loops in perpendicular magnetic field $H_\perp \le$ 15 kOe. Strong paramagnetic contribution from $Gd_3Ga_5O_{12}$ substrate has been subtracted. The specific FR θ_F [deg/μm] is found to depend on the oxygen ambient pressure used for the growth (for the value, refer to the Table).

suggesting that this material is promising for MO applications.

In summary, epitaxial Ce-substituted YIG films have been grown on GGG (111) single crystal using Nd:YAG pulsed laser deposition technique. Oxygen ambient pressure used for the growth has been found to control crystallinity, magnetic, and magnetooptic properties as well as the stoichiometry of the PLD made films. Monotonic increase of the lattice parameter with a decrease of oxygen pressure suggests higher incorporation of Ce ions in YIG lattice at lower pressures. In addition, from the microprobe analysis using EDS, Fe concentration in the films is also found to decrease greatly with an increase of the oxygen ambient pressure. 50 mTorr of oxygen pressure has been found to be optimum to gain epitaxial Ce:YIG film quality indicated by the narrowest FWHM = 0.06° of the (444) rocking curve, the maximum saturation magnetization of 1255 G together with the minimum coercivity of 35 Oe, and the maximum Faraday rotation of 1.78 deg/μm at 633 nm. By optimizing the processing parameters of PLD technique, it is expected that the specific FR can be increased, further reducing its optical path length and hence the size of the devices.

ACKNOWLEDGEMENT

The authors would like to thank Dr. Birger Emmoth for the assistance with the RBS measurements.

REFERENCES

1. M. Guillot, H. Le Gall, J. M. Desvignes, and M. Artinian, J. Appl. Phys., **81**, 5432 (1997).
2. You Xu, Jie Hui Yang, Xi Juan Zhang, Phys. Rev. B, **50**, 13428 (1994).
3. T. Sekijima, T. Funakoshi, K. Katabe, K. Tahara, T. Fujii, Jpn. J. Appl. Phys., **37**, 4854 (1998).
4. Hyonju. K., A. M. Grishin, K. V. Rao, S. C. Yu, R. Sbiaa, H. Le Gall, IEEE Trans. Magn., Vol. **35**, 3163 (1999).
5. L. Vegard, Z. Phys. **5**, 17 (1921).
6. C.-R. Cho, Alex Grishin, Appl. Phys. Lett. **75**, 268 (1999), J. Appl. Phys. **87** (May 1, 2000).

Ferroelectrics and
Related Materials

JUNCTION PROPERTIES OF METAL/SrTiO₃ SYSTEMS

Takashi SHIMIZU, and Hideyo OKUSHI

Electrotechnical Laboratory, 1-1-4 Umezono, Tsukuba, Ibaraki 305-0045 Japan

ABSTRACT

Electrical properties of Nb-doped $SrTiO_3$ (STO:Nb) Schottky barrier (SB) junctions have been investigated in detail for a comprehensive understanding of metal/oxide interfaces. Using a high-purity ozone surface treatment, rectification ratio over 9th order of magnitude has been successfully obtained, while without the surface treatment, anomalous large reverse bias leak currents were observed in the current-voltage characteristic of the junctions. The X-ray photoelectron spectroscopy (XPS) shows that carbon contamination which adsorbed the STO:Nb surface in air, induces surface states in the band gap of the STO:Nb, which probably originate the large reverse bias leak currents of the metal/STO:Nb junctions. Thus we present importance of surface treatment for oxides to obtain controllability and reproducibility of the electrical properties of the oxide devices. Photocapacitance spectroscopy has been performed to investigate deep levels due to bulk defects and impurities in the Au/STO:Nb junctions. The photocapacitance spectra clearly indicate existence of the deep levels in the Au/STO:Nb and the concentration of the deep levels were of the order of $10^{13} \sim 10^{15}$ cm^{-3}. These values are too low to affect the Fermi level pinning at the interface if the deep levels exist in the near surface region of the bulk STO:Nb. We have shown some interesting electrical properties, characteristic of the SB junction of the dielectric oxide compared with that of the conventional semiconductor's. The schematic band diagram of the Au/STO:Nb junction with the intrinsic low permittivity layer at the interface has been proposed, which explains all the characteristic electrical properties. Considering the chemical trend of the SB height (SBH) estimated from the J-V results, we have pointed out the importance of the metal reactivity for understanding the formation mechanism of the SBH.

INTRODUCTION

Recently, great efforts have been made to deposit oxide thin films directly on silicon and to apply their dielectric properties to ultra-large-scale-integrated-circuit (ULSI) technology. Since transition-metal oxides including titanium, tantalum and niobium are considered to be promising materials for the applications, studies of electrical properties of their junction devices including those of interface states, interface layer, and deep levels due to defects or impurities are important to obtain reliability and reproducibility of the junction devices. In this paper,

Mat. Res. Soc. Symp. Proc. Vol. 623 © 2000 Materials Research Society

preliminary results of junction properties of metal/Nb-doped SrTiO$_3$ systems, including extrinsic origin of interface states, characteristic junction properties originating from their dielectric properties and deep levels observed using the photocapacitance method, are presented.

EXPERIMENTAL

The Schottky barrier (SB) junctions were fabricated on 0.01 wt.% Nb-doped SrTiO$_3$ (STO:Nb) (001). To prevent effect of grain boundary and/or low crystallinity, Verneuil grown single crystal substrates were used. They were as-polished substrates chemically etched by concentrated HNO$_3$ acid for 2 minutes or so-called "stepped substrates" chemically etched by buffered HF. Each substrates were annealed in flowing O$_2$ at 1000 ℃ for 1 hour and at 400 ℃ for 4 hours to obtain the substrate surface having well-ordered surface steps and atomically flat terraces [1][2]. Before deposition of metals for the SB junction in ultra-high vacuum chambers, cleaning procedure using high-purity ozone was performed in the same chambers, in which substrate surfaces were annealed in the high purity ozone atmosphere (the ozone flux was about 10^{16} molecules-cm^{-2}) at 400-500 ℃ for 20-40 minutes (we call this procedure as "ozone cleaning"). The pressure of the chamber during the ozone cleaning was 1-3 $\times$ 10^{-3} Pa. The reflection high energy electron diffraction (RHEED) and atomic force microscopy (AFM) measurements showed that the surface reconstruction and the surface morphology are almost unchanged after the ozone cleaning [3].

After the ozone cleaning, the substrates were cooled down spontaneously in the ozone atmosphere to 50-100 ℃ at which metal electrodes were deposited for fabricating metal/STO:Nb junctions. The electrodes for the SB junctions were deposited *in situ* through the stainless steal mask of holes soon after stopping introduction of the high-purity ozone gas. Electrode metals for the SB junctions were mainly Au, while Cu, Ag and Pt electrodes were also used. Electrodes for the Ohmic contacts were Al and Ti deposited backside of the STO:Nb substrates to form SB diodes. Although guide-ring electrodes to prevent edge effect in electrical measurements were not able to be deposited, *in-situ* preparation will be advantageous to prevent the STO:Nb surface from mechanical damages compared with preparation methods using photo lithography technique, which requires chemical etching and/or ion etching of the STO:Nb surface.

The electrical measurements were performed in a vacuum chamber using tungsten probes. Current density-voltage (*J-V*) and capacitance-voltage (*C-V*) characteristics were measured using a picoampere meter and an impedance analyzer, respectively. The measurement frequencies of the *C-V* measurements were between 100 Hz and 1 MHz. Using a single-grating monochrometer, monochromatic light was irradiated on Au/STO:Nb SB junctions, to study photocurrent and photocapacitance characteristics. The schematic of the measurement system is shown in Fig. 1. The photocurrents were measured using the picoampere meter under 0 bias voltage. The photocapacitances were measured at room temperature using the impedance analyzer at reverse bias voltage of –4 V with a measurement frequency of 1 MHz. A Si photo detector showed that the incident photon flux has a slight photon energy dependence (within 2.1 %) in the range between 1.24 and 1.70 eV in which the photocurrent measurements were performed. The photo detector also showed gradual decrease of the incident photon flux (~45 % at 3.65 eV photon energy) in the range between 1.55 eV and 3.65 eV for the photocapacitance measurements. It is noted that this photon energy dependence of the photon flux does not affect steady state photocapacitance spectra. It is also noted that since the bandgap of the $SrTiO_3$ is reported to be about 3.3 eV at room temperature [4], the photon energy in the photocapacitance study in this paper was almost above the midgap energy of the $SrTiO_3$.

RESULTS AND DISCUSSION

Control of SB height using the ozone cleaning

The rectification properties of metal/$SrTiO_3$ junctions are drastically improved when the junctions were fabricated after the ozone cleaning as reported previously [1,2,5]. The rectification ratio over 9th order of magnitudes has been successfully obtained for the first time because the reverse bias leak currents of the junctions are drastically supressed to become lower than detection limit of the measurement system. The ideality factor n and the SB height (SBH)

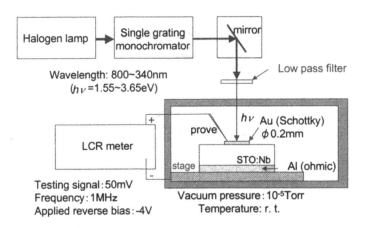

Figure 1. Schematic view of the measurement system.

are estimated from the forward J-V characteristics using the thermionic-emission theory, which is described under the condition of $V > 3kT/q$ by the following equation [6]:

$$J = A^*T^2 \exp(-\frac{q\Phi}{kT})\exp(\frac{qV}{nkT}) \cdot \qquad (1)$$

Here A^* is the Richardson constant, T is the temperature, q is the Coulomb charge, Φ is the SBH, k is the Boltzmann constant, V is applied bias voltage, respectively. The Richardson constant A^* is estimated as $156 \text{ Acm}^{-2}\text{K}^{-2}$.

Figure 2 shows the ideality factor n vs. SBH of Au/STO:Nb junctions fabricated on a bare "stepped" STO:Nb substrate without the ozone cleaning and a "stepped" STO:Nb substrate with the ozone cleaning after 1000 °C O_2 annealing. As shown in the fig. 2, distribution of both the ideality factor n and the SBH becomes narrower when the ozone cleaning was applied to the STO:Nb surface, which indicates that reliability and reproducibility of the junction properties have been improved. The SBH also increases when the ozone cleaning was applied and the rectification ratio in the J-V characteristics was drastically improved from about 4th order to over 9th order of magnitude [1,2]. It is noted that smaller ideality factors of the junctions without the ozone cleaning does not mean better junction properties, because the forward J-V characteristics of the junctions scarcely have the log-linear region. Therefore, as for the junctions without the ozone cleaning, the ideality factors shown in the fig.2 mean the smallest limits of the ideality factors estimated from the steepest tangent line of the forward J-V curves.

To investigate the origin of the drastic improvement of the electrical properties, surface

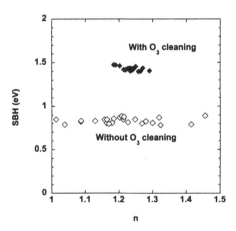

Figure 2. The ideality factor n vs. SBH of Au/0.01 wt.% Nb-doped SrTiO₃ (001) junctions fabricated with and without O₃ cleaning.

electronic states of the STO:Nb were investigated using the X-ray photoelectron spectroscopy (XPS). Figure 3 shows the XP spectra around C 1s peak of the STO:Nb surfaces before and after the ozone cleaning [7]. The XPS showed that the C 1s XP peak, clearly observed in the STO:Nb surface before the ozone cleaning, is diminished after the ozone cleaning. This indicates that the origin of the large reverse bias leak currents of the Au/STO:Nb junctions is arisen from the carbon contamination of the substrate surface. Moreover, as shown in fig.4, the gap states observed in the valence band region of the STO are diminished after the ozone cleaning. This is direct evidence that the carbon contamination of the surface induces surface states of the STO:Nb which affect the electrical properties of fabricated metal/STO:Nb junctions. Since the ozone cleaning is effective for eliminating the carbon contamination, the metal/STO:Nb junction prepared with the ozone cleaning should show intrinsic electrical properties of the junction without any extrinsic interface states. It is noted that almost the same results were obtained for the Nb-doped TiO_2 substrates. These indicate that the improvement of the electrical properties by eliminating carbon contamination will be the general results for the junction devices based on oxide-semiconductors. It is interesting that the oxide surface is not so stable to adsorbates. The junction properties of oxide-semiconductors seem to be sensitive to the carbon contaminants. These results suggest that one of the reason for the leakage currents observed in several reported dielectric and/or ferroelectric thin films may be due to the carbon contamination in the films.

Since insufficient oxidation of organometal source in chemical vapor deposition (CVD) technique will introduce carbon contamination to deposited films, it may be advantageous to

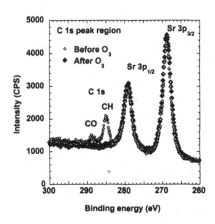

Figure 3. The XPspectra of the STO:Nb before and after the O_3 cleaning.

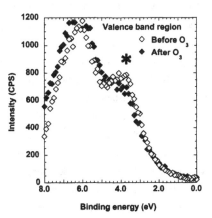

Figure 4. The XPspectra near the valence band of the STO:Nb before and after the O_3 cleaning.

use reactive oxidizing agent to prepare the oxide films by CVD. In case of the pulsed laser deposition and sputtering techniques, it should be taken care that target material and introduced oxygen gas should contain organic impurities as low as possible. Using high-purity oxidizing agent, such as high-purity ozone gas, is considered to be more and more important when fabricating of device quality oxide films.

Deep level characterization of the Au/STO:Nb junctions

In conventional semiconductors, characterization of deep levels due to defects and/or impurities is important for junction devices, because they act as trap centers and/or recombination centers for free carriers. Moreover, electronic states near surface region of a semiconductor may cause the Fermi level pinning phenomenon when concentration of the deep levels is higher than about 10^{10} cm^{-2} [10]. Therefore it will be also important to characterize the deep levels in the dielectric oxides when we consider junction properties of the metal/dielectric oxides as well as when they are applied to ULSI technology. Although deep level transient spectroscopy (DLTS) is known to be one of the suitable method to investigate the deep levels in the conventional semiconductors, it is difficult to use it to investigate the deep levels in the dielectric oxides. Since the capacitance of these oxides reflects the temperature-dependence and electric field-dependence of the electrostatic permittivity, transient capacitance in the DLTS measurement also includes these effects beside with the transient property of the space charge density in the depletion region. Although isothermal capacitance transient spectroscopy (ICTS) measurement was found to be more suitable for the oxides when such temperature dependence was considered [8,9], it is also difficult to investigate the Au/STO:Nb junctions by the conventional ICTS measurements since they show anomalous transient curves even in the isothermal conditions, which might be induced by the electric field dependence of the permittivity. We have adopted the photocapacitance method to the Au/STO:Nb junctions, which is performed under constant electrical field and the isothermal conditions.

Figure 5 shows typical steady-state photocapacitance spectra of the Au/STO:Nb in the photon energy range between 1.55 eV and 3.65 eV. The ΔC in the fig. 5 represents the difference of the capacitance from that of at 1.55 eV photon energy. It is noted that a half of the bandgap energy of the STO (3.2~3.3 eV at room temperature) is about 1.6 eV. Observed photocapacitance spectra clearly indicates the existence of the deep levels in the Au/STO:Nb junctions. Since rapid increase of photocapacitance above around 3.3 eV is probably due to band gap excitation of the STO, there observed four features in a photocapacitance spectrum which is considered to be due to deep levels in the Au/STO:Nb junctions: that is capacitance increase in the range of 1.7~2.0 eV, 2.3~2.5 eV and 2.5~3.0 eV and capacitance decrease in the range of 2.0~2.2 eV. Considering that capacitance increase corresponds to existence of deep levels below mid gap energy and that capacitance decrease corresponds to that above mid gap energy, we attribute the four features in the photocapacitance spectra mainly to four deep levels as shown in fig. 6. We denote these four deep levels as E_{T1}(1.0~1.2 eV from the conduction band edge), E_{T2} (1.7~2.0 eV), E_{T3} (2.3~2.5 eV) and E_{T4}(2.5~3.0 eV). When deep level concentrations of the E_{T1}~E_{T4} are estimated from $2\Delta C/C$ using space charge densities which are roughly estimated from the C^{-2}-V characteristics, they are 10^{13}~10^{14} cm^{-3} (E_{T1}, E_{T2}, E_{T3}) and 10^{14}~10^{15} cm^{-3} (E_{T4}), respectively. It is noted from the estimation that the deep level concentrations are much lower than doping concentration of Nb and are too low to make Fermi level pinning at the interface. However, when these deep levels also exist in undoped STO films, they may affect the electrical properties of the junction devices as charge trap centers and/or recombination centers.

Intrinsic low permittivity layer at metal/STO:Nb interface

There is a possibility that surface and/or interface of the dielectric oxide has different dielectric property from the bulk, since interdipole interaction is reduced at the surface and/or interface [11]. Moreover, atomic rearrangement of the oxide at the surface and/or interface will cause different lattice properties from the bulk [12-15], which will originate the different dielectric property. Thus, both macroscopic and microscopic model suggest that intrinsic interface layer, which show different dielectric property from the bulk STO, exists at the metal/STO:Nb interface. Since such interface layer will affect the electrical properties of the

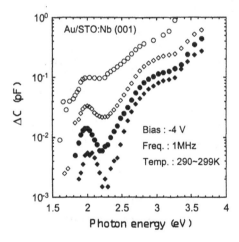

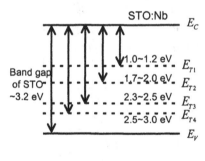

Figure 5. Typical photocapacitance spectra of the Au/STO:Nb junctions.

Figure 6. Schematic drawing of deep levels in a Au/STO:Nb junction.

oxide junction, it will be more and more important to determine whether such interface layer actually exists and to understand the nature of the interface layer if exists for controlling electrical properties of the oxide based junction devices.

The Au/STO:Nb (001) junctions which were fabricated on the ozone cleaned surface were investigated in detail to clear out intrinsic junction properties of metal/STO:Nb systems [2]. Figure 7 and figure 8 show temperature dependencies of J-V and photocuurent characteristics of a Au/STO:Nb (001) SB junction. The SB heights estimated from the J-V characteristics using the thermionic-emission theory are also shown as arrows in the fig.8. According to the Fowler theory, a straight line should be obtained in the square root of the photoresponse plotted as a function of photon energy and the SBH should be directly given as the extrapolated value on the energy axis. As shown in the fig. 8, the threshold energy of the photoresponse seems to approximately coincide with the SBH estimated from the J-V results. It is noted that both J-V and photocurrent characteristics indicate that the SBH of the Au/STO:Nb junction decreases with temperature decreases. In case of C-V characteristics, different properties from the conventional SB junctions were observed as shown in fig.9. The C-V characteristics show temperature dependence, which reflects the temperature dependence of the permittivity of the STO. The C^{-2} plot deviates from the linear relation to the voltage especially at low temperatures, which reflects the electric-field dependence of the permittivity of the STO. These results are characteristic junction properties of metal/dielectric oxide systems. The flat band voltage, which is estimated from the intersection of C^{-2} vs. V plot with the voltage, is found to be temperature

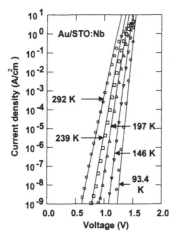

Figure 7. Temperature dependence of J-V characteristics of a Au/STO:Nb junction. Solid lines represent simulated J-V characteristics.

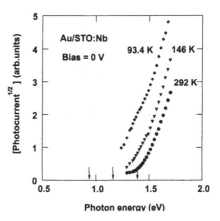

Figure 8. Photocurrent spectra of a Au/STO:Nb junction. Arrows indicate corresponding SBHs estimated from J-V characteristics at 292, 146, 101 K.

independent as shown in the fig. 9, while the *J-V* results indicate temperature dependence of the SBH as described above.

It is found that when the intrinsic low permittivity layer at metal/STO:Nb interface is considered, as schematically shown in fig. 10, all these results can be explained quantitatively. In the fig. 10, the thickness of the layer is noted as δ_i, the relative permittivity is noted as ε_i. The space charge densities within the interface layer and in the depletion layer are assumed to be 0 and constant (denoted as N_d), respectively. We assumed the relationship between E and P in the depletion region as $E = \alpha P + \beta P^3$ according to the Devonshire theory. It is noted that $\varepsilon_B(T) = (\alpha \varepsilon_0)^{-1}$ in the limit of zero electric field, where $\varepsilon_B(T)$ is the temperature-dependent relative permittivity expressed by Barret's equation as

$$\varepsilon_B(T) = \frac{M}{(T_1/2)\coth(T_1/2T) - T_0} . \qquad (2)$$

Here $M = 9.0 \times 10^4$ K, $T_0 = 38$ K, $T_1 = 84$ K and T is the temperature of the sample. For the simulation of the *J-V* characteristics, tunneling probability of the currents through the intrinsic low permittivity layer was assumed as unity and the thermionic-emission theory without series resistance of the junction was used.

When fitting parameters of $N_d = 8.08 \times 10^{23}$ m^{-3}, $\beta = 5.66 \times 10^8$ Vm^{5}C^{-3}, and $\delta_i/\varepsilon_i = 10.4$ Vm^{2}C^{-1} are used, simulated *J-V* and C^{-2}-*V* characteristics well coincide with measurement

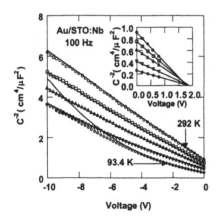

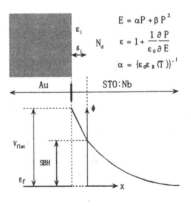

Figure 9. Temperature dependence of C^{-2}-*V* characteristics of a Au/STO:Nb junction. Solid lines represent simulated C^{-2}-*V* characteristics.

Figure 10. Schematic model of a Au/STO:Nb junction with an intrinsic low permittivity layer at the interface.

results, as shown by solid lines in fig. 7 and the fig. 9. The agreement between measured and simulated electrical properties proves the model shown in the fig. 10. Thus, it is confirmed from the electrical properties that the intrinsic low permittivity layer actually exists at the interface between Au and STO:Nb.

Chemical trends of SBH of the Au/STO:Nb junctions

Formation mechanism of the SBH is still unclear yet. The J-V characteristics of various metal/STO:Nb junctions indicate an order of SBH(Au) > SBH(Pt) > SBH(Ag) > SBH(Cu) > SBH(Ti), SBH (Al) as shown in fig.11, while work functions Φ_m of the metals are reported to be $\Phi_m(Pt) > \Phi_m(Au) > \Phi_m(Cu) > \Phi_m(Ti) > \Phi_m(Al) > \Phi_m(Ag)$ [6][16]. Since the sequence of the SBH seems to be rather related to Pauling's electronegativity, reactivity of the metal (may be reactivity to oxygen) probably correlates with the SBH of the junctions. Therefore, both the Fermi level position and the reactivity of the metals should be considered to understand the formation mechanism of SBH of the metal/oxide junctions.

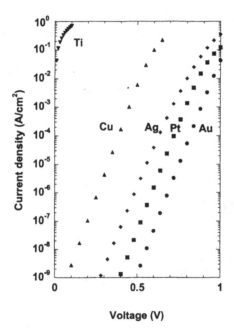

Figure 11. Chemical trends of J-V characteristics
of the metal/STO:Nb junctions.

CONCLUSIONS

Electrical properties of STO:Nb SB junctions have been investigated in detail for a comprehensive understanding of metal/oxide interfaces. Using a high-purity ozone surface treatment, rectification ratio over 9th order of magnitude has been successfully obtained, while without the surface treatment, anomalous large reverse bias leak currents were observed in the current-voltage characteristics. The XPS shows that carbon contaminants, adsorbed on STO:Nb surface in air, induce surface states in the band gap of the STO:Nb, which probably originate the large reverse bias leak currents of the metal/STO:Nb junctions. Thus we present importance of surface treatment for oxides to obtain controllability and reproducibility of the electrical properties of the oxide devices. Using these high quality metal/STO:Nb junctions, firstly photocapacitance spectroscopy has been performed to investigate deep levels. Deep levels in the Au/STO:Nb junctions are of the order of $10^{13} \sim 10^{15}$ cm^{-3} and these densities are too low to arise Fermi level pinning. Secondly we have shown some interesting electrical properties, characteristic of the SB junction of the dielectric oxide compared with that of the conventional semiconductor's. In order to explain all the characteristic properties of the metal/STO:Nb junctions mentioned in the text, the schematic band diagram of the Au/STO:Nb junction with the intrinsic low permittivity layer at the interface has been proposed. Considering the chemical trend of the SBH estimated from the $J\text{-}V$ results, we have pointed out the importance of the metal reactivity for understanding the mechanism of the SBH formation.

ACKNOWLEDGMENTS

The authors would like to thank Messrs. N. Shinozaki, N. Gotoh, Y. Usui and T. Nakagawa for their technical assistance. We would like to acknowledge Dr. K. Kurokawa for the XPS measurement and to Dr. K. Arai and Dr. H. Nonaka for their help to this work.

REFERENCES

1. T. Shimizu, H. Okushi, Appl. Phys. Lett. **67**, 1411 (1995).
2. T. Shimizu, H. Okushi, J. Appl. Phys. **85**, 7244 (1999).
3. Almost the same junction properties were observed for both types of substrates, while the RHEED patterns and the AFM images showed slightly different surface reconstruction and surface morphologies.
4. D. Goldschmigt, H. L. Tuller, Phys. Rev. **B 35**, 4360 (1987).
5. T. Shimizu, N. Gotoh, N. Shinozaki, H. Okushi, Appl. Surf. Sci. **117/118**, 400 (1997).

6. E. H. Rhoderick, R. H. Williams, *Metal-Semiconductor Contacts*, 2nd ed., (Oxford, New York, 1988).

7. Cleaning condition of the STO:Nb substrate was slightly different from the conventional ozone cleaning that described in the experimental section. In case of the STO:Nb, whose XP spectra are shown in fig. 3 and fig. 4, substrate surface was cleaned at room temperature under 800 Pa O_3 with ultraviolet (UV) light. However, results of the XP spectra are almost the same as those of the conventional ozone cleaning, which has been already published in: T. Shimizu, H. Okushi, Bull. Electrotech. Lab. **63**, 5 (1999).

8. H. Okushi, Y. Tokumaru, Jpn. J. Appl. Phys. **20**, 261 (1980).

9. H. Okushi, Phil. Mag. **B52**, 33 (1985).

10.   J. Bardeen, Phys. Rev. **71**, 717 (1947).

11.   C. Zhou, D. M. Newns, J. Appl. Phys. **82**, 3081 (1997).

12.   N. Bickel, G. Schmidt, K. Heunz, K. Müller, Phys. Rev. Lett. **62**, 2009 (1989).

13.   T. Hikita, T. Hanada, M. Kudo, M. Kawai, J. Vac. Sci. Technol. **A11**, 2649 (1993).

14.   J. Prade, U. Schröder, W. Kress, F. W. de Wetts, A. D. Kulkarni, J. Phys.: Condens. Matter **5**, 1 (1993).

15.   V. Ravikumar, D. Wolf, V. P. Dravid, Phys. Rev. Lett. **74**, 960 (1995).

16.   W. Mönch, *Semiconductor Surfaces and Interfaces*, 2nd ed. (Springer, 1978).

THIN Na$_{0.5}$K$_{0.5}$NbO$_3$ FILMS FOR VARACTOR APPLICATIONS

C.-R. CHO, J.-H. KOH, A. GRISHIN, S. ABADEI*, P. PETROV*, S. GEVORGIAN*
Condensed Matter Physics, Department of Physics, Royal Institute of Technology, Stockholm, S-100 44, Sweden

* Department of Microelectronics, Chalmers University of Technology, S-412 96, Göteborg, Sweden

ABSTRACT

Single phase Na$_{0.5}$K$_{0.5}$NbO$_3$ (NKN) thin films have been pulsed laser deposited on SiO$_2$/Si(001) wafers and LaAlO$_3$(001) and MgO(001) single crystals. Radio frequency (up to 1 MHz) and microwave (up to 50 GHz) dielectric spectroscopy studies have been carried out to characterize thin NKN films for electrically tunable microwave device applications. Films on single crystal oxide substrates showed tunabilities as high as 30-40 % at 40 V bias and dissipation factor of 0.01-0.02 at 1 MHz. The films on Si substrates showed low dielectric losses of < 0.01, and low leakage currents. Dielectric properties of ferroelectric films on Si substrates at low frequencies are greatly influenced by the depletion capacitance and the resistance inserted by semiconductor substrate. Microwave frequency measurements for NKN film on Si wafers yield more than 10 % tunability at 50 GHz and loss tanδ <0.1 at 10 GHz.

INTRODUCTION

Because of their outstanding dielectric properties, perovskite ferroelectric thin films attract strong renewed interest for microwave applications, such as frequency and phase agile materials for electronically steered antennas, components for signal storing and processing, remote sensing, and communication applications. Films in the paraelectric phase, such as (Ba,Sr)TiO$_3$ and incipient ferroelectrics SrTiO$_3$ and KTaO$_3$ have long time been regarded as superior candidates to ferroelectric films in polar phases such as Pb(Zr,Ti)O$_3$ and BaTiO$_3$ since they possess significantly lower microwave losses. However, recent results on magnetron sputtered Na$_{0.5}$K$_{0.5}$NbO$_3$ (NKN) films on LaAlO$_3$ (LAO) [1] and pulsed laser deposited NKN films on SiO$_2$/Si wafers [2,3] have demonstrated high tunabilities with low losses tanδ <0.01, although NKN films were strictly in the polar, ferroelectric phase. In this paper, we report on the dielectric properties of NKN films on single crystal oxide substrates and Si(001) substrates with ultra-thin SiO$_2$ buffer layers at radio frequencies and microwaves.

EXPERIMENTAL PROCEDURES

Highly resistive (ρ ~ 10 kΩ×cm) and lowly resistive (ρ ~ 0.01 Ω×cm) n-Si(001) wafers with ultra-thin SiO$_2$ buffer layers and LaAlO$_3$(001) (LAO) and MgO(001) single crystals were used as substrates. A KrF excimer laser (Lambda PhysiK-300, λ = 248 nm, pulse width of 25 nm) was used to ablate stoichiometric NKN target at the energy density of 4-5 J/cm^2 and repetition rate of 15 Hz. NKN films on various substrates were grown at 650 °C in an ambient oxygen partial pressure around 400 mTorr. The deposition was followed by in-situ annealing at 400 °C in 500 Torr oxygen for 30 min. Film thickness was controlled by deposition time and deposition rate, which was calibrated by Atomic Force Microscope (AFM). To measure dielectric properties, interdigital capacitor (IDC) structures were defined by photolithographic patterning. Both lift-off of 0.5 μm thick Au/Cr layer (for the films on oxide substrates) and ion milling in 0.4 μm thick Au/Ti layer (for the films on Si wafers) were used to define upper metal electrodes. Radio frequency (up to 1 MHz) C-V and $\tan\delta$-V characteristics were measured using *HP4284A* and *HP4285* LCR meters, while the microwave frequency (up to 50 GHz) dielectric spectroscopy was performed by *HP8510C* vector network analyzer. I-V measurements were done by *HP 4145B* semiconductor parameter characteriograph. The surface morphology of deposited films was analyzed by Digital Instruments multimode Scanning Probe Microscope (SPM) used in the tapping mode.

RESULTS AND DISCUSSION

Figs. 1 **a-c** show x-ray diffraction (XRD) patterns of NKN films on: **a** - Si(001), **b** - LAO(001), and **c** - MgO(001) substrates. The diffraction patterns, shown in logarithmic scale, indicate all the films are single phase, despite the highly volatile components Na and K. We have explained that the congruent transfer of material from the NKN target to the substrate can be achieved at sufficiently high ambient oxygen pressure through the process of complete thermalization of plasma plume and subsequent forming of the shock-wave. [3] NKN films on Si wafer exhibit high *c*-axis orientation with anomalously intensive "quadrupled" superlattice structure. This can be attributed to strong Jahn-Teller distortions of NbO₆ octahedra in NKN perovskite unit cell. [4] The NKN films on Si(001) substrates with various thickness of ultra-thin SiO₂ buffer layers show crystalline properties similar to that seen in Fig.1a, and were discussed in detail elsewhere. [5]

The films deposited onto different oxide substrates (see Figs. 1b and 1c) show quite different diffraction patterns, although both substrates have similar crystalline structures. The films deposited on LAO substrate are

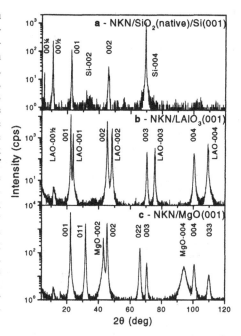

Fig. 1 Patterns of CuK_α x-ray diffraction (XRD) for NKN films on different substrates. **a** - 0.9 μm thick film on SiO₂(native)/Si(001) wafer, **b** - 0.65 μm thick film on a LAlO₃(001) substrate, **c** - 0.65 μm thick film on a MgO(001) substrate. The *hkl* Bragg peaks indicate the NKN reflections.

exclusively oriented in the [001] direction, while the film deposited on MgO substrate showed mixed [001] and [011] orientation. The reason of these different preferential orientations could be found in film-substrate lattice mismatch, since both of the films have been deposited under the same process conditions. The orthorhombic NKN bulk ceramic has lattice constants of $a = b = 4.02$, $c = 3.93$ Å at room temperature [6], while pseudocubic LAO and cubic MgO single crystals have lattice constant of 3.787 Å $[(a_{NKN} - a_{LAO})/a_{LAO} = +$ 6.15 % mismatch] and 4.213 Å $([(a_{NKN} - a_{MgO})/a_{MgO} = -4.58$ % mismatch), respectively. Therefore, if one assumes that the NKN films grow in the epitaxial cube-on-cube relationship on both single crystals, then NKN films should experience compressive stress in the lateral direction on LAO and tensile stress on MgO substrate. The θ-2θ XRD patterns of NKN films on MgO substrates (Fig. 1c) imply that the relaxation of the tensile stress occurs through developing considerable amount of [011]-oriented crystallites. Furthermore, φ-scan XRD patterns (not shown) indicate both [001]- and [011]-oriented crystallites have a strong in-plane texture in strict relation with the MgO substrate. Bi-axial texture in NKN/MgO film structures has significant implication on the nature of NKN film growth, and will be reported elsewhere.

Low frequency C-V and $tan\delta$-V characteristics of the NKN films on oxide substrates were traced at 1 MHz and are shown in Figs. 2a and 2b. NKN films on LAO and MgO substrates have the same thickness of 270 nm as well as the same dimensions of the upper IDC electrodes (5 pairs of 1.0 mm long fingers with 4 μm gap). The capacitance of the NKN/LAO IDC is 3.54 pF at zero bias voltage (C_0) and 2.46 pF at 40 V bias voltage, hence the tunability (1- C_{40V}/C_0) was calculated to be as high as 30.5 % at 40 V. For the case of the MgO substrate, $C_0 = 3.30$ pF and $C_{40V} = 1.99$ pF, thus the tunability is about 39.70 % at 40 V, which is a little bit higher then that of NKN/LAO capacitor. Slightly opened C-V hysteresis loops indicate that both films are in the ferroelectric state. The dissipation factors for both films are also different. In the case of MgO, the $tan\delta$ vs. V characteristics are almost reversible and very symmetric, while this is not true for LAO. The asymmetry and the irreversibility of $tan\delta$ - V characteristics in NKN/LAO

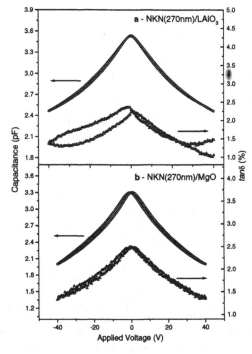

Fig. 2 C-V and $tan\delta$ - V characteristics @ 1 MHz for 4 μm slot IDC structures: **a** - 270 nm thick NKN film on LAO substrate, **b** - 270 nm thick NKN film on MgO substrate.

capacitor could be attributed to the excess of monovalent Na^+ and K^+ ions and/or oxygen O^{2-} vacancies. Since the compressive stress in the NKN/LAO films is not released through the formation of the [011]-phase, it can cause the appearance of vacancies of the most slightly bonded ions. Nevertheless the dissipation factors for both films are rather low and decrease with the applied bias voltage: at zero bias, $tan\delta$ is 2.3 % for both LAO and MgO substrates, and it decreases to 1.0 % for LAO and to 1.4 % for MgO substrate, respectively.

Fig. 3a shows C-V and $tan\delta$ - V characteristics measured at 1 MHz for a 0.9 μm thick NKN film IDC prepared on a highly resistive Si wafer. A schematic representation of the IDC structure and a simplified equivalent circuit are shown in Figs. 3b and 3c. As shown in Fig. 3a, the tunability is abnormally high, approximately 89 % at 15 V of bias voltage. The dissipation factor is 57 % at zero bias voltage, increases with increasing bias voltage, and reaches the value of 200 % at 15 V. These large changes of the real and imaginary parts of the dielectric permittivity in an applied electric field are attributed to charge injection processes.

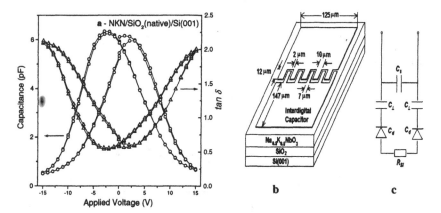

Fig. 3 **a** - C-V and tan δ - V characteristics of a 0.9 μm thick NKN film on a Si wafer with 4 μm slot interdigital capacitor (IDC) structure similar to that shown in **b**, **b** - schematic representation of the IDC structure used for the characterization of NKN films on Si substrates. **c** - equivalent electric circuit to the IDC shown in **b**.

As we have reported earlier, the dissipation factor is approximately linearly proportional to the serially connected resistance of the bottom electrode (in our case R_{Si}). [7] As the bias voltage increases, R_{Si} also increases, since the carriers in the semiconductor are accumulated or depleted with the bias voltage. The abnormally high tunability can also partly come from the Si wafer. The depletion capacitance (C_d in Fig. 3c) of the highly resistive Si substrate decreases with the bias voltage since the depletion depth increases, and C_d finally becomes much smaller than $C_\perp$, which is the main component of the NKN capacitor. [2] Therefore, the serially connected, small C_d considerably diminishes the measured capacitance with increasing bias voltage. The reproducible C - V and $tan\delta$ - V hysteresis loops, clearly seen in Fig. 3a, indicate that both R_{Si} and C_d are strongly influenced by the polarization switching in NKN the film. Thus, it is confirmed once again that NKN films on Si substrates are in the ferroelectric state.

Fig. 4 depicts the frequency dependencies of the capacitance and the dielectric loss of the NKN IDC (dimension are as shown in Fig. 3b) on a highly conductive (phosphorous concentration $N_D > 10^{19}$/cm^3) Si (001) substrate with a 40 Å thick SiO$_2$ buffer layer. In the case of such a highly doped Si substrate, the Debye length is very short, so that the difference between the measured flat-band capacitance and the real capacitance insulator (1.3 μm NKN/40 Å SiO$_2$) was found to be less than 1 %, while the resistivity of the Si substrate was less than 0.05 Ω×cm. Therefore, we deduced the dielectric permittivity ε' from the measured capacitance neglecting C_d and R_{Si}. The dielectric permittivity was found to decrease slightly (by 6 %) from 114.0 to 107.2 in the frequency range from 1 kHz to 1 MHz and can be straightened in log f scale: ε' = 114.0 - 1.8 log (f/1 kHz). In contrast to the case of the highly resistive Si substrate, the measured loss, a major barrier for high-frequency applications of ferroelectric materials, is very low. The loss decreases slightly with increasing frequency up to 10 kHz and is then almost invariant up to 1 MHz at a level of approximately 0.9 %. Considering the increasing contribution of the resistance of the Si substrate (R_{Si}) to the measured tanδ [7] with increasing frequency, the loss from the NKN film should be even lower. Such a low loss is an indirect indication of large domain sizes and the right stoichiometry of PLD-made NKN films, since domain wall motion and charge transfer due to the loss of constituents are the main sources of electrical losses in ferroelectric films. Furthermore, it reveals that there are little oxygen interdiffusion, chemical reaction, and/or structural defects at the NKN/SiO$_2$ interface, which would result in an "interfacial dead layer" considered as the main reason of the degradation of properties in thin dielectric films. [8] In spite of all the above explanations, separate studies of intrinsic loss mechanisms in NKN films are required to fully understand such low losses in ferroelectric film that are strictly in the polar state.

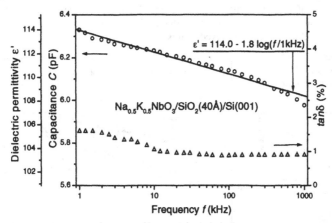

Fig. 4 Frequency dependence of dielectric permittivity and *tan* δ for Au/Ti/Na$_{0.5}$K$_{0.5}$NbO$_3$(1.3μm)/SiO$_2$(40Å)/Si(001) thin film interdigital capacitor structure.

We investigated the surface morphology of films deposited onto SiO$_2$ (40 Å)/Si(001) by using a Scanning Probe Microscope (SPM). Grains with areas up to several μm^2 are visible in the 8 × 8 μm^2 scan depicted in Fig.5. These grains are orders of magnitude larger than those of epitaxial films on oxide substrates. [9] Moreover, the film surface has been found to be extremely smooth, the average roughness in the area shown in Fig. 5 is 6.0 Å (corresponding to 1.5 times the height of the NKN unit cell). No particulates were observed in the film, though particulates are common to many films prepared by PLD. [10] We attribute these SPM analysis results to strong

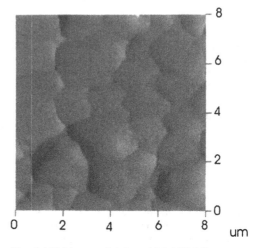

Fig. 5 AFM image of 1.3μm thick NKN film on SiO$_2$(40Å)/Si(001) substrate

interactions between NbO$_6$ octahedra, the high volatility of Na and K, and the polar faces in the A^{+1}B^{+5}O$_3$$^{-2}$ type perovskite NKN unit cell [3]. The interaction between NbO$_6$ octahedra results in the "quadrupled" structure shown in Fig.1a, and could be one of the reasons for the large grains. The high volatility enables Na and K adatoms to migrate along the film surface despite the relatively high ambient gas pressure of about 400 mTorr and the moderate process temperature of 650 °C, while alternating net charges in adjacent adatom layers promote the "layer by layer" growth mode. Nevertheless, more theoretical work is required to fully understand and interpret the observed results.

Fig. 6 shows *I-V* characteristics of an NKN IDC (shown in Fig. 3b) on a highly resistive Si wafer. Three successive *I-V* curves, which coincide well, show the hysteresis caused by the polarization switching of the NKN film. The ferroelectric material retains charges on its surfaces when it is polarized. When the direction of polarization is switched, a switching current is required to store oppositely signed charges on the surfaces. [11] An NKN IDC on a lowly

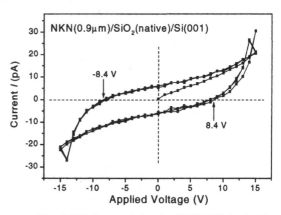

Fig. 6 *I-V* characteristics for NKN IDC (as in Fig. 3b) on high resistive Si wafer.

resistive Si substrate also showed similar *I-V* hysteresis loops with a low leakage current level. Recently, we observed maximum conductance at the center of the polarization switching field in a Au/NKN/SiO$_2$/Si structure [5], which corresponds to these results. Currents flowing in opposite direction to the bias field are observed in low voltage regimes: from 0 V to -8.4 V for negative and from 0 V to +8.4 V for positive bias voltages. This clearly indicates that the polarization switching is the dominant conducting mechanism in the low voltage regime. Although the *I-V* hysteresis loop is widely opened, low leakage currents (of the order of pA) are measured. The small leakage currents could be partly attributed to the surface barriers onto silicon-insulator interface that are equivalent to back-to-back connected diodes C_d in Fig. 3c. A leakage current might also flow along the grain boundaries [12] - the large size of the grains (see Fig. 5) would then explain its low level. The right stoichiometry of the deposited film could be the reason for the reproducibility of the *I-V* loop: excess ions or ion vacancies could act as charge carriers and result in irreversible *I-V* characteristics.

NKN films on SiO$_2$/Si substrates are promising not only for varactors but also for other high frequency devices, such as surface acoustic wave (SAW) devices and piezoelectric transducers, for the following reasons. Firstly, bulk ceramic NKN shows a high maximum radial coupling coefficient of 0.39 and a piezoelectric coefficient d_{33} of 160, which is one order of magnitude higher than that of other candidates such as ZnO. [13] The exclusively polar axis oriented film texture (see Fig.1a) with only small mosaic broadening [14], which was observed in NKN films on SiO$_2$/Si substrates, promises enhanced piezoelectricity. [15, 16] Moreover, high piezo-activity is also expected at high frequencies up to 10 GHz (see the results on high frequency polarizability later). Secondly, the moderate dielectric permittivity and the low loss (see Fig.4), the low leakage current and the high breakdown voltage (see Fig.6) are ideal for these applications. Thirdly, the surfaces are smooth and free of particulates in an area comparable with the area of a device, therefore, the attenuation of acoustic waves propagating along the film surface (in the case of SAW) and normal to the film surface (in the case of Bulk Acoustic Waves (BAW)) should be low. The attenuation caused by a rough surface is known to be a major limitation of PLD deposited piezoelectric films for applications in acoustic wave devices. [17,18] Ferroelectric films on (ultra-thin)SiO$_2$/Si substrates could be incorporated into integrated circuits and are compatible with the Si micromachining process. Recent reports on the propagation of surface acoustic waves (SAW) in metal/piezoelectric/semiconductor structures favor thinner buffer layers to obtain higher SAW phase velocities as well as maximum coupling coefficients κ^2 [19], while a thermally grown thin SiO$_2$ layer could be used as an effective etch stop layer in the Si etching process.

Even though, dielectric properties of NKN film on Si wafer are crucially influenced by substrate property, electrically tunable material on Si substrate has significant technical importance for realization of Silicon Microwave Monolithic Integrated Circuits (SiMMIC). To characterize NKN(0.9μm)/SiO$_2$(native)/Si(001) thin film structure at microwave frequencies, IDCs with straight slots of 2 μm gap were photolithographically defined in Au/Ti layer by ion milling. The frequency dependence of NKN planar structure capacitance at zero and 10 V DC bias voltages has been measured by sweeping the frequency f in the range 45 MHz to 50 GHz using the vector network analyzer. To compute the capacitance and effective loss $\tan\delta$ from measured complex reflection coefficient, S_{11}, we used the following expression for complex impedance:

$$Z = r - i\frac{1}{2\pi f C} = Z_o \frac{1+S_{11}}{1-S_{11}} \ ,$$

where r and C are resistance and capacitance of equivalent to DUT circuit, and $Z_0 = 50\ \Omega$. The accuracy of the measurement is estimated better than 5 %. Fig. 7a shows the capacitance while Fig. 7b shows the deduced tunability vs. f. The capacitance at zero bias and tunability steeply decrease with the frequency increase up to several GHz. Most probably this frequency dispersion is caused by Maxwell relaxation in silicon wafer. In a wide frequency band, 10 to 50 GHz, low dielectric dispersion has been observed. In this high frequency range the capacitance dispersion is predominantly due to NKN film for the following reasons. First, the thickness of the Au/Ti electrodes ($\sim 0.5\ \mu$m) is the order of skin depth, thereby no substantial dispersion may be expected due to the skin effect. The dielectric (Maxwell) relaxation frequency for the silicon, $f_M = 1/(2\pi\tau_M) = 1/(2\pi\varepsilon_{Si}\varepsilon_o\rho_{Si})$ is about 15 MHz (τ_M is the relaxation time). Thus, there is no dielectric dispersion in silicon substrate. Furthermore, since the resistance and the dielectric permittivity of NKN film are much larger than that of silicon, the interfacial (Maxwell-Wagner) frequency may be approximated by $f_{MW} = t/(2\sqrt{3}\pi g\rho_{Si}\varepsilon_o\varepsilon_{NKN})$. For the given geometry and material parameters $f_{MW} \sim 0.25$ MHz and the interfacial relaxation also does not contribute in the dispersion at frequencies above 10 GHz. Measured tunability is decreased to several percent at around 15 GHz and slightly increases with frequency reached about 10 % at 50 GHz at 10 V bias.

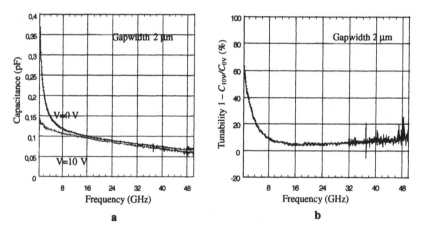

Fig. 7 **a** - the capacitance of 0.9 μm thick NKN planar capacitor at zero and 10 V bias at microwave frequencies; **b** - associated tunability, deduced from the measurement shown in **a**.

Figs. 8 show C-V and tan δ- V characteristics of NKN/Si IDC measured at 10 GHz. More than 10 % tunability and loss tan $\delta \sim 0.1$ is observed. It is somehow unexpected to observe two typical maxima in bias dependent capacitance at about $\pm$ 10 V in Fig. 8a. Maxima in bias dependent dielectric permittivity (capacitance) are typically observed in C-V characteristics in ferroelectric state at low frequencies. This indicates that ferroelectric activity (thereby piezoelectric activity as well) in NKN film remains at frequency as high as

10 GHz. After one cycle of voltage sweep, both of capacitance and tan δ are reduced which have not appeared in epitaxial NKN films on oxide substrates (see Figs. 2). One possible explanation for this phenomenon is the neutralization of entrapped alkali ions (Na^+ and K^+) on the NKN/SiO_2 and/or SiO_2/Si interface by injected electrons from n-Si substrate. [5] Nevertheless, more than 10 % tunability and low loss of ~ 0.1 observed in these not yet optimized capacitor structure is quite promising for applications in microwave devices.

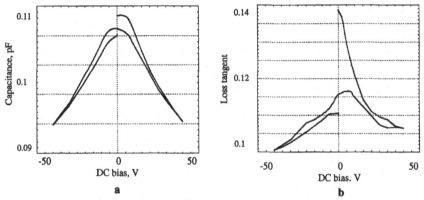

Fig. 8 **a** - *C* - *V* and **b** - *tanδ* - *V* characteristics of NKN/Si IDC at 10 GHz.

SUMMARY AND CONCLUSIONS

Dielectric properties of highly *c*-axis oriented NKN thin films on $SiO_2/Si(001)$ wafers and epitaxial films on $LaAlO_3(001)$ and MgO(001) single crystals have been studied for electrically tunable microwave device applications. *C-V* and *tanδ - V* characteristics at radio frequencies (up to 1 MHz) measured for the NKN film 4 µm gap interdigital capacitor structures yield 30-40 % tunabily at 40V bias and loss *tanδ* as low as 0.023. NKN film on low resistive Si wafer showed low loss < 0.01, while the film on high resistive Si substrate showed abnormally high loss and tunability. These substrate dependencies have been dedicated to the depletion capacitance, C_d, and inserted resistance, R_{Si} in Si substrates. Large grain size and with very smooth film surface, reproducible *I-V* hysteresis loops with very low leakage current were observed in NKN films on SiO_2/Si substrate. Dielectric spectroscopy of NKN films on Si wafers performed at microwave frequencies showed more than 10 % tunability up to 50 GHz under 10 V bias and less than 10 % loss *tanδ* at 10 GHz, which indicate NKN films on Si to be a promising candidate for SiMMICs devices.

ACKNOWLEDGMENT

This research was funded by the Swedish Agency NUTEK.

REFERENCES

[1] X. Wang, U. Helmersson, S. Olafsson, S. Rudner, L. Wernlund, S. Gevorgian, Appl. Phys. Lett. **74**, 927 (1998).

[2] C.-R. Cho, J.-H. Koh, A. Grishin, S. Abadei, S. Gevorgian, Appl. Phys. Lett. **76**, 1971 (2000).

[3] C.-R. Cho, A. Grishin, J. Appl. Phys., **87**, 4439 (2000).

[4] A. M. Glazer, Acta Crystallogr. Sect. A: Cryst Phys., Diffr., Theor. Gen. Crystallogr. **A31**, 756 (1975).

[5] C.-R. Cho, A. Grishin, MRS Spring 2000, San Francisco, April 24-28, 2000.

[6] G. Shirane, R. Newnham, R. Pepinsky, Phys. Rev. **96**, 581 (1954).

[7] C.-R. Cho, S. I. Khartsev, A. Grishin, T. Lindbäck, Mat. Res. Soc. Symp. Proc., **574**, 249 (1999).

[8] C. Zhou, D. M. Newns, J. Appl. Phys. **82**, 3080 (1997).

[9] M. J. Dalberth, R. E. Stanber, J. C. Price, C. T. Rogers, D. Galt, Appl. Phys. Lett. **72**, 507 (1998).

[10] G. Koren, A. Gupta, R. J. Baseman, M. I. Lutwyche, R. B. Laibowitz, Appl. Phys. Lett. **55**, 2450 (1989).

[11] Y. Watanabe, Phys. Rev. B. **59**, 11257 (1999).

[12] H. Fujisawa, M. Shimizu, T. Horiuchi, T. Shiosaki, K. Matsushige, Appl. Phys. Lett. **71**, 416 (1997).

[13] L. Egerton, D. M. Dillon, J. Am. Ceram. Soc. **42**, 438 (1959).

[14] C. -R. Cho, A. Grishin, 12th International Symposium on Integrated Ferroelectrics (ISIF 2000), Aachen, March 12-15, 2000.

[15] X. Du, J. Zheng, U. Belegucu, K. Uchino, Appl. Phys. Lett. **72**, 2421 (1998).

[16] J. G. E. Gardeniers, Z. M. Rittersma, G. J. Burger, J. Appl. Phys. **83**, 7844 (1998).

[17] H. Du, D. W. Johnson, Jr., W. Zhu, J. E. Graebner, G. W. Kammlott, S. Jin, J. Rogers, R. Willett, R. M. Fleming, J. Appl. Phys. **86**, 2220 (1999).

[18] M. -A. Dubois, P. Muralt, Appl. Phys. Lett. **74**, 3032 (1999).

[19] M. Wu, W. -C. Shin, Jpn. J. Appl. Phys. **36**, 2192 (1997).

PYROELECTRIC COEFFICIENT OPTIMIZATION
THROUGH GRAIN SIZE CONTROL IN (Ba,Sr)TiO$_3$ THIN FILMS

LAWRENCE F. SCHLOSS AND EUGENE E. HALLER
Department of Materials Science and Engineering, UC Berkeley and
Materials Science Division, Lawrence Berkeley National Laboratory, Berkeley, CA 94720

ABSTRACT

Thin Ba$_{0.2}$Sr$_{0.8}$TiO$_3$ films were pulsed laser deposited at laser pulse repetition rates of 1, 5 and 20 Hz in order to assess the effect of grain size on pyroelectric properties. Scanning electron microscopy reveals an increase in columnar grain size with decreasing laser pulse rate, as expected. Substrate bending measurements show an increase in tensile stress with increasing grain size, suggesting some form of stress relaxation has occurred. X-ray diffraction studies reveal all films to be epitaxial and oriented, while Rutherford Backscattering Spectroscopy finds no significant differences in film composition. The films grown at 5 and 20 Hz display a suppressed dielectric permittivity as a function of temperature, while the film deposited at 1 Hz displays an unusually sharp and large dielectric permittivity peak. A variation on current theory concerning phase transitions of stressed ferroelectric thin films is proposed to explain these results.

INTRODUCTION

Thin Ba$_{1-x}$Sr$_x$TiO$_3$ films are being developed as the active layer in a passively-cooled pyroelectric detector to be utilized aboard Earth-orbiting satellites in their 90 K environment. For this application, the Ba/Sr ratio has been tailored so that the bolometric pyroelectric coefficient, which is proportional to the change in the film's dielectric permittivity with temperature, is a maximum near 90 K. Bulk KTa$_{1-x}$Nb$_x$O$_3$ crystals with Curie temperatures in this range display a bolometric sensitivity, or normalized change in capacitance with temperature $(1/C*dC/dT)$, of 10-15 %/K.[1] Thin Ba$_{1-x}$Sr$_x$TiO$_3$ films with similar Curie temperatures, however, display sensitivities of only 1 %/K. This suppression of Curie-Weiss behavior in thin films is widely observed and mechanisms involving interfacial capacitance, film thickness and stress have been proposed in explanation.[2,3,4]

THEORETICAL CONSIDERATIONS

One model proposed by Pertsev et al.[5] attributes the Curie-Weiss deviation to the effect of the two-dimensional clamping at the film-substrate interface. Their theory applies to ferroelectric thin, epitaxial films grown on cubic substrates, and calculates that the thermodynamically stable phases differ significantly from a freestanding crystal. In the specific case where the thermal expansion mismatch produces a two-dimensional, in-plane tensile stress in BaTiO$_3$ films, the first ferroelectric phase transition upon cooling is predicted to be a paraelectric to ferroelectric, orthorhombic phase with an in-plane spontaneous polarization. This phase change, which occurs above the theoretical Curie temperature, is preferred because the tensile stress can be relaxed by the orthorhombic unit cell distortion. Further, Pertsev et al.[6] calculate that this phase transition will be accompanied by modified Curie-Weiss behavior. The dielectric permittivity will increase more rapidly with temperature above the transition, but the transition will display only a small permittivity peak instead of a discontinuity. Since more

119

stress increases this modified slope, one could, in a limited way, increase a film's bolometric sensitivity by increasing the film's tensile stress.

Pertsev *et al.*'s theory only considers crystals with a single domain. Yet, domain structures of 90° twins have been predicted and observed in ferroelectric thin films. Indeed, the same two-dimensional tensile stress which can be accommodated by an in-plane orthorhombic distortion can also be accommodated by in-plane, 90° twinned tetragonal distortions.[7] Further, the tetragonally distorted unit cell volume is larger than the pseudocubic orthorhombic unit cell.[8] This suggests that the tetragonal transition may be preferred in films under greater stress. In this case, the phase transition should be accompanied by a large dielectric peak since in-plane tetragonal transitions display such peaks in bulk crystals.[9]

Yet, this transition may not appear in thin films due to the small columnar grain size. The number of 90° domains has been observed to diminish significantly in bulk ceramics as the grain size is reduced below 1 μm.[10] In the case of columnar grains, it is the column diameter which would be the limiting factor for the formation of an in-plane domain structure. Thus, the proposed in-plane, 90° twinned tetragonal phase transition may only appear in large-grained films under large tensile stress.

In the materials system of $Ba_{1-x}Sr_xTiO_3$ thin films deposited on $SrTiO_3$ (STO), tensile thermal mismatch stresses will be generated. By depositing films of various grain sizes, we have attempted to create a film with a large grain size and in-plane tensile stress. In this way, we hoped to observe the in-plane tetragonal phase transition and its associated increase in bolometric sensitivity.

EXPERIMENT

Thin Film Growth

$Ba_{0.2}Sr_{0.8}TiO_3/La_{0.5}Sr_{0.5}CoO_3$ (BST/LSCO) thin films were pulsed laser deposited onto single crystal (001) exact-cut STO substrates in a 300 mTorr oxygen ambient. Ablation occurred via a 248 nm KrF excimer laser with an energy density of 2.4 J/cm^2. The substrate temperature was set to 680 °C while the target to substrate distance was approximately 5 cm. The LSCO layers were deposited at a laser pulse repetition rate of 5 Hz while the BST layers were grown at 1, 5, and 20 Hz. Total pulse count was varied to achieve a BST thickness of 4000 Å.

Characterization

Film stress was calculated by measuring substrate bending both pre- and post-deposition using a Tencor Flexus FLX2320 laser interferometer stress gauge. Surface morphology was examined using a LEO Gemini 1550 field-emission scanning electron microscope (SEM). Film compositions were calculated from Rutherford backscattering spectroscopy (RBS) measurements utilizing a 1.95 MeV He^+ beam with a silicon detector positioned at 165° with respect to the beam. X-ray diffraction (XRD) θ-2θ scans were performed on a Siemens 4000 high resolution Cu K_α x-ray diffractometer, equipped with a 4-circle goniometer and a 4-crystal Ge monochrometer. Temperature-dependent capacitance-voltage measurements were performed using an evacuated liquid nitrogen-capable dewar and a HP 4277 LCZ meter with a 60 mV_{p-p}, 10 kHz test signal. Top contacts were 200 μm diameter circles of Au/Ti electron-beam deposited through a shadow mask. The LSCO bottom electrode was accessed through a 1500 μm diameter contact pad connected in series to the capacitor of interest.

RESULTS

Grain Size

The grain size of the film grown at 1 Hz is estimated to be around 40 nm based upon the scanning electron micrograph shown in Figure 1. The films grown at 5 and 20 Hz had grain sizes smaller than the resolution limit of the SEM. This increase in grain size with decreasing laser pulse rate was expected since the increased time between pulses allows for more Ostwald ripening of the BST islands prior to coalescence.

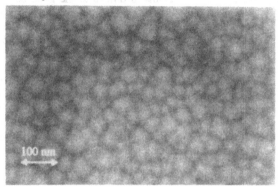

Figure 1. Scanning electron micrograph of the BST film grown at 1Hz

Film Stress

According to Hoffman,[11] grain coalescence generates tensile stress at the grain boundaries. Nix and Clemens[12] calculated that the average tensile film stress depends inversely on the square root of the grain size. That is, in the absence of relaxation, as grain size decreases, film tensile stress increases. The dependence of film stress on laser pulse repetition rate is plotted in Figure 2.

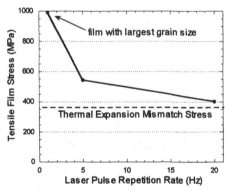

Figure 2. Film stress as a function of laser pulse repetition rate

It is apparent that some kind of relaxation has occurred in the films deposited at 5 and 20 Hz, since the film stress is shown to decrease with decreasing grain size. One possibility is the creation of misfit dislocations. The film grown at 20 Hz initially should have generated the largest tensile stress. Since all films are approximately the same thickness, one would expect this film to generate the largest number of misfit dislocations and experience the most relaxation.

Nix and Clemens suggest another relaxation mechanism. Once the tensile grain boundaries have formed, atoms arriving at the growth surface will be attracted, and thus adhere preferentially, to the grain boundaries. As atoms fill the "gaps", the film's tensile stress will decrease. By this mechanism, Nix and Clemens suggest that a higher surface diffusion rate allows a greater proportion of arriving atoms to diffuse to the grain boundaries, resulting in greater relaxation.

One should also be able to apply this theory to films with equivalent surface diffusion rates but different grain sizes. Among the films grown at 1, 5 and 20 Hz, the film grown at 20 Hz can be expected to have the smallest grain size and thus should have the highest film stress prior to relaxation. However, the distance between grain boundaries is smallest, too. Thus, for equivalent surface diffusion rates, a larger portion of the arriving atoms could migrate to the grain boundaries. In addition, Chang et al.[13] observed that the thickness at which coalescence of $YB_2C_3O_{7-x}$ islands occurred varied inversely with laser pulse rate. It is expected, therefore, that the BST film grown at 20 Hz would coalesce earliest, leaving more deposition time for relaxation. It appears in the cases of the films deposited at 5 and 20 Hz that the balance in grain size between grain boundary stress and relaxation is dominated by the relaxation mechanism.

Film Composition and Orientation

RBS results show no significant differences in film composition. XRD scans reveal the films to be both epitaxial and oriented. Unfortunately, the close lattice match between the BST and the STO substrate, along with the tensile strain, cause the BST and STO peaks to overlap significantly, making a determination of the exact BST lattice parameters difficult.

Bolometric Sensitivity

Comparisons of the BST bolometric sensitivity dependencies on temperature and laser pulse rate are shown in Figure 3.

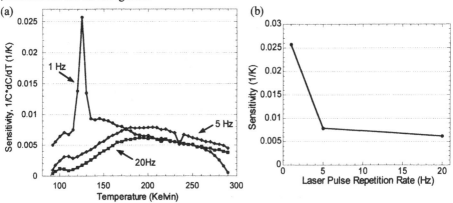

Figure 3. Sensitivity (a) vs temperature and (b) vs laser pulse rate

While the films deposited at 5 and 20 Hz display similar curves with no prominent peak, the film grown at 1 Hz displays an extremely sharp sensitivity peak at 125 K. The peak sensitivity of the film deposited at 1 Hz is 225% larger than that grown at 5 Hz. For reference, given the Ba/Sr ratio of 2/8 in these films, the theoretical Curie temperature is predicted to be 120 K.

The films grown at 5 and 20 Hz show very similar sensitivities, though the film deposited at 5 Hz displays a slightly higher sensitivity at all temperatures. Pertsev et al.'s theory predicts such an increase in the bolometric sensitivity with increasing stress. The prominent sensitivity peak occurring at 125 K in the film grown at 1 Hz, however, is not predicted by their theory. The theory suggests that the in-plane two-dimensional tensile stress distorts the unit cell into an orthorhombic shape, leading to stress relaxation when, at the transition temperature, such a distortion becomes stable. In effect, the energetic barrier at the transition is reduced, allowing the transition to occur more easily. With increasing stress, the stress relaxation increases, leading to an increase in the transition temperature. If the sensitivity peak observed in the film grown at 1 Hz, which has the largest in-plane tensile stress, were associated with the paraelectric-orthorhombic transition, it should occur above the sensitivity peaks in the films grown at 5 and 20 Hz.

A more detailed look at the theory finds a prediction of a transition from the orthogonal phase to a phase with an out-of-plane polarization component that should show a dielectric peak. However, this transition's temperature should increase with decreasing stress, and a similarly shaped peak is not observed in the films grown at 5 or 20 Hz. Thus, the measured sensitivity peak observed in the film grown at 1 Hz does not correspond to any predicted transition, suggesting that an alternate transition has occurred. A paraelectric-tetragonal transition with 90° in-plane twinned domain structure is a likely candidate since it would be accompanied by a sharp sensitivity peak. Still, the grain diameters in this film are only 40 nm, and it is unclear whether a ferroelectric domain structure could exist. It is possible that the grains themselves act as domains, with the domain structure created by the twinning of neighboring grains.

CONCLUSIONS

Thin pyroelectric films deviate from the Curie-Weiss behavior commonly observed in bulk ferroelectric crystals. Theory suggests that this is the result of a preferential paraelectric to in-plane orthogonal phase transition due to the two-dimensional, in-plane tensile stress generated from thermal expansion mismatch between the film and substrate. The theory, however, does not include the possibility of ferroelectric domain structure. Thus, a second transition to an in-plane, 90° twinned tetragonal transition may also occur under the same stress conditions, assuming the film's grain size is large enough to support domain boundaries. Unlike the transition to the orthogonal phase, a transition to this tetragonal phase should be accompanied by a large sensitivity peak. Due to the larger unit cell distortion, this alternative transition may be preferred in films with large tensile stresses. BST thin films were pulsed laser deposited with identical growth parameters except the laser pulse repetition rate. The film grown at 1 Hz has the largest grain size, measured the largest stress, and displayed a sharp permittivity peak as a function of temperature. This peak increased the BST bolometric sensitivity more than 225% over the other films.

ACKNOWLEDGMENTS

The authors would like to thank Dr. Kin Man Yu for performing the RBS measurements. This work was performed with support from an interagency agreement between the Department of Energy and NASA under contract #A53228D. A portion of this research benefited from the use of the UCB Integrated Materials and Microfabrication Laboratories.

REFERENCES

1. Hilary Baumann Cherry, MS Thesis, UC Berkeley, 1993.

2. C. Basceri, S. K. Streiffer, A. I. Kingon, R. Waser, J. Appl. Phys. **82,** 2497-2504 (1997).

3. C. Zhou and D. M. Newns, J. Appl. Phys. **82,** 3081-3088 (1997).

4. T. M. Shaw, Z. Suo, M. Huang, E. Liniger, R. B. Laibowitz, J. D. Baniecki, Appl. Phys. Lett. **75,** 2129-2131 (1999).

5. N. A. Pertsev, A. G. Zembilgotov, A. K. Tagantsev, Phys. Rev. Lett. **80,** 1988-1991 (1998).

6. N. A. Pertsev, A. G. Zembilgotov, S. Hoffman, R. Waser, A. K. Tagantsev, J. Appl. Phys. **85,** 1698-1701 (1999).

7. B. S. Kwak, A. Erbil, J. D. Budai, M. F. Chisholm, L. A. Boatner, B. J. Wilkens, Phys. Rev. B **49,** 14865-14879 (1994).

8. H. F. Kay and P. Vousden, Phil. Mag. **40,** 1019 (1949).

9. W. P. Merz, Phys. Rev. **75,** 687 (1949).

10. G. Arlt, D. Hennings, G. de With, J. Appl. Phys. **58,** 1619-1625 (1985).

11. B. Hoffman, Thin Solid Films **34,** 185-190 (1976).

12. W. D. Nix and B. M. Clemens, J. Mater. Res. **14,** 3467-3473 (1999).

13. C. C. Chang, X. D. Wu, R. Ramesh, X. X. Xi, T. S. Ravi, T. Venkatesan, D. M. Hwang, R. E. Muenchausen, S. Foltyn, N. S. Nogar, Appl. Phys. Lett. **57,** 1814-1816 (1990).

THE EFFECTS OF Mg DOPING ON THE MATERIALS AND DIELECTRIC PROPERTIES OF $Ba_{1-x}Sr_xTiO_3$ THIN FILMS

M.W. Cole*, P.C. Joshi*, R.L Pfeffer**, C.W. Hubbard*, E. Ngo*, M.H. Ervin*, M.C. Wood*
* U.S. Army Research Laboratory, Weapons and Materials Research Directorate, Aberdeen Proving Ground, MD 21005
** Physics Dept., Rurgers University, Piscataway, New Jersey 08854

ABSTRACT

We have investigated the dielectric, insulating, structural, microstructural, interfacial, and surface morphological properties of $Ba_{0.60}Sr_{0.40}TiO_3$ thin films Mg doped from 0 to 20 mol%. A strong correlation was observed between the films structural, dielectric and insulating properties as a function of Mg doping. Non textured polycrystalline films with a dense microstructure and abrupt film-Pt electrode interface were obtained after annealing at 750°C for 30 min. Single phase solid solution films were achieved at Mg doping levels up to 5 mol%, while multiphased films were obtained for Mg doping levels of 20 mol%. Decreases in the films dielectric constant, dielectric loss, tunability and leakage current characteristics were paralleled by a reduction in grain size as a function of increasing Mg dopant concentration. Our results suggest that Mg doping serves to limit grain growth and is thereby responsible for lowering the dielectric constant from 450 to 205. It is suggested that Mg behaves as an acceptor-type and is responsible for the doped films low dielectric loss and good leakage current characteristics. Performance-property trade-offs advocates the 5 mol% Mg doped $Ba_{0.60}Sr_{0.40}TiO_3$ film to be an excellent choice for tunable microwave device applications.

INTRODUCTION

$Ba_{1-x}Sr_xTiO_3$ (BST) is a promising material for tunable microwave device applications such as electronically tunable mixers, delay lines, filters, capacitors, oscillators, resonators and phase shifters. The tunability of this material arises because it is possible to change its dielectric constant with application of an electric field [1]. Bulk ceramic BST phase shifters, in a microstrip geometry, have been demonstrated at 5-10 GHz [2]. However, the relatively high loss tangent of these materials, especially at microwave frequencies, have precluded their use in phase shifter applications. Recently, the dielectric properties of theses bulk materials have been improved, that is, the loss tangents were reduced to less than 0.006 at 10 GHz [3]. This reduction in loss tangent was achieved by the addition of MgO to form BST/MgO bulk ceramic composites. Utilization of these BST/MgO materials as phase shifting elements in this bulk ceramic form is still quite limited due to the large voltages, on the order of ≥ 1000 V, needed to bias these bulk materials in a microstrip geometry [4]. However, fabrication of this BST/MgO based material in the thin film form, reduces the needed bias voltages to less than a 100 V, which is compatible with the voltage requirements of present semiconductor based systems [4, 5]. Additionally, the thin film material regime allows high frequency device operation (>15 GHz) thereby enhancing the S/N ratio, and direct integration opportunities with other semiconductor components and devices. In this work we report an investigation of the dielectric, insulating, structural, microstructural, interfacial, and morphological properties of $Ba_{0.60}Sr_{0.40}TiO_3$ thin films as a function of Mg dopant concentration from 0 to 20 mol%. The properties were measured and correlated in order to evaluate the long-term reliability and to define the trade-offs between film structure, composition, dielectric loss, tunability, and insulating characteristics for tunable device applications.

Mat. Res. Soc. Symp. Proc. Vol. 623 © 2000 Materials Research Society

EXPERIMENTAL

Undoped and Mg doped (5, 20 mol%) $Ba_{0.6}Sr_{0.4}TiO_3$ films were fabricated via the metalorganic solution decomposition (MOSD) technique. Barium acetate, strontium acetate, and titanium isopropoxide were used as precursors to form BST. Acetic acid and 2-methoxyethanol were used as solvents and magnesium methoxide was employed as the dopant precursor. The precursor films were spin coated onto Pt-coated silicon substrates. Crystallinity was achieved via post-deposition annealing at 750°C for 30 min. in an oxygen ambient. The films were characterized for structural, microstructural, compositional, dielectric and insulating properties. X-ray diffraction, was employed to assess film crystallinity and phase formation. Field emission scanning electron microscopy (FESEM) and atomic force microscopy (AFM) were employed to assess film microstructure and surface morphology. Cross-sectional transmission electron microscopy (X-TEM), combined with energy dispersive spectroscopy (EDS) analysis, was used to detail the film microstructure, compositional uniformity, phase formation, and the nature of the film-substrate interface. Rutherford backscattering spectroscopy (RBS) was employed to determine film composition. Elemental distribution within the film and across the film-Pt interface was determined by Auger electron spectroscopy (AES). The dielectric and insulating measurements were conducted on the films in the metal-insulator-metal (MIM) capacitor configuration. The film Capacitance (C_p) and dissipation factor (tan δ) were measured with an HP 4194A impedance/gain-phase analyzer. The insulating properties of the films were evaluated via I-V measurements.

RESULTS AND DISCUSSION

Table I summarizes the dielectric and insulating properties of the 0, 5, and 20 mol% Mg doped BST films at 100 MHz. Table I shows that the amount of Mg dopant has a strong influence on the dielectric and insulating properties of BST thin films. Specifically the dielectric and insulating properties decrease with increasing Mg doping concentration. However, in order to fully assess the usefulness and reliability of these films for device applications the influence of the Mg content on the film structure, microstructure, surface morphology, compositional uniformity, and nature of the film-substrate interface must be determined. X-ray diffraction was utilized to assess film structure and crystallinity. Figure 1 displays the x-ray diffraction patterns of the Mg doped (0 – 20 mol%) BST thin films annealed at 750°C for 30 min. The x-ray results demonstrated that all films possessed a non-textured polycrystalline structure. There was no apparent change in peak intensity resultant of the Mg doping indicating that all films were well developed at this annealing temperature and time. The full-width-at-half-maximum (FWHM) of the most intense diffraction peaks increased with increasing Mg content. This peak broadening is indicative of a decrease in grain size [6]. The x-ray results showed that the 0 and 5 mol% Mg doped films were single phase. In contrast, the x-ray pattern of the 20 mol% Mg doped film exhibited an extra peak at 2θ =33.01°. Additional peaks in the x-ray data are indicative of secondary or complex phase formation [7], and in this instance is most likely related to excess Mg. A secondary phase, representative of MgO, should exhibit an intense peak at 2θ =37° (2.413Å); however, the presence of a peak in this region would not be evident due to masking by the Pt peak from the bottom electrode. In order to delineate the nature of this complex phase formation noted in the x-ray results, selected area electron diffraction (SAD) experiments were performed on all films. The electron diffraction analyses supported the x-ray results by demonstrating that all films were polycrystalline with no preferred crystallographic orientation. The electron diffraction patterns taken from the 0, 5, and 20 mol% Mg doped BST films consisted of reflections corresponding to the d-spacings of 2.80Å, 2.29Å, 1.98Å, 1.77Å, 1.40Å, and 1.32Å which matched the (100), (111) (200) (210), (220) and (221) reflections of $Ba_{.60}Sr_{.40}TiO_3$. However, the SAD results for the 20 mol% Mg doped film differed markedly from the 0 and 5 mol% Mg doped films, in that, additional diffraction data were present in the SAD pattern. The d-spacings of the additional rings, 1.21Å, 1.05Å, and 0.95A, matched the (222), (400), and (331) reflections of MgO. Thus, based on the SAD results, the 20 mol% Mg doped film was not a single phase film and the secondary phase was determined to be MgO. Therefore, from these experimental results the x-ray peak existing at $2\theta = 33.01°$ in figure 1 is concluded to be resultant of a second phase, MgO, within the film.

Table I. Summary of the effects of Mg doping on the dielectric and insulating properties of $Ba_{0.60}Sr_{0.40}TiO_3$ thin films. Tunability and I_L were measured at E = 200 kV/cm.

Mg (mol%)	ε_r	tan δ	Tunability (%)	I_L (A)
0	450	0.013	28.1	1.0×10^{-10}
5	386	0.007	17.2	6.5×10^{-11}
20	205	0.009	7.9	2.0×10^{-11}

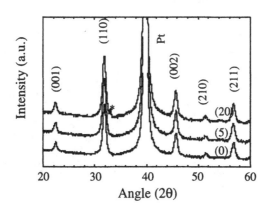

Fig. 1. X-ray diffraction patterns of the Mg doped (0 – 20 mol%) BST annealed films.

Energy dispersive spectroscopy (EDS) nanoanalysis via X-TEM, RBS and AES elemental depth profiles were employed to determine the films compositional uniformity, interdiffusion and quality of the film-electrode interface. The 0 and 5 mol % Mg doped films showed no appreciable compositional variation, as a function of vertical and horizontal EDS probe position. The EDS analyses detected Ba, Sr, Ti, and O peaks, representative of the BST film composition. In contrast, the EDS analysis of the 20 mol% Mg doped film demonstrated significant compositional variations with respect to probe position within the film. The EDS spectra for various probe points differed primarily in the presence or absence of the Mg signal. This compositional heterogeneity is supportive of a multiphased film. The RBS results for the undoped and 20 mol% Mg doped BST films are displayed in figure 2. The sharp edges of the peaks, indicate that the interface between the film and substrate was sharp and that, within the resolution of RBS, no diffusion took place between the film and substrate. A small peak in the RBS spectrum is noted between channels 240 and 300 (figure 2b). This small peak is due to Mg and its presence in the RBS spectrum further confirms the Mg signal noted in the EDS data.

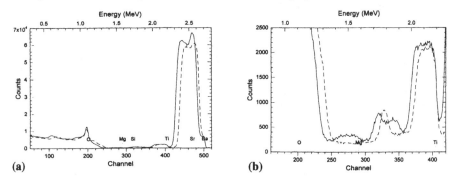

Fig. 2. The RBS spectra for the undoped BST film (solid line) and the 20 mol% Mg doped BST film (dotted line). Fig. 2(b) is an enlargement spectrum (2a) of between channels 120 - 440.

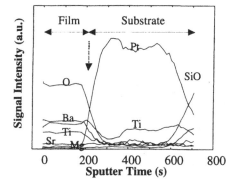

Fig. 3. The AES elemental depth profile for the 5 mol% Mg BST film.

Fig. 4. Plan-view FESEM image of the 20 mol% Mg doped BST film

The AES profile (figure 3) revealed a relatively sharp interface with little interdiffusion of constituent elements between the dielectric film and Pt electrode. The depth profile also revealed that each element component of the film possessed a uniform distribution from the film surface to the interface of the Pt electrode substrate. This data substantiates the fact that the film and the platinized-silicon substrate maintain chemical and thermal stability at processing temperatures up to 750°C (the annealing temperature). No impurities were observed in the AES profile, which without doubt, contributed to the films excellent dielectric properties and low leakage current.

Contaminants provide a negative impact on the films dielectric and leakage current properties. This impurity free film composition was also observed in the EDS analysis. The films surface morphology was evaluated via plan-view FESEM and AFM. The films were crack free and the surface roughness was smooth to within 20Å. Observation of the films surface morphologies showed a dependence of grain size on composition. The 0 and 5 mol% Mg doped films displayed a granular uniform structure, however, this was not the case for the higher doped film. A non-uniform, bimodal grain size distribution was clearly representative of the 20 mol% Mg doped film (fig. 4). A bimodal grain size distribution is common for multiphased materials [8].

Fig. 5. Field emission SEM (images of the 0, 5, and 20 mol% Mg doped $Ba_{0.60}Sr_{0.40}TiO_3$ films.

Figure 5 displays the FESEM cross-sectional images of the 0, 5, and 20 mol% doped BST films. The X-FESEM micrographs clearly delineate the BST film/Pt electrode/SiO₂/Si wafer structure. No amorphous layer or voiding was observed at the interface between the BST film and the bottom Pt electrode. This defect free, and structurally abrupt interface bodes well for the excellent mechanical integrity and good adhesion characteristics of the BST film-Pt coated Si substrate at all doping levels. The microscopy results demonstrated that the doped and undoped BST films possessed a well crystallized, and dense, void free microstructure. This microstructure is most likely responsible for the low leakage current of these films since poor electrical behavior is usually attributed to existence of an amorphous phase between grain boundaries, high defect densities, voids, and poor quality of the film/electrode interface [9]. All films were polycrystalline and were composed of granular multi-grains randomly distributed throughout the film thickness. Dark field TEM analyses revealed that the films grain size decreased from 36 nm to 25 nm as the Mg dopant increased from 0 to 5 mol%, respectively. A decrease in ε_r paralleled this reduction in grain size, which is in agreement with previous findings reported in the literature [8, 10]. The 20 mol% Mg doped film possessed a bimodal grain size distribution with grain sizes averaging 21 nm for the smaller grains and 60 nm for the larger grains. This bimodal grain size distribution, in combination with the x-ray, SAD and EDS results strongly supports a multiphased, non solid-solution film structure.

CONCLUSIONS

Our results have shown that Mg doping dramatically changed the film microstructure, dielectric and insulating properties. Specifically, the Mg doping reduced the grain size, dielectric loss, dielectric constant, and leakage current but at the same time reduced the dielectric tuning.

At Mg doping levels from 0 to 5 mol% the films were single phase and at the Mg concentration of 20 mol% the film was multiphased. The compositional and microstructural non-uniformity of multiphased films is detrimental to long term reliability and processing-property reproducibility. Therefore, the trade-offs between film structure, composition, dielectric loss, leakage current, and tunability must be considered. Our results suggest the 5 mol% Mg doped BST film to be the most suitable for high performance, reproducible, reliable, and tunable device applications. The 5 mol% Mg doped BST film was single phase with low dielectric loss (tan δ =0.007), low I_L (6.5×10^{-11} at applied field of 200 kV/cm) and possessed the necessary ϵ_r of 386 for impedance matching purposes. The tunability, 17.2% at 200 kV/cm, was much lower than tunable devices demand, however the overall applied voltage translates to only ~5.2 V. Thus, the film tunability can be easily elevated toward the necessary 50 % mark as desired for high frequency phase shifter applications, by applying higher field strength. However, the I_L at this elevated applied field must also be low for this to be useful in device applications.

REFERENCES

1. J. S. Horwitz, J. M Pond, B. Tadayan, R. C. Y. Auyeung, P. C. Dorsey, D. B. Christy, S. B.Qudri, and C. Muller, Mat. Res. Soc. Symp. Proc. **361**, 515 (1995).
2. W. Drach, T. Koscica, R. W. Babbitt, L. Sengupta, E. Ngo, S. Stowell, and R. Lancto, Proc. of the 9th IEEE Symp. on Ferroelectrics **79** (1994).
3. L. C Sengupta, E. Ngo and J. Synowczynski, Integrated Ferroelectrics **17**, 287 (1997).
4. V. K. Varadan, D. K. Ghodgaonkar, V. V. Varadan, J. F. Kelly, and P. Glikerdas, MicrowaveJ. **35**, 116 (1992).
5. R. W. Babbit, T. E. Koscica, and W. C. Drach, Microwave J. **35**, 63 (1992).
6. Sharmistha Lahiry, Vinay Gupta, and K. Sreenivas, Proceedings of the 11th IEEE International Symp. Applications of Ferroelectrics 129 (1998).
7. Nak-Jin Seong and Soon-Gil Yoon, Integrated Ferroelectrics **21**, 207 (1998).
8. Robert Tsu, Hung-Yu Liu, Wei-Yung Hsu, Scott Summerfelt, Katsuhiro Aoki, and Bruce Gnade, Mat. Res Soc. Symp. Proc. **361**, 275 (1995).
9. H. Tabata, T. Kawai, S. Kawai, O. Nurata, J. Fujoka, and S. Minakata, Appl. Phys. Lett. **59**, 2354 (1991).
10. C. S. Hwang, O. S. Park, C. S. Kang, H.-J. Cho, H.-K. Kang, S. T. Ahn, and M. Y. Lee, Jpn. J. Appl. Phys. 1 **34**, 5178 (1995).

INFLUENCE OF LaNiO₃ AS AN ELECTRODE ON THE PROPERTIES OF FERROELECTRIC OXIDES

R. Katare, M. Vedawyas and A. Kumar[1]
Department of Electrical and Computer Engineering, University of South Alabama, Mobile, AL 36688
1 Department of Mechanical Engineering and Center for Microelectronics Research, University of South Florida, Tampa, FL 33620. Email : akumar1@eng.usf.edu
(Corresponding author)

ABSTRACT
An electrode plays an important role in realising a ferroelectric thin film as a potential memory device. We have investigated LaNiO₃ (LNO) as a potential electrode material and evaluated the ferroelectric properties of oxide materials like strontium bismuth tantalate (SBT) and barium titanate(BT). We have successfully deposited epitaxial films of LNO on Pt coated Si(100) and LaAlO₃ (LAO) substrates using the pulsed excimer laser deposition technique. We are able to grow high quality SBT and BT films on top of this LNO layer. The X-ray diffraction revealed the epitaxy of the LNO, SBT and BT films. The ferroelectric properties of SBTand BT were investigated using the RT66A ferroelectric tester.

INTRODUCTION

Recently much attention has been attracted in the fabrication of ferroelectric capacitors for high speed non-volatile random access memories (NVRAM)[1,2], high storage capacity of the insulator for memory cells of dynamic random access memories (DRAM)[3,4] and for integrated micro-mechanical devices. In the random access memory applications (DRAMs and NVRAMs) both the conducting and insulating properties of the oxides are utilized. In such applications a ferroelectric material is sandwitched between two conducting electrodes. The reliability of a ferroelectric capacitor depends on the bottom and top electrodes and recent research has attracted conducting oxides as the promising electrodes[4-8]. The oxide electrodes contribute in controlling the long term properties of ferroelectric capacitors. The top and bottom metal electrodes have been shown to detiorate the switched polarization fatigue and leakage currents, a reliability problem hindering the progress of integrated ferroelectric devices, which are related to the quality of electrode/ferroelectric interface. The successful utilization of oxide electrodes in memory devices depend on factors such as : their chemical inertness with the substrates and ferroelectric material at the deposition temperature, their ability to maintain the post-deposition desired compositional and electrical properties and structural compatibility with the perovskite phase of ferroelectric layer. To overcome the problems of degradation, these factors have prompted the use of RuO₂, LaSrCo₃, SrRuO₃, etc. as the promising electrodes for the metal /insulator/metal structure since they act as a sink for oxygen vacancies and hence improve the interface quality[4-8]. Among the various oxide electrodes, perovskite electrodes are more attractive since they fall in the similar category of those of ferroelectric insulating

Mat. Res. Soc. Symp. Proc. Vol. 623 © 2000 Materials Research Society

materials and hence no lattice mismatch is seen, thus facilitating the smooth growth of electrode/ferroelectric/electrode structures. In quite a few cases it is shown that the perovskite electrodes enhance the phase stability of the ferroelectric materials.

In this paper we report our investigation on $LaNiO_3$ as an electrode to evaluate the ferroelectric properties of SBT and $BaTiO_3$. We have fabricated $LaNiO_3$/ferroelectric/$LaNiO_3$ structures on substrates like Pt coated Si (100) and $LaAlO_3$ (100) by the pulsed laser ablation technique. The ferroelectric properties have been evaluated by utilizing RT66A ferroelectric test instrument.

EXPERIMENTAL

Thin films of $LaNiO_3$ (LNO), $SrBi_2Ta_2O_9$ (SBT) and $BaTiO_3$ (BT) were fabricated utilizing the pulsed excimer laser ablation technique. An excimer KrF laser with a wavelength of 248 nm was focussed onto the respective targets to produce a fluence of 2 J/cm^2. The $LaNiO_3$ films were deposited at 750°C with oxygen ambient pressure maintained at 200 mTorr during deposition. The substrate temperature for SBT growth was maintained at 600°C and the deposition carried out in 250 mTorr oxygen ambient. The BT films were deposited at 750C in an oxygen ambient of $2X10^{-3}$ Torr. The laser pulse repetition rate was maintained at 5 Hz for all the deposition cases. LNO and SBT were cooled in 760 Torr oxygen ambient, while the BT films were cooled in the oxygen ambient at which they were deposited. The electrode/ferroelectric/electrode structures were fabricated to evaluate the ferroelectric properties. The capacitor structures were grown on Pt coated Si(100) and $LaAlO_3$ (100) (LAO) substrates. The bottom electrodes were approximately 1500 Åthick and the ferroelectric layers were about 2500 Å.

The phases of the films deposited were identified by X-ray diffraction technique with CuKa radiation. The hysterisis loop and the fatigue of the ferroelectric structures were measured using the RT66A ferroelectric tester.

RESULTS AND DISCUSSION

$LaNiO_3$ is a metallic oxide having a provskite structure as that of most ferroelectric materials[9,10]. $LaNiO_3$ has an excellent crystallographic compatibility with several perovskite type materials (super-conducting as well as insulating). It is rhombohedral metal oxide with a lattice constant of 0.383 nm, and a surface resistivity of the order of 250 $\mu\Omega$ cm . There are quite a few reports where $LaNiO_3$ has been utilized as an electrode to evaluate the ferroelectric properties of PZT and BST[3,5,11]. We have utilized LNO as an electrode to evaluate the ferroelectric properties of SBT and BT.

As described in the previous section we have grown LNO on Pt coated Si(100) and LAO substrates and used as bottom and top electrodes to evaluate the ferroelectric properties of SBT and BT. The structural perovskite phase in the deposited films were determined by the X-ray diffraction.

Fig. 1 shows the X-ray diffraction spectra of LNO films grown on Pt coated Si(100) and LAO substrates at 725°C. The films are epitaxial and highly oriented. Only LNO (100) and LNO (200) peaks at $2\theta = 23°$ and 47° are seen in the X-ray spectrum. We dia not see formation of any impurity phase of the LNO films. The room temperature surface resistivity of these LNO films were of the order of 100 $\mu\Omega$ cm. Such low surface resistivity of LNO electrodes reduces the distortion of measured ferroelectric properties.

Since the lattice constants of LAO and LNO are close to each other, the (00l) peaks of LAO and LNO could not be separated in the X-ray pattern. Nevertheless the low surface resistivity of the LNO films confirms the formation of crystalline LNO phase on LAO substrates. Thus our X-ray patterns for LNO are consistent with those reported in literature.

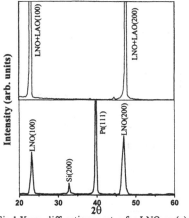

Fig.1 X-ray diffraction spectra for LNO on (a) Pt coated Si(100)and (b) LaAlO3 (100) substrates.

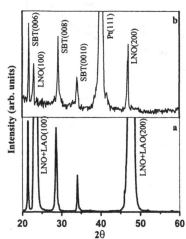

Fig.2 X-ray diffraction spectrum for LNO/SBT/LNO structure on (a)LaAlO$_3$ and (b)Pt coated Si(100)

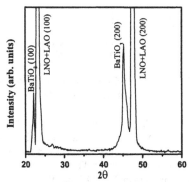

Fig.3 X-ray diffraction spectrum for LNO/BaTiO$_3$/LNO structure on LaAlO$_3$ (100) substrates.

Once the LNO films were successfully grown on LAO and Pt coated Si(100) substrate, we deposited SBT and BT films on the LNO layers and then fabricated LNO/ferroelectric/LNO structures. Fig. 2 shows the X-ray pattern for the SBT films grown on LNO layers on Pt coated Si (100) and LAO substrates at a substrate

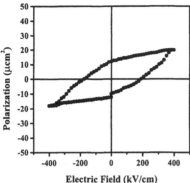

Fig.4 Hysterisis loop of LNO/SBT/LNO on Pt coated Si.

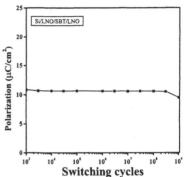

Fig.5 Fatigue curve for LNO/SBT/LNO **capacitor on Pt coated Si.**

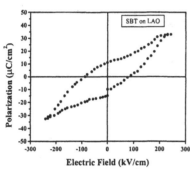

Fig.6 Hysterisis loop of a LNO/SBT/LNO capacitor on LAO substrate.

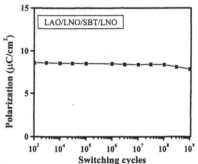

Fig.7 Fatigue curve for LNO/SBT/LNO capacitor on LAO.

temperature of 600°C. The SBT films were highly oriented along the c-axis. In Fig.2 no phases other than SBT (006), SBT (008) and SBT (0010) are seen, thus confirming the high quality of the grown SBT films. The sharpness of the peaks in the spectra suggests the high crystalline nature of the deposited films. The SBT films were highly insulating with the room temperature resistance of more than 20 MΩ. Thus the SBT films grown are of high quality.

Fig. 3 shows the X-ray spectrum for BT films grown on LNO layers on the LAO substrates. These films are also of high quality in nature as envisaged from Fig. 3. The films are highly oriented with sharp crystalline peaks along the a-axis. BT (100) and BT(200) peaks can be distinctively seen in Fig. 3. The BT films too were highly insulating at room temperature.

The LNO/ferroelectric/LNO structures were then tested for their ferroelectric behavior by the RT66A ferroelectric tester. Measurements performed include; hysterisis loop (polarization versus electric field) and fatigue characterization (retained polarization versus switching cycles). Fig. 4 shows the hysterisis loop of SBT ferroelectric capacitor fabricated on Pt coated Si substrates using the pulse voltages of ± 10 V. The remnant polarization (P_r) is about 13 $\mu C/cm^2$ and the saturation polarization (P_s) is about 20$\mu C/cm^2$. From the fatigue measurements, it could be seen that the SBT capacitors showed no polarization degradation upto 10^8 switching cycles. The fatigue characteristics were performed using ± 5V, 100 kHz square pulses. Fig. 6 and Fig. 7 show the hysterisis loop and fatigue characteristics of SBT capacitors fabricated on LAO substrates. These SBT capacitors showed similar behaviour as that of fabricated on Si substrates. The remnant polarization was measured to be 11 $\mu C/cm^2$ and the saturation polarization was 32 $\mu C/cm^2$. The fatigue characteristics showed a little improvement as can be seen from Fig.7. The polarization is retained to fairly better amount upto 10^9 switching cycles.

Our SBT films are highly oriented along the c-axis and it is reported that the remnant polarization values are low for c-axis oriented SBT films. Thus our results are consistent with those reported in literature[12,13].

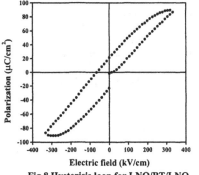

Fig.8 Hysterisis loop for LNO/BT/LNO on LAO substrate.

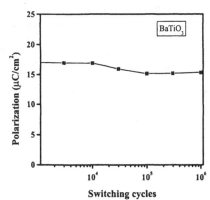

Switching cycles

Fig.9 Fatigue curve for LNO/BT/LNO capacitor.

Fig. 8 and Fig. 9 show the hysterisis and fatigue characteristics for a BT capacitor fabricated on LAO substrates with the LNO electrodes, measured using pulse voltages of ± 8 V. The remnant polarization was measured to be around 20 $\mu C/cm^2$ and the saturation polarization was about 90 $\mu C/cm^2$. These values are well within the requirements of a ferroelectric memoryand agree with the available literature[14,15]. From Fig. 9 it is evident that the BT capacitor fairly retains its polarization upto 10^6 switching cycles, but beyond 10^6 cycles the capacitor undergoes fatigue and the capacitor deteriorates.

CONCLUSION

Capacitor structures of SBT and BT are successfully fabricated with LNO as the top and bottom electrodes . It is shown that the films deposited by the pulsed laser ablation technique are of high quality and highly oriented. SBT and BT capacitor structures with LNO electrodes showed improved ferroelectric characteristics.

REFERENCES:

1. J.F. Scott and C.A.P. DeAraujo; Science **246**, 1400 (1989)
2. A. Kumar, M.R. Alam, A. Mangiaracina and M. Shamasuzzoha; J. Elec. Mater. **26**,1331 (1997)
3. C.M. Wu and T. B. Wu; Jpn. J. Appl. Phys. **36**, 1164 (1997)
4. B. Nagraj, T. Sawhney , S. Perusse, et al; Appl. Phys. Lett. **74**, 3194 (1999)
5. C.H. Lin, B.M. Yen, H.C. Kuo and H. Chen, J. Mater. Res. **15**, 115 (2000)
6. J. J. Lee, C.L. Thio and S.B. Desu; J. Appl. Phys. **78**, 5073 (1995)
7. R. Ramesh, H. Gilchrist, T. Sands, et al; Appl. Phys. Lett. **63**, 3592 (1993)
8. C. M. Foster, G.R. Bai, R. Csencsits, T. Vetrone et al; J. Appl. Phys. **81**, 2349 (1997)
9. H. Ichinose, Y. Shiwa and M. Nagano; Jpn. J. Appl. Phys. **33**, 5903 (1994)
10. K.P Rajeev, G.V. Shivashankar and A.K. Raychaudhuri; Solid State Commun. **79**, 591 (1991)
11. M.J. Shyu, T.J. Hang and T.B.Wu; Mater. Lett. **23**, 221 (1995)
12. C. Bae, J.K. Lee, S.H. Lu and H.J. Jung; J. Vac. Sci. Technol. **A 17**, 2957 (1999)
13. H. M. Yang, J.S.Luo and W.T.Lin; J. Mater. Res. **12**, 1145 (1997)
14. K. Abe, S. Komatsu, N. Yanase, K. Sana and T. Kawakuba; Jpn. J. Appl. Phys. **36**, 5846 (1997)
15. T. Hayashi, N.Ohji, K. Hirihara, T. Fukunaga and H. Maiwa; Jpn. J. App. Phys. **32**, 4092 (1993)

HETEROEPITAXIAL BARIUM HEXAFERRITE FILMS ON (111) MAGNESIUM OXIDE SUBSTRATES

STEVEN A. OLIVER*, SOACK DAE YOON**, IZABELLA KOZULIN**, MING LING CHEN**, XU ZUO** and CARMINE VITTORIA**
* Center for Electromagnetic Research, Northeastern University, Boston MA 02115.
** Department of Electrical and Computer Engineering, Northeastern University, Boston MA 02115.

ABSTRACT

High quality films of barium hexaferrite ($BaFe_{12}O_{19}$) were deposited by pulsed laser ablation onto MgO (111) substrates. In contrast to previous films deposited onto c-plane sapphire (Al_2O_3), these films were expected to have compressive biaxial stress, and indeed showed no indications for either cracking or delamination to a film thickness of 32 μm. All films were found to be highly c-axis textured by both x-ray diffraction measurements and magnetization results. The saturation magnetization ($4\pi M_s$ = 4.2 kG) and uniaxial anisotropy field (H_A = 16 kOe) values for these films approach bulk values. Ferrimagnetic resonance measurements on a calcined 3 μm thick film show a narrow linewidth ($\Delta H \sim 100$ Oe) for the uniform resonance mode. The properties of these films approach those required for self-biased millimeter wavelength devices.

INTRODUCTION

Hexaferrite materials having the hexagonal magnetoplumbite structure are the material of choice for incorporation into planar nonreciprocal devices that can be integrated with semiconductors and dielectrics into microwave integrated circuits [1]. These materials have good saturation magnetization values, high dielectric constants and low dielectric loss tangents, and low magnetic losses [2]. Most importantly, these materials show a very large uniaxial magnetocrystalline anisotropy with the magnetic easy axis coincident with the crystallographic c-axis. Thus, when samples of the hexaferrites are well aligned they can be used as permanent magnets, which is a property that can be used to good advantage in microwave devices since the presence of the strong uniaxial anisotropy field along the c-axis can be used to self-bias the ferrimagnetic resonance field at which moments uniformly precess within the material to frequencies of over 30 GHz [3]. In contrast, the spinel ferrites and garnets that are commonly used in nonreciprocal devices have small values for their cubic magnetocrystalline anisotropy, and thus self-resonate at frequencies of 1 GHz or less. Therefore, devices made from the latter materials require large external magnets to operate at higher frequencies, while devices made from the hexaferrites can operate without external magnets.

In order to be viable for planar nonreciprocal devices, the hexaferrite materials must be obtained as films, where these films must be epitaxial and must have a thickness of 50 - 100 microns, depending on the device operating frequency and design. To date, researchers have concentrated on depositing films of the prototypic hexaferrite material barium hexaferrite ($BaFe_{12}O_{19}$) onto sapphire (Al_2O_3) substrates by pulsed laser deposition techniques. However, it has proven impossible to obtain thick films of these materials in usable form for devices, due to the substantial thermally-induced biaxial tensile stress in the $BaFe_{12}O_{19}/Al_2O_3$ system that causes films having thicknesses above ~15 microns to crack and delaminate upon cooling from the growth temperature [1].

Mat. Res. Soc. Symp. Proc. Vol. 623 © 2000 Materials Research Society

Another promising substrate for this application is magnesium oxide (111), which has a face centered cubic (fcc) structure such that the continuity of close-packed oxygen planes across the interface serves as a template for an (001) $BaFe_{12}O_{19}$ film. Here the room temperature lattice parameter of MgO is slightly larger than that of $BaFe_{12}O_{19}$, with a lattice mismatch of -1% [4][5], and, most importantly, the $BaFe_{12}O_{19}$ film will be under compression since the thermal expansion coefficient of MgO is greater than that of $BaFe_{12}O_{19}$ [6][7]. Here we present results on the growth and characterization of epitaxial $BaFe_{12}O_{19}$ (001) films on MgO (111) substrates for application in planar microwave devices.

EXPERIMENT

The barium hexaferrite films characterized here were deposited onto optically polished MgO (111) substrates by pulsed laser ablation deposition using a Compex 205 KrF excimer laser ($\lambda = 248$ nm) at an energy density of 4 -5 J/cm^2 and a repetition rate of 50 Hz. For the thicker films discussed here, three 2 in. diameter targets of pressed and sintered $BaFe_{12}O_{19}$ were used in a rotating target carrousel, such that they could be rotated and rastered to maximize target surface usage. The ambient oxygen pressure was set to the optimal pressure (20 mTorr) found for growing thick $BaFe_{12}O_{19}$ films on Al_2O_3 (001) substrates.[1][8] The substrate temperature was set at 925°C, where both an *in situ* halogen lamp and a conductive heater were used to maximize the film surface temperature throughout the deposition. Films were deposited having thicknesses of from ~1 micrometer to over 30 micrometers, with selected films being characterized by x-ray diffraction, magnetometry, and ferrimagnetic resonance measurements.

RESULTS

Our initial depositions have demonstrated that barium hexaferrite films can be deposited onto MgO substrates to a thickness of almost twice that attainable on sapphire without fracture in either film or substrate, and with the film retaining good c-axis texture and excellent magnetic properties. Figure 1 shows an electron micrograph cross-section of a 32 μm $BaFe_{12}O_{19}$ film on a MgO substrate, where the sample was fractured along the cleave planes of the substrate, which are not perpendicular to the surface. Here, the torn film is shown to be very dense, with surface outgrowths of order a few microns in height scattered on the top surface.

Figure 1 Cross-section of a 32μm thick epitaxial $BaFe_{12}O_{19}$ / MgO film.

The film texture and c-axis lattice parameters were found from x-ray diffraction measurements, with the results for two films of thickness 14 μm and 28 μm being shown in Figure 2. Both films display excellent c-axis texture, as shown by the presence of all of the dominant (00n) diffraction peaks in the spectra, with only a few peaks corresponding to non-c-axis oriented crystal planes being apparent in the thicker film [4][5]. It has been noted that diffraction patterns taken on films having a thickness near 30 μm show a considerable reduction in intensity compared to that of the 15 μm films, especially in the intensity of the higher order reflections.

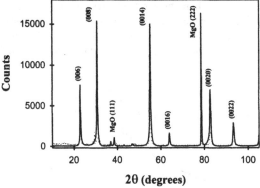

Figure 2 Diffraction patterns are shown for a 14 μm film (solid line) and a 28 μm film (x 5 - dashed line).

Rocking curves were taken on the (008) diffraction peak of these films to appraise the lattice dispersion within the sample. Figure 3 shows the result for the same 14 μm film as shown in Figure 2, where 2Θ was fixed at 30.663⁰, and reveals an asymmetric peak having a full-width-half-maximum value of 1.24⁰. Similar measurements on the thicker films show very broad lines having widths up to 4⁰. All of these results may arise from strain effects. However, they may also

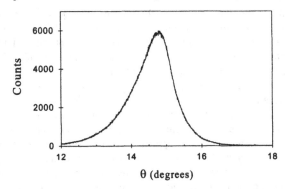

Figure 3 Rocking curve results for a 14 μm film.

arise from the displacement of the film surface away from the diffractometer focusing circle either due to the surface outgrowths that were noted in Figure 1, or due to substrate curvature.

The film magnetic properties were measured by magnetometer measurements where the applied magnetic field (H) lay either in the film plane (parallel) or along the normal to the surface (perpendicular). Figure 4 shows results for the same 32 μm film as shown in Figure 1. A distinct difference in the hysteresis curves is noted for the two orientations, as expected for a near-epitaxial permanent magnet material. In particular, the hysteresis along the film normal shows excellent square loop behavior, where the curve has been skewed from the vertical because of the film demagnetizing field. In contrast, the data taken with H in the film plane has a linear approach to saturation due to the torque required to overcome the strong uniaxial anisotropy field (H_A) and

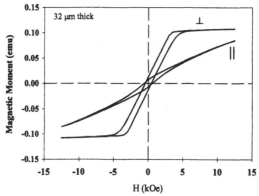

Figure 4 Hysteresis curves are shown for a 32 μm film
with H in the film plane (∥) and normal to the plane (⊥).

rotate the magnetization into the film plane. Here, the available 13 kOe field is not strong enough to fully rotate the magnetization into the film plane. A mean value for the saturation magnetization of $4\pi M_s = 4.2$ kG was found from the saturation moments and volumes of six $BaFe_{12}O_{19}$ films having thicknesses above 20 μm, in good agreement with bulk values [2]. The coercive field (H_c) values of these films are quite low, typically between 600 Oe to 900 Oe, and indicate that the film microstructure is relatively homogeneous with few domain pinning centers. However, this low coercive field value is not in accords with that desired for devices, because these films demagnetize too readily and are not good permanent magnets in their present condition. Thus, to ensure good device performance either the film microstructure must be modified after deposition to increase H_c, or an applied magnetic field of order 4.5 kOe must be applied to saturate the film, as shown by the normal oriented hysteresis curve of Figure 4.

The microwave magnetic characteristics were determined from ferrimagnetic resonance (FMR) measurements using a shorted waveguide technique at frequencies of from 40 GHz to 60 GHz, where H was oriented along the film normal. In general, the FMR linewidths of thicker films had values of $\Delta H \sim 0.6 - 0.7$ kOe, while those of thinner films (~ 3 μm) were of order 0.2 kOe. Since earlier results on $BaFe_{12}O_{19}$ films on c-plane sapphire (Al_2O_3) have shown distinct differences in the attainable FMR linewidth value for films deposited at high (300 mTorr) and low (20 mTorr) oxygen pressures, pieces of the 3 μm film were calcined for various lengths of time and temperatures of from 900°C to 1000°C to examine changes in the FMR linewidth. Figure 5 shows an FMR spectrum for a 3 μm film annealed for two hours at 1000°C. Here a single sharp resonance corresponding to the uniform mode is seen at an applied field of ~ 9 kOe. The FMR linewidth of this mode is ~100 Oe, which is approximately three times larger than the FMR linewidth value found for a thin (<0.5 μm) $BaFe_{12}O_{19}$ film deposited onto Al_2O_3 at 300 mTorr [9]. An extrapolation of the resonance field position of this mode to zero frequency indicated that this film has a self-resonant frequency of 32.7 GHz, which yields a uniaxial anisotropy field value of H_A = 16 kOe. This value is less than those found for bulk materials [2], but is in good agreement with values found for films deposited onto sapphire [1][8].

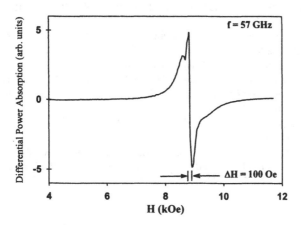

Figure 5 A 57 GHz FMR spectra is shown for a 3 μm film after calcining at 1000°C for two hours.

CONCLUSIONS

It has been shown that $BaFe_{12}O_{19}$ films deposited onto MgO (111) substrates can be grown to a much greater thickness than films deposited onto Al_2O_3 substrates, without sacrificing the excellent film magnetic and microwave magnetic properties. Such films show excellent c-axis texture, and may be epitaxial. The effects of strain may be apparent in the broadening of the rocking curve and also in ferrimagnetic resonance spectra. Altogether, the characteristics of these films match most of the requirements needed to make hexaferrite films viable for usage in nonreciprocal microwave integrated circuits.

ACKNOWLEDGMENT

This research was supported by the Office of Naval Research and the Defense Advanced Research Projects Agency under the 1996 Multidisciplinary University Research Initiative.

REFERENCES

[1] P.C. Dorsey, D.B. Chrisey, J.S. Horwitz, P. Lubitz and R.C.Y. Auyeung, IEEE Trans. Magn. **30**, 4512 (1994).

[2] H. Kojima, *Ferromagnetic Materials, Vol. 3*, edited by E.P. Wohlfarth (North-Holland Publishing Co., New York, 1982).

[3] S.A. Oliver, P. Shi, N.E. McGruer, P.M. Zavracky and C. Vittoria, IEEE Microwave Theory and Techniques, in press 2000.

[4] International Centre for Diffraction Data (ICDD) card number 43-1022.

[5] International Centre for Diffraction Data (ICDD) card numbers 39-1433 and 43-0002.

[6] *CRC Materials Science and Engineering Handbook*, edited by J.F. Shackelford, pp. 308 - 314 (CRC Press, Boca Raton, 1994).

[7] P.C. Dorsey, S.B. Qadri, J.L. Feldman, J.S. Horwitz, P. Lubitz, D.B. Chrisey and J.B. Ings, J. Appl. Phys. **79**, 3517 (1996).

[8] S.A. Oliver, M.L. Chen, I. Kozulin and C. Vittoria, J. Magn. Magn. Mat.**213**, 326 (2000).

[9] S.R. Shinde, R. Ramesh, S.E. Lofland, S.M. Bhagat, S.B. Ogale, R.P. Sharman and T. Venkatesan, Appl. Phys. Lett. **72**, 3443 (1998).

GROWTH, CHARACTERIZATION AND ELECTRICAL PROPERTIES OF PZT THIN FILM HETEROSTRUCTURES ON SILICON BY PULSED LASER DEPOSITION

SOMA CHATTOPADHYAY[#], ALEX KVIT[**], V. SREENIVASAN[#], A.K. SHARMA[**], C. B. LEE[##], AND J. NARAYAN[#]
[#]NSF Center for Advanced Materials and Smart Structures, Department of Electrical Engineering, North Carolina Agricultural and Technical State University, NC 27411; [**]North Carolina State University, Materials Science Dept., Raleigh, NC 27695 – 7916.
[##]Department of Electrical Engineering, North Carolina Agricultural and Technical State University, 551 Mcnair Hall, NC 27411.

ABSTRACT

Epitaxial thin films of $PbZr_{0.52}Ti_{0.48}O_3$ (PZT) have been synthesized successfully on $SrRuO_3/SrTiO_3/MgO/TiN/Si$ heterostructures by pulsed laser deposition. The films were single phase and had (001) orientation. The deposition parameters were varied to obtain the best epitaxial layer for each of the compounds. Transmission electron microscopy indicated good epitaxy for the entire heterostructure and sharp interfaces between the epilayers. Dielectric and P-E hysteresis loop measurements were carried out with evaporated Ag electrodes. The dielectric constant for the films was found to be between 400-450. The value of saturation polarization P_s was between 55-60 $\mu C/cm^2$ and the coercive field E_c varied from 60-70 kV/cm. Integration of PZT films with silicon will be useful for future memory and micromechanical devices.

INTRODUCTION

Recently, there has been an extensive research on oxide based ferroelectric materials since their ferroelectric and dielectric properties have potential applications in microelectronic circuitry, which includes nonvolatile memory elements, microsensors, capacitors for dynamic random access memories, surface acoustic wave (SAW) devices and optical waveguides. PZT is a perovskite-type ferroelectric material with excellent piezoelectric, dielectric and ferroelectric properties. The synthesis, processing and electrical properties of PZT thin films have been studied intensively since it can be used for SAW delay lines, ferroelectric field effect transducers, pyroelectric sensors, radiation-hard nonvolatile random access memories and dynamic random access memories [1,2]. Potential applications of nonvolatile ferroelectric random access memories include smart cards, high-speed telecommunications, RF tags, etc. [3]

There have been ongoing efforts to integrate ferroelectric thin films with existing silicon technology to fabricate reliable nonvolatile memories. Conventionally, polycrystalline ferroelectric thin films such as PZT are grown on silicon substrates with Pt as the bottom electrode: $Pt/Ti/SiO_2/Si$ has been used very commonly [4,5]. Pt has been used because of its good metallic properties and its high oxidation resistance. However, PZT capacitors with Pt top and bottom electrodes tend to show a strong loss of switchable polarization. Conducting oxide electrodes have been used to overcome the problem of fatigue in nonvolatile memory capacitors using PZT [6]. PZT films with superior properties have been synthesized successfully with $SrRuO_3$ and $(La,Sr)CoO_3$ as bottom electrodes [7,8]. Earlier, we have reported the successful growth of epitaxial thin films of PZT on $YBa_2Cu_3O_{3-\delta}/SrTiO_3/MgO/TiN/Si$ heterostructure [9] as epitaxial films have a low density of scattering centers such as dislocations and low high angle grain boundaries and hence, possess better properties than the polycrystalline films [10].

In this communication, we report the growth of epitaxial PZT thin films on silicon substrates by pulsed laser deposition (PLD) with TiN, MgO, $SrTiO_3$ (STO) and $SrRuO_3$ (SRO)

143

as the intermediate layers. The growth conditions have been optimized so as to obtain perfect epilayers for each of the compounds. The films have been characterized by X-ray diffraction (XRD) and transmission electron microscopy (TEM) and were found to possess good electrical properties.

EXPERIMENTAL

The composition of PZT that has been used is $Pb(Zr_{0.52}Ti_{0.48})O_3$ since the composition near the morphotropic phase boundary has been known to exhibit good ferroelectric properties [11]. The films were grown by PLD. A KrF excimer laser (Lambda Physik) was operated at 248 nm for the purpose of ablation. During all ablation experiments, the laser pulse width was 25 ns and the pulse repetition rate was 10 Hz for all the compounds. The Si substrates were cleaned with acetone and methanol for 15 min each, by ultrasonication and were then dipped in 10% HF for a minute. They were mounted on a heater block parallel to the target at a distance of 5.0 cm. The targets of TiN, MgO, STO, SRO and PZT were 99.9% pure and stoichiometric. The substrate temperature was controlled to within ±5°C with a Eurotherm temperature controller. The laser beam was scanned over the target by mounting it on a stepper motor controlled scanner to prevent local heating. Ablation was carried out in vacuum for TiN and in oxygen partial pressures for the other oxides. An oxygen rich atmosphere was maintained in the chamber (240 mTorr) to obtain proper stoichiometry for the SRO and PZT films and at the end of each run the chamber was flooded with oxygen and then the films were cooled slowly. After the deposition of SRO, the films were covered at the corner with a piece of silicon to expose the bottom electrode. The growth conditions for each of the films were optimized with respect to: (a) substrate temperature, (b) substrate- target distance, (c) oxygen pressure inside the chamber, (d) pulse repetition rate and (e) the laser pulse energy.

The crystallinity and the phase purity of the films were checked by a Rigaku X-ray diffractometer. Rocking curves of the (001) reflection of PZT were carried out to estimate the quality of the film. The microstructure of the different epilayers in the film was studied by TEM. Silver electrodes of 200 μm diameter were deposited on top of PZT by electron beam evaporation to serve as the top electrode and SRO was used as the bottom electrode. The thickness of the silver dots was 150 nm. Dielectric hysteresis behavior was studied using a Radiant Technology RT66A loop tracer. The capacitance and the loss tangent of the films were measured at 100 kHz and at a oscillation voltage of 1 volt using a HP 4192A LCZ meter in a standard parallel plate geometry.

RESULTS AND DISCUSSIONS

Growth and Microstructure

Figure 1 shows the XRD pattern of the five layered film deposited on silicon. The film is free from impurity phases that occur due to reactions at the interfaces and shows perfect (001) orientation. XRD pattern was noted for each of the films after the deposition. The optimized growth parameters for each of the epilayers have been listed in Table 1. The full width at half maximum (FWHM) of the rocking curve for these samples is between 0.6°-0.7° indicating the high quality of the single crystal films. The growth of TiN on Si occurs through domain matching epitaxy where four lattice constants of TiN match with three of Si. This happens since the lattice mismatch between TiN (a=0.424 nm) and Si (0.54301) is 24.6% whereas the domain mismatch is only 4.03%. The lattice mismatch between TiN and MgO (0.4216 nm) is 0.56% and that between MgO and STO (a=0.3905) is 7.7%. The lattice mismatch between STO and SRO (a=0.391 nm) is 0.0023% and that between SRO and PZT (a=0.404 nm) is 3.1%. Hence, all the other layers grow by lattice matching epitaxy.

Table 1. The optimized growth conditions for the different epilayers grown on silicon:

Compound	Growth Temperature (°C)	Laser Energy (mJ)	Chamber Pressure (Torr)	Substrate-Target distance (cm)	Pulse Repetition Rate (Hz)
TiN	540	240	7×10^{-7}	5	10
MgO	600	240	2×10^{-5}	5	10
SrTiO$_3$	750	200	1.5×10^{-3}	5	10
SRO	700	240	2.6×10^{-1}	5	10
PZT	650	220	2.4×10^{-1}	5	10

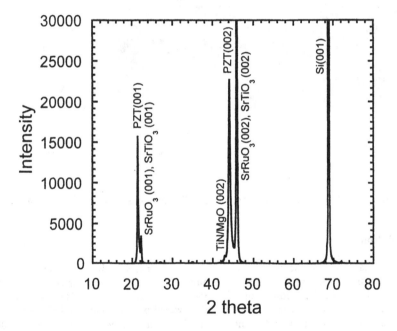

Figure 1. X-ray diffraction pattern of the PZT film grown on the silicon substrate with TiN, MgO, STO and SRO as the epilayers between the substrate and PZT.

Figure 2 shows the cross-sectional TEM image of a five layer thin film heterostructure comprising PZT/SRO/STO/MgO/TiN on silicon substrate. The interfaces are very sharp. The thickness of the mutilayer was about 1.3 μ and the typical thickness of a PZT film was 0.45 μ for 30 mins deposition. Figure 3(a) shows the selective aperture diffraction pattern (SADP) of the superimposed Si substrate ([011] zone axis) and all the films. Although SAD pattern is complicated, a good epitaxial relationship for the whole structure is evident. Fig. 3(b) shows the SAD pattern of the superimposed PZT, SRO and STO layers. In few places, the lattice was observed to be rotated by ~ 70° and such weak spots have been marked by squares in the figure. The SADP for the Si/TiN/MgO/STO layers have been found to be epitaxial and have been reported earlier [9]. The (111) planes of Si are parallel to the (111) planes of TiN and MgO which in turn are aligned with the (111) planes of the perovskite oxides deposited on top of MgO.

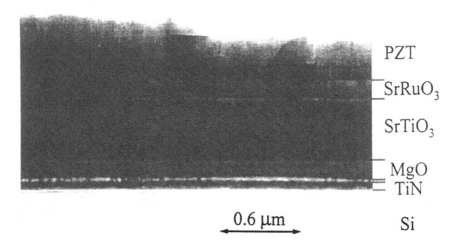

Figure 2. Cross-sectional bright field image of the PZT film whose X-ray diffraction has been shown in figure 1.

Electrical Properties

The capacitance and the loss tangent was measured for each of the silver electrodes on the top of the film with respect to the SRO bottom electrode. The capacitance of the sample holder was subtracted in each measurement to obtain the corrected value of the capacitance of the films. The dielectric constant for the PZT films varied between 400 and 450 for the films deposited at 650°C and the loss tangent for the PZT films varied between 0.06 and 0.08.

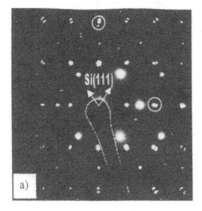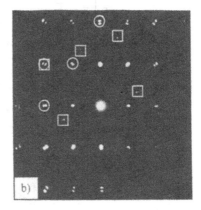

Figure 3. (a) SAD pattern of superimposed Si substrate ([011] zone axis) and all the films. The areas of superimposed diffraction spots are marked. A good epitaxial relationship for the whole structure is evident from this image. (b) SAD pattern of superimposed PZT, SRO and STO layers. Circles mark areas where epitaxial relationship is good. Weak spots are marked by squares corresponding to few features on TEM images where rotation of the lattice by and angle ~70° are observed.

Figure 4 shows the hysteresis loop for the best PZT film. The value of the spontaneous polarization (P_s) for this film is 58.3 $\mu C/cm^2$ and the value of the remnant polarization (P_r) is 30.2 $\mu C/cm^2$. These values are comparable to those obtained previously by others [12,13]. The high values obtained by us are possibly due to the excellent epitaxial quality of the PZT films. The coercive field (E_c) for this film is 70 kV/cm. The electrical properties have been found to be the best for the films deposited at 650°C. Films deposited at 700°C and higher temperatures have shown low values for the electrical parameters and are not very epitaxial in nature possibly due to lead loss taking place in them. Films deposited at temperatures lower than 650°C also showed low values for the spontaneous and remnant polarization. It would be interesting to use SRO as the top electrode for better dielectric and hysteresis properties and to study fatigue effects in these films.

CONCLUSION

In conclusion, we have been successful in growing the five layer epitaxial heterostructure PZT/SRO/STO/MgO/TiN on silicon. The films are (00l) oriented, free from impurities and are epitaxial in nature. HRTEM studies on these films have shown sharp interfaces between each of the bilayers. The films have also exhibited very good electrical properties.

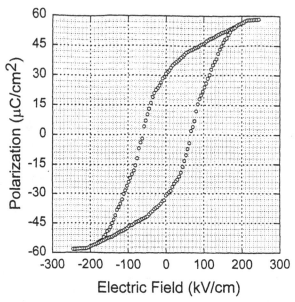

Figure 4. The hysteresis loop for the best PZT film deposited on the SRO/STO/MgO/TiN/Si heterostructure.

ACKNOWLEDGEMENT

This research was performed with the NSF Grant for the Center of Advanced Materials and Smart Structures.

REFERENCES

1) R. Takayama and Y. Tomita, J. Appl. Phys. **65** p. 1666 (1989)
2) K.R. Udaykumar, P. Schuele, J. Chen, S. Krupanidhi and L. Cross, J. Appl. Phys. **77** p. 3981 (1995)
3) Y. Shimada *et al.*, Integrated Ferroelectrics **11** p. 229 (1995)
4) Tetsuo Kumazawa *et al.*, Appl. Phys. Lett. **72** p. 608 (1998)
5) L.H. Chang and W.A. Anderson, Thin Solid Films **303** p. 94 (1997)
6) H.N. Al-Shareef, A.I. Kingon, X. Chen and O. Auciello, J. Mater. Res. **9** p. 2960 (1996)
7) S. Sadashivan *et al.*, J. Appl. Phys. **83** p. 2165 (1998)
8) C.B. Eom et al., Appl. Phys. Lett. **63** p. 2570 (1993)
9) Soma Chattopadhyay et al. (accepted for MRS Symposium Proc.- Fall 1999)
10) J.S. Speck, A. Seifert, W. Pompe and R. Ramesh, J. Appl. Phys. **76** p. 477 (1994)
11) M.J. Haun, E. Furman, S.J. Jang and L.E. Cross, Ferroelectrics **99** p. 12 (1989)
12) K.B. Lee, S. Tirumala and S.B. Desu, Appl. Phys. Lett. **74** p. 1484 (1999)
13) C.R. Cho, L.F. Francis and D.L. Polla, Materials Lett. **38** p. 125 (1999)

UV RADIATION EFFECTS ON THE SOL-GEL PROCESSING OF FERROELECTRIC PZT THIN FILMS

K. S. Brinkman, R. W. Schwartz, J. Ballato
Department of Ceramic and Materials Engineering, Clemson University
Clemson, SC 29634-0907
bob.schwartz@ces.clemson.edu

ABSTRACT

Sol gel solutions have been modified with hydrogen peroxide to improve the durability of photo-irradiated films to water and acidic solvents for photo-patterning. The solutions used for film fabrication are aqueous based and contain acetylacetonate (acac). UV-Vis absorption studies indicate that peroxide modifies the acetylacetonate ligand (in this case the zirconium precursor) creating a new absorbing species at longer wavelength which also affects the response of the acac ligand to UV radiation. Precursor modification and UV treatments have also been shown to impact the texture and improve the microstructure of resulting films. Depth profiling by radio-frequency glow discharge atomic emission spectroscopy indicates reduction in the carbon to hydrogen ratios of films crystallized after exposure to UV radiation.

INTRODUCTION

Thin films of ferroelectric materials such as lead zirconate titanate (PZT), which are based on the ABO_3 perovskite structure, are being investigated for a number of applications such as capacitors, ferroelectric random acces memory devices (FERAM) [1], infrared detectors, and microelectromechanical systems (MEMS) [2]. As for synthesis, aqueous based "chelate" routes for the preparation of these films have been under investigation because of their ease of solution preparation (avoid the use of "Schlenk" line techniques), use of less toxic solvents, and lower crystallization temperatures [3,4]. However, aging of the solutions also has been reported and such effects impact the final properties of films prepared by these processes. To date the mechanistic pathways of such effects undetermined [5]. Hydrogen peroxide initially was used in these systems as a surface wetting agent to improve film quality, and subsequently was shown to have an impact on the aging, microstructure, and electrical properties as well [3,5]. The use of these modifying ligands to tailor processing behavior of ferroelectric thin films is currently under investigation.

In addition to low temperature processing, the fine patterning of ferroelectric films is essential for device application [6]. The use of "chelate" solution modifiers such as β-diketonates (such as acetylacetonate, acac) in sol-gel processes, which lead to photosensitive solutions has been reported for single component (Zr and Ti) as well as multicomponent systems (PZT, BST) [6,7,8]. This paper reports on the investigation of β-diketone (acac) use in aqueous based PZT solutions modified with hydrogen peroxide, as well as its application to fine patterning and its impact on the structural evolution and properties of resulting films.

EXPERIMENTAL

The PZT(53/47) films with the chemical formula $Pb_{1.1}(Zr_{.53}Ti_{.47})O_3$ were fabricated by a chemical solution deposition process (CSD). The chemistry of the multicomponent solution was based on an aqueous acetate PZT system used by Webb [4] and Francis [5]. A 0.8M PZT solution was prepared by combining titanium isopropoxide (TIP) with glacial acetic acid (HOAc) (15:1 molar ratio HOAc:TIP) in a dry nitrogen glove box. After 10 minutes stirring, the TIP/HOAc solution was then combined with distilled water (32:1 H_2O: TIP). The hydrolyzed precursor formed a white precipitate, which was dissolved under ultrasonic agitation for 45 minutes. Zirconium acetylacetonate, $Zr(OC(CH_3)CHCOCH_3)_4$, and lead acetate trihydrate, $Pb(OCOCH_3)_2 \cdot 3H_2O$, powders were then added to the TIP/HOAc/water solution and refluxed under constant stirring for approximately 1 hour. The PZT solution then was split in two portions with one being modified with hydrogen peroxide, H_2O_2 (addition of peroxide corresponds to 10% of the PZT solution volume).

Single component mixtures were synthesized similarly to the multicomponent solutions. The single component Ti solution was prepared by the same procedure as performed in the PZT synthesis. Pb solutions were prepared by dissolving precursor powders in acetic acid/water (same ratio as in the PZT solution) and heating to 100°C while stirring. Zr solutions were prepared from the starting alkoxides and modified with acac (1:1 Zr to acac) according to Tohge [6]. Solutions were modified by additions (in volume %) of peroxide where indicated.

Films for microstructure and crystal structure determination were prepared on Pt coated silicon substrates (Pt/Ti/SiO$_2$/Si) by spin coating at 3000 rpm for 30 seconds. Following deposition, solvent removal and organic pyrolysis was achieved by heat treatment to ~300°C or 100°C for 5 and 10 minutes respectively on a hot plate. Multiple layers (where indicated) were deposited and pyrolyzed followed by firing at higher temperatures. The films of PZT and PZTH (peroxide modified) specimens were then fired at 700°C for 30 minutes at a ramp rate of 50°C/min. Films for UV-Vis studies were spin coated at 2000 rpm for 30 seconds on a quartz substrate and dried at 50°C for 1 hour (for single component) and 100°C for 10 minutes (for PZT to mimic patterning process of films; conditions were experimentally determined by the solubility of the dried film in solvent before irradiation). Film thickness was obtained using a stylus profilometer. UV treatments were performed using an Oriel high pressure Hg lamp at an intensity of 180mJ/cm^2. Patterning of zirconia (with and without peroxide modification) films was performed according to Tohge by drying the film at 50°C for 1 hour before UV treatment and subsequent leaching in acidic solvents. Investigation of PZT patterning was accomplished by drying the film at 100°C for 10minutes (experimentally determined) followed by subsequent UV treatments and leaching.

The elemental depth profiles of the sol-gel prepared PZT and PZTH modified layered systems were performed on a Jobin-Yvon (Division of Instruments, SA, Edison, NJ) 5000 RF rf-GD-AES spectrometer system, which has been described in detail previously [9,10]. X-ray diffraction was performed on Scintag XDS2000. FESEM was performed on a Hitachi S4700. UV-Vis spectra were obtained on a Perkin Elmer Lamda 900 spectrometer.

RESULTS AND DISCUSSION

Preliminary experiments indicate that peroxide additions to this system (10% by volume) improve the durability of photo-irradiated regions to acidic solvents. The UV-Vis spectra of the aqueous PZT solution spun onto quartz is shown in Figure 1. The spectrum shows a shoulder around 300nm. This absorption peak has been observed to decrease with UV radiation exposure. Differences in the absorption response (area between the absorption peak before and after UV treatment) of the films treated with UV were observed in specimens prepared with and without addition of hydrogen peroxide. These differences are hard to resolve in the multicomponent spectrum, therefore single component specimen spectra were investigated. Figure 1 shows the PZT; Pb, Ti, and Zr spectra in an attempt to deconvolute the relative contributions of each cationic species in the multicomponent spectrum. From this data we see that the shoulder is due

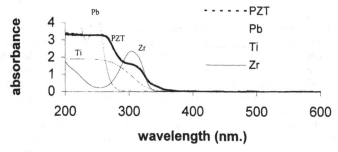

Figure 1. UV-vis of single and multicomponent spectra.

in fact to the zirconium precursor (acac absorption) and that the main UV cutoff below 300nm is due to the a combination of the Zr and Ti precursors. The lead does not seem to play a role where the absorption differences are observed.

Peroxide was added to single component solutions in order to investigate the observed difference in PZT absorption response to precursor modification. In addition, changes in the physical characteristics (precipation, color change, reactions) were observed. No changes were observed in the Pb system, while the Ti and Zr systems showed evidence of chemical modification. The Ti system changed color from clear to yellow/orange as has been previously reported [5]. The absorption spectra was not significantly different but it should be noted that the peroxide modified solution gelled before the unmodified solution. The zirconium system showed a slight color change from yellow to slightly darker yellow, while the absorption spectra showed significant changes. UV-Vis spectra obtained from freshly prepared peroxide modified solutions showed an increase in absorption at wavelengths near 350 nm. The intensity of this absorption peak increased with age (Figure 3 shows the spectra of 1 month old solution).There also was observed a difference in the response of the acac peak to UV radiation, as evidenced by the 2:1 decrease of the Zr solution compared with the 3:1 decrease of the acac peak in peroxide modified solutions. The identity of this peak (defect structure of acac) and possible kinetic modifications to the acac response with peroxide additions currently are under investigation. Absorption spectra of acac solutions modified with peroxide indicate a reaction creating species that absorb at longer wavelengths. Studies on the oligomeric differences caused by precursor modification in these single component systems and multicomponent systems are underway to identify the species responsible for this change in optical properties.

With regard to the patterning of the single component Zr system the impact of peroxide

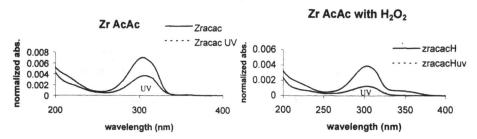

Figure 2 UV-vis of zirconium acac film   Figure 3. UV-Vis of peroxide modified zirconium

modification are inconclusive. Although differences are observed in the absorption spectra, no dramatic improvements are currently seen in the application of this modifier to the patterning of these materials. While the addition of peroxide has been shown to improve the durability of photo-irradiated areas in aqueous PZT containing acetylacetonate, there are still problems regarding pattern definition and the quality of the remaining film after leaching that must be addressed. Nevertheless, this result provides evidence that small amounts of organic free modifiers can alter the nature of the oligomeric species and affect the processing behavior of sol-gel solutions. Understanding these mechanisms may provide for further technical advances in the field of patterning and tailoring of film properties by solution precursor modification.

The addition of precursor modifiers has been shown to affect the properties of materials prepared by chemical solution deposition techniques [11]. Figure 4 shows that the peroxide modifications also lead to differences in the texture of the final crystalline films. The peroxide suppresses the (100) diffraction peak (leaving the (110) peak unaffected) in this case leading to a (111) film texture. The integrated intensities of the (100) and (111) peaks in films were normalized by those obtained from powder samples to quantify texture. It should be noted that texture differences resulting from peroxide addition have also been seen in other material systems such as strontium titanate [12]. This gives evidence of a chemical approach to texturing without the use of seed layers or other electrode technologies. Figure 5 gives additional evidence that peroxide modifies the aging behavior of solutions resulting in more stable film properties over a time period of a month for the multicomponent PZT solutions. Improvements in the electrical properties of PZT films with peroxide additions have been previously reported to result from the elimination of the pyrochlore phase. We have not investigated this effect yet, however we have observed the elimination of "rosette" type structures associated with pyrochlore which are known to degrade optical and electrical properties.

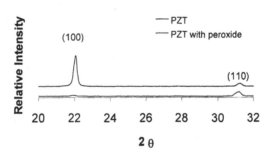

Figure 4. X-ray diffraction of PZT films heated to 700°C

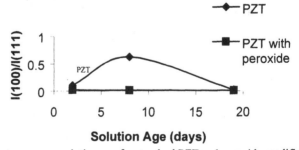

Figure 5. Film texture versus solution age for standard PZT and peroxide modified films.

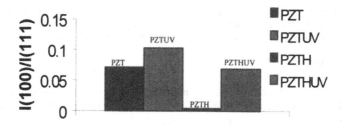

Figure 6. Film texture vs. precursor modification and UV treatment.

Figure 6 indicates differences in texture for both peroxide modified films and those treated with UV irradiation. We see that the peroxide in this case again supresses the formation of the (100) diffraction peak. These films were dried at 100°C, exposed to UV, then heat treated to 700°C. The effect of UV treatments before crystallization is to increase the (100) texture of the films in both peroxide modified and standard PZT solutions, although the peroxide modified seems to have a greater change in texture after UV exposure.

Compositional depth profiling of these films by radio frequency glow discharge atomic emission spectroscopy (rf-GDAES) was used to investigate the effects of precursor modification and UV treatments on the elemental content of these multicomponent films. Monitoring the carbon and hydrogen content of films during processing provides information on pyrolysis behavior. Additionally residual organic material in crystalline compounds may degrade the electrical properties. We have observed differences in the C and H signal intensities between the PZT and PZT modified with peroxide. The "spikes" at the surface may be attributed to plasma instability and sample contamination [13]. The PZT with peroxide films have more carbon relative to hydrogen; this has been previously seen with TGA of vacuum dried powders where the addition of peroxide delays the pyrolysis to a higher temperature. Pretreatment with UV radiation (after drying at 100°C) before crystallization at 700°C results in a decrease in the C/H ratio of the films in both the standard and peroxide modified films. It should be emphasized that these are not stoichometric ratio's of C/H, but the representative relative amounts without

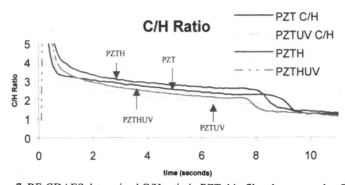

Figure 7. RF-GDAES determined C/H ratio in PZT thin films heat treated at 700°C

standards. Standards need to be employed to accurately assess the actual amounts of carbon and hydrogen. Nevertheless, qualitative comparisons may be made and these results seems to indicate elimination of residual organics in crystalline material with UV pretreatments.

CONCLUSIONS

The results of these investigations indicate peroxide additions enhance the durability of photo-irradiated areas to water and acidic solvents in aqueous based PZT with acetylacetonate absorbing ligands present. Investigations into the single component solutions show that peroxide reacts with acac, and creates new species in the zirconia acac system which absorb at a longer wavelength. The applicability of this process is currently limited due to problems with pattern definition and film quality after leaching. Studies of the effects of peroxide additions indicate improvements in microstructure and texture control through solution precursor modification. Ultraviolet radiation also has been shown to affect the processing of these materials producing films with different texture and carbon to hydrogen content.

ACKNOWLEDGEMENTS

This work was supported by the Center for Advanced Engineering Fibers and Films (CAEFF).

REFERENCES

1. T. Otsuki and K. Arita, *Integrated Ferroelectrics*, **17** (1-4), 31 (1997).
2. J. H. Kim, L. Wang, S. M. Aurn, L. Li, Y. S. Yoon, and D. L. Polla, *Integrated Ferroelectrics*, **15** (1-4), 325 (1997).
3. C. T. Lin, L. Li, J. S. Webb, R. A. Lipeless and M. S. Leung, *Integrated Ferroelectrics*, **3**, 333 (1993).
4. C.T. Lin, B. W. Scanlan, J. D. McNeill, J. S. Webb, and Li Li, *J. Mater. Res.*, **7**, 2546 (1992).
5. J. S. Wright and L. F. Francis, *Mat. Res. Soc. Symp. Proc.*, **433**, 357 (1996).
6. N. Tohge, K. Shinmou, T. Minami, *J. of Sol-Gel Science and Technology*, **2**, 581 (1994).
7. Y. Nakao, T. Nakamura, K. Hoshiba, K. Sameshima, A. Kamisawa, *Jpn. J. Appl Phys.*, **32**, 4141 (1993).
8. N. Soyama, G. Sasaki, T. Atsuki, T. Yonezawa, K. Ogi, Jpn. *J. Appl. Phys*, **33**, 5268 (1994).
9. M. Parker, M. L. Hartenstein, R. K. Marcus, *Anal. Chem.*, **68**, 4213 (1996).
10. M. L. Hartenstein, R. K. Marcus, *J. Anal. At. Spectrom.*, **12**, 1027 (1997).
11. R. W. Schwartz, R. A. Assink, D. Dimos, M. B. Sinclair, T. J. Boyle, C. D. Bucheit, *Mat. Res. Soc. Symp Proc.* **361**, 377 (1995).
12 H. Dobberstein, unpublished results.
13. A. Anfone, private communication.

FERROELECTRIC Na$_{0.5}$K$_{0.5}$NbO$_3$/SiO$_2$/Si THIN FILM STRUCTURES FOR NONVOLATILE MEMORY

C.-R. CHO, A. M. GRISHIN
Condensed Matter Physics, Department of Physics, Royal Institute of Technology, Stockholm, S-100 44, Sweden.

ABSTRACT

Highly *c*-axis oriented Na$_{0.5}$K$_{0.5}$NbO$_3$ (NKN) films have been prepared on thermally grown thin SiO$_2$ template layer onto Si(001) wafer by pulsed laser deposition technique. X-ray diffraction θ-2θ-scan data show multiple-cell structuring along the polar axis in NKN films grown onto SiO$_2$ with thickness up to 45 nm. On the other hand, the film deposited onto amorphous ceramic (Corning) glass is a mixture of slightly *c*-axis oriented NKN and pyrochlore phases, while the film onto Pt(111)/Ti/SiO$_2$/Si(001) shows perfect [111] orientation. This implies small amount of SiO$_2$ crystallites distributed in amorphous silica matrix inherits Si(001) orientation and serves as a key factor in highly oriented growth of NKN films. Au upper electrodes have been defined on the top of NKN(270nm)/SiO$_2$/Si structures to investigate Metal-Ferroelectric-Insulator-Silicon (MFIS) diode characteristics for Field Effect Transistor (MFIS-FET) nonvolatile memory applications. *C-V* measurements yield memory window of 4.14 V at 10 V of gate voltage.

INTRODUCTION

Silicon, the essential material of the semiconductor industry, is one of the most desirable substrate. Recent years there have seen extensive studies on the synthesis of perovskite materials: high temperature superconductors, giant and colossal magnetoresistors, and ferroelectrics for integration with semiconductor devices. One of the main difficulties in fabrication of high quality multi-component oxide film directly onto Si substrate lies in interdiffusion of cations through the film-substrate interface, though there is a favourable pseudoepitaxial affinity between the size of the edge of Si cubic cell and the diagonal of the perovskite unit cell in (*ab*) plane. Therefore a lot of materials such as yttrium stabilized zirconia (YSZ), CeO$_2$, MgO, and SrTiO$_3$ have been introduced as a buffer layer between Si substrate and perovskite films. [1] However, the fabrication of a buffer layer of exceptional quality has turned out a very difficult problem. Thermally grown SiO$_2$ film seems to be the ideal buffer layer since it shows a remarkably small number of electronic defects [2], hence superior performances, i.e. capacitance-voltage (*C-V*) and current-voltage (*I-V*) characteristics are expected in SiO$_2$/Si structure based devices. Nevertheless, only few studies on crystalline relation between perovskite thin film and SiO$_2$/Si substrates have been reported. In this article, we present highly *c*-axis oriented NKN thin films on thermally grown ultra-thin SiO$_2$ layer onto Si(001) wafers, and memory properties of Au/NKN/SiO$_2$/Si (MFIS) structure.

Mat. Res. Soc. Symp. Proc. Vol. 623 © 2000 Materials Research Society

EXPERIMENTAL PROCEDURES

Thin SiO_2 layers with various thickness from native to 45 nm have been thermally grown onto Si(001) substrates. The details of oxidation processes are described elsewhere. [3] A KrF excimer laser (Lambda physik-300, λ = 248 nm, pulse width of 25 nm) has been used to ablate stoichiometric NKN target with the energy density of 4-5 J/cm^2 and repetition rate of 15 Hz. NKN films onto SiO_2/Si substrates have been grown at the substrate temperature of 650 °C in a ~ 400 mTorr ambient oxygen partial pressure. The deposition was followed by in-situ annealing at 400 °C in 500 Torr oxygen for 30 min. Au both top and backside electrodes ($\varnothing$ =0.55 mm) were deposited by thermal evaporation at room temperature.

RESULTS AND DISCUSSION

Figs.1 show x-ray diffraction (XRD) θ-2θ scans for NKN films onto SiO_2 buffer layers of various thickness. High degree of preferential c-axis orientation is clearly seen in logarithmic scale. There are no pronounced reflections except (00l) peaks for NKN films onto native, 4 nm, and 10 nm thick SiO_2 buffer layers. Further increase of SiO_2 thickness up to 45 nm causes the appearance of non c-axis oriented NKN textures, such as NKN-102 and

-310. Nevertheless the intensity of NKN-310 Bragg reflection is more than ten times smaller than that of NKN-00½. Strong superlattice NKN-00¼ and -00½ reflections, which come from the periodic distortion of NbO_6 octahedrals in NKN perovskite unit cell [4], are observed in all samples.

After the prediction of a boundary layer of microcristallites for a thermally grown oxide a number of groups reported on epitaxial interfacial SiO_2 layers in thermally grown, native and wet oxides. The epitaxial oxide is twofold symmetric and had been believed to be disordered into amorphous silica thicker than ~ 7 Å. However, comprehensive investigations of thermally oxidized thin films using crystal truncation rod (CTR) x-ray scattering suggested that SiO_2 films are not purely amorphous but many small crystallites, so called *pseudocrystoballites*, are distributed within amorphous matrix, maintaining an epitaxial relation with the Si substates. [5] Differential x-ray reflection (DRX)

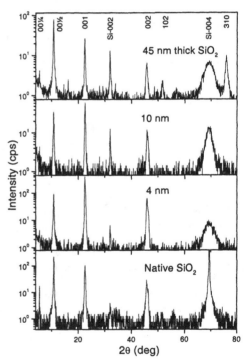

Fig. 1 The X-ray diffraction (XRD) θ-2θ scans in CuK_α radiation for NKN/SiO_2/Si(001) structure with various thickness of SiO_2 buffer layers

easurements [6] indicated the pseudocrystoballites especially populate the SiO₂/Si interface and their density decreases exponentially inside the film, and evaluated to be less than a few % even very near to the interface. There is a close coincidence between Si cubic cell edge size (5.43 Å) and the size of the face diagonal of NKN perovskite cell (5.67 Å, 4.4 % mismatch).

Based upon above considerations, we suggest nano-scale pseudocrystalites inherit crystallographic orientation of Si wafer and facilitate pseudoepitaxial growth of NKN film, seeding the preferential orientation of NKN textures at the early nucleation and coalescence stage. This conclusion is encouraged by the x-ray θ-2θ scans (Figs. 2**a** and 2**b**) for NKN films grown at the same process conditions onto ceramic glass and Pt(111)/Ti/SiO₂/Si(001) wafers. In comparison with Fig. 1, the x-ray diffraction pattern of NKN film on amorphous ceramic (Corning) glass (Fig. 2**a**), which has same stoichiometry as thermally grown SiO₂ layer, shows worse crystallinity, much higher background radiation and pyrochlore phases. On the other hand, the film deposited onto Pt(111) template layer on silicon substrate (see Fig. 2**b**) shows high degree of [111]-axis orientation. These results indicate that the orientation of NKN crystallites at some degree is influenced by substrate texture, although the phenomenon of self-assembling of NKN films in [001]-direction [7] remains the leading factor.

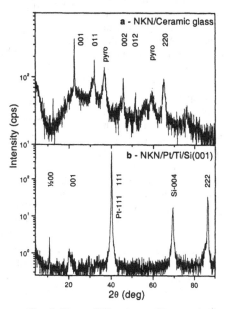

Fig. 2 X-ray diffraction θ-2θ scan for NKN films: **a** - 1.3 μm thick NKN film deposited onto amorphous ceramic (Corning) glass; **b** - 0.9 μm thick Pt/Ti layer onto SiO₂/Si(001) substrate.

Superior ferroelectric performance of NKN films, which have been reported earlier [7], and strong polar axis orientation on SiO₂/Si substrate challenged us to study memory effects in the Metal-Ferroelectric-Insulator-Si (MFIS)-diode structure. We deposited 270 nm thick NKN films onto 7, 10, and 20 nm thick SiO₂ layers, which were grown onto highly doped (phosphorous concentration $N_D > 10^{19}/cm^3$) Si(001) substrates. Capacitance-voltage (C-V) and Conductance-voltage (G-V) characteristics of Au/NKN(270nm)/SiO₂/Si/Au capacitors have been measured by *HP4284A* LCR meter at the frequency of 1 MHz. The signal voltage and bias voltage sweeping rate were fixed at 10 mV rms and 0.2 V/sec, respectively.

Fig.3 shows the C-V and G-V curves for 270 nm thick NKN film capacitor fabricated onto 10 nm thick SiO₂ buffer layer. Counterclockwise C-V hysteresis loop is obtained and the memory window reached to 4.14 V at 10 V bias. Simultaneously performed G-V tracing has close correlation with the C-V measurement. At the accumulation regime (point **a** in Fig.3), the conductance saturates to 1×10^{-4} S and does not show any anomalies up to 10 V bias. At the descending branch of the C-V hysteresis loop, the conductance remains almost invariable until 3.5 V, and then starts to increase and reached maximum value of 1.3×10^{-4} S at the steep segment of *C-V* characteristic (point **b** in Fig. 3). After that the conductance decreases conspicuously and saturates to 5.5×10^{-5} S accompanied with the saturation of the capacitance (segment **c** in Fig. 3). At the ascending branch, the conductance showed similar behavior: reaches the maximum value near the transition point **d** (Fig. 3) and saturates again. The abrupt increase in conductance at the steep segments of the *C-V* characteristic can be attributed to the charge injection when the

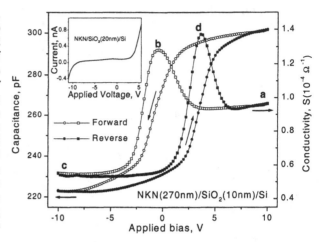

Fig. 3 Hysteresis in the capacitance-voltage (*C-V*) and conductance-voltage (*G-V*) characteristics for Au/NKN(270nm)/SiO₂(10nm)/Si(001) thin film structure. Inset shows typical current–voltage (*I-V*) curve for Au/NKN(270nm)/SiO₂(20nm)/Si (001) structure with the upper electrode area of 2.37×10^{-3} cm². The positive direction of the applied voltage is designated for the positive biasing of the top Au electrode with respect to the ohmic contact on the back side of Si wafer.

ferroelectric film switches the polarization direction. There is a close coincidence between forward and reverse maximum conductance voltages: V_{for} (G_{max}) = - 0.56 V , V_{rev} (G_{max}) = + 3.60 V and two switching voltages: V_{for} = - 0.60 V and V_{rev} = + 3.55 V. The higher conductance at the accumulation state than that in the inversion state can be also explained in the model of MFIS-diode structure. All the features of *G-V* curve in Fig. 3 indicate that eventual *C-V* hysteresis is caused by polarization switching in NKN film, not from other reasons, such as a drift of excess ions in ferroelectric film, since their motion is irreversible and depends on the polarity of the starting bias. [3] The inset in Fig. 3 shows *I-V* curve for Au/NKN(270nm)/SiO₂(20nm)/Si/Au structure. Typical diode behavior has been observed and extremely small leakage current (< 100 pA) has been measured in MFIS-diode with the upper electrode area of 2.37×10^{-3} cm² in the range - 10 to + 3 V (current density < 4.2×10^{-8} A/cm²). There is no considerable difference in *I-V* curves for 7 and 10 nm thick SiO₂ buffer layer compared to the inset in Fig.3.

Fig. 4 shows normalized *C-V* curves measured for MFIS-diodes with various thickness of SiO₂ layer. These curves clearly illustrate two results: firstly, as the SiO₂ thickness increases the *C-V* curves move to the forward direction. The center voltage of hysteresis *C-V* loop is - 0.94 V for 10 nm, + 0.19 V for 10 nm, and + 1.51 V for 20 nm thick SiO₂ films, respectively. Secondly, the squareness of the *C-V* loop improves with the oxide thickness increase. Ideally, the center of two voltage edges of memory window V_{FB} (flat-band voltage) in MFIS-diode is estimated from the metal work function and the electron affinity of the semiconductor. Experimentally, V_{FB} often deviates from this estimate due to the existence of the electronic surface states. [8] In our specific case, all the capacitors were made up of the same material and using the same processes except the SiO_2 layer thickness. Thus, we are persuaded to ascribe the eventual voltage shift to the entrapped alkali ions, i.e. Na^+ and K^+, on the NKN/SiO_2 or SiO_2/Si interfaces. The trapped positive charges create built-in electric field: in negative (reverse) direction if alkali ions trapped onto NKN/SiO_2 interface and in positive (forward) direction if they trapped onto the SiO_2/Si interface. Correspondingly, the built-in electric field causes the shift of the C-V loop towards the reverse/forward voltage direction. Most probably, this interdiffusion of the alkali occurs during the deposition, since high temperature and high bombardment energy of evaporants in laser ablated plasma plume are the main attributes of the PLD process. It is naturally to suggest the performance of SiO_2 as a diffusion barrier gets worse for thinner layers, hence charge trapping for ultra-thin SiO_2 layers occurs at NKN/SiO_2 interface (negative loop offset) while positive loop offset for thicker SiO_2 layers is caused by charge trapping onto the SiO_2/Si interface. The trapped charge also impedes the field generated by polarization switching, hence causes the decrease of the switching rate (reduces the squareness) of the C-V curve.

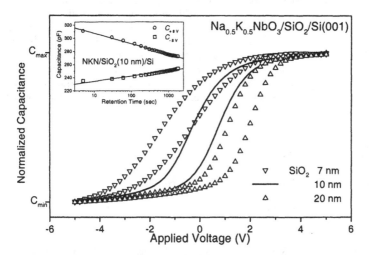

Figure. 4. Normalized C-V hysteresis loops for MFS-diode with 7, 10, and 20 nm thick SiO_2 buffer layers. The inset shows the time dependent capacitance (retention capability) after applying programmable voltage of + 8 V (○ symbols) and − 8 V (● symbols) respectively.

To characterize retention capability, we studied the decay of the remnant polarization measuring the low signal capacitance (100 kHz, 50 mV rms) of the MFIS-diode in the time domain time after poling the ferroelectric capacitor with + 8 V and - 8 V bias voltage. Although, the retention is a major barrier for realization of MFIS-FETs memory, good retention properties are not numerously reported, even though several groups observed an extension of retention time by applying appropriate bias. [9] Applying bias to improve retain performance is completely impropriate, since it destroys a main concept of nonvolatile ferroelectric memory. Therefore, we characterized retention time without applied bias. For MFIS-diodes with 7 nm and 20 nm thick SiO_2 buffer layers two remnant capacitances $C_{+8\,V}$

and C_{-8V} approached to each other and merged in a few minutes, while the MFIS-diode with 10 nm SiO_2 buffer layer possesses much better retention characteristics (see the inset to the Fig. 5). The process of the polarization decay has a typical "creep" behavior indicated by logarithmic dependence of "stored" capacitance on time: C_{+8V} $(t) = 323.2 - 15.8 \times \log (t /1$ sec), and C_{-8V} $(t) = 226.1 + 8.2 \times \log (t /1$ sec). According to above relationships, it takes more than 10^4 sec before two state of capacitance become equal. From above results, we can conclude that memory retention properties are more likely related to the ferroelectric hysteresis behavior of V_{FB} rather than interface states.

SUMMARY AND CONCLUSIONS

In summary, nearly perfect oriented "quadrupled c-axis" NKN films have been prepared by pulsed laser deposition technique on thin SiO_2 layers thermally grown onto Si(001) wafers. Through the comparison with NKN films fabricated onto the amorphous ceramic glass and [111]-oriented Pt template layer, we suggest a small amount of distributed epitaxial nano-scale pseudocrystoballites play an important role in oriented NKN film growth. We first investigate NKN film for MFIS-FETs type nonvolatile memory. NKN films onto ultrathin (up to 20 nm) SiO_2 layers show highly c-axis oriented superlattice structure. 4.14 V memory window has been obtained in Au/NKN(270nm)/SiO_2(10nm)/Si MFIS-diode structure at 10 V of gate voltage. The improvement of squareness of hysteresis C-V loop as well as its shift towards forward voltage direction with the buffer layer thickness increase support the model of alkali ions trapping at the interface of SiO_2/Si. It should be emphasized that the improvement of memory retention is directly related to the flat band voltage V_{FB}.

ACKNOWLEDGMENTS

This research was funded by the Swedish Agency NUTEK.

REFERENCES

[1] X. D. Wu, A. Inam, M. S. Hegde, B. Wilkens, C. C. Chang, D. M. Hwang, L. Nazar, T. Venkatesan, S. Miura, S. Matsubara, Y. Miyasaka, N. Shohata, Appl. Phys. Lett., **54**, 754 (1989) and references therein.
[2] S.C. Witczak, Solid state electron., **35**, 345 (1992).
[3] C.-R. Cho, J.-H. Koh, A. Grishin, S. Abadei, S. Gevorgian, Appl. Phys. Lett. **76**, 1761 (2000).
[4] M. Ahtee, A. W.Hewat, Acta Cryst. **A34**, 309 (1978).
[5] Y. Iida, T. Shimura, J. Harada, S. Samata, Y. Matsushita, Surf. Sci., **258**, 235 (1991).
[6] I. Takahashi, S. Okita, N. Awaji, Y. Sugita, Physica B., **245**, 306 (1997).
[7] C.-R. Cho, Alex Grishin, Appl. Phys. Lett., **75**, 268 (1999).
[8] Y. Watanabe, Phys. Rev. B., **59**, 11257 (1999).
[9] S.-B. Xiong, S. Sakai, Appl. Phys. Lett., **75**, 1613 (1999).

HYDROSTATIC AND BIAXIAL STRAIN IN $Ba_xSr_{1-x}TiO_3$ FILMS GROWN BY PULSED LASER DEPOSITION

C. M. Carlson *, P. A. Parilla **, T. V. Rivkin **, J. D. Perkins **, D. S. Ginley **
*Department of Physics, University of Colorado, Boulder, CO 80309-0390
**National Renewable Energy Laboratory, 1617 Cole Blvd., Golden, CO 80401

ABSTRACT

We grew $Ba_xSr_{1-x}TiO_3$ (BST) films on MgO single crystal substrates by pulsed laser deposition (PLD). We report the in-plane (a) and out-of-plane (c) lattice parameters of BST films deposited in a range of O_2 deposition pressures [$P(O_2)$], as measured by asymmetric rocking curve diffraction. As $P(O_2)$ increases, the films' biaxial strain changes from compression ($a < c$), to cubic ($a = c$), and then to tension ($a > c$). Furthermore, both a and c are larger than the lattice constant for bulk BST of the same composition. This indicates the presence of a hydrostatic strain component in addition to the biaxial component. From the measured lattice parameters, we calculate the total residual strain in terms of biaxial and hydrostatic components. We also examine the effects of a post-deposition anneal. Characterizing residual strain and understanding its origin(s) are important since strain affects the dielectric properties of BST films and thereby the properties of devices which incorporate them.

INTRODUCTION

$Ba_xSr_{1-x}TiO_3$ (BST) films have been heavily investigated for their potential use in tunable microwave and radio-frequency (RF) devices [1,2]. The basic mechanism which lends BST, and other ferroelectric materials, to these applications is the decrease of their dielectric constant ($\varepsilon/\varepsilon_0$) with applied electric field, i.e., "tuning." The tuning effect can be used in different ways to make devices such as tunable resonators, tunable filters, delay lines, phase shifters, steerable antennas, etc.. The main materials issue is to maximize the tuning and minimize the dielectric loss. One of the film properties that has been shown to affect the dielectric properties is strain [3], either residual or externally applied. Therefore, the characterization of residual strain is important for understanding both its origin and its contribution to the overall dielectric and ferroelectric response of BST and other dielectric materials.

EXPERIMENT

Film Growth, Post-Deposition Annealing, and Typical Properties

The BST films were grown by PLD using a KrF ($\lambda = 248$ nm) laser running at 10 Hz and focused to a fluence of 3 J/cm^2 at the $Ba_{0.4}Sr_{0.6}TiO_3$ target. The MgO(100) substrates were rinsed in isopropyl alcohol and then attached with Ag paint to a heater. During growth, the heater was maintained at 775 °C and held 8.5 cm directly in front of and parallel to the target. Each growth occurred in a constant O_2 ambient between 40 and 250 mTorr. After deposition, the samples were cooled in ~400 Torr O_2. Post-deposition anneals were done at ~1100 °C in flowing O_2 at ambient pressure in a quartz tube furnace for 5 hours. The composition of the films was measured by Rutherford Backscattering Spectroscopy (RBS) using 2 MeV He ions. According to RBS, the deposition pressure [$P(O_2)$] had little or no effect on the films' (Ba,Sr) composition. The average composition of $Ba_xSr_{1-x}TiO_3$ was $x = 0.39 \pm 0.02$, which is consistent with the target

161

composition, $x = 0.40$. The thickness of the films was between 2700 and 3700 Å as measured by profilometry.

The as-deposited BST/MgO films are epitaxial and show high structural quality [1]. Figure 1 shows $\theta/2\theta$ and pole figure diffraction scans for a typical film. The $\theta/2\theta$ scan (Fig. 1a) shows only $(h00)$ peaks indicating a single out-of-plane texture, and the pole figure (Fig. 1b) confirms the in-plane orientation of the epitaxial film.

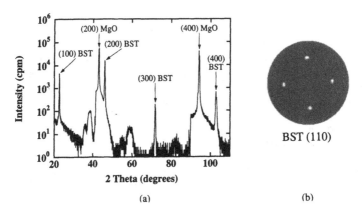

(a)   (b)

Fig. 1. (a) $\theta/2\theta$ and (b) pole figure x-ray diffraction scans for laser-ablated BST/MgO film showing single epitaxial orientation.

Annealing improves the crystalline perfection and the surface morphology [1]. The width (FWHM) of (200) rocking curves decreases (0.68° to 0.31° for $P(O_2) = 65$ mTorr film) after the anneal, as does that of (110) phi scans (1.15° to 0.91°). Annealing also decreases the surface roughness of the BST films as measured by atomic force microscopy (AFM). Figure 2 shows AFM images of an as-deposited (< 10 Å rms) and annealed film (3 Å rms) grown in $P(O_2)$ = 65 mTorr. After annealing the granular morphology is no longer visible.

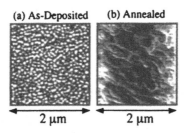

Fig. 2. AFM images of (a) as-deposited and (b) annealed BST/MgO films.

As described elsewhere [1], these films can have large dielectric constants of over 6000 with tuning greater than 65% under applied fields of 5.7 V/μm. Figure 3 shows the temperature dependence of $\varepsilon/\varepsilon_0$ for such a film. The anneal increases the maximum dielectric constant and shifts this maximum to lower temperatures. Our search for the cause of this shift lead us to consider the effects of strain and to make the measurements reported here.

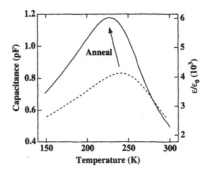

Fig. 3. The 1 MHz capacitance and $\varepsilon/\varepsilon_0$ vs. temperature for as-deposited (dashed line) and annealed (solid line) BST/MgO. The arrow indicates the change upon annealing.

Asymmetric Rocking Curve Diffraction

The in-plane (a) and out-of-plane (c) lattice parameters were measured by asymmetric rocking curve diffraction [4], which uses the splitting of the (224) Bragg reflections from the BST film and the MgO substrate. This robust technique accounts for many possible sources of error including substrate miscut, film tilt relative to the substrate, and the additional tilt of asymmetric planes due to tetragonal distortion. Asymmetric rocking curve diffraction is the technique of choice even for much more closely lattice matched systems such as III-V strained-layer superlattices. The measured splitting of the film and substrate peaks in the two geometries shown in Fig. 4 depends on two independent parameters, $\delta\theta$ (true Bragg splitting) and $\delta\phi$ (geometric splitting due to tetragonal distortion).

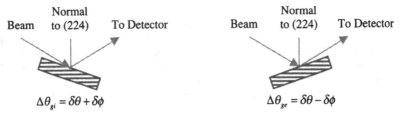

Fig. 4. Grazing incidence (gi) and grazing exit (ge) geometries for asymmetric rocking curve diffraction. The equations show dependence of splitting on $\delta\theta$ and $\delta\phi$ in either geometry.

With these measurements, $\delta\theta$ and $\delta\phi$ can be determined and then used to calculate both c and a as indicated in Eqs. 1 [4], where $\theta = \theta_{MgO} + \delta\theta$ and $\phi = \phi_{MgO} + \delta\phi$.

$$a_{BST} = \frac{\lambda}{2\sin\theta\sin\phi}\sqrt{h^2 + k^2}, \quad c_{BST} = \frac{l\lambda}{2\sin\theta\cos\phi} \tag{1}$$

Figure 5 shows the (224) rocking curves measured in the grazing incidence geometry for a BST/MgO sample grown in $P(O_2) = 85$ mTorr both as-deposited and after annealing. The Cu Kα peaks were fitted to a pseudo-Voigt function including constrained Kα_1 and Kα_2 components. The splitting ($\Delta\theta_{gi}$ in Fig. 5) is then used with that from the grazing exit scan to

163

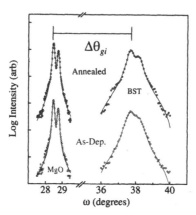

Fig. 5. Typical grazing incidence (224) rocking curves illustrating the splitting ($\Delta\theta_{gi}$) used in lattice parameter calculations. These rocking curves were made on samples grown in 85 mTorr.

determine the lattice parameters using Eqs. 1. A potential pitfall of this technique in our particular situation is that the entire $K\alpha_1$-$K\alpha_2$ doublet peak may not be measured due to the peak width and the relatively small area of our detector. If not corrected, the subsequent curve fitting gives erroneous peak positions. This was remedied by combining the rocking curves taken at several detector positions to ensure that the entire peak was measured.

RESULTS

Lattice Parameters

Using the method outlined above, we calculate the in-plane (a) and out-of-plane (c) lattice parameters of BST/MgO films grown in a range of $P(O_2)$ values (40–250 mTorr). Figure 6 gives these lattice parameters for both as-deposited and annealed films. As $P(O_2)$ increases, the as-deposited films go from biaxial compression ($a < c$), to cubic ($a = c$) at $P(O_2) = 85$ mTorr, and then to biaxial tension ($a > c$). A qualitatively similar dependence is found for the annealed films, except that the films originally deposited in $P(O_2) = 40$ mTorr are now cubic. Note that all lattice parameters are larger than that of bulk ceramic BST of the same composition. This indicates the presence of a hydrostatic strain component in addition to the biaxial component.

Hydrostatic and Biaxial Strain Components

The total residual strain can be decomposed into hydrostatic and biaxial components using the elastic compliance tensor. The hydrostatic component (U_{hyd}) affects all three axes of the unit cell identically, while the biaxial component (U_{bi}) affects the two in-plane axes (a & b) directly and the out-of-plane axis (c) indirectly by means of Poisson's ratio. The relation between this decomposition and the Cartesian one is

$$U_x = U_y \equiv U_{xy} = \frac{\Delta a}{a_0} = U_{hyd} + U_{bi}, \quad U_z = \frac{\Delta c}{a_0} = U_{hyd} + \left(\frac{2s_{12}}{s_{11} + s_{12}}\right)U_{bi}, \tag{2}$$

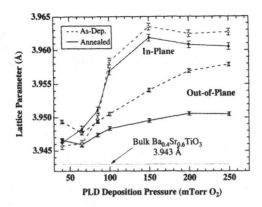

Fig. 6. In-plane and out-of-plane lattice parameters of BST/MgO films vs. deposition pressure.

where a_0 is the bulk lattice constant and the s_{ij} are elements of the elastic compliance tensor such that $s_{12} < 0$, $s_{11} + s_{12} > 0$, and $s_{11} + 2s_{12} > 0$. This decomposition helps in understanding the effects of each strain component since the potential sources of hydrostatic stress (interstitials, oxygen vacancies, etc.) are physically different from those of biaxial stress (substrate interactions, etc.).

Figure 7 shows the results of the strain decomposition along with the corresponding stress values (σ's), which are indicated by the right vertical axes. The values of s_{11} and s_{12} used in the calculations were estimated from those of BaTiO$_3$ and SrTiO$_3$ using Vegard's law [3]. Annealing reduces the hydrostatic component but increases the biaxial tension, which may be the influence of the substrate's larger lattice parameter. The dependence of both the biaxial and hydrostatic strain is consistent with the effects of energetic ion bombardment during growth, i.e., "ion peening [5]." This hypothesis will be explored in detail

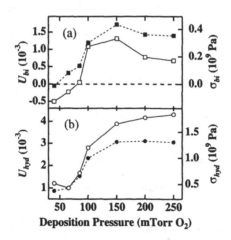

Fig. 7. The P(O$_2$)-dependence of (a) biaxial strain and stress, and (b) hydrostatic strain and stress. Solid and dashed lines denote as-deposited and annealed films respectively.

in a subsequent publication [6], but is supported by RBS data that indicates an increasing excess Ti concentration with increasing deposition pressure. This is consistent with increasing Ti interstitial defects, which hydrostatically expand the lattice. Oxygen vacancies are another source of hydrostatic strain [7], which may explain its observed reduction after annealing.

CONCLUSIONS

Using a robust x-ray diffraction technique, we have measured the in-plane and out-of-plane lattice parameters of BST films grown epitaxially on MgO substrates by PLD. From the lattice parameters, we then calculated the residual hydrostatic and biaxial strain in the films. We found that there is a deposition pressure at which the biaxial strain changes from compressive to tensile. By careful selection of the deposition pressure, it was possible to eliminate the biaxial strain in both the as-deposited and annealed films. Measurements of these films' dielectric properties, which are currently underway, will provide insight into the effects of residual strain.

ACKNOWLEDGMENTS

This work was done at NREL under DOE contract no. DE-AC36-99G010337. We wish to thank Spectral Solutions Inc. (SSI) for the use of their annealing furnace and microwave test equipment. We also thank Dr. Charles Barbour of Sandia National Labs for the RBS analysis.

REFERENCES

1. C. M. Carlson, T. V. Rivkin, P. A. Parilla, J. D. Perkins, D. S. Ginley, A. B. Kozyrev, V. N. Oshadchy and A. S. Pavlov, Appl. Phys. Lett. **76**, p. 1920 (2000).

2. C. M. Carlson, T. V. Rivkin, P. A. Parilla, J. D. Perkins, D. S. Ginley, A. B. Kozyrev, V. N. Oshadchy, A. S. Pavlov, A. Golovkov, M. Sugak, D. Kalinikos, L. C. Sengupta, L. Chiu, X. Zhang, Y. Zhu and S. Sengupta, to appear in *Materials Issues for Tunable RF and Microwave Electronics*, edited by Q. Jia, F. A. Miranda, D. E. Oates, and X. X. Xi (Mater. Res. Soc. Symp. Proc., Pittsburgh, PA 2000).

3. S. K. Streiffer, C. Basceri, C. B. Parker, S. E. Lash and A. I. Kingon, J. Appl. Phys. **86**, p. 4565 (1999).

4. D. K. Bowen and B. K. Tanner, *High Resolution X-ray Diffractometry and Topography*, Taylor & Francis, Bristol, PA, 1998.

5. D. W. Hoffman and J. A. Thornton, J. Vac. Sci. Technol. **20**, p. 355 (1982).

6. C. M. Carlson, P. A. Parilla, T. V. Rivkin, J. D. Perkins and D. S. Ginley, Submitted to Appl. Phys. Lett. (2000).

7. W. J. Kim, W. Chang, S. B. Qadri, J. M. Pond, S. W. Kirchoefer, D. B. Chrisey and J. S. Horwitz, Appl. Phys. Lett. **76**, p. 1185 (2000).

Light Scattering from Pulsed Laser Deposited BaBi$_4$Ti$_4$O$_{15}$ Thin Films

R.K. SONI[*][+], ANJU DIXIT[*], R. S. KATIYAR[*], A. PIGNOLET[**], K.M. SATYALAKSHMI[**], and D. HESSE[**]

*Physics Department, University of Puerto Rico, San Juan, Puerto Rico, PR 00931, USA
**Max-Planck-Institut für Mikrostrukturphysik, Weinberg 2, D-06120 Halle/Saale, Germany

ABSTRACT

Light scattering investigations are carried out on BaBi$_4$Ti$_4$O$_{15}$ (BBiT) which is a member of the Bi-layer structure ferroelectric oxide with n = 4. The BBiT thin films, thickness ~ 300 nm, were grown on epitaxial conducting LaNiO$_3$ electrodes on epitaxial buffer layers on (100) silicon by pulsed laser deposition. Micro-Raman measurements performed on these films reveal a sharp low-frequency mode at 51 cm^{-1} along with broad high-frequency modes corresponding to other lattice vibrations including TiO$_6$ octahedra. No temperature dependence of the low frequency mode is seen while a weak dependence of the broad high frequency vibrations are observed in the mixed oriented regions. Raman polarization carried out at room temperature indicates that the prominent modes have A$_{1g}$ and E$_g$ symmetries in the BaBi$_4$Ti$_4$O$_{15}$ thin films.

INTRODUCTION

The Bi-based layer-structured ferroelecetric oxides belong to the Aurivillius family and are described by the general formula $(Bi_2O_2)^{2+}(A_{n-1}B_nO_{3n+1})^{2-}$. The high fatigue resistance to polarization switching of these oxides is believed to be due to their unique crystal structure, where n perovskite-like oxygen octahedra are interleaved between $(Bi_2O_2)^{2+}$ layers [1]. BBiT undergoes ferroelectric phase transition from high temperature tetragonal phase to orthorhombic or puedo-orthorhombic phase at T$_c$ = 420°C [2]. The high temperature tetragonal phase is possibly J4/mmm point group while the point group of the low-temperature phase is not known. As the orthorhombic distortion is small, a puedo-tetragonal unit cell with lattice parameters of a$_t$ = 3 .86 Å and c$_t$ = 41.8 Å is commonly used to describe the structural and lattice dynamical properties of BBiT. Recent reports on morphology and structure investigations have shown that the pulsed laser deposited (PLD) films consisted of c-axis oriented as well as mixed (110), (100)- and (001) oriented regions with rectangular and equiaxed crystalline grains [3].

The dielectric and ferroelectric properties of single crystalline c-axis oriented Bi$_4$Ti$_3$O$_{12}$ (BiT) with n = 3 and BaBi$_4$Ti$_4$O$_{15}$ (BBiT) with n = 4 have been reported [3-6]. Single crystalline BBiT exhibits characteristic anisotropic dielectric behavior of layered perovskites with high dielectric constant in the ab-plane and low dielectric constant in the c-plane [6]. It has been shown that the macroscopic as well as microscopic ferreoelectric properties of a planer capacitor with BBiT thin film are extremely sensitive to the crystalline orientation of the film [3,4]. The regions of BBiT film having mixed a$_t$- and c$_t$- axis orientation show saturated ferroelectric hysteresis loops and resistance to polarization fatigue up to 10^8 cycles whereas regions with c$_t$-axis orientation and a smooth surface morphology show a linear polarization curve. Ferroelctric properties of Bi-layer oxides also display an interesting interplay between orientation of spontaneous polarization and the number n of pervoskite-like layers. The films with odd n, such as BiT (n=3), exhibits spontaneous polarization along c-axis while those with even n exhibit spontaneous polarization along a- or b-axis [5,7].

The lattice dynamical properties of Bi-layer structure oxides are not well understood. It is desirable to explore the light scattering properties of well oriented and characterized higher member (n>3) of the Bi-layer oxide family. Light scattering from single crystal BiT (n=3) has already been reported [8]. Here, we study micro Raman scattering in epitaxially grown BBiT with n=4 by pulsed laser deposition emphasizing the effect of film surface morphology, light polarization and temperature.

EXPERIMENTAL

The epitaxial 300 nm thick BBiT layer was grown by pulsed laser deposition (PLD) technique using 248 nm line of a KrF excimer laser at a pulse repetition rate of 5 Hz with a laser pulse energy density of 2 J/cm^2. The conducting 150 nm thick epitaxial LNO film, deposited on an epitaxial buffer layer stake consisting of CeO_2/YSZ on (100) oriented crystalline silicon, was used as a template to favor epitaxial growth of the BBiT film. Details of film deposition and its structural and electrical characterization have been described elsewhere [3].

The Raman measurements were carried out using a Jobin-Yvon T64000 Raman microprobe and detected by liquid nitrogen cooled CCD detector. The 514.5 nm radiation with power of 30 mW from a Coherent Innova 99 Ar$^+$ laser was focused to 2μm diameter area using a Raman microprobe with 80X eyepiece. Typical values of the slit widths were 200 μm for the entrance slit, 27 mm for the first intermediate slit, and 200 μm for the second intermediate slit. The spectral resolution of the Raman system with 1800 gr/mm grating and one inch CCD was less than 1 cm^{-1}. The peak position and full width at half maximum (FWHM) of Raman lines were determined by fitting a phonon function with appropriate bose factor using commercially available PeakFit program. A mini-cryostat from MMR based on Joule-Thomson effect was used for temperature measurements in the range 70-550 K with temperature stability better than 1K. A microscope compatible resistance heating stage was used for temperature above 550 K.

RESULTS

Fig.1 shows scanning electron micrograph of a BBiT film. The SEM images show high density of mostly rectangular-shaped microcrystallites that are arranged with long-axis along two orthogonal directions on the substrate plane. In some other regions of the film the crystallite density is smaller and the surface has a smoother morphology. Similar surface morphology was also seen on BBiT films examined by atomic force microscopy [3]. The AFM images also revealed that the crystallites are protruding out of the smooth surface from the region that inhabits rectangular-shaped crystallites. It was further inferred from the TEM cross-section (XTEM) images that the crystallites protruding out of the surface are a$_t$-oriented whereas the smooth surface regions are entirely consisting

Fig. 1 Scanning electron micrograph of a BBiT film showing rectangular crystallites arranged on the surface with long axis along two orthogonal directions.

of c-axis oriented crystallites. Furthermore, the XTEM images indicate that the BBiT films have bimodal sublayer structure. The lower sublayer, directly on the LNO electrode, consists entirely of c-axis oriented crystallites while the upper sublayer consists predominantly of a_t-axis oriented crystallite submerged in a c_t-axis oriented matrix.

Fig. 2 shows normalized Raman spectra taken from three different regions of the BBiT film, the normalization is done with the intensity of 521 cm^{-1} line of the silicon substrate. Apart from small intensity variation, the Raman spectra from these regions have common general features. These regions have different surface morphology as seen in the SEM images. From the surface with smoother surface morphology, the Raman scattering efficiency is larger, Fig 2 (c), than from those surfaces with high density of crystallites and rough surface morphology Fig 2 (a) and (b). For the psuedo-tetragonal symmetry (D_{4h}) the contribution to the light scattering in the Y(ZZ)Y configuration comes from phonons of A_{1g} symmetry while in the Y(XZ)Y configuration gives contribution from phonons of E_g symmetry.

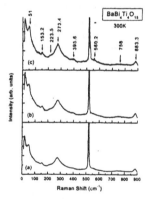

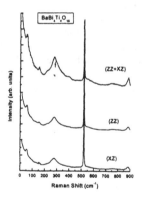

Fig.2 Normalized Raman spectra from a BBiT film taken from regions with smooth surface (a) to increasing surface roughness (b) and (c).

Fig. 3 Polarized Raman spectra from a BBiT film. The light propagation is parallel to the plane of the surface.

In Fig. 3 we shows the polarization dependence of Raman spectra from BBiT. The propagation direction of the light does not coincide with the crystal axis, therefore leakage of phonons of different symmetries is possible. The figure shows that the participating phonons primarily have A_{1g} and E_g symmetries.

The Raman lines are generally very broad with the exception of low frequency line at 51 cm^{-1}. Low frequency mode around 58 cm^{-1} has been observed in many Bi-layer perovskite oxides such as $SrBi_2Ta_2O_9$ (n=2), $Bi_4Ti_3O_{12}$ (n=3), and ($MBi_4Ti_4O_{15}$) (n=4), M= Ca, Pb, Sr and Ba [2,9]. Common to all these layered oxides is $(Bi_2O_2)^{2+}$ layer, the interleaved perovskite-like layers mass and chemical composition seem to have little effect on the low frequency mode. It is therefore conceivable to assign the low frequency mode to the vibration Bi-O in the $(Bi_2O_2)^{2+}$ layer. Comparing with bismuth based cuparate superconductors [10], which has a similar layer structure, the eigen vectors corresponding to the low frequency mode may arise from the bismuth displacements perpendicular to the $(Bi_2O_2)^{2+}$ layer with A_{1g} symmetry.

The main features of the spectrum above 150 cm^{-1} and located at about 275, 560 and 883 cm^{-1} arise from the vibration of TiO_6 octahedra. These bands exhibit a systematic shift with the number n of perovskite-like layers and on cation substitution with in the octrahedra in Bi-layered oxide materials. With increasing n, the modes corresponding to the vibration of oxygen atoms shift towards higher frequencies for an identical BO_6 octahedra. When the B site is replaced by a lighter mass cation, the frequency again shifts to higher values as a result of lower reduced mass for a given member of Bi-layer family. These trends and assignments of prominent A_{1g} and E_g modes are summarized in the Table-I. The symmetry representations customarily have been referred to the high temperature tetragonal phase. Based on these trends we assign higher frequency vibrational modes in BBiT.

The mode at 883 cm^{-1} is an A_{1g} mode due to symmetric stretching of TiO_6 octahedra. Corresponding situation for $Bi_4Ti_3O_{12}$ (n=3) single crystal is an expected low value of 845 cm^{-1}. The other modes at 273 cm^{-1} and 560 cm^{-1} are E_g modes involving apical oxygen atoms of the octahedra in the xy-plane and are two-fold degenerate. Their corresponding shift with respect to BiT (n=3) is smaller than A_{1g} modes. The E_g modes

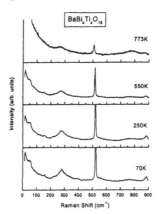

Fig. 4 Temperature dependent Raman spectra from a BBiT film.

170

are expected to show splitting due to orthorhombic distortion particularly at lower temperatures. All the observed Raman modes are broader compared to BiT owing to high structural and chemical disorder in BBiT.

Temperature dependent Raman spectra are shown in Fig. 4. The low frequency mode at 51 cm^{-1} is insensitive to the temperature variation. This is a common feature observed in many Bi-layered oxides even near the phase transition temperatures. Both A_{1g} and E_g modes exhibit frequency shift towards lower side with increasing temperature. Such a shift is expected from increased lattice constant at higher temperature. No clear splitting of the E_g mode at 273 cm^{-1} was observed though some structure around this mode is seen at 70K. The oscillator strength of A_{1g} and E_g modes decreases continuously while a broad feature at 750 cm^{-1} gains strength with temperature. At 773K, the diminishing strength of A_{1g} mode at 874 cm^{-1} and appearance of a well defined mode at 780 cm^{-1} indicates a possible onset of structural phase transition in BBiT.

Table-I Raman frequencies of prominent A_{1g} and E_g modes in Bi-layer oxides with n=2, 3 and 4 at room temperature.

Symmetry	Assignment	n = 2 $Bi_2SrTa_2O_9$ Ref. 11	n = 3 $Bi_4Ti_3O_{12}$ Ref. 8	n = 4 $BaBi_4Ti_4O_{15}$ Present work
A_{1g}	Bi z-axis vibration	59	---	51
E_g	Bi x(y)-axis vibration	130	---	---
E_g	B, x(y)-axis vibration	163	---	153
A_{1g}	B, z-axis vibration	210	224	223
E_g	O, x(y)-axis vibration	242	260	273
E_g	O, z-axis vibation	370	409	395
E_g	O, x(y)-axis vib.in phase	427	---	---
E_g	O,x(y)-axis vib.out of phase	516	530	560
A_{1g}	O, z-axis vib. in phase	600	611	---
A_{1g}	O,z-axis vib.out of phase	811	845	883

CONCLUSIONS

We have carried out Micro-Raman measurements on epitaxially grown $BaBi_4Ti_4O_{15}$ (BBiT) grown on epitaxial conducting $LaNiO_3$ electrodes on (100) silicon by pulsed laser deposition. The Raman spectrum reveals a sharp low-frequency mode at 50 cm^{-1} along with broad high-frequency modes mainly corresponding to the lattice vibrations of TiO_6 octahedra. Both c-axis oriented regions and a-axis oriented regions have broad Raman features that are insensitive to the crystallographic orientations. The polarization dependence shows that these features have A_{1g} and E_g symmetries. Tentative mode assignments were based on the trends in Raman spectrum from Bi-layer oxides with lower n values. No temperature dependence of low frequency modes is seen upto 773 K the mixed oriented regions.

ACKNOWLEDGMENTS

This work is supported by DAAG55-98-1-0012, DAAD 19-99-1-0362, and NSF-DMR-980175 grants.

+ Permanent address: Physics Department, Indian Institute of Technology, New Delhi 110016, India.

REFERENCES

1. E.C. Subarao, J. Phys. Chem. Solids **23**, 665 (1962).
2. S. Kojima, R Imaizumi, S. Hamazaki and M. Takashige, Jap J. of Appl. Phys. **33**, 5559 (1994).
3. K.M. Subalakshmi, M. Alexe, A. Pignolet, N.D. Zakharov, C. Harnagea, S. Senz and D. Hesse, Appl. Phys. Lett. **74**, 603 (1999).
4. C. Harnagea, A. Pignolet, M. Alex, K.M. Subalashmi, D. Hesse and U. Gosele, Jap. J.Appl. Phys. **38**, L1255 (1999).
5. R. Ramesh, A. Inam, W.K. Chan, B. Wilkens, K. Myens, K. Remschning, D.L. Hart, and J.M. Tarascon, Science **252**, 944 (1991).
6. S.K. Kim, M. Miyayama,, and H. Yanagida, J. Ceram. Soc. Jan. **102**, 722 (1994).
7. H. Tabata, M. Hamada, and T. Kawai, Mat. Res. Soc. Symp Proc. **401**, 73 (1996).
8. P.R. Graves, G. Hua, S. Myhra, and J.G. Thompson, J. of Solid State Chemistry **114**, 112 (1995).
9. M.P. Moret, R. Zallen, R.E. Newnham, P.C. Joshi, and S.B. Desu, Phys. Rev. B57, 5715 (1998).
10. Z.V. Popovic, C. Thomsen, M. Cardona, R. Liu, G. Stanisic, R. Kremer, and W. Konig, Sold. State Commun. **66**, 965 (1988).
11. R. Liu, Proc. of the American Physical Society Meeting 1997.

LATTICE DEFECTS IN EPITAXIAL Ba$_2$Bi$_4$Ti$_5$O$_{18}$ THIN FILMS GROWN BY PULSED LASER DEPOSITION ONTO LaNiO$_3$ BOTTOM ELECTRODES

N.D. ZAKHAROV, A.R. JAMES [+], A. PIGNOLET, S. SENZ, AND D. HESSE
Max-Planck-Institut für Mikrostrukturphysik, Weinberg 2, D-06120 Halle, Germany
[+] now with Materials Res. Lab., Penn State University, University Park, PA, USA

ABSTRACT

Epitaxial, ferroelectric Ba$_2$Bi$_4$Ti$_5$O$_{18}$ films grown on LaNiO$_3$/CeO$_2$/ZrO$_2$:Y$_2$O$_3$ epitaxial layers on Si(100) are investigated by cross-section high-resolution transmission electron microscopy (HRTEM). The films are perfectly oriented and consist of well-developed grains of rectangular shape. The grain boundaries are strained and contain many defects, especially a new type of defect, which can be described as a staircase formed by repeated lattice shifts of $\Delta \approx c/12 \approx$ 4.2 Å in the [001] direction. This repeated shift results in seemingly bent ribbons of stacked Bi$_2$O$_2$ planes, involving, however, individual Bi$_2$O$_2$ planes which remain strongly parallel to the (001) plane. These defects contain an excess of bismuth. Other defects found in the grain interior include mistakes in the stacking sequence originating from the presence of single, well-oriented, non-stoichiometric layers intergrown with the stoichiometric Ba$_2$Bi$_4$Ti$_5$O$_{18}$ film matrix.

INTRODUCTION

Ferroelectric thin films of bismuth-layer compounds like SrBi$_2$Ta$_2$O$_9$ gain more and more importance in view of their relevance to the development of non-volatile ferroelectric random access memories [1]. They are advantageous over usual perovskite materials in that they do not suffer from fatigue [2,3]. Though memories made of polycrystalline SBT films are already in use, the integration of Bi-layer type ferroelectric films into the silicon technology remains a challenge. The growth of *epitaxial* Bi-layer films on Si substrates is significant, because they allow fundamental and applied studies to be performed in order to probe the properties of these unconventional materials. As part of our studies of the structure-property relationships of various Bi-layer type ferroelectric films, like Bi$_4$Ti$_3$O$_{12}$, SrBi$_2$Ta$_2$O$_9$, or BaBi$_4$Ti$_4$O$_{15}$ [4-6], we have prepared epitaxial Ba$_2$Bi$_4$Ti$_5$O$_{18}$ films on LaNiO$_3$ electrodes on CeO$_2$/YSZ-buffered Si(100) and studied their ferroelectric and dielectric properties [7,8]. Here we report on cross-section HRTEM investigations of these films, with a strong emphasis on the lattice defects found.

Ferroelectricity in the compound Ba$_2$Bi$_4$Ti$_5$O$_{18}$ had been reported as early as in 1962 [9]. According to this work, the compound crystallizes in the tetragonal space group *I4/mmm* with the lattice parameters $a_t = 3.88$ Å and $c_t = 50.3$ Å. Though the ferroelectric phase might indeed be orthorhombic, here we are keeping to that structure. To our knowledge, there are no reports on epitaxial Ba$_2$Bi$_4$Ti$_5$O$_{18}$ films in the literature, except preliminary information given in our two recent papers [7,8]. HRTEM investigations of lattice defects in bismuth-layer type perovskites have been performed rather rarely. Suzuki et al. [10] revealed twin domains and wavy c/6 translational boundaries in (001)- and (116)-oriented SrBi$_2$Ta$_2$O$_9$ films. In our Ba$_2$Bi$_4$Ti$_5$O$_{18}$ films, similar types of defects involving shifts by $\Delta \approx c_t/12$ along [001] are present, as described below.

EXPERIMENTAL

The YSZ and CeO$_2$ buffer layers, the LaNiO$_3$ (LNO) electrode layer, and the ferroelectric Ba$_2$Bi$_4$Ti$_5$O$_{18}$ thin film were all deposited by pulsed laser deposition (PLD) in a UHV system.

173

All layers were deposited in one single run without breaking the vacuum. A KrF excimer laser ($\lambda = 248$ nm) was used at a repetition rate of 5 or 10 Hz and pulse energy densities on the target around 3 J/cm^2. Deposition temperatures were 700 °C (YSZ) and 675 °C (CeO$_2$, LNO, and Ba$_2$Bi$_4$Ti$_5$O$_{18}$), and the depositions were performed in an oxygen atmosphere of 10^{-2} mTorr for YSZ, 300 mTorr for LNO and 100 mTorr for CeO$_2$ and Ba$_2$Bi$_4$Ti$_5$O$_{18}$. Film thicknesses were 110 nm (YSZ), 70 nm (CeO$_2$), 170 nm (LNO), and 180 nm (Ba$_2$Bi$_4$Ti$_5$O$_{18}$), respectively. Further details of the deposition procedure can be found in Refs. [7,8].

After deposition, the samples were cut into several pieces used for the electrical measurements [7], X-ray diffraction, scanning electron microscopy, and scanning force microscopy analyses [8], and for the present HRTEM investigations, respectively. Cross-section samples were prepared along (100)$_t$ and (110)$_t$ planes using the method of Ref. [11], and the HRTEM investigations were performed in a Jeol 4000 EX microscope. (Index t indicates the tetragonal description of the unit cell of Ba$_2$Bi$_4$Ti$_5$O$_{18}$).

RESULTS

Morphology and orientation of the films

SEM images revealed that the Ba$_2$Bi$_4$Ti$_5$O$_{18}$ films consist of grains of almost rectangular shape ('tiles') with lateral dimensions between 0.2 and 0.5 µm. The edges of the tiles were all parallel to one of two mutually perpendicular orientations indicating the epitaxy [7,8]. TEM plan-view images (not shown here) revealed that the tile boundaries are strained and contain many defects. By X-ray diffraction Θ–2Θ scans and Φ–Ψ pole figures the perfect epitaxy of all the layers of the overall film system was proven [7,8]. From these investigations, the following orientation relationships were deduced (index 'pc' indicates the pseudo-cubic indexing of LNO):

Ba$_2$Bi$_4$Ti$_5$O$_{18}$ (001) || LNO (100)$_{pc}$ || CeO$_2$ (100) || YSZ (100) || Si(100);
Ba$_2$Bi$_4$Ti$_5$O$_{18}$ [100]$_t$ || LNO [100]$_{pc}$ || CeO$_2$ [110] || YSZ [110] || Si[110].

Fig. 1 shows shows the entire Si/YSZ/CeO$_2$/LNO/Ba$_2$Bi$_4$Ti$_5$O$_{18}$ film system, revealing that all the interfaces are plane and that the individual films of the system have homogeneous thicknesses and morphologies. While the LNO film consists of very narrow columns, the Ba$_2$Bi$_4$Ti$_5$O$_{18}$ film reveals the large tiles known from the SEM images. The tile bounda-

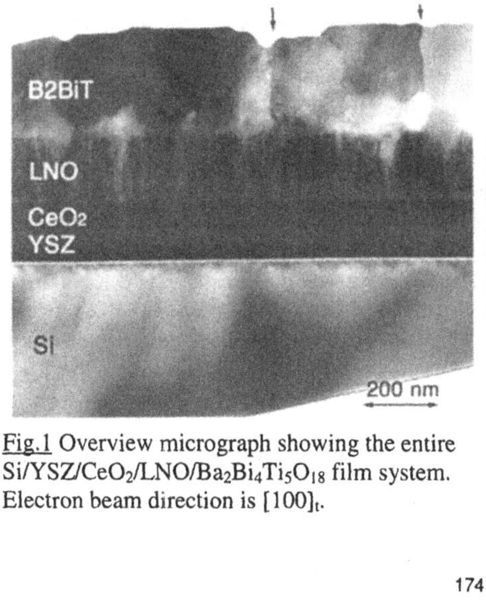

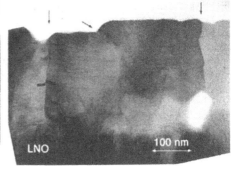

Fig.1 Overview micrograph showing the entire Si/YSZ/CeO$_2$/LNO/Ba$_2$Bi$_4$Ti$_5$O$_{18}$ film system. Electron beam direction is [100]$_t$.

Fig. 2 Cross-section HRTEM overview micrograph of one tile. Horizontal lines represent the Bi$_2$O$_2$ layers of Ba$_2$Bi$_4$Ti$_5$O$_{18}$.

ries (see arrows) are rather flat, extend from the bottom to the top of the $Ba_2Bi_4Ti_5O_{18}$ film and most of them are approximately perpendicular to the film plane.

Defects at the tile boundaries

Fig. 2 (see previous page) shows an enlarged view of the tile boundaries shown in Fig. 1. In the tiles, a pattern of almost horizontal dark and bright lines is visible, with a separation of 2.5 nm. This separation obviously corresponds to the distance of $\frac{1}{2} c_t = 25.15$ Å between two Bi_2O_2 layers in the unit cell. Due to the overall epitaxial (001) orientation of the $Ba_2Bi_4Ti_5O_{18}$ film, the Bi_2O_2 layers are parallel to the (001) plane and thus extend horizontally in the tile interior. Near the tile boundary, however, the Bi_2O_2 layers seem to be bent upwards, as indicated by a pair of curved black ink lines on each side of the left tile boundary. Fig.3 shows a HRTEM micrograph from an area near the two curved ink lines of Fig. 2. Here, the Bi_2O_2 layers manifest themselves as sharp black lines running parallel to the horizontal edges of the image and being embedded in a bright surrounding, which results in a broad bright ribbon with a white/black/white/black/white fine structure. Near the left and right rims of the image, these bright ribbons run parallel to the

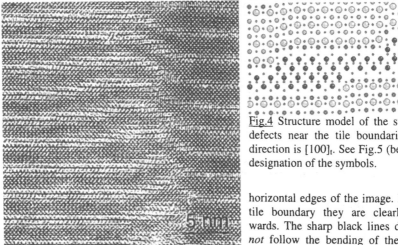

Fig.3 Detail of Fig. 2. The Bi_2O_2 planes, visualized as sharp black lines, remain strongly parallel to the (001) plane, even in the bent regions of the bright ribbons.

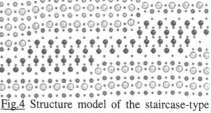

Fig.4 Structure model of the staircase-type defects near the tile boundaries. Viewing direction is $[100]_t$. See Fig.5 (below) for the designation of the symbols.

horizontal edges of the image. But near the tile boundary they are clearly bent upwards. The sharp black lines do, however, *not* follow the bending of the bright ribbons, but rather keep running horizontally. Obviously, the Bi_2O_2 layers on the atomic scale *remain parallel* to the overall (001) crystal plane even in those regions, where the bright ribbons bend upwards. This apparent bending of the ribbons is caused by a repeatedly occurring small shift of the Bi_2O_2 layers in the [001] direction, forming an overall staircase-like pattern. The magnitude of the shift is close to 4.2 Å $= c_t/12$. As a consequence of the shift of the Bi_2O_2 layers, the perovskite blocks are also shifted in the [001] direction. Strictly speaking the term 'shift' is not correct, because no discrete shift along a well-defined habit plane occurs. Rather two or three adjacent 'steps' of the staircase are overlapping. A model of this staircase-like overlap was derived from the structure model of $Ba_2Bi_4Ti_5O_{18}$ (shown in Fig. 5, below) and is displayed in Fig. 4. The locally occurring overlap of two or more *adjacent* Bi_2O_2 layers seen in Fig. 4 is equivalent to a local bismuth excess, *i.e.* it indicates a local deviation of the composition from the overall stoichiometry of the $Ba_2Bi_4Ti_5O_{18}$ film. Hence, the tile boundaries are bismuth-rich.

Structure images along [100] and [110] directions

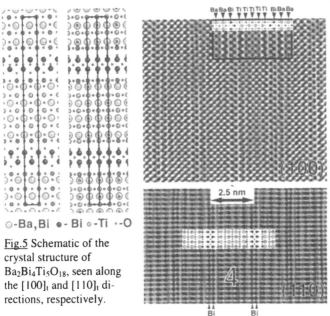

Fig.6 HRTEM structure images (noise-filtered) of the $Ba_2Bi_4Ti_5O_{18}$ film, seen along the $[100]_t$ and $[110]_t$ directions, respectively. For the inset, see the text.

O-Ba,Bi •-Bi o-Ti ··-O

Fig.5 Schematic of the crystal structure of $Ba_2Bi_4Ti_5O_{18}$, seen along the $[100]_t$ and $[110]_t$ directions, respectively.

Fig. 5 shows the crystal structure of $Ba_2Bi_4Ti_5O_{18}$, seen along the $[100]_t$ and $[110]_t$ directions, respectively. This schematic has been calculated using the atom positions given in Ref. [9]. Four perovskite units with the overall composition $([Ba_2Bi_2]Ti_5 O_{16})^{--}$ are stacked between two $(Bi_2O_2)^{++}$ double layers. The latter extend along the (001) planes of the crystal structure with a separation of ½ $c_t = 25.15$ Å. Experimental structure images along $[100]_t$ and $[110]_t$ are shown in Fig. 6, together with inserted structure images and a simulated image (insert on top of the upper image). Note that Fig. 6 is rotated by 90° with respect to Fig. 5. There is a perfect correspondence between the simulated and the experimental images.

Defects in the interior of the tiles

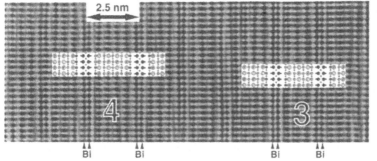

Fig.7 HRTEM cross-section image of a single layer with only three perovskite blocks between the Bi_2O_2 layers inserted into the $Ba_2Bi_4Ti_5O_{18}$ film matrix. Beam direction is $[110]_t$.

TEM plan-view and cross-section overview images (not shown) revealed a much lower density of lattice defects inside the tiles compared to the tile boundaries. The prevailing type of defects inside the tiles were mistakes in the stacking sequence originating from the intergrowth of single, well-oriented, non-stoichiometric layers within the stoichiometric $Ba_2Bi_4Ti_5O_{18}$ film matrix. In particular, layers having only three perovskite blocks (instead of four) between the Bi_2O_2 layers were found. This is equivalent to the composition $BaBi_4Ti_4O_{15}$. This phase is also ferroelectric, and epitaxial films of it have been grown as reported in Ref. [4]. On the right of Fig. 7 such a single $BaBi_4Ti_4O_{15}$ layer is shown embedded in the $Ba_2Bi_4Ti_5O_{18}$ surrounding. The inserted computed structure schematics of $Ba_2Bi_4Ti_5O_{18}$ (left) and $BaBi_4Ti_4O_{15}$ (right) demonstrate that the single layer is characterized by three (instead of four) perovskite blocks.

In those places of the lattice, where the intergrown layers end, characteristic transition structures form. An example is shown in Fig. 8. Here, layers with three perovskite blocks, designated by '3' and located left and right outside the image, transform into a $Ba_2Bi_4Ti_5O_{18}$ layer having four perovskite blocks, designated by '4'. The step-like transition region is characterized by a structure (Fig. 9), which is most similar to the individual steps of the staircase-like defects in the tile boundaries as already shown in Figs. 3 and 4. Again, a bismuth-rich region characterized by a shift of $\Delta \approx c_t/12$ along [001] is formed.

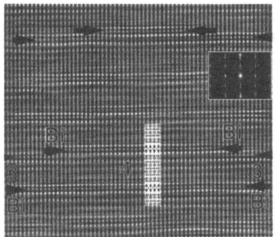

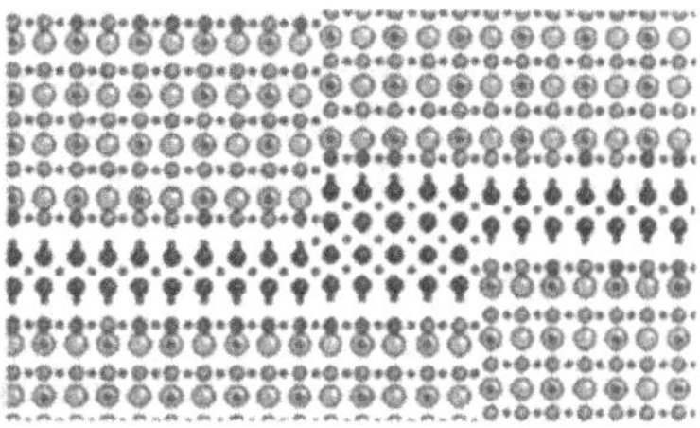

Fig.9 Schematic of the transition region between layers of different number of perovskite blocks between the Bi_2O_2 layers. Electron beam direction is $[1\bar{1}0]_t$. For the meaning of the symbols, see Fig.5.

Fig.8 Transition regions between $BaBi_4Ti_4O_{15}$ and $Ba_2Bi_4Ti_5O_{18}$ layers seen along the $[1\bar{1}0]_t$ direction. Arrows indicate the Bi_2O_2 layers.

DISCUSSION AND CONCLUSIONS

The observed defects are all characterized by mistakes in the stacking sequence corresponding to lattice shifts along the [001] direction by an amount of $\Delta \approx c_t/12$. In this respect they are similar to the $c/6$ translational boundaries observed by Suzuki et al. in epitaxial $SrBi_2Ta_2O_9$ films [10]. However, the schematic Figures 4 and 9 demonstrate that, at least in our case, these defects are carrying an excess of bismuth. In particular the tile boundaries seem to be Bi-rich. What can be the origin of this Bi excess?

It is known that an (accidental or adjusted at-will) excess of Bi during the growth of a Bi-containing complex oxide thin film is either desorbing from the surface of the growing film, or is being driven towards the final film surface by the moving free surface of the film. This behaviour of excess Bi is a consequence of its high volatility, and it is known from the growth of high-T_C superconducting Bi-Sr-Ca-Cu-O thin films. Though PLD is generally considered to provide a stoichiometric transfer of the target to the growing thin film, a Bi excess may have occurred due to non-optimal deposition parameters.

Considering the facts that our $Ba_2Bi_4Ti_5O_{18}$ films consist of individual rectangular-shaped, epitaxial grains (tiles) and that these grains are in fact columns extending from the bottom to the top of the film, it is reasonable to assume that during an early stage of film growth these grains were separated islands, *i.e.* islands that were not in contact with each other. If now a Bi ecxess was indeed present during film growth, it may have been driven to the surface of the growing islands, *i.e.* both to their upper surface *and to their side walls.* Consequently, under the Bi-rich conditions, the defects forming during later coalescence at the tile boundaries have not only to accommodate different stacking sequences, originating from the individually grown separate islands, but have also to comply with the bismuth excess. The staircase-like defects of Figs. 3 and 4 are obviously an answer to these two very different requirements: They are able to accommodate differences in the layer stacking sequence between two former islands, and simultaneously they bind the excess bismuth by incorporating it into the defect structure, keeping the perfect epitaxial orientation.

The intergrowth of layers having the "wrong" number of perovskite blocks between the Bi_2O_2 layers is a phenomenon already known from epitaxial films of the high-T_C superconductors of the Bi-Sr-Ca-Cu-O system, *cf.*, e.g., Ref. [12]. The transition regions between layers of different number of perovskite blocks to our knowledge have, however, not been visualized before.

REFERENCES

1. J.F.Scott, *Ferroelectric Memories*, Springer, Heidelberg, 2000.

2. Symetrix Corporation, International Patent H01L27/115, 21/320529/92 (1992).

3. C.A. Paz de Araujo, J.D. Cuchiaro, L.D. McMillan, M.C. Scott, and J.F. Scott, Nature **374**, 627 (1995).

4. K.M. Satyalakshmi, M. Alexe, A. Pignolet, N.D. Zakharov, C. Harnagea, S. Senz, and D. Hesse, Appl.Phys.Lett. **74**, 603 (1999).

5. A. Pignolet, C. Schäfer, K.M. Satyalakshmi, C. Harnagea, D. Hesse, and U. Gösele, Appl.Phys. A **70**, 283 (2000).

6. C. Harnagea, A. Pignolet, M. Alexe, D. Hesse, and U. Gösele, Appl.Phys. A **70**, 261 (2000).

7. A.R. James, A. Pignolet, D. Hesse, and U. Gösele, J.Appl.Phys. **87**, 2825 (2000).

8. A.R. James, A. Pignolet, S. Senz, N.D. Zakharov, and D. Hesse, Solid State Commun. **114**, 249 (2000).

9. B. Aurivillius and P.H. Fang, Phys.Rev. **126**, 893 (1962).

10. T. Suzuki, Y. Nishi, M. Fujimoto, K. Ishikawa, and H. Funakubo, Jpn.J.Appl.Phys. **38**, L1261 and L1265 (1999).

11. C. Traeholt, J.G. Wen, V. Svetchnikov, A. Delsing, and H.W. Zandbergen, Physica C **206**, 318 (1993).

12. N.D. Zakharov, D. Hesse, J. Auge, H.-G. Roskos, H. Kurz, H. Hoffschulz, J. Dreßen, H. Stahl, and G. Güntherodt, J.Mater.Res. **11**, 2416 (1996).

HYDROTHERMAL PREPARATION OF Ba(Ti,Zr)O₃ THIN FILMS FROM Ti-Zr METALLIC ALLOYS ON SILICON SUBSTRATE

Chang-Tai XIA, V. M. FUENZALIDA, and R. A. ZARATE
Universidad de Chile, Facultad de Ciencias Físicas y Matemáticas
Departamento de Física, Casilla 487-3, Santiago, Chile. cxia@cec.uchile.cl

ABSTRACT

Ba(Ti,Zr)O₃ thin films were grown hydrothermally on silicon substrates coated with sputtered a Ti-34 at.%Zr metallic alloy thin film. To ensure the formation of Ba(Ti,Zr)O₃ under the hydrothermal conditions at 150 °C, the concentration of the Ba(OH)₂ had to be greater than 0.25 M. Preliminary capacitance measurement revealed a dielectric constant of 200 in Ba(Ti,Zr)O₃ films of approximately 320 nm. The formation mechanism is discussed.

INTRODUCTION

The processing of electronic ceramics is a rapidly evolving area of research driven by a variety of applications such as nonvolatile memories, dynamic random access memories, electro-optic and electro-mechanic devices, among others [1-3]. The high dielectric constant of cubic (Ba,Sr)TiO₃ makes it a prospective material for applications where no ferroelecticity is required, e.g., volatile memories and integrated capacitors [1]. A recent study by Hoffmann and Waser [4] reported that Ba(Ti,Zr)O₃ is a potentially better candidate as a high permittivity dielectric for future high-density DRAMs than the widely studied (Ba,Sr)TiO₃.

Hydrothermal and hydrothermal-electrochemical methods, which were first developed by Yoshimura and coworkers [5,6] in 1989, are increasingly being used for the preparation of ceramic thin films [7-12]. These methods offer the advantages of simple instrumentation, high film crystallinity, low preparation temperature, and high purity. Moreover, the fabrication of the BaTiO₃ microstructures by the combination of hydrothermal and lift-off techniques has been reported [13].

EXPERIMENTAL PROCEDURE

The titanium-zirconium alloy film was deposited onto n-type (100) silicon wafer polished on one side under a sputtering pressure using DC sputtering with 1 Pa Ar under a DC mode (CrC-150 Sputtering system, Plasma Sciences Inc., Virginia, USA). The prepared structures were of the form Pt(500nm)\W-Ti(5nm)\Si(100)\W-Ti(5nm)\Pt(500nm)\Ti-Zr, where the W-Ti acted as an adherence layer. The alloy sources with a purity of 99.5 % were purchased from Pure Tech Inc. (New York, USA).

The alloy-coated silicon substrates were hydrothermally reacted in a 450-ml stainless-steel autoclave (Büchi AG, Switzerland). The hydrothermal treatment consisted in submerging the substrate in a 150 ml Ba(OH)₂ solution in a 300-ml Tefelon beaker, prepared from high purity Ba(OH)₂·8H₂O with a strontium content below 0.024 % (Solvay Bario e Derivati S.p.A, Italy). The autoclave was sealed and heated up to the working temperature of 150 °C in 40 min., where the system was maintained for a given period of time. In order to study the evolution of the morphology at the early growth stages, a few samples were kept over the solution during the heating up and cooling down periods. This was performed by submerging the substrate into the solution only when the working temperature of 150 °C was achieved, and pulling it out of the

179

solution after the hydrothermal treatment, using a Pt string attached to the sample holder. After the system was cooled, the samples were rinsed in boiling deionized water, acetone, and dried by blowing N_2.

The products were characterized by x-ray powder diffraction using CuKα radiation at 40 kV and 30 mA and a scan rate of 2θ 1.2 °/min(Siemens D5000). The film morphology was characterized using a JEOL JSM-25SII scanning electron microscope (SEM). Film thickness and rugosity were measured with a TENCOR Alpha step 500 profile meter. To facilitate the measurement steps were fabricated during the sputtering of Ti-Zr onto the silicon substrates.

The electrical measurements were performed on metal/insulator/metal (MIM) structures fabricated by depositing an array of 2.0-mm-diameter Au spots through a shadow mask by evaporation onto the Ba(Ti,Zr)O₃ film surface (top electrode), whereas the Pt under the film was the bottom electrode. The capacitance measurements were performed at room temperature using an Impedance Analyzer (Keithley 590 CV).

RESULTS AND DISCUSSION

Phase Identification

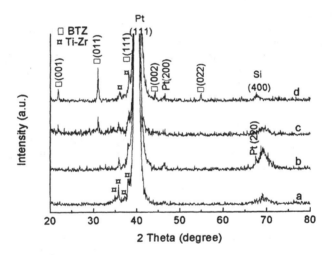

Fig. 1. XRD patterns of the Ti-Zr alloy coated on silicon substrates after hydrothermal treatments in Ba(OH)₂ aqueous solutions: (a) untreated; (b) 0.25 M, 8 h; (c) 0.5 M, 2 h; (d) 0.5 M, 4 h.

After hydrothermal treatments the smooth and mirrorlike surface of the films exhibited colors varying from slight yellow, blue, and purple, to dark-gray as the treatment progressed in time. Figure 1 shows the XRD patterns of the samples after different hydrothermal treatments. Cubic crystalline Ba(Ti,Zr)O₃ films with no preferred orientations formed after the hydrothermal treatment. At lower Ba(OH)₂ concentration (0.25 M), even with 8 hours treatment, the film was only slightly crystallized, as shown in the curve b where only the (011) reflection could be detected in the XRD pattern. This is consistent with the results of Alvarez et. al. [14] reporting that the Ti-Zr alloy film treated hydrothermally in 0.25 M Ba(OH)₂ at 150 °C exhibited only partial reaction.

Using a higher Ba(OH)$_2$ concentration of 1 M, a relatively short reaction time is sufficient to grow a highly crystallized film. However, in this case, the film is more prone to be contaminated with BaCO$_3$, which is the product from the reaction of the residual solution on the film and atmospheric CO$_2$. A concentration of 0.5 M seems to be appropriate for the hydrothermal treatment.

Microstructure Evolution

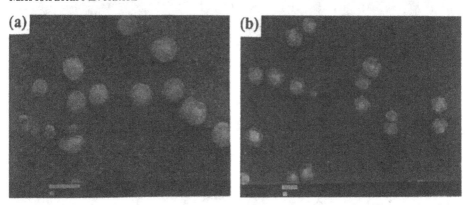

Fig.2. Scanning electron micrographs of the films hydrothermally prepared from 0.25 M Ba(OH)$_2$ at 150 °C for different treatment times: (a) 2 h; (b) 8 h. White bar = 1 μm.

Figure 2 shows the scanning electron micrographs of the films prepared in 0.25 M Ba(OH)$_2$ with different treatment times. The reaction is only slight at this dilute concentration, and increasing the treatment time has no significant influence neither on the particle morphology nor on the number of nuclei. This is consistent with the XRD results, and also with the well-known fact that the formation of BaTiO$_3$ at low temperature requires highly alkaline solutions (pH≈14), as indicated by the phase stability diagram of the Ba-Ti-H$_2$O system [11,15] and results from other authors [16]. The calculated pH value of a 0.25 M Ba(OH)$_2$ solution is about 13.7, assuming that the Ba(OH)$_2$ is fully disassociated. The actual value must be a bit smaller due to the fact that additional deionized water (about 10 ml) was added between the autoclave wall and the Teflon beaker to ensure a good thermal conductivity between the autoclave wall and the Teflon beaker. During the experiment and due to the different vapor pressures, some of this water migrates to the solution. It should also be pointed out that the pH of the alkaline solutions typically decreases with temperature due to the temperature dependence of the dissociation constant of water. The actual value in higher temperatures can be calculated from Debye-Hückel Theory [17].

Figure 3 shows the scanning electron micrographs of the films prepared in 0.5 M Ba(OH)$_2$ with different treatment times, depicting the early evolution of the Ba(Ti,Zr)O$_3$ nuclei (Fig. 3a, 5 min). After 2 hours of reaction, there are enough Ba(Ti,Zr)O$_3$ nuclei to completely cover the surface (Fig. 3b). However, further interaction between the substrate and the solution is still occurring. This is the dissolution-recrystallization process that generally occurring during the hydrothermal preparation of ceramic powders [18,19]. This can be deduced from the observation of the dissolution (Fig. 3c) and the recrystallization (Fig. 3d) of the Ba(Ti,Zr)O$_3$ crystallites.

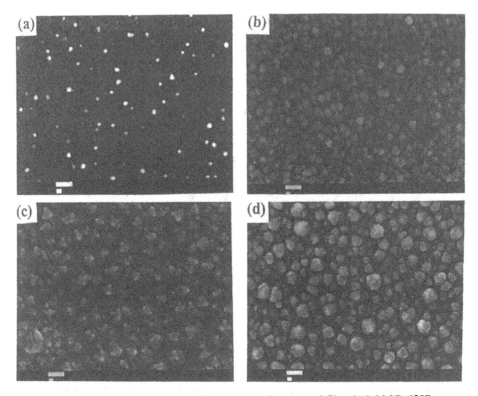

Fig. 3. Scanning electron micrographs of hydrothermally prepared films in 0.5 M Ba(OH)$_2$ at 150 °C for different duration: (a) 5 min; (b) 2 h; (c) 3.5 h; (d) 4 h. White bar = 1 μm.

Thickness Evolution

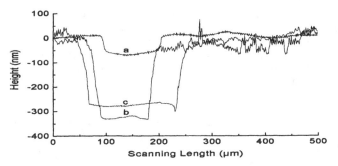

Fig. 4. Scan profiles around fabricated steps of Ti-Zr alloy coated on silicon substrate (a) before and after hydrothermal treatment in 0.5 M Ba(OH)$_2$ at 150 °C for (b) 4 h and (c) 8 h.

Figure 4 shows scan profiles around the fabricated steps of Ti-Zr coated on a silicon substrate before and after the hydrothermal treatments. The untreated Ti-Zr film was of ca. 70 nm thick with an RMS roughness ca. 5 nm, the film after 4 hours treatment in 0.5 M Ba(OH)$_2$ at 150 °C was of ca.320 nm thick with a RMS roughness ca.30 nm, while the film after 8 hours treatment under the same condition was of ca. 280 nm thick with a RMS roughness ca. 18 nm. The thickness and roughness decreased in a film treated for 4 hours to one treated for 8 hours could be a result of further dissolution-recrystallization as has been mentioned in the former section.

Electrical Properties

Preliminary electrical measurement revealed a dielectric constant of ca. 200 for the ca. 320 nm hydrothermal Ba(Ti,Zr)O$_3$ film fabricated in 0.5 M Ba(OH)$_2$ at 150 °C for 4 h. One reason for the relatively low dielectric constant may be the incompleteness of the reaction as indicated by XRD (Fig.1) and by the AES depth profile, which will be detailed elsewhere [20]. The other reason should be the unavoidable incorporation of water or OH$^-$ in the film, as observed in hydrothermal BaTiO$_3$ films [21,22]. Several groups reported that a proper annealing could improve the electrical properties of the latter [23-25].

Growth Mechanism

The kinetic studies necessary to determine the rate-limiting step(s) specific to the formation of Ba(Zr,Ti)O$_3$ films have not yet been performed. However, based on a knowledge of the corrosion behavior of titanium [26,27] and zirconium [28], the hydrothermal synthesis of BaTiO$_3$ [19,29,30], BaZrO$_3$ [8], and the current work, a film formation mechanism is proposed that accommodates both the literature and this data. It is proposed that the formation of Ba(Ti,Zr)O$_3$ films is controlled by three distinct, but interrelated, processes:
(1) The rapid formation of either a (Ti, Zr) oxide or a (Ti, Zr) hydrous oxide film, whose crystallinity and composition depend on the alloy composition, the solution chemistry, the atmosphere, and the reaction temperature.
(2) The nucleation of fine barium zirconate titanate crystallites on the surface of the amorphous, metastable Ti,Zr oxide film, under highly alkaline conditions, and their further growth to constitute an insulating, polycrystalline film of barium zirconate titanate.
(3) Concurrent with 2, a dissolution-recrystalization process should occur simultaneously. However, at the early stage of the reaction, the nucleation and growth is the controlling step, while at the later stage of the reaction, the dissolution is the controlling step.

SUMMARY

It has been demonstrated that the Ba(Ti,Zr)O$_3$ thin films could be prepared from the Ti-Zr metallic alloys on the silicon substrate. For the Ti-34 at.% Zr alloy, a 0.25 M Ba(OH)$_2$ aqueous solution was found to be too dilute to ensure the formation of Ba(Ti,Zr)O$_3$ at 150 °C. The relatively low dielectric constant (ca. 200) of a ca. 320 nm Ba(Ti,Zr)O$_3$ fabricated in 0.5 M Ba(OH)$_2$ for 4 hours has been mainly attributed to the incompleteness of the reaction and the incorporation of OH$^-$ and H$_2$O in the film.

The formation mechanism was proposed as (1) Formation of the (Ti,Zr) oxide or hydrous oxide film; (2) nucleation and growth of the Ba(Ti,Zr)O$_3$ crystallites; and (3) dissolution-recrystallization of the Ba(Ti,Zr)O$_3$ crystallites.

ACKNOWLEDGMENTS

Work financed by Chilean Government under awards FONDECYT 3970012 and 1970310. The analysis facilities were provided by Fundación Andes through grants c-12510 and c-12776. The high purity $Ba(OH)_2$ is a courtesy of Solvay-Sabed, Italy. We are also grateful to Prof. R.E. Avila of the Comisión Chilena de Energía Nuclear for the use of the electrical measurement facilities. One of us (C.-T. Xia) greatly appreciate the fruitful discussions with Dr. M.E. Pilleux and Dr. J.G. Lisoni.

REFERENCES

1. O. Auciello, J. F. Scott, and R. Ramesh, Physics Today **51,** 22 (1998).
2. J. F. Scott and C. A. Paz de Araujo, Science **246,** 1400 (1989).
3. G. H. Haertling, J. Am. Ceram. Soc. **82,** 797 (1999).
4. S. Hoffmann and R. M. Waser, Integrated Ferroelectrics **17,** 141 (1997)
5. M. Yoshimura, S.-E. Yoo, M. Hayashi, and N. Ishizawa, Jpn. J. Appl. Phys. **28,** L2007 (1989).
6. N. Ishizawa, H. Banno, M. Hayashi, S.-E. Yoo, and M. Yoshimura, Jpn. J. App. Phys. **29,** 2467 (1990).
7. M. E. Pilleux, C. R. Grahmann, and V. M. Fuenzalida, Appl. Surf. Sci. **65/66,** 283 (1993).
8. V. M. Fuenzalida and M. E. Pilleux, J. Mater. Res. **10,** 2749 (1995).
9. A. T. Chien, J. S. Speck, and F. F. Lange, J. Mater. Res. **12,** 1176 (1997).
10. H. Ishizawa and M. Ogino, J. Mater. Sci. **31,** 6279 (1996).
11. P. Bendale, S. Venigalla, J. R. Ambrose, E. D. Verink Jr, and J. H. Adair, J. Am. Ceram. Soc. **76,** 2619 (1993).
12. E. B. Slamovich and I. A. Aksay, J. Am. Ceram. Soc. **79,** 239 (1996).
13. T. Hoffmann, T. Doll, and V. M. Fuenzalida, J. Electrochem. Soc. **144,** L292 (1997).
14. A. V. Alvarez and V. M. Fuenzalida, J. Mater. Res. **14,** 4136 (1999).
15. M. M. Lencka and R. E. Riman, Chem. Mater. **5,** 61 (1993).
16. E. B. Slamovich and I. A. Aksay, J. Am. Ceram. Soc. **79,** 239 (1996).
17. H. Galsater, pH Measurement, VCH, Weinbeim, Germany, 1991.
18. W. J. Dawson, Am. Ceram. Soc. Bull. **67,** 1673 (1988).
19. C.-T. Xia, E.-W. Shi, W.-Z. Zhong, and J.-K. Guo, J. Eur. Ceram. Soc. **15,** 1171 (1995).
20. C.-T. Xia, V. M. Fuenzalida, and R. A. Zarate, unpublished results.
21. C.-T. Xia, E.-W. Shi, W.-Z. Zhong, and J.-K. Guo, Chin. Sci. Bull. **40,** 2002 (1995).
22. J. G. Lisoni, F. J. Piera, M. Sánchez, C. F. Soto, and V. M. Fuenzalida, Appl. Surf. Sci. **134,** 225 (1998).
23. R. R. Bacsa, J. R. Dougherty, and L. J. Pilione, Appl. Phys. Lett. **63,** 1053 (1993).
24. K. Kajiyoshi, Y. Sakabe, and M. Yoshimura, Jpn. J. Appl. Phys. **36,** 1209 (1997).
25. A. T. Chien, X. Xu, J. H. Kim, J. Sachleben, J. S. Speck, and F. F. Lange, J. Mater. Res. **14,** 3330 (1999).
26. L. Young, *Anodic Oxide Films* (Academic Press, New York, 1961), p. 253-57.
27. A. R. Prusi and L. D. Arsov, Corros. Sci. **33,** 153 (1992).
28. M. Pourbaix, *Atlas of Electrochemical Equilibria in Aqueous Solutions,* 2nd ed.(National Association of Corrosion Engineers, Houston, TX, 1974), p. 213-29..
29. W. Hertl, J. Am. Ceram. Soc. **71,** 879 (1988).
30. A. T. Chien, J. S. Speck, and F. F. Lange, A. C. Daykin, and C. G. Levi, J. Mater. Res. **10,** 1784 (1995).

PHASE TRANSFORMATIONS IN SOL-GEL PZT THIN FILMS

D.P. Eakin, M.G. Norton and D.F. Bahr
Mechanical and Materials Engineering, Washington State University, Pullman, WA 99164-2920

ABSTRACT

Thin films of PZT were deposited onto platinized and bare single crystal NaCl using spin coating and sol-gel precursors. These films were then analyzed using in situ heating in a transmission electron microscope. The results of in situ heating are compared with those of an ex situ heat treatment in a standard furnace, mimicking the heat treatment given to entire wafers of these materials for use in MEMS and ferroelectric applications. Films are shown to transform from amorphous to nanocrystalline over the course of days when held at room temperature. While chemical variations are found between films crystallized in ambient conditions and films crystallized in the vacuum conditions of the microscope, the resulting crystal structures appear to be insensitive to these differences. Significant changes in crystal structure are found at 500 °C, primarily the change from largely amorphous to the beginnings of clearly crystalline films. Crystallization does occur over the course of weeks at room temperature in these films. Structural changes are more modest in these films when heated in the TEM then those observed on actual wafers. The presence of Pt significantly influences both the resulting structure and morphology in both in situ and ex situ heated films. Without Pt present, the films appear to form small, 10 nm grains consisting of both cubic and tetragonal phases, whereas in the case of the Pt larger, 100 nm grains of a tetragonal phase are formed.

INTRODUCTION

Lead zirconate titanate (PZT) at compositions near the morphotropic phase boundary, around $Pb(Zr_{0.52}Ti_{0.48})O_3$, are readily formed using sol-gel or solution based deposition [1,4]. These deposition methods rely upon depositing the material in a solution containing the proper precursor elements via spin coating, with subsequent heat treatments on the substrate – film system. Three heat treating steps are often used: a low temperature heat (≈100 °C) is sometimes used to remove solvent from the film; a higher temperature pyrolosis step (≈350 °C) removes the organic components; and a high temperature (≈700 °C) heat treatment to form the perovskite crystal structure.

Two materials dominate the literature in terms of substrates for PZT thin films deposited via solution methods: platinum and ruthenium oxide (RuO_2) coated silicon wafers. It has been suggested that platinum [5] assists in the formation of the perovskite crystal structure, either through a lattice match that would encourage epitaxy or the formation of secondary phases which behave in a similar manner. Additionally, the chemical composition of the film has been shown to be very important if forming the perovskite crystal structure [6]. Two analytical techniques have been used in other studies [7,8] to characterize these materials, transmission electron microscopy (TEM) and x-ray diffraction (XRD). While there have been studies of phase transformations during the crystallization process using XRD, an in situ TEM heating of the pyrolyzed material would allow both crystal structure and morphology to be directly monitored in these thin films, in much the same way which in situ TEM has been used to track the crystallization process of barium titanate thin films made via sol-gel processing [9]. Therefore, this study was undertaken to examine the phase transformations which occur during the crystallization of PZT thin films with and without the presence of a Pt electrode. Of particular interest is the ability to perform in situ TEM heating experiments. As the vacuum conditions in

185

the TEM differ significantly from those in ambient of PbO environments (used by some researchers to promote perovskite formation [6]), the differences between *in* and *ex situ* heat treatments will be examined.

EXPERIMENTAL PROCEDURE

Two substrate systems were chosen for PZT deposition, a platinized NaCl single crystal, and a bare NaCl single crystal. The single crystal was cleaved from a large single crystal to form sections approximately 1 cm square and 2 mm thick. Platinum was then deposited using DC magnetron sputtering to thicknesses of approximately 50 nm. A PZT solution containing the proper stoicheometry of $Pb(Zr_{0.52}Ti_{0.48})O_3$ and containing an excess of 10 mol % Pb was formed using the 2-MOE route in previously published methods [1,10]. The excess Pb is present to enhance the formation of the tetragonal, rather than cubic, crystal structure [6]. This solution was spun onto the salt crystals at a rate of 4000 rpm for 15 seconds. Both sets of crystals were then heated to 90 °C to evaporate solvent and 350 °C to pyrolyze the film. This process was repeated twice to build up PZT films approximately 250 nm thick. The films were removed from the salt using dissolution and lift off techniques, and captured on Au TEM grids.

One set of TEM grids was then subjected to the same heat treatment used for processing wafers; they were placed in a furnace at 700 °C for 15 minutes at temperature, then removed and allowed to air cool. A second set of grids was used to examine the phase transformations in the TEM. A Philips CM 200 was used to perform the *in situ* heating of the grids with both a platinum and no platinum substrate. Samples were then heated using a Phillips Heating Stage grid holder. The current was controlled through a current source to produce the resistive heating. Current was slowly increased as the temperature was increased. Bright field images, diffraction patterns, and EDS analysis was taken at temperatures between 300 °C and 700 °C, at 100 °C intervals. Once the sample temperature reached 700 °C, the sample was held at that temperature for 30 minutes to observe any changes. Bright field images and selected area diffraction patterns were taken at the previously mentioned temperatures. The bright field images provided a higher contrast image, assisting in recognizing grain nucleation and growth. The lattice spacings were evaluated from the diffraction patterns. As the patterns all contain multiple grains, no effort was made to determine individual grain orientation. Energy Dispersive Spectroscopy (EDS) was used to determine relative amounts of each element in the thin films. Multiple readings were taken from all of samples to observe qualitative changes in film composition with the environment of the heat treatment.

RESULTS

As deposited pyrolyzed films

The as deposited film without Pt (the bright field image in Figure 1a with the corresponding diffraction patterns in Figure 1b an 1c) shows an interesting time dependent behavior. When films were examined two days after deposition, primarily amorphous films are observed. However, when the same film is examined three weeks later, a significant amount of crystallization has occurred. Figure 1c shows the diffraction pattern of a similar film in Fig. 1a after three weeks of storage at room temperature. Obviously, it is possible for the PZT to crystallize at room temperature. Most of the films examined in this study were tested approximately 5 days after deposition, and therefore show some evidence of crystallization in the as deposited conditions. However, in these films the lattice spacings were not exactly the same as tetragonal PZT [11], as shown in Table I. EDS spectra of these films are shown in Table II.

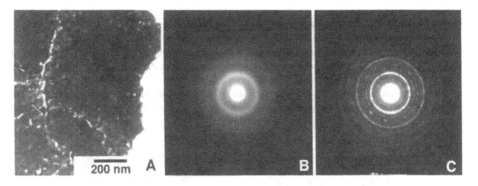

Figure 1. (a) Bright field image of as-spun PZT after three weeks at room temperature. (b) DP after 4 days at room temperature (c) DP after 3 weeks at room temperature of same film.

measured lattice spacings from these experiments are shown in Table I. Additional planes are not included for clarity, as are any peaks corresponding solely to Pt without a similarly spaced PZT family of planes. The EDS spectra of these films were analyzed using semi-quantitative methods, as shown in Table II.

Ex situ heated films

Films that were heat treated in a furnace open to the ambient atmosphere were analyzed both with and without the Pt underlayer. Figures 2 and 3 show these films in bright field and the corresponding diffraction patterns. The lattice spacings corresponding to the diffraction patterns shown in these figures are given in Table I, while the composition determined via EDS spectra are tabulated in Table II.

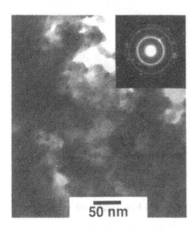

Figure 2. Bright field and DP of PZT film without Pt film after ex situ heating.

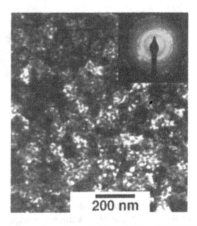

Figure 3. Bright field and DP of PZT film with Pt film after ex situ heating.

In situ heating – film morphology and phase transitions

As deposited films heated to 400 °C over the course of 30 minutes show very little evidence of any phase transformations or any significant morphological changes. Beginning at 500 °C, the initial ring patterns which are indicative of mainly amorphous materials start to show evidence of developing long range crystalline order, as shown in Figure 4. Once 700 °C is achieved, images at the initial time as well as 10, 20 and 30 minutes at temperature were recorded. Representative changes are shown in Figure 5. The resulting lattice spacings and compositions are given in Tables I and II respectively.

Figure 4. Diffraction patterns at (a) 500 °C and (b) 700 °C of PZT film on Pt. Note the formation of spots in (a) marked with the arrow, which prior to this temperature was amorphous.

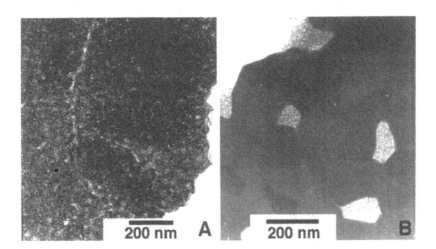

Figure 5. Bright field images of PZT film at 700 °C during *in situ* heat treatment (a) without Pt and (b) with Pt. Note, 5(a) is the same region shown in Fig. 1(a).

Table I. Measured Lattice Spacings (nm)

Reference lattice spacings [11] T- Tetragonal, C- Cubic	As Deposited no Pt	As Deposited with Pt	Ex situ heat treatment no Pt	Ex situ heat treatment with Pt	In situ @ 700°C no Pt	In situ @ 700°C with Pt
0.404 PZT{001} T				0.406		
	0.387		0.385		0.396	
0.289 PZT{101} T		0.289		0.293		0.298
0.285 PZT{110} T			0.286		0.281	
0.243 PZT {331} C	0.248		0.247		0.243	
0.235 PZT{111} T				0.237		0.231
0.228 Pt {111} T	0.223		0.228	0.228	0.228	.228
0.207 PZT{002} T						
0.201 PZT{200} T 0.197 Pt {200)			0.194	0.195	0.193	0.197

Table II. Semi-quantitative EDS analysis of films. Atomic ratio of (Pb+O) / (Ti+Zr)

Sample	As Spun	Ex situ heat treatment	In situ heat treatment
No Pt film	5.92±3.55	2.54±0.26	1.77±0.09
Pt film present	4.21±1.45	1.71±0.29	1.51±0.76

DISCUSSION

Three significant observations have been made during this study. First, the PZT films can crystallize after three weeks held at room temperature. Traditionally this process has been carried out via heat treating, which clearly hastens the process. However, this suggests that an activation energy could be determined for crystallization of PZT from an amorphous thin film by monitoring changes with time and temperature. The amorphous rings present in films spun onto Pt films are at the same lattice spacing as the primary {101} tetragonal PZT, 0.289 nm. When films are deposited on salt and then lifted off, the amorphous rings do not begin at the same lattice spacings. As the amorphous films have been heated to moderate (≈350 °C) temperatures for 10 minutes to pyrolyze the films, it is likely that the substrate may be influencing the structure of the overlayer during this processing step. Therefore, the presence of Pt does indeed play a significant role in the formation of the desired crystal structure in the PZT thin film.

The second observation demonstrates some of the limitations of this method. Unlike the reports of samples which were studied after processing wafers [12], we do not see evidence of perovskite forming from a pyrochlore type matrix. This may be an effect of the heat treatment occurring in vacuum, which allows for substantial Pb and O loss. Therefore, caution should be used in comparing these structures to those formed directly on wafers.

The final observation is the change at ≈500 °C in which the amorphous ring at 0.289 nm begins to show evidence of crystallization. This is similar to the temperature at which transformations were observed in x-ray studies [2]. As much smaller crystals are formed in the current study than in previous experiments [12], it would appear that some of the variations between ambient and vacuum based heating results in a tighter nucleation rate distribution (i.e. more crystallization at a given temperature or time) and therefore finer and more uniform grains.

CONCLUSIONS

- *In situ* TEM can be used to study the transformations from amorphous to crystalline PZT thin films fabricated by sol-gel techniques by utilizing relatively simple lift off sample preparation.
- PZT thin films can crystallize at room temperature over the course of weeks to nano-crystalline structures, resulting in very fine, randomly oriented grains.
- Rapid crystallization in the TEM occurs at temperatures similar to those observed in x-ray studies of heat treatments in ambient environments.
- A thin Pt film used as the substrate greatly influences both the as-pyrolyzed amorphous film and the subsequently crystallized film. Lattice spacings corresponding to a perovskite PZT structure occurred in the presence of Pt.

ACKNOWLEDGEMENTS

The authors wish to acknowledge the support of the National Science Foundation and the REU site program under grant #9876937. The assistance of Mr. T.B. Myers, Mr. P. Banerjee, and Prof. A. Bandyhopadhay of Washington State University for solution fabrication is greatly appreciated. DFB whishes to thank the Washington Technology Center for partial support of this work.

REFERENCES

1. K.D. Budd, S.K. Dey, and D.A. Payne, Br. Ceram. Proc., **36**, 107 (1985).
2. Y.L. Tu and S.J. Milne, J. Mater. Res., **10**, 3222 (1995).
3. D.L. Polla and L.F. Francis, Annu. Rev. Mater. Sci., **28**, 563 (1998).
4. B.A. Tuttle and R.W. Schwartz, MRS Bull., **21**, 49 (1996).
5. A. Seifert, F.F. Lange, And J.S. Speck, J. Mater. Res., **10**, 680 (1995).
6. M.J. Lefervre, J.S. Speck, R.W. Schwartz, D. Dimos, and S.J. Lockwood, J. Mater. Res., **11**, 2076 (1996).
7. B.A. Tuttle, T.J. Headley, B.C. Bunker, R.W. Schwartz, T.J. Zender, C.L. Hernadez, D.C. Goodnow, R.J. Tissot, J. Micael, and A.H. Carim, J. Mater. Res., **7**, 1876 (1992).
8. C.J. Kim, D.S. Yoon, J.S. Lee, C.G. Choi, and K. No, J. Mater. Res., **12**, 1043 (1997).
9. M.C. Gust, N.D. Evans, and M.L. Mecartney, Proc. Microscopy and Microanalysis 1995, 258, (1995).
10. D.F. Bahr, J.C. Merlino, P. Banerjee, C.M. Yip, and A. Bandyopadhyay, Proc. Mater. Res. Soc., **546**, 153 (1999).
11. JCPDS data files card number 33-0784.
12. C.C. Hsueh and M. L. Mecartney, J. Mater. Res., **6**, 2208 (1991).

ELECTRIC AND MAGNETIC PROPERTIES OF EPITAXIAL $Fe_{2-x}Ti_xO_{3+\delta}$ FILMS

T. Fujii*, K. Ayama*, M. Nakanishi*, M. Sohma**, K. Kawaguchi**, and J. Takada*
*Department of Applied Chemistry, Okayama University, Okayama 700-8530, Japan,
tfujii@cc.okayama-u.ac.jp
**National Institute of Materials and Chemical Research, Tsukuba 305-0046, Japan

ABSTRACT

Solid solution films of the α-Fe_2O_3-$FeTiO_3$ series are one of the candidates for noble half-metallic oxides. They were epitaxially formed on α-Al_2O_3(001) single crystalline substrates by O_2-reactive evaporation method. The $Fe_{2-x}Ti_xO_{3+\delta}$ films prepared at higher T_S=973 K and with larger Ti content $x \geqq 0.4$ had the ilmenite structure with $R\overline{3}$ symmetry. Other films at lower T_S or with smaller x possessed the corundum structure with $R\overline{3}c$. Only the films with $R\overline{3}$ symmetry had large ferrimagnetic moments, though the observed spontaneous magnetization was less than half of the ideal value expected from the fully ordered structure. Room temperature resistivity of intermediate composites dropped to $10^{-1}\Omega$cm due to the formation of the mixed valence states between Fe^{2+} and Fe^{3+}. However the Fe^{2+} content of the films was rather small as compared with stoichiometric $Fe_{2-x}Ti_xO_3$. The Ti-rich films had large oxygen nonstoichiometry of about δ=0.3.

INTRODUCTION

Half-metallic ferromagnetic materials have attracted attention due to their technological potentials for use in magnetic random access memory (MRAM) and a tunnel magnetoresistance (TMR) sensors. Solid solutions of α-Fe_2O_3-$FeTiO_3$ (hematite-ilmenite) series are known to have interesting magnetic and electric properties [1, 2]. Though the end members of this series are antiferromagnetic insulators, the intermediate compositions between them are half-metallic ferrimagnets. The compositions are expressed as $Fe^{3+}_{2-2x}Fe^{2+}_xTi^{4+}_xO_3$, where x is the mole fraction of ilmenite. The crystal structure consists of an hcp O^{2-} framework with cations occupying octahedral sites. The ferrimagnetism appears only when the crystal has a space group of $R\overline{3}$, where the cations are arranged into two nonequivalent layers along the c-axis. One is a Ti^{4+}-rich layer and another is a Fe^{2+}-rich layer. The mixed valence states between Fe^{2+} and Fe^{3+} gives anisotropic conductivity within the c-plane [3].

However it was very difficult to prepare the ferrimagnetic solid solutions films [4]. The complete solid solution at high temperature has the antiferromagnetic structure with a space group $R\overline{3}c$, where all cation layers are equivalent. While at low temperature, the binary phase diagram in α-Fe_2O_3-$FeTiO_3$ shows a miscibility gap [5]. Carefully controlled heating and oxygen pressure should be required to prepare stoichiometric solid solution films. We have recently succeeded in preparing well-crystallized epitaxial $Fe_{2-x}Ti_xO_{3+\delta}$ films by activated reactive evaporation [6]. However the spontaneous magnetization of sample films was a little inferior to the ideal value reported in the bulk form [1]. Structures, stoichiometries, and magnetic and electric properties of the epitaxial $Fe_{2-x}Ti_xO_{3+\delta}$ (x=0.0~1.0) films are investigated as a function of the compositions and the preparation conditions.

Mat. Res. Soc. Symp. Proc. Vol. 623 © 2000 Materials Research Society

Table 1. Preparation conditions of epitaxial $Fe_{2-x}Ti_xO_{3+\delta}$ films.

Apparatus	AORE system	UHV system
Base pressure (Pa)	10^{-3}	10^{-7}
Oxygen pressure (Pa)	$2.6 - 5.3 \times 10^{-2}$	$2.0 - 4.0 \times 10^{-4}$
Substrate temperature (K)	773 - 973	773
Evaporation rate (Å/s) Fe	0.5	0.5
Ti	0 - 0.5	0 - 0.5
RF / radical power (W)	200	300
Film thickness (Å)	3000	2000

EXPERIMENT

Sample films were deposited on α-Al_2O_3(0001) single crystalline substrates by using two different O_2-reactive evaporation systems. One is an activated O_2-reactive evaporation (AORE) system where oxygen plasma was broadly generated over the evaporation pass. Another is an ultra high vacuum (UHV) system with radical O_2-beam source. The UHV system was equipped with *in situ* x-ray photoelectron spectroscopy (XPS). In both systems high purity Fe and Ti metals were evaporated individually by electron-beam-guns to control the Fe/Ti evaporation rate ratio. During deposition activated oxygen was introduced into the chamber to keep certain pressures (P_{O2}). Substrate temperature (T_S) of the UHV system was fixed to 773 K, while the T_S of the AORE system was heated up to 973 K. Detailed deposition conditions of each system were summarized in Table 1. All deposited films were characterized by x-ray diffraction (XRD) techniques using Cu Kα radiation. Magnetic and electronic properties of deposited films were examined by a vibrating sample magnetometer (VSM), [57]Fe conversion electron Mössbauer spectroscopy (CEMS), *in situ* XPS using Mg Kα x-rays and a dc four-probe resistivity tester. Chemical compositions of deposited films were analyzed by energy dispersive x-ray spectroscopy (EDX).

RESULTS AND DISCUSSION

Crystal Structures

Because $Fe_{2-x}Ti_xO_3$ has the same hcp O^{2-} framework as α-Al_2O_3, the films could be epitaxially formed on the substrates. All deposited films in certain deposition conditions showed monophasic XRD patterns. Epitaxial growth of $Fe_{2-x}Ti_xO_3$ on α-Al_2O_3 was confirmed by XRD pole figure measurements. Fig.1 shows

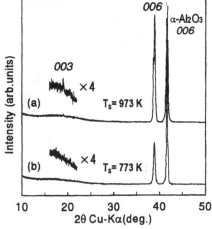

Fig. 1.   XRD patterns for $Fe_{1.54}Ti_{0.46}O_3$ films prepared at two different substrate temperatures; a) 973 K, b) 773 K.

typical XRD patterns of the films prepared by using the AORE system at high and low T_S's of 973 and 773 K, respectively. The XRD pattern for the film of T_S=773 K had only one reflection indexed as $Fe_{2-x}Ti_xO_3$ *006*. However the pattern of T_S=973 K had an additional small peak at 2θ=19.08°, which could be indexed as $Fe_{2-x}Ti_xO_3$ *003*. Advent of the *003* reflection indicates that the $Fe_{2-x}Ti_xO_3$ film had the structure with an ordered array of the Ti^{4+}-rich layer and the Fe^{2+}-rich layer along the c-axis. The disordered cation array with $R\bar{3}c$ symmetry annihilates all *00l* reflections except for $l=6n$ conditions. While the ordered array with $R\bar{3}$ symmetry shows the additional reflections with $l=3n$ [7].

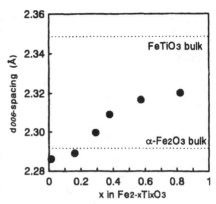

Fig. 2. d_{006}-spacings of epitaxial $Fe_{2-x}Ti_xO_{3+\delta}$ films as a function of the Ti content.

d_{006}-spacings of epitaxial $Fe_{2-x}Ti_xO_3$ (001) films are potted in Fig. 2 as a function of the Ti content x. The dashed lines indicate the d_{006}'s of α-Fe_2O_3 and $FeTiO_3$ bulk crystals, respectively. With increasing the Ti^{4+} content, the d_{006} values increased linearly for $x<0.5$ and then reached the ceiling at $x>0.5$. However it was reported that the lattice parameters of $Fe_{2-x}Ti_xO_3$ bulks were linearly changed according with the Vegard's law [5]. The ceiling of the measured d_{006}-spacings of our films at $x>0.5$ could be one of the proofs of formation of nonstoichiometric $Fe_{2-x}Ti_xO_{3+\delta}$ films.

$Fe_{2-x}Ti_xO_{3+\delta}$ films on α-Al_2O_3(001) had the α-Fe_2O_3-$FeTiO_3$ solid solution phase epitaxially formed on the substrates. However the symmetry of the deposited films, judging from the *003* reflection, strongly depended upon both x and T_S. Only the films prepared at higher Ts=973 K and with larger Ti content, $x\geq0.4$ had the ordered structure with $R\bar{3}$ symmetry. Other films with smaller x or at lower T_S possessed the disordered structure with $R\bar{3}c$ symmetry.

Magnetic and Electric Properties

Room- and low-temperature spontaneous magnetization of $Fe_{2-x}Ti_xO_{3+\delta}$ films prepared at T_S=973 K are plotted in Fig. 3 as a function of the Ti content x. Large magnetizations were observed in the intermediate compositions ranging on $0.4\leq x<1.0$. The formation range of ferrimagnetic films was fully consistent with that of the ordered structure with $R\bar{3}$ symmetry examined by XRD. Difference in the net magnetic moments between the Ti-rich and the Fe-rich layers produced ferrimagnetism. However the observed spontaneous magnetization for ferrimagnetic films were fairly smaller than the theoretical values (360 emu/cm³ at x=0.5) expected from the perfectly ordered $R\bar{3}$ structure [1]. This suggested that the octahedral cations of the films did not arranged perfectly into the two nonequivalent layers. Moreover the oxygen nonstoichiometry should also influence the magnetic properties.

Fig. 4 shows the dependence of room temperature electric resistivity of $Fe_{2-x}Ti_xO_{3+\delta}$ films on the Ti content. The films examined here had the disordered structure with the $R\bar{3}c$ symmetry. The resistivity of both $FeTiO_3$ and α-Fe_2O_3 films was relatively high (over 10^1 Ωcm). For the films with intermediate compositions, $Fe_{2-x}Ti_xO_{3+\delta}$, resistivity went down to 10^{-1} Ωcm

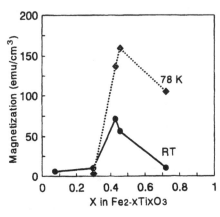 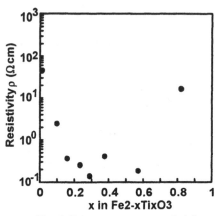

Fig. 3. Room- and low-temperature spontaneous magnetization of $Fe_{2-x}Ti_xO_{3+\delta}$ films as a function of the Ti content.

Fig. 4. Room temperature resistivity of $Fe_{2-x}Ti_xO_{3+\delta}$ films as a function of the Ti content.

at about $x=0.3$. The decreasing resistivity of the solid solution films suggested the formation of mixed valence states between Fe^{2+} and Fe^{3+}. However the observed conductivity was still small as compared with other mixed valence oxides like Fe_3O_4. Disordered cation arrays in octahedral interstices and oxygen nonstoichiometry could affect the properties of $Fe_{2-x}Ti_xO_{3+\delta}$ films to increase the resistivity.

Stoichiometries

Stoichiometries of $Fe_{2-x}Ti_xO_{3+\delta}$ solid solution films prepared by the UHV system were examined *in situ* by XPS and *ex situ* by CEMS. XPS Fe 2p core-level spectra exhibit so-called shake-up satellite structures, which are very sensitive to the electronic structure of iron ions. The Fe 2p spectra of various $Fe_{2-x}Ti_xO_{3+\delta}$ films are shown in Fig. 5 as a function of Ti content. It is known that the broad satellite centered at about 720 eV is characteristic of octahedral Fe^{3+} ions, while the one due to octahedral Fe^{2+} is at about 715 eV. The XPS spectra of $Fe_{2-x}Ti_xO_{3+\delta}$ films looked similar to those of nonstoichiometric $Fe_{3-\delta}O_4$ films from γ-Fe_2O_3 to Fe_3O_4 [8]. With increasing the Ti content, the intensity of Fe^{3+} satellites decreased and that of Fe^{2+} increased. The binding energy of the Fe 2p main lines was also shifted gradually. The substitution of Ti^{4+} ions for Fe^{3+} brought the mixed valence states between Fe^{3+} and Fe^{2+}. However the content of produced Fe^{2+} ions seemed to be rather small than that of the stoichiometric composition of $Fe_{2-x}Ti_xO_3$. The Fe^{2+}/Fe^{3+} ratio in $Fe_{2-x}Ti_xO_{3+\delta}$ films was determined quantitatively by CEMS.

In Fig. 6, the room temperature CEMS spectra of $Fe_{2-x}Ti_xO_{3+\delta}$ films are shown as a function of Ti content. In bulk forms the Curie temperature (T_C) decreases linearly with increasing x from 948 K in α-Fe_2O_3 to 55 K in $FeTiO_3$, crossing room temperature at about $x=0.73$. The CEMS spectra of the samples films clearly indicated that the T_C crossed room temperature at about $x=0.75$ as well as the bulk. The spectra of a magnetically split sextet below T_C changed to paramagnetic doublets above T_C. We have deconvoluted the paramagnetic spectra by using two asymmetric doublets; one is an octahedral Fe^{3+} component and another is

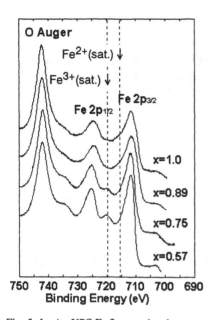

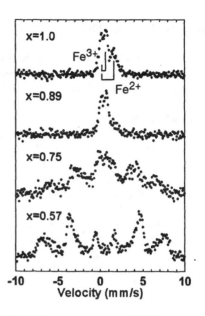

Fig. 5. *In situ* XPS Fe 2p core-level spectra of epitaxial $Fe_{2-x}Ti_xO_{3+\delta}$ films as a function of the Ti content.

Fig. 6. Room temperature CEMS spectra of epitaxial $Fe_{2-x}Ti_xO_{3+\delta}$ films as a function of the Ti content.

an octahedral Fe^{2+} component. Asymmetry of the doublets resulted from the geometric relationship between the irradiated γ-ray direction and the electric field gradient axis of epitaxial crystals. The number of oxygen nonstoichiometry δ was calculated to assume the fixed valence states of Ti^{4+} and O^{2-}. The fitted results are summarized in Table 2. These epitaxial $Fe_{2-x}Ti_xO_{3+\delta}$ films we prepared had the compositions with large nonstoichiomery of δ=0.3. There was no report to form such largely nonstoichiometric $Fe_{2-x}Ti_xO_{3+\delta}$ crystals in bulk forms. The excess oxygen should influence the structural, magnetic and electric properties of the films as discussed above.

Table 2. Results of CEMS analysis of paramagnetic $Fe_{2-x}Ti_xO_{3+\delta}$ films. IS, QS, and δ stand for isomer shift, quadrupole splitting, and number of oxygen nonstoichiometry, respectively.

Sample	Site	IS (mm/s)	QS (mm/s)	Area ratio (%)	δ
x=0.89	Fe^{2+}	0.85	1.68	23	0.31
	Fe^{3+}	0.42	0.60	77	
x=1.0	Fe^{2+}	0.87	1.45	40	0.30
	Fe^{3+}	0.39	0.58	60	

ACKNOWLEDGEMENT

The authors acknowledge Dr. S. Prasanna for her cooperation in UHV film deposition. This work was financially supported by grant-in-aid from Ministry of Education, Science, Sports and Culture, Japan and Mazda Foundation's Research Grant.

REFERENCES

1. Y. Ishikawa, J. Phys. Soc. Jpn. **17**, p. 1835 (1962).
2. Y. Ishikawa, J. Phys. Soc. Jpn. **13**, p. 37 (1958).
3. A. K. Mukerjee, Ind. J. Phys. 38, p. 10 (1964).
4. Z. Dai, P. Zhu, S. Yamamoto, A. Miyashita, K. Narumi, and H. Naramoto, Thin Solid Films **339**, p. 114 (1999).
5. D.H. Lindsley, Rev. Mineral. **3**, p. L-1 (1976).
6. T.Fujii, K. Ayama, M. Nakanishi, and J. Takada, J. Magn. Soc. Jpn. **22**, S1, p. 206 (1998).
7. N.F.M Henry and K. Lonsdale, *International Tables for X-ray Crystallography*, vol. 1, Birmingham, Kynoch Press, 1969.
8. T. Fujii, F.M.F. de Groot, G.A. Sawatzky, F.C. Voogt, T. Hibma, K. Okada, Phys. Rev. B **59**, p. 3195 (1999).

Transparent Conductors

FUNDAMENTAL ADVANCES IN TRANSPARENT CONDUCTING OXIDES

Timothy J. Coutts, David L. Young, and Xiaonan Li
National Renewable Energy Laboratory, Golden, CO, USA

ABSTRACT

Increasingly large-volume markets for large-area, flat-panel displays and photovoltaic panels are likely to be established in the early years of the next century and transparent conducting oxides (TCOs) of improved opto-electronic properties will be required to enable some of these applications to be realized. Our work is focusing on improving both the fabrication-limited properties of the materials (extrinsic), and materials-limited properties (intrinsic). The emphasis on achieving improved electrical and optical properties hinges on achieving higher electron mobility via intrinsic and/or extrinsic properties. To this end, we have investigated the properties of several TCOs including cadmium oxide, tin oxide, zinc oxide, cadmium stannate and zinc stannate. These may be deposited by chemical vapor deposition (CVD) or sputtering and we hope to establish the capability to fabricate compounds and alloys in the cadmium oxide, tin oxide, zinc oxide ternary phase diagram.

The properties of the materials have been investigated using a wide variety of techniques including high-resolution electron microscopy, atomic force microscopy and X-ray diffraction, as well as Mössbauer, Raman and UV/visible/NIR spectroscopies. We have measured the transport properties (conductivity, Hall, Seebeck and Nernst coefficients) and have obtained the effective mass, relaxation time, Fermi energy, and scattering parameter. This information has been obtained as a function of carrier concentration, which depends on the deposition and annealing procedures. We have found that the mobilities of free-electrons in the cadmium-bearing compounds are greatly superior to those in the other materials, because they have much longer electron relaxation times. In the case of cadmium oxide, there is also great benefit from a much lower effective mass. We are gaining a clearer understanding of the fundamental microscopic attributes needed for TCOs, which will be required in more-demanding, and rapidly emerging, applications.

INTRODUCTION

Transparent conducting oxides have been used for various applications over the last 30-40 years and the level of their understanding, as well as their performance, has always been adequate. However, it has become increasingly apparent recently, that this situation has changed or, at very least, will change very soon in the next generation of devices and applications.[1]

The three applications that may be anticipated are photovoltaic modules, flat-panel displays, and architectural windows. The first two applications demand TCOs of high conductivity and excellent optical transmittance to minimize power loss and to minimize power consumption, respectively. The third application requires a low absorptance in the visible wavelength range and a low emittance in the infrared range, rather than a high conductivity. However, the two sets of attributes are inseparable. A fourth application (really an extension of the third) is that of 'smart' windows. These are based on electrochromic materials, such as WO_3, that change from being transmissive to absorptive when a voltage is applied across them. It is too soon to speculate on prospective market sizes of 'smart' windows, because the technology is not sufficiently advanced yet. The attraction of these materials is that they may be used to minimize

Mat. Res. Soc. Symp. Proc. Vol. 623 © 2000 Materials Research Society

solar gain in hot climates (thus minimizing air-conditioning) and/or to minimize heat-loss from buildings in cold climates. Part of their development depends on reducing the time taken to change state from transmissive to absorptive, which depends, to some extent, on reducing the sheet resistance of the TCO.

The market, in terms of the area of coated glass required annually (or in financial terms), for each of these applications is large now but has the potential to increase greatly. The efficiency of thin-film photovoltaic modules is, optimistically, 10%. The peak annual insolation is 1 k watt m^{-2}.[2] At present, the U.S. has approximately 700 G watts of installed electricity generating capacity.[3] For photovoltaics to be of national significance, one may speculate that their production rate must be at least 1-G watt annually and, based on these estimates, at least 4 square miles of photovoltaic panels must be manufactured annually. Production rates of this magnitude are not out of the question. The market for photovoltaics is increasing annually at approximately 25% and it is expected that this will persist for many years to come.[4] If thin-film photovoltaics command an increasing share of the market, as is expected, and given that the three leading thin-film candidates all require a TCO in their construction, we may confidently predict that the volume of TCO-coated glass will increase substantially. The glass-coatings industry manufactures at least 16 square miles per annum, although not all of this involves a TCO coating.[5] However, American Float Glass manufactures approximately 2 square miles of tin oxide-coated glass annually.[6] Not all of this is for photovoltaics but the point is that the manufacturing capacity does exist. Indeed, photovoltaics are seen as one of the major future product lines by the glass industry. If the industry meets the cost-target of 1 $ watt^{-1}, set by the Department of Energy, then the equivalent revenue will be $1Bn per annum based on our estimate of 1 G watt per annum.[4] However, it appears likely that this will grow substantially when, and if, thin-film photovoltaics become significant.

The market for flat-panel displays is already large and is also likely to grow to multiple billions of dollars per annum. Estimates of the future market for flat-panel displays are also impressive. Predictions suggest that the existing market size is approximately $18 Bn per year, and this is expected to increase to $22 Bn per year by 2002 and to $27 Bn per year by 2005.[7] Higher performance TCOs with lower sheet resistance and superior optical properties will accelerate the achievement of these markets in two ways. Firstly, much of the power consumed by laptop computers is due to the screen. Reducing the sheet resistance would therefore provide a longer battery lifetime. Secondly, the realization of large flat-screen televisions depends on reducing the sheet resistance of the TCOs.

Similar comments may be made about energy conserving windows (either low emissivity or electrochromics), and they too are likely to gain in importance as the emphasis on conservation increases. Based on these points, we can see that TCOs are commercially important, as well as being scientifically interesting!

THE STRATEGY

A generic TCO has a high conductivity and an excellent optical transmittance in the visible spectrum. Typically, TCOs transmit freely between wavelengths of about 0.35-1.5 μm, both of these values being related to their concentration of free electrons[1]. At the same time, they must have a sheet resistance, for present-day applications such as those mentioned above, of less than

---

[1] The long wavelength cut-off of transmittance is established by the onset of reflection at the plasma edge. The short wavelength limit depends on the Burstein-Moss effect, which is caused by the unavailability of the states at the bottom of the conduction band, when the material is degenerate. This has the effect of increasing the minimum energy required for a band-to-band transition beyond that of the fundamental bandgap.

10 Ω/□. A film with a resistivity of 5×10^{-4} Ω cm and a thickness of 0.5 μm, meets this requirement. In the research environment, resistivities of as low or lower than 10^{-4} Ω cm have been achieved. This corresponds to a sheet resistance of 2 Ω/□. However, values of at least half this will be required for each of the above applications. The sheet resistance of a material is equal to its resistivity divided by its thickness. Although this could be reduced simply be increasing the thickness, this is not acceptable because this would cause additional optical absorption. The transmittance needs to be at least 85% for most applications. Free electrons give the materials their conductivity but they are also responsible for the absorption of light. This presents an inevitable compromise facing TCOs. Maxwell's equations show that the optical and electrical properties of materials are inter-related.

Many authors have commented that the only remedy to this is to increase the free-carrier mobility in TCO materials.[8, 9] Merely increasing the free-carrier concentration worsens optical absorbance, by increasing the height of the free-carrier absorption band and by moving it further into the visible part of the spectrum. This is illustrated in figures 1a and 1b. Figure 1a shows that increasing the carrier concentration, while keeping the mobility constant at 100 cm² V⁻¹ s⁻¹, increases the height of the absorption band, even though its width decreases due to an increase in the conductivity. Figure 1b shows that increasing the mobility while keeping the carrier

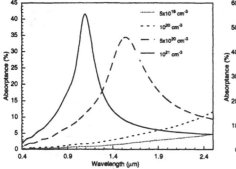 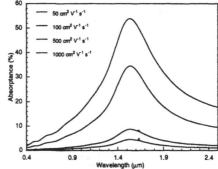

Figure 1a. Modeled variation of free-carrier absorptance with wavelength. The mobility was 100 cm² V⁻¹ s⁻¹ and the carrier concentration was treated parametrically. The effective mass was taken as 0.3 m_e and the high frequency permittivity was 4.

Figure 1b. Modeled variation of free-carrier absorptance with wavelength. The carrier concentration was 5×10^{20} cm⁻³ and the mobility was treated parametrically. The effective mass was 0.3 m_e and the high frequency permittivity was 4.

concentration constant at 5×10^{20} cm⁻³, is a better alternative because the height of the absorption band decreases. In addition, figure 1a shows that the absorption band-edge moves into the visible part of the spectrum. This worsens the optical properties for all of the major applications. In figure 1b, the wavelength of the absorption band does not change significantly. Of course, no TCO has a mobility of 1000 cm² V⁻¹ s⁻¹ but the point is that increasing the mobility improves both the electrical and optical properties.

To improve the mobility of a particular semiconductor, it is necessary to increase its carrier relaxation time—the interval of time between successive randomizing collisions with various defects, lattice vibrations, impurities etc. Some of these may be considered intrinsic to the material (e.g. phonon and ionized impurity scattering) and there may be little or nothing that can be done to lessen their effect. Other scattering mechanisms may be regarded as extrinsic and it may be possible to reduce their effect by improving crystallinity through improved fabrication methods. We assert it is unlikely that there will be significant improvements in the properties of materials that may be considered to be 'conventional' (e.g. tin oxide, indium tin oxide, zinc oxide) through application of the latter approach. These materials have been extensively researched over many years, with films having been deposited by many methods and using a wide variety of deposition properties. Indeed, the modern literature is replete with papers discussing the properties of TCOs that were equaled or exceeded many years ago.

Perhaps the limitation of investigations into TCOs to date has been the relatively small number of materials researched. Given that mobility is critical to the performance of TCOs, a better approach may be to study materials that have an intrinsically high mobility because of low-effective-mass free carriers.[9] This may give an intrinsic advantage, irrespective of extrinsic limitations and, because of this, there is clearly a recognition amongst international researchers that new, and improved, materials are soon certain to be required. This is evidenced by the fact that there are now significant research and development efforts into TCOs in Japan, and the US. Much of the work has involved cations from the transition metals with a d^{10} electronic structure. These are described by figure 2a showing a hexahedron of phase space with the relevant binary oxides at the apexes.

Each of the binary compounds has been used commercially and each has attractions, with the exception of CdO, that has a bandgap that is too low for applications using visible light. In our

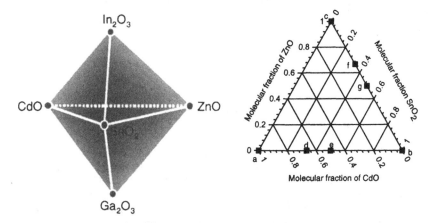

Figure 2a. Hexahedron of phase space of relevance to TCOs. All the materials in this volume are under investigation by various groups in the world.[10, 11]

Figure 2b. The base plane of figure 2a. The point compounds a-g signify CdO, SnO$_2$, ZnO, Cd$_2$SnO$_4$, CdSnO$_{3+}$, Z n$_2$SnO$_4$, and ZnSnO$_3$

own work, we have focused on the base plane of the hexahedron consisting of ZnO, SnO$_2$, and CdO, as shown in figure 2b. We have approached the problem from a fundamental point-of-view and we are attempting to probe the electronic properties of key materials more deeply than has been typical. We have focused on the spinel-structured materials cadmium stannate (CTO or Cd$_2$SnO$_4$) and zinc stannate (ZTO or Zn$_2$SnO$_4$) and have obtained encouraging results. The optical properties of CTO have proved superior to most other TCOs, due to the high mobility,

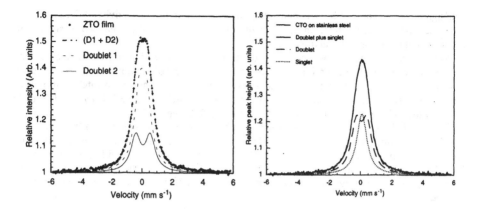

Figure 4a. Mössbauer spectrum of a ZTO film.

Figure 4b. Mössbauer spectrum of a CTO film.

the octahedral site, under the assumption that the recoil-less fraction for the two species is equal. This implies that the arrangement of the tin and zinc cations was approximately random, with maximum entropy.[14] The presence of the two doublets, alone, implies that the film must exhibit at least a partially inverse form. There was no isomer shift, which means that the tin is in the Sn^{4+} state. Figure 4b shows the Mössbauer spectrum for a film of CTO deposited on a stainless steel substrate. This exhibits the need for a singlet and a doublet contribution, again implying the presence of the inverse phase. In this case, the ratio of the areas of the two peaks was 39:61, again suggesting that the film was intermediate between normal and inverse, assuming that the recoil-less fraction was equal for the two species.

We used the *method of four coefficients* to characterize the electrical properties of the two materials. The technique has been described elsewhere.[15] It is, perhaps, the only method for measuring the effective mass, relaxation time[2], scattering parameter[3], and the Fermi energy of highly disordered polycrystalline thin films. To obtain all this information, it is necessary to measure the conductivity, the Hall, Seebeck and Nernst coefficients and we shall now discuss the results for these.

Figure 5 shows the mobility and resistivity of CTO films as a function of the carrier concentration. The mobility reaches a value of approximately 80 cm^2 V^{-1} s^{-1} and remains in the range 50-80 cm^2 V^{-1} s^{-1} over the whole range of concentrations. It is relatively unusual for the mobility not to decrease with carrier concentration and we shall offer an explanation for this later. The resistivity decreases monotonically with carrier concentration to a value of approximately 1.1×10^{-4} Ω cm which, for a film thickness of 0.5 μm, corresponds to a sheet resistance of 2-3 Ω/□. This is typical of the best results obtained in research but is much lower than typical commercial TCO values. The optical properties of CTO are also superior to those of other TCOs such as SnO_2, which has been exploited in CdTe solar cells. The excellent optical

---

[2] The relaxation time gives the time taken for the electron distribution to return to its equilibrium value after having been perturbed.

[3] The scattering parameter relates the relaxation time to the energy of the charge carriers, electrons in this case. The value of the scattering parameter gives an indication of the dominant scattering mechanism. This is invaluable in ascertaining the likelihood of being able to improve the film mobility further, or whether a fundamental limit has been reached.

although there has not yet been a commanding improvement in its conductivity. It has, however, been successfully incorporated in the structure of CdTe solar cells and has led to significant improvement in performance and reproducibility.[12]

We have also remained aware of issues of cadmium toxicity and have studied zinc stannate as a possible replacement for CTO. In our view, the issue of cadmium toxicity is greatly exaggerated when it is part of a compound. For example, we have shown separately that films of CTO are highly stable when heated in vacuum over a long period of time.[9] They are also stable in HCl vapor after exposure at elevated temperature. Etching of CTO requires the use of HF acid. Nevertheless, we shall discuss our work on both of these materials later in this paper. We also note that improving the carrier relaxation time will increase the mobility. Hence, learning how to make films of improved structural and crystallographic properties will also be beneficial. This attacks the 'extrinsic' part of the problem. In the present work, the films were made using radio frequency sputtering and post-deposition annealing; both of which have been described previously.

RESULTS

The deposition and annealing procedures, described previously, give single-phase material, which is essential to the achievement of the best properties.[9] The quality of the intra-grain material is excellent, as evidenced by high-resolution electron microscopy.[13] However, the spinel phase may be either normal or inverse, which will almost certainly play a role in the properties of the material.[14] In the normal form, the group IV cations are located on the tetrahedral sites and the group II cations on the octahedral sites. However, the relative sizes of the cations and their charge influences the actual locations of the two cations. Bulk CTO and ZTO are both claimed to be of the inverse form, in which the group IV cations are found on 50% of the octahedral sites, while 50% of the group II cations occupy all the tetrahedral sites.[14] The structure factors of cadmium and tin are almost identical for both X-rays and neutrons and it is not straightforward to determine which of the forms is most likely. However, ZTO does not suffer from this problem and it is possible to determine the form from the X-ray diffraction spectrum.

Figure 3 shows the XRD spectrum of a film of ZTO, the sputtering target, and the calculated relative peak heights for the normal and inverse forms. It is clear from this that both the film and the target are very close to the inverse form, as claimed in the literature. Figure 4a shows the Mössbauer

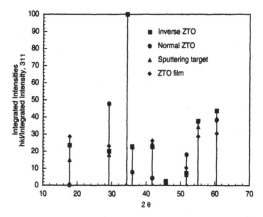

Figure 3. X-ray diffraction spectrum of a ZTO film and a ZTO sputtering target. The calculated spectra for the normal and inverse spinel forms are also shown.

spectrum for the same ZTO film. The dots show the measured data and the solid line shows the best fit to the data. This had the best goodness of fit parameter for a model consisting of two doublets. The relative areas of the peaks imply that approximately 74% of the tin cations were in

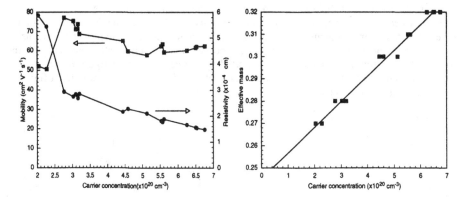

Figure 5. Resistivity and mobility of CTO films as a function of the carrier concentration.

Figure 6. Effective mass of conduction electrons in CTO as a function of the carrier concentration.

and electrical properties both stem from the higher mobility of CTO. Figure 6 shows the effective mass of electrons in CTO as a function of the carrier concentration. The uncertainty in the effective mass is approximately 0.015 m_e, where m_e is the mass of free carriers. The linear increase in the effective mass (actually the density of states effective mass) indicates clearly that the conduction band is non-parabolic. This conclusion is the reverse of our previously published data that appeared to suggest that the conduction band was indeed parabolic. We no longer believe this to be the case. The result is important because it influences subsequent modeling of the transport properties for different scattering mechanisms.

This is illustrated in figures 7a and 7b, which show the Seebeck and Nernst coefficients

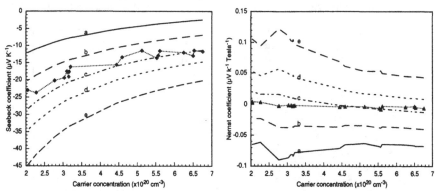

Figure 7a. Seebeck coefficient of CTO films as a function of carrier concentration. The letters a-e signify scattering due to acoustic phonons, neutral impurities, optical phonons, screened ionized impurities, and ionized impurities, respectively.

Figure 7b. Nernst coefficient of CTO films as a function of carrier concentration. The letters a-e signify the same scattering mechanisms as shown in figure 7a.

respectively, as functions of carrier concentration. The experimental data are shown by the erratic line near the center of the family of modeled curves. These were modeled for each of a series of scattering mechanisms. The indication is that optical phonons are primarily responsible for scattering the electrons, although we are not yet able to eliminate other mechanisms completely. Scattering by phonons is supported by the data shown in figure 8. In this, the resistivity increases with temperature while the mobility decreases. This was for a film of a carrier concentration of 5×10^{20} cm^{-3}.

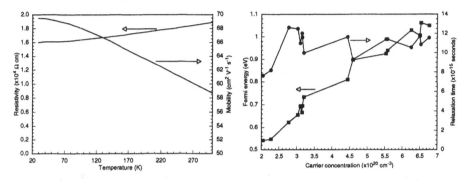

Figure 8. Variation of the resistivity and mobility of a high-carrier concentration CTO film with absolute temperature.

Figure 9. Variation of Fermi energy and relaxation time of electrons in CTO films as a function of their carrier concentration.

In figure 9, we show the position of the Fermi energy with respect to the minimum in the conduction band. The figure also shows the relaxation time, which is approximately 10^{-14} seconds, over the full range of carrier concentrations. The Fermi energy increases to over 1 eV, indicating that the optical gap must be approximately 3.9-4.2 eV, the fundamental bandgap being

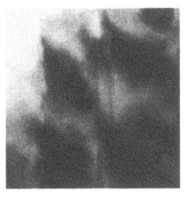

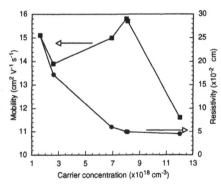

Figure 10. High resolution electron micrograph of a ZTO film. The field of view is approximately 30×30 nm.

Figure 11. Mobility and resistivity of ZTO films as a function of their carrier concentration.

in the range 2.8-3.1 eV. We have confirmed the value of the optical gap using spectroscopic ellipsometry.

Figure 10 shows a high-resolution electron micrograph of a thin film of ZTO. Although lattice fringes can be clearly seen, they are not parallel over the entire field of view, as was the case for CTO. Although this was an unannealed film, it is typical of micrographs taken on annealed films. Clearly the state of crystallinity of the film is poorer than that of CTO and we believe that this is largely responsible for the differences in the mobilities of the two materials.

Figure 11 shows the mobility and resistivity for a series of ZTO films as a function of carrier concentration. Several important points emerge from this figure. Firstly, the carrier concentration range is much less than that of the CTO films discussed earlier. We have not yet learned how to increase the concentration to the same levels as the CTO films. The mobility is much less than the CTO films over the entire range of carrier concentrations. Not surprisingly, the resistivity of the ZTO films does not decrease below 5×10^{-2} Ω cm, i.e. a factor of 500 times greater than that of CTO.

Figure 12 shows the effective mass and relaxation time of electrons in ZTO films as a function of carrier concentration. The effective mass is significantly less than that of electrons in CTO, which may be due to the much lower carrier concentration. The lower carrier concentration ensures that the Fermi level is much nearer the bottom of the conduction band, where the effective mass may be expected to be less than higher in the band. The relaxation time is also less than that of CTO by a factor of ten. The combined effect of these quantities leads to the reduction of five times in the mobility of ZTO compared with that of CTO.

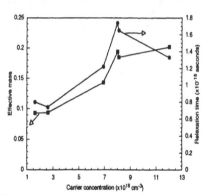

Figure 12. Effective mass and relaxation time of electrons in ZTO films as a function of their carrier concentration.

CONCLUSIONS AND FUTURE DIRECTIONS

We have used novel electron transport methods to determine the fundamental properties of two of the new TCOs of interest to our group. From these we have been able to determine the effective mass of the carriers, their Fermi energy, the likely scattering mechanism, and the relaxation time. We have shown that the two spinels considered are partially inverse. For CTO, this can only be determined using Mössbauer spectroscopy but, for ZTO, both this technique and X-ray diffraction provide a clear indication of the inverse nature of the film material. This aspect of the structure of these films may well be related to their behavior as TCOs and to the source of carriers, particular in CTO. The quality of the CTO grains appears to be excellent as indicated by lattice fringing. In this material, the Fermi level is so high in relation to the bottom of the conduction band that it is unlikely that grain boundaries scatter charge. Consequently, we observe long relaxation times, which leads to the high mobilities observed. In ZTO, however, scattering occurs within grains due to the poorer crystalline order. The carrier concentration is also much lower than in CTO, the implication of which is that grain boundary scattering may also be worse. Although the effective mass of the electrons in ZTO is half that in CTO, this is more than offset by the much shorter relaxation times. Consequently, we observe significantly lower mobilities for ZTO.

The probable dominant scattering mechanism in CTO appears to be optical phonons although we are unable to be certain of this yet. Further work is required at lower carrier concentrations to determine if this mechanism changes. We shall also investigate the effect of temperature in greater detail. If indeed the mobility is limited by a phonon-related mechanism, then we may have reached the mobility limit in CTO. However, this has yet to be established. Future work will focus on learning how to anneal the ZTO to the same degree of quality as the CTO. For the work to be relevant to industry, fabrication techniques must be developed that can be applied in large volume, with short deposition times.

The choice of directions for future TCOs is dictated not only by the extent to which it is possible to achieve a high degree of crystal perfection, but also by the fundamental properties of the free carriers. All other things being equal, lighter free electrons would cause the free-carrier absorption band to move towards the visible range of wavelengths and, consequently, the optical properties of the TCO would deteriorate. However, a lighter effective mass would cause the mobility to increase and there may not be much change in the figure-of-merit[4]. Conversely, an increase in the carrier effective mass would move the absorption band away from the visible and would thereby improve the optical properties. However, the conductivity would suffer because the mobility would decrease. Improving the carrier relaxation time, is the only unambiguous course of action. This does not move the position of the absorption band but an increase in its magnitude decreases the height of the band, while also improving the conductivity. Hence, this must be the first course of action in the development of future TCOs. However, it must also be remembered that the effective mass, carrier concentration and relaxation time are not independent quantities. Heavier carriers may also lead to a decrease in the relaxation time, as may a higher carrier concentration.

---

[4] The figure-of-merit is often defined as the ratio of the electrical conductivity to the optical absorption coefficient, at a wavelength in the visible range.

REFERENCES

[1] T. J. Coutts, T. O. Mason, J. D. Perkins, et al., in *Photovoltaics for the 21st. Century*, edited by V. J. Kapur, R. D. McConnell, D. Carlson, G. P. Ceasar and A. Rohatgi (The Electrochemical Society, Inc., Seattle, Washington, 1999), Vol. Proceedings Volume 99-11, p. 274.

[2] C. Riordan, in *Advances in Solar Energy*, edited by K. W. Böer (American Solar Energy Society, Inc., Boulder, Colorado, 1992), Vol. 7, p. 211–238.

[3] Edison Electric Institute, in *Statistical Yearbook of the Electric Utility Industry* (Edison Electric Institute, Washington, DC, 1999), Section 1, p. 7.

[4] www.nrel.gov/ncpv,, (Facilitated by the National Center for Photovoltaics (NCPV) for the U.S. PV Industry (Prepared by Energetics, Incorporated, Columbia, Maryland), 1999).

[5] R. H. Hill and S. J. Nadel, *Coated glass Applications and Markets* (BOC Coating Technology, Fairfield, California, 1999).

[6] C. Corning,, "Annual manufacture of coated glass", personal e-mail communication received by T. J. Coutts, NREL, Golden, CO, 2000.

[7] C. Bright,, Annual market for TCO-coated glass for flat-panel displays", personal e-mail communication received by T. J. Coutts, NREL, Golden, CO, 2000.

[8] G. Haacke, in *Annual Review of Materials Science*, edited by R. A. Huggins, R. H. Bube and R. W. Roberts (Annual Reviews, Inc., Palo Alto, California, 1977), Vol. 7, p. 73.

[9] T. J. Coutts, X. Wu, and W. P. Mulligan, Journal of Electronic Materials **25**, 935 (1996).

[10] D. R. Kammler, D. D. Edwards, B. J. Ingram, et al., in *195th. Meeting of the Electrochemical Society–Photovoltaics for the 21ˢᵗ Century*, edited by V. K. Kapur, R. D. McConnel, D. Carlson, G. P. Ceasar and A. Rohatgi (The Electrochemical Society, Pennington, NJ, Seattle, WA, 1999), Vol. 99–11, p. 68.

[11] T. Minami, T. Kakumu, K. Shimokawa, et al., Thin Solid Films **317**, 318 (1998).

[12] X. Wu, P. Sheldon, Y. Mahathongdy, et al., in *NCPV Photovoltaics Program Review–Proceedings of the 15th. Conference*, edited by M. Al-Jassim, J. P. Thornton and J. M. Gee (Am. Inst. Phys.,, Denver, Colorado, 1998), Vol. 462, p. 37.

[13] T. J. Coutts, D. L. Young, X. Li, et al., J. Vac. Sci. Technol. to be published (2000).

[14] L. A. Siegel, Journal of Applied Crystallography **11**, 284 (1978).

[15] D. L. Young, T. J. Coutts, and V. I. Kaydanov, Rev. Sci. Inst. **71**, 462 (2000).

In$_2$O$_3$ Based Multicomponent Oxide Transparent Conducting Films Prepared by R.F. Magnetron Sputtering

Tadatsugu Minami, Toshihiro Miyata, Hidenobu Toda and Shingo Suzuki
Electron Device System Laboratory, Kanazawa Institute of Technology,
7-1 Ohgigaoka, Nonoichi, Ishikawa 921-8501, Japan

ABSTRACT

Transparent and conductive thin films using new multicomponent oxides consisting of a combination of different In$_2$O$_3$ based ternary compounds have been prepared on room temperature substrates by r.f. magnetron sputtering. Transparent and conductive (Ga,In)$_2$O$_3$-MgIn$_2$O$_4$, (Ga,In)$_2$O$_3$-Zn$_2$In$_2$O$_2$, (Ga,In)$_2$O$_3$-In$_4$Sn$_3$O$_{12}$, Zn$_2$In$_2$O$_5$-In$_4$Sn$_3$O$_{12}$ and Zn$_2$In$_2$O$_5$-MgIn$_2$O$_4$ films were prepared over the whole range of compositions in these multicomponent oxides. The electrical and chemical properties of the resulting films could be controlled by varying the composition in the target. The resistivity, band-gap energy, work function and etching rate of the resulting multicomponent oxide films ranged between the properties of the two ternary compound films. This paper also presents a discussion of a significant spatial distribution of resistivity found on the substrate of the multicomponent oxide films as a function of composition. The resistivity distribution is attributable to the oxygen concentration on the substrate surface rather than the bombardment effect of high energy particles.

INTRODUCTION

Transparent conducting oxide (TCO) thin films using binary compounds such as SnO$_2$, In$_2$O$_3$ and ZnO are in practical use.[1] Various dependendent application properties are required for TCO films as the demand for transparent electrodes of various optoelectronic devices, transparent resistor and window coatings is increased. As a result, the use of these films is often limited in specialized applications. In order to resolve these problems, new TCO materials have been extensively investigated in recent years. New ternary compounds, such as Zn$_2$SnO$_4$,[2] CdSb$_2$O$_6$,[3,4] MgIn$_2$O$_4$,[5,6] ZnSnO$_3$,[7,8] GaInO$_3$,[9,10] Zn$_2$In$_2$O$_5$,[11-13] and In$_4$Sn$_3$O$_{12}$,[14,15] have been reported as promising materials for TCO films. The use of ternary compounds composed of binary compounds may allow an improvement in the attainable properties seen in conventional TCO films using binary compounds such as In$_2$O$_3$, ZnO and SnO$_2$. For example, ZnSnO$_3$ films exhibit higher chemical stability than ZnO films as well as lower resistivity than SnO$_2$ films.[7,8] Zn$_2$In$_2$O$_5$ films exhibit lower resistivity than ZnO and In$_2$O$_3$ films, and the band-gap energy of Zn$_2$In$_2$O$_5$ is lower than either ZnO or In$_2$O$_3$.[11-13] GaInO$_3$ [9] and Zn$_2$In$_2$O$_5$ [11] films exhibit refractive indices of 1.65 and about 2.4, relatively lower and higher than the refractive index of about 2.0 seen in conventional TCO films using binary compounds such as In$_2$O$_3$, ZnO and SnO$_2$. As described above, ternary compound TCO films which have novel properties as well as the advantages of the binary compound TCO films may be a promising material for TCO films suitable for specialized applications.

In addition, multicomponent oxides consisting of a combination of different metal oxides such as binary or ternary compounds have recently attracted much attention as new materials for transparent conducting oxide (TCO) films.[16] Transparent conducting multicomponent oxide films may exhibit properties which are suitable for specialized application as a result of changes in physical properties brought about by controlling the composition. In this paper, we describe the preparation of transparent conducting thin films

211

using various multicomponent oxides consisting of a combination of different In_2O_3 based ternary compounds. $(Ga,In)_2O_3$-$MgIn_2O_4$, $(Ga,In)_2O_3$-$Zn_2In_2O_5$, $(Ga,In)_2O_3$-$In_4Sn_3O_{12}$, $Zn_2In_2O_5$-$In_4Sn_3O_{12}$ and $Zn_2In_2O_5$ $MgIn_2O_4$ system thin films have been prepared.

EXPERIMENTAL

Films were prepared by conventional r.f. planar magnetron sputtering using powder targets. A mixture of ZnO (purity, 99.99%), MgO (purity, 99.99%), In_2O_3 (purity, 99.99%), Ga_2O_3 (purity, 99.99%) and SnO_2 (purity, 99.99%) powders calcined at 1000℃ in an argon (Ar) atmosphere for 5 hours was used as the target: stainless steel holder, diameter of 80 mm. Substrates of Corning 7059 glass were placed parallel to the target surface at a distance of 35 mm. Sputtering deposition was carried out at sputter gas pressures of 0.2 to 2.0 Pa in a pure Ar gas or a mixture of Ar and oxygen (O_2) gases with a d.c. power of 40 W. The O_2 gas content in the Ar+O_2 gas atmosphere was varied from 0 to 8%. Substrate temperatures were room temperature (RT). Although substrates at RT were not intentionally heated, surface temperatures reached about 160℃ after a sputter deposition of approximately 30 min. The deposition rate was dependent on the powder composition; the deposition rates for ZnO, MgO, In_2O_3, Ga_2O_3 and SnO_2 powders were about 15, 10, 10, 8, 8 nm/min, respectively.

Film thickness was measured using a conventional surface roughness detector with stylus. The thickness of films deposited in this work ranged from 340 to 490 nm. The composition of deposited films was measured by energy dispersive X-ray spectroscopy (EDX). The EDX of the deposited multicomponent oxide films showed that the composition (the metal element content, or the atomic ratio) in the films was approximately equal to that in the target. The crystalline structure of the deposited films was investigated by X-ray diffraction; a conventional X-ray unit with a copper anode was used. Electrical resistivity and Hall mobility were measured using the van der Pauw method. Optical transmission through the film was measured in the visible wavelength range, 300 to 800 nm. The work function of the films was determined from the wavelength dependence of photoemission of electrons using ultraviolet photoelectron spectroscopy (Model AC-1 Riken Keiki Co. Ltd.).

RESULTS AND DISCUSSION

Transparent conducting $MgIn_2O_4$ and $In_4Sn_3O_{12}$ films prepared on substrates at RT by r.f. magnetron sputtering were always polycrystalline.[2,16] In contrast, $(Ga,In)_2O_3$ and $Zn_2In_2O_5$ films prepared on RT substrates were amorphous;[10-13] however, these films prepared on substrates at 350℃ were polycrystalline,[10-13] identified as $(Ga,In)_2O_3$ and $Zn_2In_2O_5$, respectively. Therefore, in the following description, the formula of the above materials are presented as ternary compounds even for materials where the films prepared at RT are amorphous.

$(Ga,In)_2O_3$-$In_4Sn_3O_{12}$ Thin Films

We reported that transparent conducting $In_4Sn_3O_{12}$ films were prepared at substrate temperatures of RT and 350℃ by r.f. or d.c. magnetron sputtering using a In_2O_3-SnO_2 target with a Sn content (Sn/(In+Sn) atomic ratio) of approximately 50 atomic%.[14,15] The preparation of transparent conducting $GaInO_3$ films was first reported by Phillips et al.[9] In addition, we reported that in Ga_2O_3-In_2O_3 system films prepared at a substrate temperature of RT and 350℃ by r.f. or d.c. magnetron sputtering, a minimum resistivity

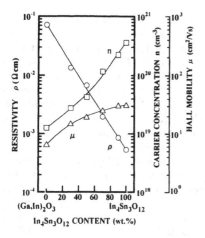

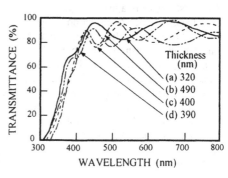

Fig.1. *Resistivity (○), Hall mobility (△) and carrier concentration (□) as functions of $In_4Sn_3O_{12}$ content for $(Ga,In)_2O_3$-$In_4Sn_3O_{12}$ films.*

Fig.2. *Optical transmission spectra for $(Ga,In)_2O_3$-$In_4Sn_3O_{12}$ films prepared with $In_4Sn_3O_{12}$ contents of 0 (a), 30 (b), 70 (c) and 100 (d) wt.%.*

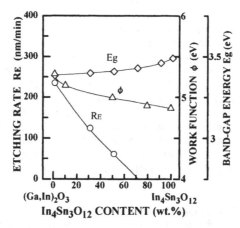

Fig.3. *Band-gap energy (◇), work function (△) and etching rate (○) as functions of $In_4Sn_3O_{12}$ content for $(Ga,In)_2O_3$-$In_4Sn_3O_{12}$ films.*

was obtained using a Ga_2O_3-In_2O_3 target with a Ga content (Ga/(In+Ga) atomic ratio) of approximately 50 at.%.[17,18] When prepared on substrates at 350℃, the deposited films was identified as $(Ga,In)_2O_3$ by X-ray diffraction analyses.[18]

The electrical properties of resistivity (ρ), carrier concentration (n) and Hall mobility (μ) as functions of the $In_4Sn_3O_{12}$ content are shown in Fig.1 for $(Ga,In)_2O_3$-$In_4Sn_3O_{12}$ films prepared at a substrate temperature of RT. The film deposition was carried out at a sputter gas pressure of 0.25 Pa with an O_2 gas content of 1% and under the optimal conditions for the preparation of both $In_4Sn_3O_{12}$ and $(Ga,In)_2O_3$ films. The electrical properties varied continuously as the $In_4Sn_3O_{12}$ content was varied from 0 to 100 wt.%. The resistivity of $(Ga,In)_2O_3$-$In_4Sn_3O_{12}$ films was varied in a range from 10^{-2} to 10^{-4} Ω cm by changing the composition. Figure 2 shows optical transmission spectra for $(Ga,In)_2O_3$-$In_4Sn_3O_{12}$ films prepared at RT using targets with different $In_4Sn_3O_{12}$ contents. An average transmittance above 85% in the visible range was obtained in all $(Ga,In)_2O_3$-$In_4Sn_3O_{12}$ films: $In_4Sn_3O_{12}$ content varied from 0 to 100 wt.%. From the transmission spectra shown in Fig.2, the band-gap energy [13,16] of $(Ga,In)_2O_3$-$In_4Sn_3O_{12}$ films is roughly estimated to have slightly increased from 3.4 to 3.5 eV as the $In_4Sn_3O_{12}$ content was increased. It should be noted that transparent and conductive thin films could be prepared over all the compositions in the $(Ga,In)_2O_3$-$In_4Sn_3O_{12}$ system.

For specialized applications, it is desirable to control the etching rate of TCO films. $(Ga,In)_2O_3$ films were more easily etched in acid solutions such as HCl than indium-tin oxide (ITO) films, which have been used for practical conventional applications. However, $In_4Sn_3O_{12}$ films were difficult to etch in acid solutions. Figure 3 shows work function (ϕ), band-gap energy (Eg) and etching rate (RE) as functions of the $In_4Sn_3O_{12}$ content for $(Ga,In)_2O_3$-$In_4Sn_3O_{12}$ films prepared on substrates at RT; work function and band-gap energy changed slightly as the composition was changed. As can be seen in Figs. 1 and 3, it was found that the work function decreased from 5.2 to 4.8 eV but the carrier concentration increased from approximately 10^{19} to 10^{21} as the $In_4Sn_3O_{12}$ content was increased from 0 to 100 wt%. It should be noted that the work function showed a tendency to decrease with increasing carrier concentration. The above relationship between work function and carrier concentration may be attributable to the theoretically expected relationship between carrier concentration and Fermi energy found in degenerated semiconductors. The etching test of films was carried out in 0.2 M HCl at 25°C. The etching rate decreased as the $In_4Sn_3O_{12}$ content was increased up to about 60 wt.%; however, films with $In_4Sn_3O_{12}$ contents above 70 wt.% were not etched in this etchant. The decrease of etching rate with increasing $In_4Sn_3O_{12}$ content is related to the increasing Sn content in the TCO films; i.e., the chemical properties of TCO films were basically determined by the kind and amount of metal elements they contain.[16] From X-ray diffraction analyses, $(Ga,In)_2O_3$-$In_4Sn_3O_{12}$ films prepared on RT substrates with $In_4Sn_3O_{12}$ contents from 0 to about 90 wt.% were amorphous. Although $In_4Sn_3O_{12}$ films prepared at RT was crystalline, the electrical, optical and chemical properties of the films, as described above, were relatively independent of the crystallographical properties of films prepared with a $In_4Sn_3O_{12}$ content ranging from 0 to 100 wt.%.

$(Ga,In)_2O_3$-$Zn_2In_2O_5$ Thin Films

We have reported that transparent conducting $Zn_2In_2O_5$ films could be prepared at substrate temperatures of RT and 350°C by r.f. or d.c. magnetron sputtering using an ZnO-In_2O_3 target with a Zn content (Zn/(In+Zn) atomic ratio) of 15-40 at.%.[11-13] The electrical properties as functions of the $Zn_2In_2O_5$ content are shown in Fig.4 for $(Ga,In)_2O_3$-$Zn_2In_2O_5$ films prepared at a substrate temperature of RT. The film deposition was carried out in pure Ar at a sputter gas pressure of 0.25 Pa and under the optimal conditions for the preparation of $Zn_2In_2O_5$ films. The electrical properties varied continuously as the $Zn_2In_2O_5$ content was varied from 0 to 100 wt.%. The resistivity of $(Ga,In)_2O_3$-$Zn_2In_2O_5$ films could be controlled in a range from 10^{-3} to 10^{-4} Ω cm by changing the composition. It was found that an average transmittance above 80% in the visible range was obtained in all $(Ga,In)_2O_3$-$Zn_2In_2O_5$ films: $Zn_2In_2O_5$ content ranging from 0 to 100 wt.%. The band-gap energy of $(Ga,In)_2O_3$- $Zn_2In_2O_5$ films is roughly

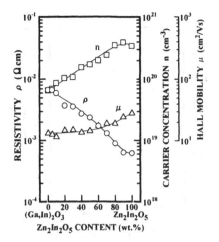 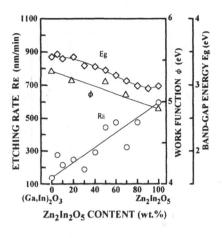

Fig.4 Resistivity (○), Hall mobility
(△) and carrier concentration (□) as
functions of $Zn_2In_2O_5$ content for
$(Ga,In)_2O_3$-$Zn_2In_2O_5$ films.

Fig.5. Band-gap energy (◇), work
function (△) and etching rate (○) as
functions of $Zn_2In_2O_5$ content for
$(Ga,In)_2O_3$-$Zn_2In_2O_5$ films.

estimated to have slightly decreased from 3.4 to 2.9 eV as the $Zn_2In_2O_5$ content was
increased. It should be noted that transparent and conductive thin films could be prepared
over all the compositions in the $(Ga,In)_2O_3$-$Zn_2In_2O_5$ system. In addition, $(Ga,In)_2O_3$-
$Zn_2In_2O_5$ films prepared at RT with a $Zn_2In_2O_5$ content that ranged from 0 to 100 wt.%
were amorphous.

Figure 5 shows work function, band-gap energy and etching rate as functions of the
$Zn_2In_2O_5$ content for $(Ga,In)_2O_3$-$Zn_2In_2O_5$ films prepared on substrates at RT. The work
function and band-gap energy in this system decreased slightly as the $Zn_2In_2O_5$ content was
increased. Figures 4 and 5 show that the work function tends to decrease with increasing
carrier concentration. The etching rate of $(Ga,In)_2O_3$-$Zn_2In_2O_5$ films was increased as the
$Zn_2In_2O_5$ content was increased; the etching test was carried out in 0.2 M HCl at 25℃. The
increase in etching rate with increasing $Zn_2In_2O_5$ content is related to the increasing Zn
content in the TCO films.[16]

$(Ga,In)_2O_3$-$MgIn_2O_4$ Thin Films

The preparation of transparent conducting $MgIn_2O_4$ films was first reported by Un'no
et al.[6] In addition, we reported that in MgO-In_2O_3 system films prepared at a substrate
temperature of RT by r.f. or d.c. magnetron sputtering, a minimum resistivity could be
obtained by using a MgO-In_2O_3 target with an In content (In/(Mg+In) atomic ratio) of
approximately 84 at.%.[19,20] The electrical properties as functions of the $MgIn_2O_4$ content
are shown in Fig.6 for $(Ga,In)_2O_3$-$MgIn_2O_4$ films prepared at a substrate temperature of RT.
The film deposition was carried out in pure Ar at a sputter gas pressure of 0.8 Pa and under
the optimal conditions for the preparation of $MgIn_2O_4$ films. The electrical properties

215

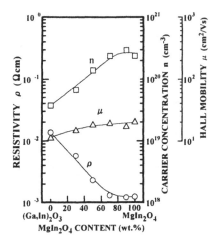

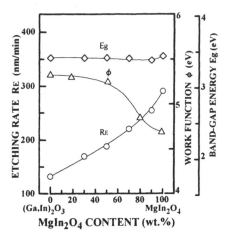

Fig.6. Resistivity (○), Hall mobility (△) and carrier concentration (□) as functions of MgIn₂O₄ content for (Ga,In)₂O₃-MgIn₂O₄ films.

Fig.7. Band-gap energy (◇), work function (△) and etching rate (○) as functions of MgIn₂O₄ content for (Ga,In)₂O₃-MgIn₂O₄ films.

varied continuously as the MgIn₂O₄ content was varied from 0 to 100 wt.%. The resistivity of (Ga,In)₂O₃-MgIn₂O₄ films could be controlled in a range from 10^{-2} to 10^{-3} Ω cm by changing the composition. It was found that an average transmittance above 80% in the visible range was obtained in all (Ga,In)₂O₃-MgIn₂O₄ films: MgIn₂O₄ content ranging from 0 to 100 wt.%. The band-gap energy of (Ga,In)₂O₃-MgIn₂O₄ films is roughly estimated to be 3.4 eV. It should be noted that transparent and conductive thin films could be prepared over all the compositions in the (Ga,In)₂O₃-MgIn₂O₄ system.

Figure 7 shows work function, band-gap energy and etching rate as functions of the MgIn₂O₄ content for (Ga,In)₂O₃-MgIn₂O₄ films prepared on substrates at RT. The band-gap energy in this system was relatively independent of the composition. The work function decreased from 5.2 to 4.7 eV as the MgIn₂O₄ content was increased. In addition, it should be noted that the decrease of work function is related to the increase of carrier concentration in a manner which is similar to the relationship between the work function and carrier concentration found in the other multicomponent oxide films described above. However, the change in work function with changes in the composition of multicomponent oxide films was not related to that of band-gap energy. The etching rate in 0.2 M HCl at 25°C could be controlled by varying the MgIn₂O₄ content. From X-ray diffraction analyses, the (Ga,In)₂O₃-MgIn₂O₄ films prepared on RT substrates with MgIn₂O₄ contents from 0 to about 90 wt.% were found to be amorphous. Although MgIn₂O₄ films prepared at RT were crystalline, the electrical, optical and chemical properties of the films, as described above, were relatively independent of the crystallographical properties of films prepared with a MgIn₂O₄ content that ranged from 0 to 100 wt.%.

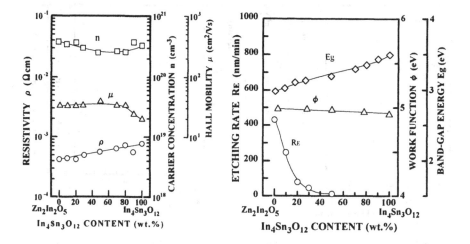

Fig.8. Resistivity (○), Hall mobility (△) and carrier concentration (□) as functions of $In_4Sn_3O_{12}$ content for $Zn_2In_2O_5$-$In_4Sn_3O_{12}$ films.

Fig.9. Band-gap energy (◇), work function (△) and etching rate (○) as functions of $In_4Sn_3O_{12}$ content for $Zn_2In_2O_5$-$In_4Sn_3O_{12}$ films.

$Zn_2In_2O_5$-$In_4Sn_3O_{12}$ Thin Films

The electrical properties as functions of the $Zn_2In_2O_5$ content are shown in Fig.8 for $Zn_2In_2O_5$-$In_4Sn_3O_{12}$ films prepared at a substrate temperature of RT. The film deposition was carried out in pure Ar at a sputter gas pressure of 1.2 Pa and under the optimal conditions for the preparation of $Zn_2In_2O_5$ films. The electrical properties varied continuously as the $In_4Sn_3O_{12}$ content was varied from 0 to 100 wt.%. The resistivity of $Zn_2In_2O_5$-$In_4Sn_3O_{12}$ films could be controlled in a range from 3 to 8×10^{-4} Ω cm by changing the composition. It was found that an average transmittance above 80% in the visible range was obtained in all $Zn_2In_2O_5$-$In_4Sn_3O_{12}$ films: $In_4Sn_3O_{12}$ content ranging from 0 to 100 wt.%. The band-gap energy of $Zn_2In_2O_5$-$In_4Sn_3O_{12}$ films is roughly estimated to have slightly increased from 2.9 to 3.4 eV as the $In_4Sn_3O_{12}$ content was increased. It should be noted that highly transparent and conductive thin films could be prepared over all the compositions in the $Zn_2In_2O_5$-$In_4Sn_3O_{12}$ system.

Figure 9 shows work function, band-gap energy and etching rate as functions of the $In_4Sn_3O_{12}$ content for $Zn_2In_2O_5$-$In_4Sn_3O_{12}$ films prepared on substrates at RT. The work function and the carrier concentration in this system were relatively independent of the composition. In contrast, the band-gap energy gradually increased as the $In_4Sn_3O_{12}$ content was increased. $Zn_2In_2O_5$-$In_4Sn_3O_{12}$ films with $In_4Sn_3O_{12}$ contents up to 60 wt.% were not etched; however, the etching rate increased as the $In_4Sn_3O_{12}$ content was decreased from 50

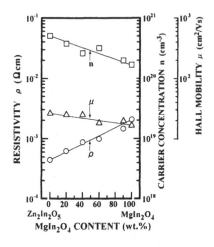

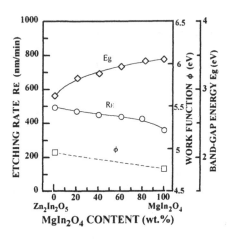

Fig.10. Resistivity (○), Hall mobility (△) and carrier concentration (□) as functions of MgIn₂O₄ content for Zn₂In₂O₅-MgIn₂O₄ films.

Fig.11. Band-gap energy (◇), work function (△) and etching rate (○) as functions of MgIn₂O₄ content for Zn₂In₂O₅-MgIn₂O₄ films.

to 0 wt.%. The decrease in etching rate with increasing $In_4Sn_3O_{12}$ content is related to not only the decreasing Zn content in the TCO films but also the increasing Sn content.[16] $Zn_2In_2O_5$-$In_4Sn_3O_{12}$ films prepared on RT substrates with $In_4Sn_3O_{12}$ contents ranging from 0 to about 90 wt.% were amorphous. As described above, the electrical, optical and chemical properties of the films were relatively independent of the crystallographical properties of films prepared with a $In_4Sn_3O_{12}$ content ranging from 0 to 100 wt.%.

Zn₂In₂O₅-MgIn₂O₄ Thin Films

The electrical properties of $Zn_2In_2O_5$-$MgIn_2O_4$ films varied continuously as the $Zn_2In_2O_5$ content was varied from 0 to 100 wt.%, as shown in Fig.10. The film deposition was carried out in pure Ar at a sputter gas pressure of 1.2 Pa. The resistivity of $Zn_2In_2O_5$-$MgIn_2O_4$ films could be controlled in a range from 10^{-3} to 10^{-4} Ω cm by changing the composition. It was found that an average transmittance above 80% in the visible range was obtained in all $Zn_2In_2O_5$-$MgIn_2O_4$ films: $MgIn_2O_4$ content ranging from 0 to 100 wt.%. The band-gap energy of $Zn_2In_2O_5$-$MgIn_2O_4$ films is roughly estimated to have slightly increased from 2.9 to 3.4 eV as the $MgIn_2O_4$ content was increased. It should be noted that transparent and conductive thin films could be prepared over all the compositions in the $Zn_2In_2O_5$-$MgIn_2O_4$ system.

Figure 11 shows work function, band-gap energy and etching rate as functions of the $MgIn_2O_4$ content for $Zn_2In_2O_5$-$MgIn_2O_4$ films prepared on substrates at RT. The band-gap energy increased as the $MgIn_2O_4$ content was increased. The work function in this system decreased slightly as the $MgIn_2O_4$ content was increased. In contrast, the carrier concentration of $Zn_2In_2O_5$-$MgIn_2O_4$ film decreased slightly as the $MgIn_2O_4$ was increased, as seen in Fig. 10. It should be noted that the relationship between the work function and carrier concentration in this system was different from the other systems described above.

This may be attributable to the work function of MgIn$_2$O$_4$, being lower than the other ternary compounds used in this work. The etching rate of Zn$_2$In$_2$O$_5$-MgIn$_2$O$_4$ films was also independent of the composition; the etching test was carried out in 0.2 M HCl at 25°C.

Spatial Resistivity Distribution on the Substrate Surface

It is well known that polycrystalline TCO films prepared on low temperature substrates by magnetron sputtering exhibit a spatial resistivity distribution on the substrate surface.[21-24] Although most of the multicomponent oxide films described above were amorphous, as evidenced from X-ray diffraction analyses, they exhibited a spatial resistivity distribution on the surface of substrates placed parallel to the target surface. For example, the spatial resistivity distribution of MgIn$_2$O$_4$-(Ga,In)$_2$O$_3$ thin films was considerably affected by the MgIn$_2$O$_4$ content in the target. Figures 12 and 13 show the electrical properties as functions of substrate surface location for MgIn$_2$O$_4$-(Ga,In)$_2$O$_3$ thin films prepared on RT substrates using targets with MgIn$_2$O$_4$ contents of 20 and 84.2 wt.% and 40 wt.%, respectively. The film depositions were carried out in pure Ar at a sputter gas pressure of 0.25 Pa and under the optimal conditions for the preparation of (Ga,In)$_2$O$_3$ films. As can be seen in these figures, the curvature of the spatial resistivity distribution inverted as the MgIn$_2$O$_4$ content was increased. The carrier concentration at a location on the substrate corresponding to the erosion pattern of the target (approximately 3 to 4 cm) changed markedly depending on the MgIn$_2$O$_4$ content. The electrical properties at locations on the substrate corresponding to either the erosion pattern (closed symbols) and the center

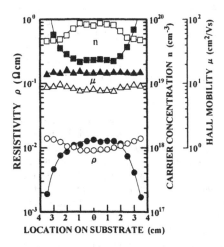 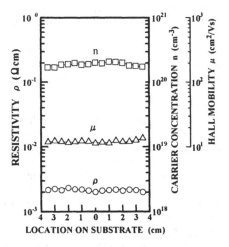

Fig.12. *Resistivity (○), Hall mobility (△) and carrier concentration (□) as functions of substrate location for (Ga,In)$_2$O$_3$-MgIn$_2$O$_4$ films prepared with MgIn$_2$O$_4$ contents of 20 (open symbols) and 84.2 wt.% (closed symbols).*

Fig.13. *Resistivity (○), Hall mobility (△) and carrier concentration (□) as functions of substrate location for (Ga,In)$_2$O$_3$-MgIn$_2$O$_4$ films prepared with a MgIn$_2$O$_4$ content of 60 wt.%.*

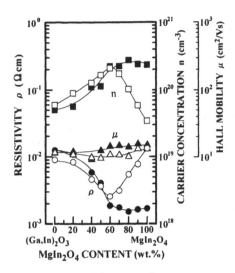

Fig.14. MgIn$_2$O$_4$ content dependence of resistivity (○), Hall mobility (△) and carrier concentration (□) at locations on the substrate corresponding to either the erosion pattern (closed symbols) or the center (open symbols) of the target.

(open symbols) of the target are shown in Fig.14 as a function of the MgIn$_2$O$_4$ content for MgIn$_2$O$_4$-(Ga,In)$_2$O$_3$ thin films prepared under the same deposition conditions as the films shown in Figs.12 and 13. The MgIn$_2$O$_4$ content dependence of the resistivity distribution is mainly related to that of carrier concentration. The above results may suggest that the spatial resistivity distribution of multicomponent TCO films is mainly affected by the amount of oxygen [25] rather than the bombardment by high energetic particles.[23,24]

CONCLUSIONS

Transparent conducting oxide films using multicomponent oxides consisting of a combination of various In$_2$O$_3$ based ternary compounds were newly developed. Transparent conducting thin films using (Ga,In)$_2$O$_3$-MgIn$_2$O$_4$, (Ga,In)$_2$O$_3$-Zn$_2$In$_2$O$_2$, (Ga,In)$_2$O$_3$-In$_4$Sn$_3$O$_{12}$, Zn$_2$In$_2$O$_5$-In$_4$Sn$_3$O$_{12}$ and Zn$_2$In$_2$O$_5$-MgIn$_2$O$_4$ were prepared over all the compositions in these multicomponent oxides. The resulting transparent conducting multicomponent oxide films prepared on room temperature substrates by r.f. magnetron sputtering are very promising for use in specialized applications, because the resistivity, band-gap energy, work function and etching rate of the films could be controlled by varying the composition in the target. It can be concluded that multicomponent oxides consisting of combinations of ternary compounds which are transparent conducting oxide film materials or transparent conductors were transparent conductors at all compositions. In addition, multicomponent oxide films prepared at RT exhibited a significant spatial distribution of resistivity on the substrate which was dependent on the deposition conditions and composition. The resistivity distribution is attributable to the oxygen concentration on the substrate surface rather than the bombardment effect of high energy particles. It is concluded that transparent conducting multicomponent oxide films exhibit properties which are suitable for specialized applications because of the changes in physical properties brought about by controlling the composition of materials.

ACKNOWLEDGMENTS

The authors wish to acknowledge Takeshi Usami, Katsushige Kobayashi and Toshimitsu Tanaka for their technical assistance in the experiments. This work was partially supported by a Grant-in-Aid for Scientific Research No.11650033 from the Ministry of Education, Science and Culture of Japan.

REFERENCES

1. H.L.Hartnagel, A.L.Dawar, A.K.Jain and C.Jagadish, "Semiconducting Transparent Thin Films" , Chap.2(Institute of Physics Publishing Bristol and Philadelphia, PA, 1995) p.22.
2. H.Enoki, T.Nakayama and J.Echigoya, Phys. Stat. Sol. (a), **129**, p.181 (1992).
3. K.Yanagawa, Y.Ohki, N.Ueda, T.Omata, T.Hashimoto and H.Kawazoe, Appl. Phys. Lett., **63**, p.3335 (1993).
4. K.Yanagawa, Y.Ohki, T.Omata, H.Hosono, N.Ueda and H.Kawazoe, Appl. Phys. Lett., **65**, p.406 (1994).
5. N.Ueda, T.Omata, N.Hikuma, K.Ueda, H.Mizoguchi, T.Hashimoto and H.Kawazoe, Appl.Phys. Lett. , **61**, p.1954 (1992).
6. H.Un'no, N.Hikuma, T.Omata, N.Ueda, T.Hashimoto and H.Kawazoe, Jpn. J. Appl. Phys., **32**, L1260 (1993).
7. T.Minami, H.Sonohara, S.Takata and H.Sato, Jpn. J. Appl. Phys. , **33**, L1693 (1994).
8. T.Minami, S.Takata, H.Sato and H.Sonohara, J. Vac. Sci. Technol. A,**13**, p.1095 (1995)
9. R.J.Cava, J.M.Phillips, J.Kwo, G.A.Thomas, R.B.van Dover, S.A.Carter, J.J.Krajewski, W.F.Peck Jr., J.H.Marshall and D.H.Rapkine, Appl. Phys. Lett. , **64**, p.2071 (1994).
10. J.M.Phillips, J.Kwo, G.A.Thomas, S.A.Carter, R.J.Cava, S.Y.Hou, J.J.Krajewski, J.H.Marshall, W.F.Peck Jr., D.H.Rapkine and R.B. van Dover, Appl. Phys. Lett., **65**, p.115 (1994).
11. T.Minami, H.Sonohara, T.Kakumu and S.Takata, Jpn. J. Appl. Phys. , **34**, L971 (1995).
12. T.Minami, T.Kakumu and S.Takata, J. Vac. Sci. Technol. A ,**14**, p.1704 (1996).
13. T.Minami, T.Kakumu, Y.Takeda and S.Takata, Thin Solid Films, **290/291**, p.1 (1996).
14. T.Minami, Y.Takeda, S.Takata and T.Kakumu, Thin Solid Films, **308-309**, p.13 (1997).
15. T.Minami, T.Kakumu, K.Shimokawa and S.Takata, Thin Solid Films, **317**, p.318 (1998).
16. T.Minami, J. Vac. Sci. Technol. A ,**17**, p.1765 (1999).
17. T.Minami, S.Takata and T.Kakumu, J. Vac. Sci. Technol. A ,**14**, p.1689 (1996).
18. T.Minami, Y.Takeda. T.Kakumu, S.Takata and I.Fukuda, J. Vac. Sci. Technol. A,**15**, p.958 (1997).
19. T.Minami, H.Sonohara, S.Takata and H.Sato, J. Vac. Sci. Technol. A,**13**, p.1095 (1995).
20. T.Minami, S.Takata, T.Kakumu and H.Sonohara, Thin Solid Films, **270**, p22 (1995).
21. T.Minami, H.Nanto, H.Sato and S.Takata, Thin Solid Films , **164**, p.275 (1988).
22. J.B.Webb, Thin Solid Films , **136**, p.135 (1986).
23. K.Tominaga, T.Yuasa, M.Kume and O.Tada, Jpn. J. Appl. Phys. , **24**, p.944 (1985).
24. S.Ishibashi, Y.Higuchi, Y.Ota and K.Nakamura, J. Vac. Sci. Technol. A, **8**, p1403 (1990).
25. K.Ichihara, N.Inoue, M.Okubo and N.Yasuda, Thin Solid Films , **245**, p152 (1994).

CONTROL OF VALENCE STATES IN ZnO BY CODOPING METHOD

T. YAMAMOTO *, H. K.-YOSHIDA **
*Electronic and Photonic Systems Engineering Department, Kochi University of Technology,
Tosayamada-cho, Kochi 782-8502, JAPAN, yamateko@ele.kochi-tech.ac.jp
**Condensed Matter Physics Department, ISIR, Osaka University, Osaka 567-0047, JAPAN

ABSTRACT

We have investigated the electronic structures of p- or n-type doped ZnO based on *ab initio* electronic band structure calculations in order to control valence states in ZnO for the fabrication of low-resistivity p-type ZnO. We find *unipolarity* in ZnO; p-type doping using Li or N increases the Madelung energy while n-type doping using Al, Ga, In or F species decreases the Madelung energy. We have proposed materials design: codoping using N acceptors and reactive codopants, Al or Ga, enhances electric properties in p-type codoped ZnO. It has been already verified by experiments using the N acceptors and Ga reactive donor codopants. We find a very weak repulsive interaction between Li acceptors and the delocalization of the Li-impurity states for Li-doped ZnO, in contrast with the case of N-doped ZnO. On the other hand, we find the compensation mechanism by the formation of O vacancy in the vicinity of the Li-acceptor sites. We propose a group VII element, F species, as a promising candidate for use of the reactive codopant as for Li-doped ZnO in order to realize low-resistivity p-type ZnO.

INTRODUCTION

Zinc oxide (ZnO) with a wurtzite structure is a wide band gap (3.436 eV at 4.2K) semiconductor which has many applications such as piezoelectronic transducers, varistors, and highly optically transparent conducting films. Recent successes in producing large-area single crystals have opened up the possible applications in short-wave length light emitting devices [1,2]. The main advantage of ZnO as the light emitter is large exciton binding energy (60 meV). ZnO is also much more resistant to radiation damage than other common semiconducting materials, such as Si, GaAs, CdS, and even GaN; thus it should be useful for space applications. In order to develop not only the optoelectronic devices but also such large-scale applications, one important issue that should be resolved is the fabrication of low-resistivity p-type doped ZnO, as well as other wide band gap semiconductors such as ZnSe and GaN. The realization of the p-type ZnO will lead to the fabrication of a unique pn junction, a key structure in the semiconductor technology.

ZnO is naturally an n-type semiconductor because of a deviation from stoichiometry due to the presence of intrinsic point defects such as O vacancies (V_O) and Zn interstitials (Zn_i). We have found a substantial decrease in the Madelung energy of n-type ZnO with the excess of Zn originated in the formation of V_O and Zn_i compared with that of stoichiometric ZnO, which causes the stabilization of ionic charge distributions in n-type ZnO, from *ab initio* electronic band structure calculations [3]. Several authors reported the fabrication of low-resistivity n-type ZnO using group III elements, [4,5] Si, a Zn-substituting species [6], or a group VII element, F [7]. On the other hand, ZnO has proven to be difficult to dope as p-type; few studies concerning the fabrication of p-type ZnO have been reported [8,9]. Recently, Minegishi et al. succeeded in realizing p-type ZnO at room temperature by chemical vapor deposition, using simultaneous codoping of NH_3 and excess Zn [10]. That study, however, showed poor reproducibility, high resisitivity of typically 100 $\Omega\cdot$cm and low carrier concentration in the order of 10^{16} cm^{-3} which are not sufficient for many applications.

Recently, we have proposed the codoping method using acceptors and reactive donors simultaneously as materials design for the fabrication of low-resistivity p-type wide band gap semiconductors such as $CuInS_2$ [11-13], GaN [14-16], ZnSe [17-19], ZnO [20] and AlN [21]. For GaN, our prediction concerning the codopant pair, Be acceptors and O reactive donors, which is one of four pairs proposed by us [14-16], was confirmed by experiments successfully conducted by German group [22]. For ZnO, we predicted two effective codopant pairs, (N, Ga) and (N, Al) [20]. Very recently, Osaka's group verified it, reporting the realization of p-type ZnO with a

223

room temperature resistivity of 2 $\Omega \cdot$cm and high hole concentration of 4×10^{19} cm^{-3} using N as acceptors and Ga as reactive donors [23]. They concluded, from X-ray photoelectron spectroscopic (XPS) measurements, that they get a 2N : 1Ga ratio in the codoped ZnO films, which is the ratio predicted by our theoretical calculations for obtaining low-resistivity p-type ZnO.

In this paper, we first study the *"unipolarity"* doping problem in ZnO crystals or films, based on the results of *ab initio* electronic band structure calculations. We investigate what hampers the use of N, an O-substituting species, or Li, a Zn-substituting one, as a practical acceptor dopant. Second, we summarize the effects of the codoping on p-type doping in wide-band-gap semiconductors in order to propose an effective solution to the *"unipolarity"* doping problem. Third, we discuss the influence of reactive donors on the incorporation of acceptors, N and Li species, and electronic structures of the p-type codoped ZnO.

METHODOLOGY

The results of our band structure calculations for ZnO crystals were based on the local-density approximation (LDA) treatment of electronic exchange and correlation [24-26] and on the augmented spherical wave (ASW) formalism for the solution of effective single-particle equations [27]. For the calculations, the atomic sphere approximation (ASA) with a correction term was adopted. For undoped ZnO crystals, Brillouin zone integration was carried out for 84-k points in an irreducible wedge and for 24-k points for doped and codoped ZnO crystals. For valence electrons, we employed outermost s, p and d orbitals for Zn atoms and s and p orbitals for the other atoms. The Madelung energy, which reflects long-range electrostatic interaction in the system, was assumed to be restricted to a sum over the monopoles.

We studied the crystal structures of doped and codoped ZnO with periodic boundary conditions by generating supercells that contain the object of interest. (1) For n-type ZnO doped with group III elements (III=Al, Ga or In), a Zn-substituting species, we replace one of the 16 sites of Zn atoms by a donor site in model supercells, as shown Fig. 1. (2) For n-type ZnO doped with a group VII element, F species, an O-substituting one, we replace one of the 16 sites of O atoms by a donor site in model supercells. (3) For p-type ZnO doped with N (Li) alone, we replace one of the 16 sites of O (Zn) atoms by an acceptor site. (4) For p-type ZnO codoped with the reactive donors using the group III elements and 2N, (ZnO:(III, 2N)), we replace two of the 16 sites of the O atoms by the N atom sites and one of the 16 sites of the Zn atoms by the donor site. (5) For p-type ZnO codoped with the reactive donors using the F species and 2Li, (ZnO:(F, 2Li)), we replace two of the 16 sites of the Zn atoms by the Li atom sites and one of the 16 sites of the O atoms by the donor site.

In this work, for codoped ZnO crystals, we decided the crystal structures under the condition that the total energy is minimized from all atomic configurations. We neglect the effects of relaxation due to dopants whose covalent radii are similar to those of substituted atoms on the lattice constants and displacement of atoms in the vicinity of the dopants because the magnitude of change in the total energy due to the relaxation, almost on the order of 10^2 meV, is smaller than those of the Madelung energy calculated below.

RESULTS AND DISCUUSION

n-Type Doping

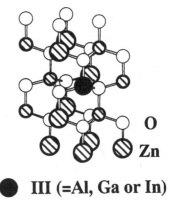

O

Zn

III (=Al, Ga or In)

Fig. 1.   Crystal structure of supercell for ZnO doped with the group III elements, Al, Ga or In.

First, we show the crystal structure of ZnO doped with the group III elements (Al, Ga and In) in Fig. 1. Next, we present the total density of states (DOS) of undoped ZnO crystals in Fig. 2(a) as a standard reference and those of ZnO doped with Al, Ga, In or F in Figs. 2(b) to 2(e), respectively. The $2s$ states at O sites are included in the calculation as valence states, but those located around -18 eV below the Fermi energy (E_F) are omitted in Fig. 2 because such electrons form a narrow band and have little interaction with other states. Energy is measured relative to E_F.

For undoped ZnO in Fig. 2(a), we find two groups in the valence band. (1) There are bands with a strong d character originating mostly from d states at Zn sites from -6.5 eV to -4.0 eV. (2) The upper valence band located above approximately - 4.0 eV originates mainly from the p states at O sites. The lowest conduction bands have a strong Zn $4s$ contribution; there are charge transfers from Zn $4s$ to O $2p$ due to the mixing between the s and p states at O sites, and the s and p states of the surrounding Zn shifts the center of gravity of the local DOS at the O sites towards lower-energy regions. This results from a remarkable difference in electronegativity, 1.79, between Zn species (1.65) and O one (3.44) by Pauling, whose percent ionic character is 55 %; It means that the chemical bonds in ZnO with a wurtzite structure have an ionic rather than a covalent character. The features of the DOS of undoped ZnO are in good agreement with those already calculated theoretically using the LDA [28, 29].

Figures 2(b) to 2(e) show that an excess electron is generated at the bottom of the conduction band. The arrows in Figs. 2(b) to 2(d) indicate the bonding states between the s states of the donor dopants and the p states of O atoms located near the dopants. For F-doped ZnO in Fig. 2(e), the arrow indicates the bonding states between the s states at the F

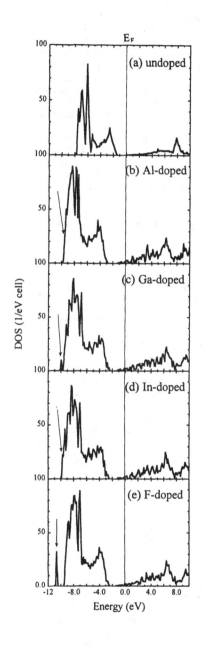

Fig. 2. Total DOS of (a) undoped ZnO, (b) Al-doped ZnO, (c) Ga-doped ZnO, (d) In-doped ZnO, and (e) F-doped ZnO crystals.

sites and the s states at the sites of Zn close to the site of the F atoms. The impurity states for n-type ZnO:Al, Ga, In or F are found to be largely delocalized near E_F from the examination of the site-decomposed DOS and the dispersion relations. This results in the same band shape as that of undoped ZnO in Fig. 2(a). Thus, the *ab initio* calculations indicate that the four n-type doped ZnO have shallow donor levels. We note that there is a sharp DOS peak, marked by arrows, due to attractive potentials except for the case of n-type Al-doped ZnO, as shown in Fig. 2(b). This means that the impurity states for the three n-type ZnO doped with Ga, In or F are localized compared with that for n-type ZnO:Al. As a result, we verify weak repulsive interactions between Al acceptors for n-type ZnO:Al compared with those for the three n-type ZnO [20], suggesting that the Al species would be most stable among the four donor dopants.

p-Type Doping

We show the total DOS of undoped ZnO crystals in Fig. 3(a) as a standard and those and site-decomposed DOS of ZnO doped with N or Li in Figs. 3(b) to 3(e), respectively. In Figs. 3(b) and 3(d), we see that a hole band is generated at the top of the valence for p-type ZnO:(N or Li): both N at O sites and Li at Zn sites would be acceptors in ZnO.

For N-doped ZnO, the examination of N-site-decomposed DOS in Fig. 3(c) reveals the largely localized impurity states at the N sites due to the strong repulsive potential. It suggests that the N at the O sites are deep acceptors. It caused by the difference in electronegativity between the N and O species. The O species has large electronegativity of 3.44 compared with 3.04 of N. In such a case, in O-Zn-N chemical bonds with a tetrahedral coordinate, the charge transfer from the cation, Zn, sites to the anion, O, sites is larger than that

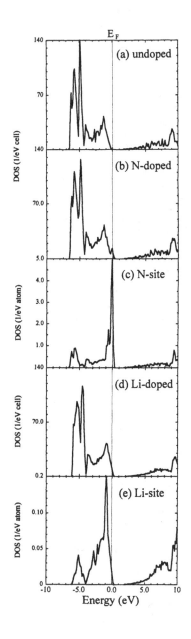

Fig. 3. Total DOS of (a) undoped ZnO, (b) total and (c) N-site decomposed DOS of ZnO doped with N and (d) total and (e) Li-site decomposed DOS of ZnO doped with Li.

from the Zn to another anion, N, sites. Then the outermost of valence orbitals, $2p$, of the N are shifted towards high energy regions, resulting in not only the deep acceptors but also the instability of N atoms in p-type ZnO:N. The control of impurity levels in the band gap is a key issue to be solved for N-doped ZnO. In other words, we must enhance the incorporation of N and lower the N acceptor levels for the fabrication of low-resistivity p-type ZnO doped with N, discussed below.

For Li-doped ZnO, Fig. 3(e) shows delocalized states at the Li sites compared with those at the N sites for N-doped ZnO. It also indicates the strong interaction between the Li and O in the vicinity of the Li sites, resulting in the shifts of p states at the Li sites towards lower energy regions. It suggests that the Li dopants are shallow acceptors in p-type ZnO:Li.. Moreover, we determined the crystal structures of both ZnO:2N and ZnO:2Li, as shown Fig. 4, in order to study the interaction between the acceptors using supercells method. They are determined by *ab-initio* electronic band structure calculations under the condition that the total energy is minimized from all atomic configurations. We find a short distance of 4.57 Å between two Li acceptors to be energetically favorable in p-type ZnO:2Li compared with that, 6.14 Å, in p-type ZnO:2N. For doping of Al, which has high solubility in ZnO [30], we found the same distance originated in weak repulsive interaction between Al donors for n-type ZnO:2Al [20]. This finding suggest high solubility of Li impurities as well as Al, in contrast with ZnSe:Li [10].

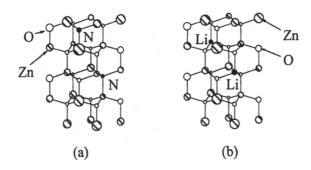

(a)   (b)

Fig. 4.   Crystal structures of (a) ZnO doped with 2N and (b) ZnO doped with 2Li.

Unipolarity

"*Unipolarity*" means that materials exhibit an asymmetry in their ability to be doped n-type or p-type. ZnO had been a good n-type conductor but had not been able to be made p-type. In our previous works, we established *unipolarity* of wide-band-gap semiconductors such as CuInS$_2$ [11-13], GaN [14-16] and ZnSe [17-19] from the view point of a change in the Madelung energy, caused by intrinsic or extrinsic doping. For example, for GaN, p-type doping using Be or Mg as acceptors gives rise to an increase in the Madelung energy while n-type doping using Si or O as donors leads to a decrease in the Madelung energy [14-16]. An increase in the Madelung energy by p-type doping results in the shifts of outermost valence states, especially p states, at anion sites in the vicinity of the dopants near the top of the valence band towards higher energy regions, leading the instability to the anion. As a result, there increases the compensation due to the formation of anion vacancies with an increase in the concentration of the acceptor dopants.

The wurtzite structure is favored by more ionic compounds. The bonding between hard acids, Al, Ga, In and Li and hard bases, N, O and F can be described approximately in terms of ionic or dipole-dipole interactions. Thus the stability of the ionic charge distribution for p- or n-

Table I. Calculated difference in the Madelung energies, ΔE_M, (units: eV) between undoped and n- or p-type doped ZnO crystals.

n-type				p-type	
Al	Ga	In	F	N	Li
-6.44	-13.72	-9.73	-1.86	+0.79	+13.56

type doped ZnO depends strongly on a change in the Madelung, ΔE_M, energy. For both p- and n-type doped ZnO, we summarized ΔE_M in Table 1.

Table I shows the occurrence of the doping problem called "*unipolarity*" of ZnO crystals: while n-type doping using Al, Ga, In or F species leads a decrease in the Madelung energy, p-type doping using N or Li causes an increase in it whereby both low-resistivity n- and p-type ZnO crystals are difficult to fabricate. For Li doping, it suggests the compensation of the formation of vacancies of the O sites in the vicinity of the Li sites, resulting in high resistivity.

From Table I, we propose the Ga species is eminently suitable for use of n-type dopants. For p-type doping using N species, N acceptor levels must be lowered, discussed the above section, with a decrease in the Madelung energy. For Li doping, it is a key issue to control valence states in order to decrease the vacancies of the O sites.

Codoping Method

In this section, we summarize the significant physics of our codoping method using acceptors and donors as reactive dopants simultaneously for the fabrication of low-resistivity of p-type wide-band-gap semiconductors. We find that the codoping method forms acceptor and donor complexes in the crystals, and contributes (i) to reduce the Madelung energies and enhance the incorporation of acceptors because the acceptor-donor attractive interaction overcomes the repulsive one between the acceptors, and (ii) to lower the energy levels of the acceptors and raise them of the donors in the band gap due to the strong interaction between the acceptors and donors with forming an acceptor-donor-acceptor complex in Fig. 5, and (iii) to increase the carrier mobility due to the short-range dipole-like scattering mechanism (long-range Coulomb scattering one is dominated in the case of doping of acceptors alone). Thus, the p-type codoped semiconductors exhibit low-resistivity with high carrier density and mobility.

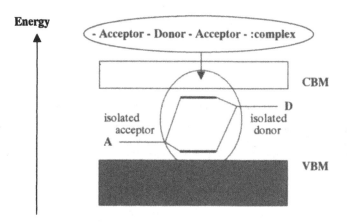

Fig. 5 The acceptor (A) level is lowered and the reactive donor (D) level is raised with the formation of acceptor-donor-acceptor complexes upon codoping.

Codoping of N and the group III elements (Al, Ga, and In)

We apply the codoping method to ZnO using N as acceptors and the group III elements, Al, Ga and In, as reactive donor codopants, in the ratio of 2 to 1. We show the crystal structure of p-type ZnO:(2N, III(=Al, Ga or In)) in Fig. 6. We determined them under the condition that the total energy is minimized from all atomic configuration in the supercell. A strong correlation between the reactive donors and N acceptors results in the formation of the complexes, including the III-N pair, which occupy nearest-neighbor sites, and a more distant N, located at the next-nearest-neighbor site in a layer close to the layer including the III-N pair.

For p-type GaN:(2Mg, O) or (2Be, O) with a wurtzite structure [14-16] and p-type ZnSe:(2N, In) [17] or (2Li, Cl(I)) [18,19] with a zincblende structure, the total energy calculations revealed that the formation of acceptor-donor-acceptor complexes which occupy nearest-neighbor sites is energetically favorable. The difference in the structure of the complexes between p-type codoped ZnO and the other codoped semiconductors, GaN and ZnSe, suggests that ZnO has a strong ionic character in chemical bonds compared with the two crystals above.

We summarized the calculated differences in the Madelung energy among p-type ZnO:N as a standard, ZnO:2N and codoped ZnO in Table II. It shows that simultaneous codoping decreases the Madelung energy of p-type codoped ZnO:(2N, III) compared with p-type ZnO doped with the N acceptor alone: the codoping enhances the incorporation of the N acceptors with the stabilization of the ionic charge distribution in p-type ZnO crystals.

It is of interest to study what happens when the complexes including acceptors and reactive-donor codopants are formed by the codoping. We show the N-site-decomposed DOS of p-type ZnO:N as a standard in Fig. 7(a) and of ZnO:(III, 2N) in Figs. 7(b) to 7(d). We note that the formation of the complexes leads to a mixed state of a hole generated at the top of the valence band originating from the two N acceptors, especially at the N acceptor site close to the reactive donor sites (dotted curve).

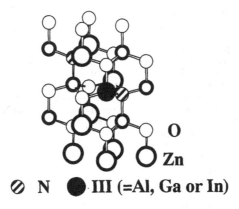

$\oslash$ N   ● III (=Al, Ga or In)

Fig. 6   Crystal structure of ZnO codoped with 2N acceptors and the group III(=Al, Ga and In) elements as reactive donor codopants.

Table II.   Calculated differences in the Madelung energy among p-type ZnO:N as a standard, p-type ZnO:2N and p-type codoped ZnO:(2N, III(=Al, Ga and In)). units:eV

N-doped alone	codoped		
2N: +0.91	(2N, Al): -4.74,	(2N, Ga): -12.06,	(2N, In): -7.79

Figures 7(b) to 7(d) show a shift of the weight of the p states at the site of N atoms (dotted curve) close to the reactive donor sites, towards lower-energy regions due to the charge transfer from the Al, Ga or In to the N atoms. It causes not only the stabilization but also delocalization of N acceptors in p-type codoped ZnO compared with p-type ZnO doped with N alone in Fig. 7(a). We note that Figures 7(b) to 7(d) show that as the polarizing power of reactive donors increases in the order of In < Ga < Al, we find that the delocalization of states of the N close to the reactive donors increases in the same order. At the same time, a strong interaction between the two N acceptors gives rise to a slight shift of the weight of the p states of the remaining N (solid curve) towards lower-energy regions, resulting in the same shift of a sharp DOS peak near the top of the valence band, as shown in Figs. 7(b)-7(d). It means that the codoping using reactive donor codopants gives rise to the stabilization of both electronic and lattice system. As an effect of the codoping, we predict that the acceptor levels in the band gap are lowered due to the strong interaction between the N acceptor and reactive donor codopants, as shown in Fig. 8. We note that the acceptor level of N, belonging to the III-N pair will be lowered very well. Thus, we concluded that Al and Ga are eminently suitable as the reactive donor codopants for the fabrication of low-resistivity p-type ZnO crystals.

Our theoretical prediction, especially the effects of the codoping of N and Ga, was verified by experiments by Osaka`s group. They reported that the codoping enhances p-type activity in terms of conductivity and carrier concentration. They realized p-type ZnO with a room temperature resistivity of 2 $\Omega \cdot$cm and high hole concentration of 4×10^{19} cm^{-3}. Moreover, from XPS data of GaN and p-type codoped ZnO, they concluded the relative ratio of Ga to N in the p-type ZnO thin films is obtained as 1 to 2.

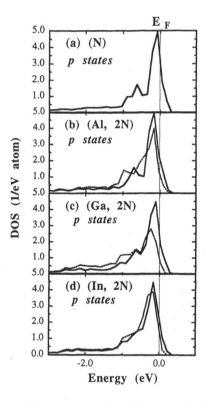

Fig. 7. Site-decomposed DOS of p-states for (a) ZnO:N, ZnO:(b) (Al,N), (c) (Ga,N) and (d) (In,N). Dotted curve indicates the DOS at theN close to the III element sites; the solid, the DOS at the sites of second-nearest-neighbor N atoms. It is after Fig. 3. of Ref. [20].

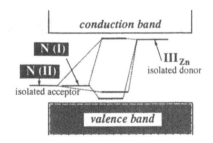

Fig. 8. A change in acceptor and donor level by the codoping. N(I) is N belonging to the III-N pair and N(II) is N, being located at the second-nearest-neighbor sites. See Fig. 6.

Codoping of N and Zn

Minegishi and coworkers reported that they realized p-type ZnO film by the simultaneous addition of NH_3 in carrier hydrogen and excess Zn in source ZnO powder [10]. In general, many oxides of transition and post-transition elements such as Zn can have quite high concentrations of defects. The defects are often associated with deviations from perfect stoichiometry, and are formed because of the relative ease of reduction or oxidation of the metal ion. Reduction can be accompanied by formation of interstitial metal atoms, or by oxygen vacancies. Thus, in this work, we assume that the excess Zn would form interstitial Zn (Zn_i) in ZnO crystals under consideration. The reduction leaves extra electron in ZnO. In other words, $ZnO:Zn_i$ behaves n-type. From this view point, the doping method used in the above report [10] is the codoping one using acceptor and donor simultaneously.

We show the crystal structure of $ZnO:Zn_i$ in Fig. 9(a), which is determined using *ab initio* electronic band structure calculations under the condition of the minimization of the total energy.

We find that the interstitial Zn is surrounded by three oxygen atoms as the first neighbors with the distance of 2.005 Å between the Zn_i and the O sites, being almost same value as that between Zn and O in host material, ZnO. We note that the Zn_i by doping of excess Zn causes a remarkable increase in the Madelung energy, 11.38 eV; the defect is unstable.

Next, we show the crystal structure of ZnO codoped with Zn_i and 2N in Fig. 9(b), which is determined using *ab initio* electronic band structure calculations under the condition of the minimization of the total energy. The calculations show that the formation of the complex including N-Zn_i-N, which occupy nearest-neighbor sites, is energetically favorable. It suggests that the growth technique using excess Zn enhances the incorporation of N species into ZnO crystals. Considering that the above complex behaves not acceptor but neutral, we must control a partial pressure of N gas in such a way that the complex including N-Zn-N-N- or a "chain-like" complex, such as N-Zn-N --- N , which occupy more distant sites from the rest of the complex will be formed. In such a case, codoping using N and excess Zn as reactive donor codopant may cause a problem of reproducibility. The reason is that the calculations show a 13.33 eV increase in the Madelung energy for intrinsic ZnO with the complex, N-Zn_i-N, compared with that for undoped ZnO with stoichiometry.

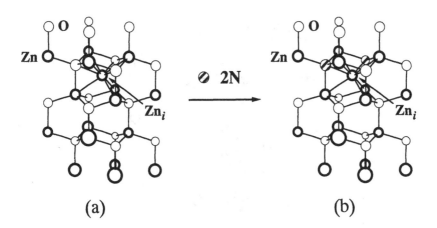

(a)   (b)

Fig. 9   Crystal structures of (a) ZnO doped with interstitial $Zn(Zn_i)$ and (b) ZnO codoped with 2N acceptors and reactive donor codopant, Zn_i.

Codping of Li and F

From our calculations, Li species is an attractive candidate for use of shallow acceptor. Recently, Onodera and co-workers reported high solubility, $Zn_{0.9}Li_{0.1}O$, of Li species into ZnO single crystals [30]. However, there has been no report of the fabrication of p-type ZnO:Li while there were reports that doping of Li increases its resistivity [31] and also shows ferroelectric nature [32,33].

Our calculations concerning the effects of the doping of Li on a change in the Madelung energy shows a remarkable increase in Table. 1. It means that Li-doping causes the instability of ionic charge distributions around the doping sites, especially at the O sites in the vicinity of the Li_{Zn} sites: as a result, there occurs the formation of donor defects, O vacancies (V_O). Then we investigate the crystal structure of ZnO: (Li, V_O) under the condition that the total energy is minimized from all atomic configurations. We show it in Fig. 10. We find that the distance between the Li and V_O sites is 3.83 Å. This means that the doping of Li acceptors gives rise to bad crystallinity due to the compensation by O vacancies. As a result, Li-doped ZnO probably exhibit high-resistive n-type behavior.

From those studies on the effects of Li-doping on the Madelung energy and crystallinity, most important key technology to fabricate low-resistivity p-type ZnO crystals using Li species as acceptors is such a way that the formation of O vacancies can be avoided without degradation of the solubility of Li impurities. Considering that Li species is hard acids with small electronegativity and has a strong polarizing effect on the valence electrons of the anion, we focus on hard bases, F species, which occupy "O-sites", to be suitable for use in reactive-codopant donors. In such a case, considering that the hard acid-base interactions are predominantly electrostatic, long-range Coulomb attractive ones, we expect the possibility of the formation of the complexes including Li-F-Li, which occupy nearest-neighbor sites, although its Madelung energy are increased due to their small charge, being approximately to a unit, under not thermal equilibrium but one of metastable states using MBE and MOCVD growth technique. Here we append that the melting points of Li_2O with a antifluorite structure and LiF with a rock salt structure are 1570° C and 848. C, respectively, which are lower than 1980° C of ZnO. This suggests little problem of segregation or precipitates of Li-O or Li-F compounds.

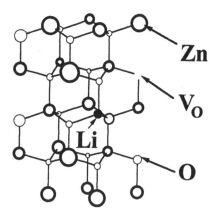

Fig. 10. Crystal structure of supercell for ZnO doped with Li and O vacancies (V_O). The distance between the Li and V_O is 3.83 Å.

Total energy calculations show that the formation of Li-F-Li complexes that occupy nearest-neighbor sites, which causes lowered (raised) acceptor (donor) levels in the band gap, is energetically favorable. Figure 11 shows the crystal structure of ZnO:(2Li, F). From the calculations, we find a small decrease of 380 meV in the Madelung energy compared with ZnO doped with 2Li alone, being, however, very larger than that for ZnO:Li, without the vacancies of O. This indicates that it is very difficult to realize p-type ZnO with the same order of carriers as p-type ZnO:(2N, Ga) or (2N, Al) crystals. Considering that Li has high solubility and the bonding between a hard acid, Li, and a hard base, F, can be described approximately in terms of ionic long-range Coulomb interactions, it may be possible to realize p-type ZnO:(2Li, F) if we control a partial pressure of O gas the under the O-rich condition as a metastable state.

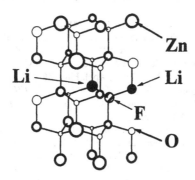

Fig. 11   Crystal structure of p-type ZnO codoped with 2Li and F species.

CONCUSIONS

The theoretical investigation of materials design for the fabrication of low-resistivity p-type ZnO crystal leads to the following conclusions. (1) We predict the *"unipolarity"* in ZnO crystals. (2) We propose Ga and Al species as suitable candidates for use of donors. (3) Simultaneous codoping using N species as an acceptor and the Ga or Al as reactive-codopant donor is very effective for materials design to fabricate low-resistivity p-type ZnO crystals. (4) For both ZnO:(2N, Zn$_i$) and ZnO:(2Li, F), there will remain problems of poor reproducibility with degradation of crystallinity due to an increase in the Madelung energy while the donor codopants, excess Zn or F, enhances the incorporation of the acceptors into ZnO crystals.

These results await experimental verification.

ACKNOWLEDGMENTS

We would like to thank Prof. T. Kawai, Prof. H. Tabata and Dr. M. Joseph for their fruitful discussion. One of authors, T. Y., thank Dr. J. Sticht for his technical support. We used ESOCS code of Molecular Simulations Inc.

REFERENCES

1. P. Yu, Z. K. Tang, G. K. L. Wong, M. Kawasaki, A. Ohtomo, H. Koinuma and Y. Segawa: Proc. 23th Int. Conf. Physics of Semiconductors, Berlin, 2, p. 1453 (1996).
2. Y. Segawa, A. Ohtomo, M. Kawasaki, H. Koinuma, Z. K. Tang, P. Yu and G. K. L. Wong: Phys. Status

Solidi (b) **202**, p.669 (1997).

3. T. Yamamoto and H. Katayama-Yoshida, unpublished.

4. T. Minami, H. Nanto and S. Takata, Jpn. J. Appl. Phys. **23**, p. L280 (1984).

5. T. Minami, H. Sato, H. Nanto and S. Takata, Jpn. J. Appl. Phys. **24**, p. L781 (1985).

6. T. Minami, H. Sato, H. Nanto and S. Takata, Jpn. J. Appl. Phys. **25**, p. L776 (1986).

7. J. Hu and R. G. Gordon: Solar Cells **30**, p. 437 (1991).

8. M. Kasuga and S. Ogawa, Jpn. J. Appl. Phys. **22**, p. 794 (1983).

9. Y. Sato and S. Sato, Thin Solid Films **281-282**, p. 445 (1996).

10. K. Minegishi, Y. Koiwai, Y. Kikuchi, K.Yano, M. Kasuga and A. Shimizu, Jpn. J. Appl. Phys. **36**, p. L1453 (1997).

11. T. Yamamoto and H. Katayama-Yoshida in *Electronic structure of p-type CuInS$_2$*, edited by D. Ginley, A. Catalano, H. W. Schock, C. Eberspcher, T. M. Peterson and T. Wada (Mater. Res. Soc. **426**, Pittsburgh, PA 1996), p. 201-206.

12. T. Yamamoto and H. Katayama-Yoshida, Solar Energy Materials and Solar Cells, **49**, p. 391 (1997).

13. T. Yamamoto and H. Katayama-Yoshida in A CODOPING METHOD IN CuInS2 PROPOSED BY ab initio ELECTRONIC-STRUCTURE CALCULATIONS, Inst. Phys. Conf. Ser. **152**, pp. 37-41 (1998). 11Th Int. Conf. Ternary and Multinary Compounds, Salford, 1997. Invited.

14. T. Yamamoto and H. Katayama-Yoshida, Jpn. J. Appl. Phys. **36**, p. L180 (1997).

15. T. Yamamoto and H. Katayama-Yoshida in *CONTROL OF VALENCE STATES BY A CODOPING METHOD IN P-TYPE GaN MATERIALS*, edited by C. A. Abernathy, H. Amano and J. C. Zolper (Mater. Res. Soc. Proc. **468**, Pittsburgh, PA 1997) p. 105-110.

16. T. Yamamoto and H. Katayama-Yoshida, Journal of Crystal Growth, **189/190**, p. 532 (1998).

17. H. Katayama-Yoshida and T. Yamamoto, phys. Stat. Sol. (b) **202**, p. 763(1997).

18. T. Yamamoto and H. Katayama-Yoshida in Role of n-type Codopants on Enhancing *p*-type Dopants Incorporation in *p*-type Codoped ZnSe, edited by (Mater. Res. Soc. Proc. **510**, Pittsburgh, PA 1998), p. 67-72.

19. T. Yamamoto and H. Katayama-Yoshida, Jpn. J. Appl. Phys. **37**, p. L910 (1998).

20. T. Yamamoto and H. Katayama-Yoshida, Jpn. J. Appl. Phys. **38**, p. L166 (1999).

21. T. Yamamoto and H. Katayama-Yoshida, Physica B, **273-274**, p. 113 (1999).

22. O. Brandt, H. Yang, H. Kostial and K. H. Ploog, Appl. Phys. Lett, **69**, p. 2707 (1996).

23. M. Joseph, H. Tabata and T. Kawai, Jpn. J. Appl. Phys. **38**, p. L1205 (1999).

24. W. Kohn and L. J. Sham: Phys. Rev. **140**, P. A1133 (1965).

25. L. Hedin and B. I. Lundquist: J. Phys. C4, p. 3107 (1971).

26. U. von Barth and L. Hedin: J. Phys. C5, p. 1629 (1972).

27. A. R. Williams, J. Kübler and C. D. Gelatt: Phys. Rev. B19, p. 6094 (1979).

28. P. Schröer, P. Krüger and J. Pollmann: Phys. Rev. **47**, p. 6971 (1993).

29. S. Massidda, R. Resta, M. Posternak and A. Baldereschi: Phys. Rev. **52**, p. R16977 (1995).

30. A. Onodera, K. Yoshio, H. Satoh, H. Yamashita, N. Sakagami, Jpn. J. Appl. Phys. **37**, p. 5315 (1998).

31. E.D.Kolb and R.A.Laudise, J. Am. Ceram. Soc. 49, p. 302 (1966).

32. A. Onodera, A. Tamaki, Y. Kawamura, T. Sawada and H. Yamashita, Jpn. J. Appl. Phys. **35**, p. 5160 (1996).

33. M. Joseph, H. Tabata and T. Kawai, Appl. Phys. Lett, **74**, p. 2534 (1999).

TRANSPARENT P- AND N-TYPE CONDUCTIVE OXIDES WITH DELAFOSSITE STRUCTURE

HIROSHI YANAGI, KAZUSHIGE UEDA, SHUNTARO IBUKI, TOMOMI HASE, HIDEO HOSONO, and HIROSHI KAWAZOE*
Materials and Structures Laboratory, Tokyo Institute of Technology, Yokohama, Japan.
*Present affiliation: R & D Center, HOYA Corporation, Tokyo, Japan.

ABSTRACT

Thin films of $CuAlO_2$, $CuGaO_2$ and $AgInO_2$ with delafossite structure were prepared on sapphire substrates by pulsed laser deposition method. The resulting $CuAlO_2$ thin films exhibited p-type conduction and the electrical conductivity at room temperature was 0.3 S cm^{-1}. $CuGaO_2$ thin films were grown epitaxially on α-Al_2O_3 (001) surface and showed p-type conduction (conductivity at room temperature = 0.06 S cm^{-1}). The optical band gap was estimated to be ~3.5 eV for $CuAlO_2$ or ~3.6 eV for $CuGaO_2$. On the other hand, the thin film of Sn doped $AgInO_2$ exhibited n-type conduction. The optical band gap and electrical conductivity at room temperature were ~4.1 eV and 70 S cm^{-1}, respectively. The recent work demonstrates the validity of our chemical design concept for p- and n-type transparent conducting oxides, providing an opportunity for realization of transparent p-n junction using delafossite-type oxides.

INTRODUCTION

Transparent conductive oxides (TCOs) such as electron doped ZnO, In_2O_3, SnO_2 are widely and practically used as transparent electrodes in flat panel displays, solar cells and touch panels. However, all conduction type of these materials is n-type, no p-type TCO was found so far. This mono-polarity is the primary origin of the restricted application of TCOs. Therefore, realization of p-type TCOs should be a milestone to expand the utilization of TCOs as transparent oxide semiconductors, because a wide variety of active functions of semiconductor devices comes from p-n junctions.

We have proposed a guideline for chemical design of p-type TCOs and found $CuAlO_2$,[1] $CuGaO_2$[2] and $SrCu_2O_2$[3] as a consequence of exploration efforts following the guideline. Further, all oxide-based transparent polycrystalline p-$SrCu_2O_2$/n-ZnO heterojunction thin film diodes exhibiting rectifying properties were successfully fabricated.[4] In order to achieve better performance, fabrication of p-n heterojunctions with same crystal structure is desirable. Transparent oxides with delafossite structure are unique materials that satisfy our working hypotheses for exploring both p- and n-type TCOs. It turned out that $CuAlO_2$ and $CuGaO_2$ delafossites are p-type TCOs, while Sn doped $AgInO_2$ delafossite is n-type TCO.[5]

Our guideline to find p- and n- type TCOs are summarized in as follows:[1, 6] The essence of guideline to find p-type TCOs is to have monovalent copper as the major constituents. The selection of Cu^+ is based on its electronic configuration and energy levels of 3d orbitals. Cuprous ion has the electronic configuration of (Ar) $3d^{10}4s^0$ (closed shell), which is free from visible coloration arising from a d-d transition commonly seen in transition metal ions. The energy level of a $3d^{10}$ state is close to that of an O $2p^6$ state. As a consequence, covalent bonding formation or hybridization of orbitals is expected between Cu $3d^{10}$ and O $2p^6$. The hybridization of the orbitals will bring large dispersion to the valence band or reduction of the localization of positive holes. Formation of a covalent bonding between Cu^+ and O^{2-} ions demands an appropriate crystal structure. The crystal structure is required to meet two requirements, one is that Cu^+-O^{2-} bonds with a moderated covalency exist, and another is to retain a wide energy gap for optical transparency. This argument can be valid also to Ag^+ instead of Cu^+.

For the n-type TCO, high mobility of carrier electron in the conduction band is required. A liner chain of edge-sharing octahedra, in which the p-block heavy cations (M^{i+}) with ns^0 electric configuration (n: the principal quantum number) occupy the central position, is preferred to satisfy this requirement. Since there is no intervening oxygen between the two neighboring M cations in the chain of edge-sharing octahedra, direct overlap between ns atomic orbitals of the neighboring M cations is possible for the p-block heavy cations. In this case, the bottom edge of conduction band is mainly composed of ns atomic orbitals of M cations. As a consequent, a large dispersion of the conduction band, which is appropriate for high mobility of electron carrier, may be expected.

Taking the above stated requirements into consideration, we selected as the candidate materials ternary noble metal oxide with delafossite structure (ABO_2). Figure 1 illustrates delafossite-type crystal structure. The delafossite is composed of an alternate stacking O-A-O

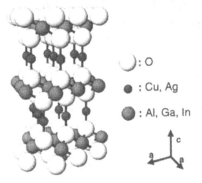

): O

● : Cu, Ag

◐ : Al, Ga, In

c
a a

FIG. 1. Crystal structure of delafossite (ABO_2). This material has layer structure composed of A (A = Cu, Ag) plane and BO_2 (B = Al, Ga, In) layer alternately stacked along c-axis.

dumbbells and BO_6 edge-sharing octahedral layer. According to our hypothesis, delafossite oxides satisfy the requirements for p- and n-type TCOs, when Cu and/or Ag as A cation and p-block heavy cations, such as Ga^{3+} and In^{3+}, as B cation are chosen. It is considered that BO_6 octahedral layers and O-A-O layers work as conduction paths for electrons and positive holes, respectively. That is to say, double oxides with delafossite structure are candidate materials for preparing p- and n-type TCOs aiming at a transparent p-n junction using the same crystal structure.

The present paper describes (1) the preparation of p-type conducting $CuAlO_2$ and $CuGaO_2$ thin films and n-type conducting $AgInO_2$ thin films by pulsed laser deposition (PLD) method and (2) the electric structure of $CuAlO_2$, which is a prototype of p-type conducting transparent oxides, probed by photoemission spectroscopy.

EXPERIMENT

Polycrystalline $CuAlO_2$ was synthesized by heating a stoichiometric mixture of Cu_2O and Al_2O_3 at 1373 K for 10 h.[7] The powder was pressed into a pellet by a cold isostatic press (CIP) at 800 kg cm^{-2} and then sintered at 1373 K for 10 h. Resulting sintered pellets were used as the target for preparation of $CuAlO_2$ thin films by PLD method.

Sintered pellets of $CuGaO_2$ for the target PLD were prepared by using solid state reactions of Cu_2O and Ga_2O_3.[8] Stoichiometric amounts of the raw materials were mixed thoroughly with methanol and calcined at 1373 K for 24 h. The material was pelletized, CIPed and sintered at 1373 K for 24 h.

$AgInO_2$ delafossite were prepared by cation exchange reaction,[9, 10] because direct preparation of $AgInO_2$ by a conventional solid state reaction of In_2O_3 with $AgNO_3$ or Ag_2O at high temperature was unsuccessful due to the precipitate of metallic silver. $NaInO_2$ with rock salt structure was first prepared by direct solid state reaction between Na_2CO_3 and In_2O_3 at 1123 K for 12 h in O_2 gas flow. Then $NaInO_2$ powder was mixed with $AgNO_3$ and KNO_3 in the molar ratio 1:1.8:0.8, and heated at 453 K for 48 h in air. Under the heat treatment, $NaInO_2$ was subjected to the cation exchange of Na^+ with Ag^+ in the melt and as a consequent $AgInO_2$ with delafossite structure was synthesized. The solidified mixture was washed with water to dissolve the remaining nitrates. The $AgInO_2$ powders were shaped into disk by CIP. The molded disks were sintered at 773 K for 24 h in O_2 gas flow.

In the preparation of targets for PLD, substitution of Sn 5 at.% with In was carried out to dope carrier electrons by using Sn doped $NaInO_2$ in the cation exchange reaction,[5] because non-doped $AgInO_2$ sintered disks were almost insulating. Since this substitution was successful to enhance the conductivity in the disks, fabrication of $AgInO_2$ thin films by PLD was performed using these disks as targets.

TABLE I Preparation condition for $CuAlO_2$, $CuGaO_2$ and $AgInO_2$ thin films.

		$CuAlO_2$	$CuGaO_2$	$AgInO_2$
Target (pellet)		$CuAlO_2$	$CuGaO_2$	Sn doped $AgInO_2$
Substrate		α-Al_2O_3 (001)	α-Al_2O_3 (001)	α-Al_2O_3 (001)
Substrate-target distance	(mm)	25	25	30
Substrate temperature	(K)	963	973	723
Oxygen pressure	(Pa)	1.3	9	13
KrF excimer laser				
Repetition frequency	(Hz)	20	20	2
Energy density	(J pulse^{-1} cm^{-2})	5.0	6.0	8.0
Post annealing	(h)	3	----------	----------

Thin films of $CuAlO_2$, $CuGaO_2$ and $AgInO_2$ were deposited on α-Al_2O_3 single crystal substrate by PLD method using KrF excimer laser (Lambda Physik ComPex 102). The base pressure before the deposition was ~10^{-6} Pa. Table I summarizes the optimized preparation conditions for each thin film. The films of $CuAlO_2$ were post-annealed at 963 K for 3 h under the deposition after deposition, and were cooled down to room temperature maintaining an atmosphere oxygen partial pressure of 1.3 Pa.

Crystalline phases in the pellets and the films were identified by X-ray diffractometer (XRD) (Cu $K\alpha$, Rigaku Rint-2500), and film thickness was measured with a stylus (Sloan DEKTAK^{3}ST). Optical transmission spectra of the films were measured with a dual beam spectrophotometer (Hitachi U-4000). Electrical conductivity was measured by the two-probe method in a temperature range 10 K ~ 300 K. Measurements of the Seebeck and Hall coefficients were carried out at room temperature.

Measurements of photoemission spectroscopy (PES) and inverse photoemission spectroscopy (IPES) were performed on the thin film samples at room temperature using an instrument built up in our laboratory.[11] In the PES measurement, two excitation lights, He II resonance radiation (40.8 eV) and Mg $K\alpha$ X-ray (1254 eV), were used for UPS and XPS, respectively. IPES spectra were measured in BIS mode by monitoring the intensity of emitting photon at 9.45 eV. These measurements were carried out under the vacuum level 1×10^{-7} ~ 5×10^{-8} Pa.

RESULTS

Figure 2 shows XRD patterns of thin films and powders of the $CuAlO_2$, $CuGaO_2$ and Sn doped $AgInO_2$. All diffraction peaks were indexed as delafossite structure except for peaks from the substrate. The $CuAlO_2$ and $CuGaO_2$ thin films have orientation c-axis. Epitaxial growth of

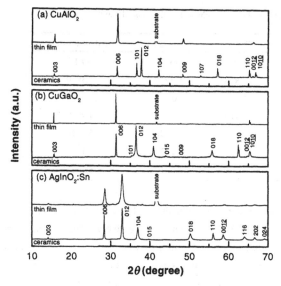

FIG. 2. XRD patterns of delafossites, CuAlO$_2$ (a), CuGaO$_2$ (b) and Sn doped AgInO$_2$ (c). Top and bottom pattern of each material are due to thin film and ceramics (for PLD target), respectively.

CuGaO$_2$ was confirmed by X-ray pole figure and in-plane measurements, and its detail will be reported elsewhere.[12]

Figure 3 shows optical transmission spectra of the CuAlO$_2$, CuGaO$_2$ and Sn doped AgInO$_2$ thin films in visible and near infrared (NIR) regions. No distinct optical absorption band was observed in the visible region. Relatively high transmittance, ~70% at 500 nm, was achieved for all films deposited under the optimized preparatory conditions. The direct allowed optical band gaps of CuAlO$_2$, CuGaO$_2$ and AgInO$_2$ thin films are listed in Table II. In the spectrum of Sn doped AgInO$_2$ thin film, a decrease in NIR (> ~2000 nm) is due to the absorption by carrier electrons.

Figure 4 shows temperature dependence of electrical conductivity in the CuAlO$_2$, CuGaO$_2$ and Sn doped AgInO$_2$ thin films. Table II summarizes optical and electrical properties of these

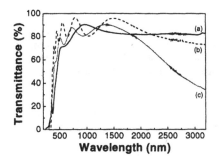

FIG. 3. Optical transmission spectra of CuAlO$_2$ (a), CuGaO$_2$ (b) and Sn doped AgInO$_2$ (c) thin films. The surface reflection loss of sapphire substrate was subtracted.

TABLE II Optical and electrical properties of $CuAlO_2$, $CuGaO_2$ and Sn doped $AgInO_2$ thin films at room temperature.

		$CuAlO_2$	$CuGaO_2$	Sn doped $AgInO_2$
Optical band gap	(eV)	~3.5	~3.6	~4.1
DC electrical conductivity	(S cm^{-1})	3.4×10^{-1}	6.3×10^{-2}	7.3×10^{1}
Activation energy	(eV)	0.22	0.13	----------
Hall coefficient	(cm^3 C^{-1})	$+2.3\times10^{-1}$	$+3.7$	-1.9×10^{-2}
Carrier density	(cm^{-3})	2.7×10^{19}	1.7×10^{18}	3.3×10^{20}
Hall mobility	(cm^{2}V^{-1}s^{-1})	0.13	0.23	1.4
Seebeck coefficient	(μVK^{-1})	$+2.1\times10^{2}$	$+5.6\times10^{2}$	-5.1×10^{1}

films at room temperature. The dc conductivities of the $CuAlO_2$ and $CuGaO_2$ thin films at room temperature were 3.4×10^{-1} S cm^{-1} and 6.3×10^{-2} S cm^{-1}, respectively. The temperature dependence of conductivities of the $CuAlO_2$ and $CuGaO_2$ films may be described by Arrehenius type behavior near room temperature and the activation energy is estimated as 0.22 eV for $CuAlO_2$ or 0.13 eV for $CuGaO_2$. On the other hand, the data on plots of log σ vs. $T^{-1/4}$ at low temperature region suggested that a variable-range hopping mechanism is dominant in this region. In the case of the Sn doped $AgInO_2$ thin film, no remarkable dependence of the conductivity on temperature was observed except for at higher temperatures near room temperature, indicating that the Fermi level of the specimen locates at above of the conduction band bottom. The Seebeck and Hall coefficients of the $CuAlO_2$ and $CuGaO_2$ were positive, indicating positive holes are undoubtedly major carriers in these thin films. On the other hand, the Seebeck and Hall coefficients of the Sn doped $AgInO_2$ was negative. Therefore we conclude that $AgInO_2$:Sn thin films are n-type conductors.

XPS, UPS, and IPES spectra of $CuAlO_2$ thin film are shown in figure 5. Fermi energy determined experimentally was set as zero on the binding energy scale in the three spectra. The

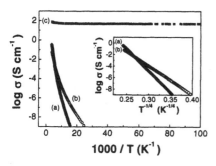

FIG. 4. Arrhenius plots of dc electrical conductivity in $CuAlO_2$ (a), $CuGaO_2$ (b) and Sn doped $AgInO_2$ thin films. Inset shows plots of log σ vs. $T^{-1/4}$ in $CuAlO_2$ and $CuGaO_2$ films.

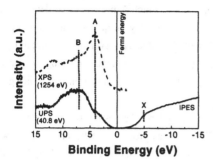

FIG. 5. PES and IPES spectra of CuAlO₂ thin film. The origin of the energy axis is Fermi level which was determined by using Au deposited on sample. Bands A, B and X are discussed in the text.

intensity of IPES was adjusted to be comparable to those of the PES spectra. The band gap, which was directly observable between the valence band edge in the PES spectra and the conduction band edge in the IPES spectrum, was ~3.5 eV. This value is in a reasonable agreement with the band gap evaluated by the optical measurement. The Fermi energy lies around the top of the valence band. This observation agrees with the fact that the specimen exhibits p-type electrical conductivity. A clear band was resolved at ~4 eV (band A) in the XPS spectrum. On the other hand, a broad band at ~7 eV (band B) and a shoulder at ~4 eV (band A) were observed in the UPS spectrum. In the IPES spectrum, a very broad band, which is indexed as X, was observed. No other band was resolved in the unoccupied state.

DISCUSSION

In this article, we described the detailed preparation procedure of p-type and n-type TCOs with delafossite structure. These materials described here, $CuAlO_2$, $CuGaO_2$ and $AgInO_2$, were found for the first time by us on the basis of our guideline. Here, we discuss the material selection toward the realization of transparent p-n junction using delafossite TCOs.

The candidate material for transparent n-type conductor with delafossite structure is only $AgInO_2$, but there are two candidates for p-type; $CuAlO_2$ and $CuGaO_2$. Both $CuAlO_2$ and $CuGaO_2$ delafossites thin films were oriented c-axis, however $CuGaO_2$ thin films were epitaxially growth in an as-deposited states.[12] This is an advantage for $CuGaO_2$ thin films to fabricate p-n junction, because epitaxial growth is expected to reduce structural imperfections at the p-n interface. The distances between neighboring oxygen ions on c-surface for $CuGaO_2$ and $AgInO_2$ are 0.298 nm and 0.328 nm, respectively, and this lattice mismatch is ~10%. On the other hand, the distances for $CuAlO_2$ and α-Al_2O_3 substrates are 0.286 nm and 0.252-0.287 nm, respectively, and the mismatch between $AgInO_2$ and $CuAlO_2$ is ~13%. Therefor, $CuGaO_2$ is expected to be more appropriate for combination than $CuAlO_2$. Furthermore, an advantage of

CuGaO$_2$ is that the oxygen pressure during deposition of CuGaO$_2$ is closed to that of AgInO$_2$. This may be effective to prevent the oxidation of p-layer, which is needed to be deposited on sapphire before the deposition of AgInO$_2$ so as to obtain good crystal quality. The advantage of CuAlO$_2$ is higher electrical conductivity than that of CuGaO$_2$. However, the conductivity of CuGaO$_2$ thin films will be enough to fabricate p-n junction to realize rectifying properties, because the conductivity of SrCu$_2$O$_2$ layer in p-SrCu$_2$O$_2$/n-ZnO heterojunction exhibiting good rectifying properties, was 10^{-3} S cm^{-1}.

Next, the electronic structure on CuAlO$_2$ is discussed on the results obtained here. The structure of the upper part of valence band of CuAlO$_2$ was probed by UPS and XPS. It was found that the relative intensity of the band A with respect to band B increased largely in the XPS spectrum in comparison with the UPS spectrum. The ratios of emission cross section of Cu 3d electron to O 2p electron for 40.8 eV excitation and for 1254 eV excitation are ~1.5 and ~30,[13] respectively. Therefore, the change in the relative intensity of band A with varying excitation energies indicates that the contribution of Cu 3d to the upper valence band is significant. In most transparent oxides, the upper valence band is almost entirely composed of O 2p orbitals and the contribution metal cation's orbitals is not significant. This makes a sharp contrast with the nature of present material. As a consequence, it is now concluded that our chemical design[1] to find a new TCO exhibiting p-type conductivity is valid in the viewpoints of electronic structure.

CONCLUSIONS

Thin films of CuAlO$_2$, CuGaO$_2$, and Sn doped AgInO$_2$ with delafossite structure were prepared by PLD method on α-Al$_2$O$_3$ (001) as a first step to fabricate transparent homo-structural p-n junction. The conclusions obtained are summarized as follows.

(1) CuAlO$_2$ and CuGaO$_2$ thin films oriented c-axis were obtained. No preferential orientation was seen for AgInO$_2$ thin films.

(2) The Seebeck and Hall coefficients of the CuAlO$_2$ and CuGaO$_2$ were positive, while those of the Sn doped AgInO$_2$ were negative. Therefore CuAlO$_2$ and CuGaO$_2$ were p-type conductor and AgInO$_2$ was n-type one.

(3) Photoemission spectroscopy revealed that the upper valence band of CuAlO$_2$ consist of Cu 3d orbitals to a large degree. This result provided a solid basis for our working hypotheses to explore p-type transparent oxides.

ACKNOWLEDGMENTS

One (HY) of the authors thanks the supports by Japan Society for the Promotion of Science and the Sasakawa Scientific Research Grant from the Japan Science Society.

REFERENCES

1. H. Kawazoe, M. Yasukawa, H. Hyodo, M. Kurita, H. Yanagi, and H. Hosono, Nature, **389**, 939 (1997).

2. H. Yanagi, H. Kawazoe, A. Kudo, M. Yasukawa, and H. Hosono, J. Electroceramics, **4**, 427 (2000)

3. A. Kudo, H. Yanagi, H. Hosono, and H. Kawazoe, Appl. Phys. Lett., **73**, 220 (1998).

4. A. Kudo, H. Yanagi, K. Ueda, H. Hosono, H. Kawazoe, and Y. Yano, Appl. Phys. Lett., **75**, 2851 (1999).

5. T. Otabe, K. Ueda, A. Kudoh, H. Hosono, and H. Kawazoe, Appl. Phys. Lett., **72**, 1036 (1998).

6. H. Kawazoe, N. Ueda, H. Un'no, H. Hosono, and H. Tanoue, J. Appl. Phys.,**76**, 7953 (1994)

7. T. Ishiguro, A. Kitazawa, N. Mizutani, and M. Kato, J. Solid State Chem., **40**, 170 (1981)

8. K. T. Jacob, and C. B. Alcock, Rev. int. Htes. Temp. et Refract., **13**, 37 (1976).

9. R. D. Shannon, D. B. Rogers, and C. T. Prewitt, Inorg. Chem., **10**, 713 (1971)

10. Y. J. Shin, J. P. Doumerc, P. Dordor, C. Delmas, M. Pouchard, and P. Hagenmuller, J. Solid State Chem., **107**, 303 (1993)

11. H. Yanagi, Master Thesis, Interdisciplinary Graduate School, Tokyo Inst. Tech. (1997)

12. K. Ueda, T. Hase, H. Yanagi, H. Kawazoe, H. Hosono, H. Ohta, M. Orita, and M. Hirano, submitted to J. Appl. Phys.

13. J. -J. Yen, *Atomic Calculation of Photoionization Cross-Sections and Asymmetry Parameters*, Gordon and Breach Science Publishers, Pennsylvania, (1993)

A Study of the Crystallization of Amorphous Indium (Tin) Oxide

David C. Paine, Eric Chason, Eric Chen, Dan Sparacin, and Hyo-Young Yeom
Brown UniversityDivision of Engineering, Box D
Providence, RI 02912

ABSTRACT

The crystallization of amorphous indium oxide thin films with zero to 9.8 wt% SnO_2 was studied using a combination of *in situ* MOSS (multibeam optical stress senor) and *in situ* resistivity measurements. We report that amorphous indium oxide deposited using electron beam evaporation undergoes crystallization in a two part process of amorphous structure relaxation followed by crystallization. MOSS measurements show that the relaxation process corresponds to a densification of the amorphous structure while crystallization results in a molar volume increase.

INTRODUCTION

Tin-doped indium oxide (ITO) is the transparent conductor of choice for a number of critical technologies including flat panel displays, EMF shielding, and solar cells. Although other transparent conductors are available such as, $SnO_2{:}F$ and $Zn(Al)O$, ITO is favored because, of the available choices, it offers the highest transmissivity of visible light combined with the lowest electrical resistivity. For high performance applications, ITO is deposited on heated substrates (300-400°C) typically using dc-magnetron sputtering of a ceramic target in a controlled oxygen ambient to achieve resistivities of 1-1.5×10^{-4} Ωcm. Flat panel display technologies currently under development require deposition of ITO films onto heat-sensitive polymer substrates and polymeric color filters that cannot survive vacuum processing temperatures above 200°C. Under these deposition conditions the optical transmissivity and electrical conductivity of ITO are severely degraded. In fact, physical vapor deposition of ITO at low substrate temperatures may result in the deposition of an amorphous material with lower conductivity and optical transmissivity than the crystalline phase. Remarkably, the amorphous material will undergo solid state crystallization at extremely low temperatures (<150°C) relative to the In_2O_3 melting point (1910°C). In this paper we report on the amorphous to crystalline phase transformation in electron beam deposited indium oxide.

Both dc-magnetron sputtering [1] and electron beam evaporation [[2], [3]] techniques are reported to deposit amorphous, partially amorphous, or crystalline indium (tin) oxide on cool (T<200°C) substrates. In all of these cases the deposited material possesses inferior electrical and optical properties. For example, deposition of thin (below 50 nm) tin-doped indium oxide (ITO) films at substrate temperatures below 150°C reportedly [4], [5] results in an all or partly amorphous material that can be crystallized via a brief low temperature anneal. There are also reports in the literature [6, 7] that suggest that even at higher substrate temperatures there are conditions where the initial layers (<5-10 nm) are deposited in the amorphous state but, as the film thickness increases, the film reverts to its crystalline form. Similarly, in dc magnetron sputtered materials, both Yi [8] and Vink [9] have reported that ITO deposited on substrates that are close to 150°C is initially amorphous. Based on these reports it appears likely that the mechanisms of the a/c-transformation during both growth and post-deposition processing plays a role in determining the ultimate microstructure of the film. In our work[10] in this area, we have chosen to study electron beam deposited material since it offers a method for producing amorphous ITO that is relatively process independent and allows target compositional changes to be easily implemented.

Crystalline indium oxide has the bixbyite structure which is described using the c-type rare earth vacancy defect-oxide crystal prototype. Bixbyite has an 80-atom unit cell with the Ia3 space group and a 1.0 nm lattice parameter in an arrangement that is based on the stacking of MO_6 coordination groups. The structure is closely related to fluorite which is a face-centered cubic array of cations with all the tetrahedral interstitial positions occupied with anions. The bixbyite structure is similar except that the MO_8 coordination units (oxygen position on the corners of a cube and M located near the center of the cube) are replaced with units that have oxygen missing from either the body or the face diagonal as shown in Figure 1. The removal of two oxygen ions from the MO_8 to form the MO_6 coordination units forces the displacement of the cation from the center of the cube. In this way, indium is distributed in two non-equivalent sites with 1/4 of the indium atoms positioned at the center of a trigonally distorted oxygen octahedron (diagonally missing O). The remaining 3/4 of the indium atoms are positioned at the center of a more distorted octahedron that forms with the removal of two oxygen atoms from the face of the octahedron. The resulting MO_6 coordination units are then stacked such that 1/4 of the oxygen ions are missing from each {100} plane.

Amorphous indium oxide is likely formed during physical vapor deposition processing when the InO_x units that evolve from the source or which form from species arriving on the growth surface are incorrectly oriented as they are incorporated into the growing film. The restricted mobility of the indium-oxygen clusters at low substrate temperatures preserves the misorientation of the coordination units and consequent bond distortion. Remarkably, the crystallization of amorphous ITO occurs rapidly at very low homologous temperatures ($T/T_m<0.19$) with 150°C being commonly cited as the temperature at which crystallization occurs rapidly.

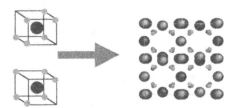

Fig. 1: Bixbyite consists of a 2x2x2 array of In-O_6 structural units to form a 80 atom unit cell with a 1 nm lattice parameter. The two configurations of In-O_6 structural units are shown.

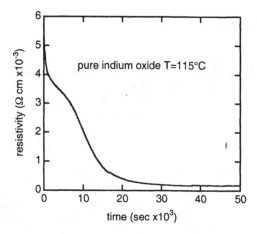

Fig. 2: Typical plot of resistivity versus time showing first order (Avrami parameter of 1) relaxation of the crystalline structure (approx. first 5000 sec) and crystallization via nucleation and growth (Avrami parameter 2-3).

EXPERIMENT

Indium oxide samples were deposited using electron beam evaporation from sintered ceramic targets containing 0, 2.5, 5, and 9.8 wt% SnO_2. The films were deposited to a thickness of 100, 200, and 300 nm onto optical-quality glass substrates at a deposition rate of about 0.8 nm/s in a chamber that was pre-pumped to a base-pressure of 3×10^{-6} Torr . No gases were introduced during deposition.

The kinetics of the crystallization process were studied using 500-μm-wide Hall-spider type structures that were fabricated using lift-off lithography. Electrical contact was made to large contact pads during isothermal annealing in air using tungsten pressure probes. The resistance of the material was monitored during anneals at temperatures ranging between 100 and 165 °C using a 47 Hz AC probe current no larger than 1 μA while the specimen voltage was monitored using a lock-in amplifier.

The evolution of film stress due to the change in volume associated with the crystallization process was studied using a multiple-beam optical stress sensor system. The details of this system are discussed elsewhere [11] Briefly, MOSS is a laser-based method of measuring wafer curvature that creates multiple, highly parallel beams, by passing a HeNe laser beam through an etalon. The parallel beams are reflected off the substrate and into a CCD camera. The change in position of the beams on the CCD during the transformation yields a measure of the change in substrate curvature which, using the Stoney equation, provides a measure of film stress.

RESULTS AND DISCUSSION

Figure 2 shows a typical plot of resistivity versus time at temperature for a pure indium oxide sample annealed at 115°C. Similar results were obtained from samples containing 0, 2.5, 5, and 9.8 wt% SnO_2 and, in every case, showed a shift to longer times at lower temperatures which is an obvious characteristic of a thermally activated process. Inspection of the resistivity curve, however, suggests that there are two processes that operate during annealing and which result in a decrease in film resistivity. The first of these can be seen in Fig. 2 as the rapid 20% decrease in resistivity that occurs in this sample (at 115°C) in the first 5000 sec. At higher temperatures the characteristic time for this initial resistance change decreases. Glancing incidence angle x-ray diffraction (GIAX) studies [10] reveal that the initial reduction in resistivity occurs without the formation of detectable amounts of crystalline material. This is shown in Fig. 3 where a set of 9.8 wt% SnO_2 samples were annealed at 120 °C while monitoring the resistance at temperature. At roughly 5000 sec intervals one of the samples was removed to allow x-ray data to be taken. The lower curve shows the broad amorphous peak that is characteristic of as-deposited amorphous ITO. This broad amorphous peak (curves b and c) persists even after the resistance has dropped by almost 20%. After about 20,000 sec, the bixbyite 222 peak, at a two-theta of 30.5°, becomes apparent in the top-most curve corresponding to point (d) on the resistance plot.

A model that accurately describes the relationship between volume fraction crystalline material and resistivity in thin film indium oxide was developed based on a 2-d modification of Landauer's 3-d model[12] for two-phase resistivity. The details are available elsewhere[10] but using x-ray diffraction to determine the volume fraction crystalline material it was shown that this model closely predicts the change is resistance as a function of volume fraction crystalline material.

Using the above two-phase resistivity model and a non-linear least squares fitting routine, the measured resistivity was used (previously reported work [10]) to show that the change in resistivity is consistent with a two step process consisting of a precrystallization relaxation of the amorphous structure followed by classical nucleation and growth. The relaxation process occurs with a first order kinetic parameter for all compositions tested (0, 2.5, 5, 9.8 wt% SnO_2) as is consistent with a point-defect mediated process. The relaxation-induced part of the resistivitivity curve is shown in Fig. 3. The subsequent crystallization process was shown to have a kinetic parameter of approximately 2 in the case of pure indium oxide and was as high as 3.7 for the indium oxide samples containing 9.8 wt% SnO_2. Plan view TEM reveals that the higher tin concentration sample has a much finer grain size (<100 nm diameter grains) than the pure indium oxide case (>1000 nm grains) suggesting that the SnO_2 plays a role in determining the nucleation of the crystalline phase.

MOSS measurements of 100, 200, and 300 nm-thick indium oxide with 9.8 wt% SnO_2 were used to monitor the effect of the transformation on film stress. The change in stress that accompanies the heating of a 200 nm thick sample at 135 °C is shown in Fig. 4. In this plot, the change in stress is plotted on the left axis while the corresponding change in resistance is plotted on the right. The ITO thin film shows an initial positive (tensile) change in stress which peaks at about 5000 sec and then decreases in the subsequent 15000 secs of the anneal. Inspection of the resistance curve in Fig. 4 shows that the first 5000 seconds corresponds to the amorphous relaxation process that precedes

248

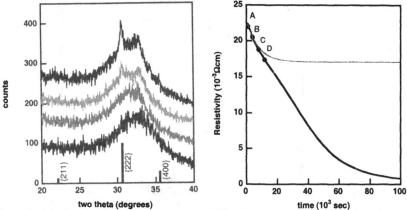

Figure 3: (a) Glancing incidence angle x-ray diffraction patterns showing (from bottom to top) indium (9.8 wt% SnO_2) oxide in the as deposited (bottom curve) condition and the change in resistivity with increasing time at temperature (b-d).

crystallization while the time scale of the stress decrease is consistent with crystallization. The initial increase in tensile stress is, we speculate, due to densification resulting from the relaxation of distorted In-O_x bonds to a more equilibrium configuration. The decrease in tensile stress that we observe is due to a volume increase associated with the crystallization of the ITO from the relaxed amorphous structure. This, initially surprising result, appears to be possible because of the arrangement of vacancies that is part of the bixbyite crystal structure described above and, in effect, suggests that that the molar volume of bixbyite is greater than that of the relaxed amorphous material.

It is, in principle, possible that the increase in volume that attends the crystallization of our ITO samples is due to the oxidation of sub-stoiciometric indium oxide. We have discounted this possibility for three reasons. First, the temperature at which these tests were performed (100-165°C) is likely too low to allow for long range oxygen diffusion. Second, measurement of the rate of crystallization in dry N_2 was similar to that observed for tests run in air. Third, the resistivity curves show a consistent drop in resistance which a variety (Hall, reflectivity, temperature dependence) of measurements all show is due to an increase in carrier density which implies an increase in either substitutional tin or doubly charged oxygen vacancy density. The study of the molar volume change as a function of processing and ambient is ongoing.

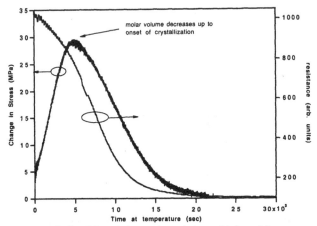

Figure 4: Change in stress (left axis) and change in resistance (right axis).

CONCLUSIONS

Using a combination of resistivity and stress measurements we have investigated the crystallization behavior of electron beam deposited indium oxide. We have shown that crystallization is preceded by a relaxation of the as-deposited amorphous structure and have shown that the relaxed structure has a lower molar volume than the as deposited state. The reduction in resistance that accompanies the relaxation process in both pure indium oxide and Sn-doped indium oxide suggests that there is a fundamental change in electronic structure during the relaxation of the amorphous structure and, kinetic analysis reveals that this change occurs with an Avarami kinetic parameter of unity which is consistent with a point defect mediated process. Crystallization was shown to occur via nucleation and growth. The kinetic parameter for the crystallization appears to increase with increasing Sn content from 2 to 3.7 . Crystallization was seen to be a thermally activated process that occurs at reasonable rates (time scale of days) at temperatures as low as 100 °C. Furthermore, the crystalline state (bixbyite) appears to have a larger molar volume than the relaxed amorphous state. Finally, the above observations highlight the extent to which the crystallization of this complex oxide is rich with important fundamental and technological issues.

ACKNOWLEDGEMENTS

This work was supported, in part, by the MRSEC at Brown University and by Arconium Corp.

REFERENCES

1.. Song, P.K.,. Jpn. J. Appl. Phys., 1998. **37(pt.1)**(4A): p. 1870-1876.
2. Shigesato, Y., S. Takaki, and T. Haranoh, J. Appl. Phys., 1992. **71**: p. 3356.
3. Oyama, T, J. Vac. Sci Tech A, 1992. **10**: p. 1682.
4. Chaudhuri, S., J. Bhattacharyya, and A.K. Pal, Thin Solid Films, 1987. **148**: p. 279-284.
5. Hoheisel, M., A. Mitwalsky, and C. Mrotzek,. Phys. Stat. Sol., 1991. **123**: p. 461-472.
6. Muranaka, S., Y. Bando, and T. Takada,. Thin Solid Films, 1987. **151**: p. 355-364.
7. Muranaka, S., *Crystallization of Amorphous In$_2$O$_3$ Films During Film Growth.* Jpn. J. Appl. Phys. Part 2, 1991. **30**(12A): p. L2062.
8. Yi, C.H.,. Jpn. J. Appl. Phys. part 2, 1995. **34**(2B): p. 244-247.
9. Vink, T.J., W. Walrave, J.L.C. Daams, P.C. Baarslag, J.E.A.M. van der Meerakker, Thin Solid Films, 1995. **266**: p. 145.
10. Paine, D.C., T. Whitson,D. Janiac, R. Beresford, C. Ow Yang and B. Lewis, J. Appl. Phys., 1999. **85**(12): p. 8445-8450.
11. E. Chason and J.A. Floro, Mat. Res. Soc. Symp. Proc. Vol 428, p.499-504(1996).
12. Landauer, R. J. Appl. Phys., 1952. **23**: p779

Low Electrically Resistive Transparent Indium-Tin-Oxide Epitaxial Film on (001) Surface of YSZ by Pulsed Laser Deposition

Hiromichi Ohta[*], Masahiro Orita[*], Masahiro Hirano[*], Hideo Hosono[*, **], Hiroshi Kawazoe[***], Hiroaki Tanji[***]

[*]Hosono Project of Transparent ElectroActive Materials, ERATO, Japan Science and Technology Corporation, KSP C-1232, 3-2-1, Sakado, Takatsu-ku, Kawasaki, 213-0012, JAPAN,

[**]Materials and Structures Laboratory, Tokyo Institute of Technology, Nagatsuta, Midori-ku, Yokohama 226-8503, JAPAN,

[**]R&D Center, HOYA Corporation, 3-3-1 Musashino, Akishima-shi, Tokyo 196-8510, JAPAN

e-mail : h-ohta@ ksp.or.jp

ABSTRACT

High quality ITO thin films were grown hetero-epitaxially on extremely flat substrate of (001) YSZ by a pulsed laser deposition technique at a substrate temperature of 600°C. The crystal orientation relationship between the film and YSZ were confirmed as ITO (001) // YSZ (001) and ITO (010) // YSZ (010), respectively, by HR-XRD and HR-TEM. The carrier densities of the films were almost equal to SnO_2 concentration in the films. That is, almost all the doped Sn^{4+} ions were activated to release electrons to the conduction band. The carrier densities of the films were enhanced up to $1.9 \times 10^{21} cm^{-3}$, while the Hall mobility showed a slight, almost linear, decrease from 55 to $40 cm^2 V^{-1} s^{-1}$ with increasing SnO_2 concentration. The low resistivity is due to larger electron mobility, which most likely resulted from good crystal quality of the films. The optical transmissivity of the film exceeded 85% at wavelengths from 340 to 780nm.

INTRODUCTION

Indium-tin-oxide (ITO), a typical transparent conductor, has been widely used as a transparent electrode for display devices such as LCDs, PDPs and organic LEDs. Because of the requirement for high transparency of the electrodes, the films become thinner in these devices. This inevitably leads to an increase in the electric resistance, provided that the electric conductivity of the film remains unchanged. Thus, many attempts have been made to increase the electrical conductivity. However, using conventional evaporation or sputtering methods, it is almost impossible to decrease the resistivity down below $10^{-4} \Omega cm$.

A few ideas have been proposed with the intention to enhance the conductivity drastically. The first is to apply new film deposition techniques to the growth of ITO films. DC magnetron sputtering[1], electron beam evaporation[2] and pulsed laser deposition[3-7] (PLD) techniques have been tried. However, ITO films having resistivity below $10^{-4}\Omega$cm have not been obtained so far. Use of new deposition techniques alone does not seem to lead to any dramatic improvement. The second one is to improve the crystal quality of ITO films. This may lead to an increase in the electron mobility due to the reduction of the scattering centers such as crystal imperfections and neutral donor centers. Use of a single crystal substrate[1-4], hopefully one lattice-matched with ITO, and a high temperature deposition process seem to be essentially important to accomplish it. The third approach is to design such a quantum structure that the donor-doped-layer is spatially separated from the carrier drifting one, which is analogous to a HEMT device. The mobility may be improved in this film structure because of the reduction of interaction between carriers and scattering centers.

We have employed a combination of the first and the second approaches. That is, an ITO film was grown on an extremely flat yittria-stabilized-zirconia (YSZ)[1-3] single crystal substrate at a temperature of 600°C by a pulsed laser deposition technique. Twice the cubic lattice constant of YSZ (1.02nm) is very close to that of ITO (1.01nm), which could make epitaxial growth of ITO possible. The PLD method is suited for oxide film preparation. Additionally, film composition is easily controlled by changing the target composition.

We have obtained epitaxially grown ITO films using this approach and the resistivity was reproducibly reduced down to 7.8 x $10^{-5}\Omega$cm.

EXPERIMENTAL

Oxygen gas was introduced into the chamber with a pressure of 1.2 x 10^{-3}Pa. YSZ (001) was used as the substrate. Surface polished SiO_2 glass was also used as a substrate for comparison. The substrate temperature was maintained at 600 °C during film deposition. An ITO ceramic target (SnO_2 concentration: $0 \sim 15$wt.%) was set at the center of the PLD chamber, and the substrate was positioned facing the target. An ITO film was deposited on a rotating substrate by focusing a KrF (λ=248nm) excimer laser onto the rotating target. The distance between the substrate and the target was 30mm. The pulsed laser with a repetition rate of 10Hz was focused onto the target with an energy density of about 6J/cm^2.

The film thickness was determined by means of a conventional stylus profilometer (Taylor-Hobson, Tallystep). An atomic force microscope (AFM, S.I.I., SPI-3700) was used to investigate surface morphology of the films. The SnO_2 concentration was analyzed by XRF measurement (RIX2100, Rigaku). Crystalline quality and orientation were analyzed by X-ray diffraction (ATX-G, Rigaku Co.), in which the ω axis was rotated in the horizontal plane and the ϕ axis was rotated azimuthally. The detector was manipulated with both the 2θ axis in the horizontal plane and the $2\theta\chi$ axis in the vertical plane. Out-of-plane crystal orientation was measured by synchronous scan of the 2θ with the ω axis. Tilting was analyzed from measurements of 2θ-fixed ω-scan (out-of-plane locking curve). In-plane crystal orientation was measured by synchronous scan of the $2\theta\chi$ and with the ϕ axis. Twisting was analyzed from measurements of $2\theta\chi$-fixed ϕ-scan (in-plane locking curve). The interface between the ITO film and the substrate was observed by high resolution TEM (JEM-2000EX, JEOL). The electrical properties were measured at room temperature by the van der Pauw method. Optical transmission and reflection spectra were measured by U-4000 spectrometer (Hitachi Co.).

RESULTS

Figure 1 shows an AFM image of a polished YSZ surface after annealed at 1350°C for 1h in air. Flat terraces and steps were clearly observed. Terrace width was about 150nm, and step increments were about 0.26nm, which was in good agreement with the distance between

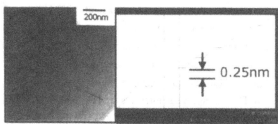

Fig. 1. AFM image of annealed substrate of YSZ (001). Flat terraces and steps were clearly observed.

two adjacent (002) planes of YSZ (0.257nm). 400nm-thick ITO films were grown on the annealed substrate of YSZ by PLD at a substrate temperature of 600°C. Figure 2 shows the AFM images of the ITO thin films deposited on YSZ and SiO₂ glass. The grain size of ITO on YSZ was about 200nm, which was larger than that of ITO on SiO₂ glass (100nm). Surface

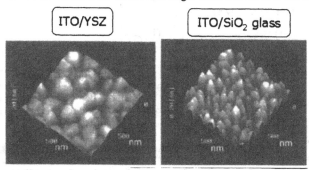

Fig. 2. AFM images of the ITO thin film deposited on YSZ and SiO₂ glass

roughness of the films on YSZ was smaller than the films grown on SiO₂ glass.

Figure 3 shows out-of-plane XRD patterns of ITO thin films deposited on (001) YSZ (Fig. 3(a)) and SiO₂ glass (Fig. 3(b)). Only intense (00l) ITO diffraction peaks were observed together with (00l) YSZ

diffraction peaks in Fig. 3(a), indicating that the crystal orientation is ITO (001) // YSZ (001). The full width at half maximum of out-of-plane locking curves of (004) ITO diffraction was 0.4°, indicating strong orientation of ITO (001) // YSZ (001). On the other hand, a sharp diffraction peak of (004) ITO with weak diffraction peaks such as (222) was observed in Fig. 3 (b), indicating a random orientation of the ITO crystals. Figure 4 shows in-plane XRD patterns of the ITO thin film grown on YSZ surface. In-plane measurements were performed around the critical angle of the X-ray total external reflection. The grazing incident angle of the X-ray was 0.3°. Only intense (080) ITO diffraction peak is observed together with (040) YSZ diffraction peak in Fig. 3(a). This indicates that crystal orientation relationship is ITO [010] // YSZ [010]. Figure 4(b) shows in-plane locking curve under the condition that $2\theta\chi$ axis was fixed at (080) ITO diffraction angle. Four peaks locating at 90° interval were observed, indicating hetero-epitaxial growth in lateral direction was taken place. Figure 5(a) demonstrates a cross sectional HR-TEM image near interface of ITO and YSZ projected <100>. Its inversed FFT image was shown in Fig. 5(b). Although several incoherencies by geometrical misfit dislocations could be observed on interface shown in Fig. 5(b), ITO and YSZ were continuously connected, proving that hetero-epitaxial growth occurs. From these observations, it is verified that high quality hetero-epitaxial ITO thin films were successfully obtained keeping crystallographic orientation of (0k0) ITO // (0k0) YSZ and (00l) ITO // (00l) YSZ.

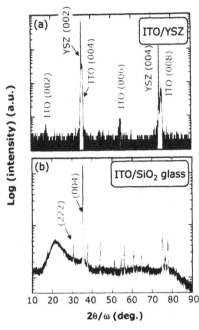

Fig. 3. Out-of-plane XRD patterns of the ITO thin films deposited on YSZ (a) and SiO_2 glass (b).

The SnO_2 concentration analyzed by XRF, carrier density ($n_{meas.}$), and its calculate value ($n_{calc.}$) from the SnO_2 concentration are listed in Table 1. The conversion efficiency of the dopant ($n_{meas.}$ / $n_{calc.}$) was

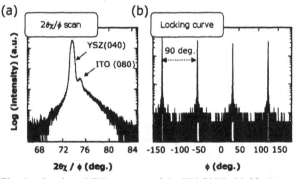

Fig. 4. In-plane XRD patterns of the ITO/YSZ. (a) $2\theta\chi/\phi$ scan and (b) locking curve of ITO (080).

almost 100%, which indicates nearly all Sn^{4+} ions release electrons into the conduction band. This is probably resulted from good crystallinitiy of the film. This also implies neutral impurities, which act as scattering center for the carrier, become very small density, leading to the high electron mobility of the film. Figure 6 summarized Hall mobility, and the electrical resistivity of ITO/YSZ and ITO/SiO_2 films as a function of SnO_2 concentration. Although Hall mobility of ITO/SiO$_2$ glass decreased drastically with increasing SnO$_2$ concentration, that of ITO/YSZ shows slight decrease almost linearly with the concentration. In both cases, carrier electrons were mainly scattered by ionized impurity such as dopant Sn^{4+} ion. Additionally, in case of ITO/SiO$_2$ glass non-homogeneity and neutral impurity would cause larger electron scattering. The electrical resistivity of ITO/YSZ decreases sharply from about $10^{-2}\Omega$cm to below $10^{-4}\Omega$cm with doping of SnO$_2$, and it saturates when SnO$_2$ exceeds 3wt.%, showing a gentle peak at 5.7wt% SnO$_2$. The lowest resistivity of 7.8 x $10^{-5}\Omega$cm was reproducibly obtained in the film containing 5.7wt.% SnO$_2$, in which the mobility was 42cm^{2}V^{-1}s^{-1} and the carrier density was 1.9 x 10^{21}cm^{-3}.

Figure 7 shows the optical transmission (T) and the reflection spectra (R) in the visible and near-IR region for the ITO/YSZ film containing 5.7wt.% SnO$_2$ (lowest resistivity specimen). In this figure, transmission and reflection due to the YSZ substrate are also shown by dotted lines. The optical transmissivity of the film exceeded 85% at wavelengths from 340 to 780nm.

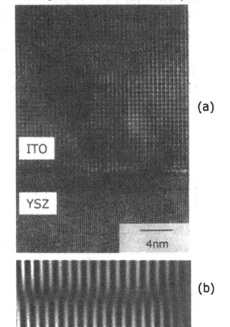

(a)

(b)

Fig. 5 (a) X-TEM image of ITO/YSZ and (b) its inversed FFT image of near interface region.

Table 1 The SnO$_2$ concentration analyzed by XRF, carrier density ($n_{meas.}$), and its calculate value ($n_{calc.}$) from SnO$_2$ concentration.

SnO$_2$ concentration in films (wt.%)	Carrier density (10^{21}cm^{-3})		Doping efficiency (%)
	$n_{meas.}$	$n_{calc.}$	
3.0	0.96	0.86	110
5.7	1.9	1.8	106
7.4	1.9	2.2	86

CONCLUSIONS

High quality ITO thin film was successfully grown hetero-epitaxially on extremely flat substrate of (001) YSZ by pulsed laser deposition method at a substrate temperature of 600°C. HR-XRD and HR-TEM measurements verify crystal orientation relationship between the film and YSZ were ITO (001) // YSZ (001) and ITO (010) // YSZ (010), respectively. Almost all the Sn^{4+} ions in the film release electrons to the conduction bands, resulted from high crystallinity of ITO. The carrier densities of the films were enhanced up to $1.9 \times 10^{21} cm^{-3}$, while the Hall mobility showed a slight, almost linear, decrease from 55 to $40 cm^2 V^{-1} s^{-1}$ with increasing SnO_2 concentration. The low resistivity is due to larger electron mobility, which was most likely resulted from high crystal quality of the films. The optical transmissivity of the film exceeded 85% at wavelengths from 340 to 780nm.

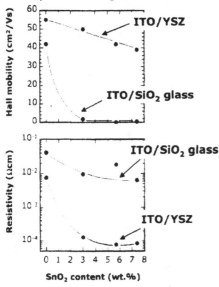

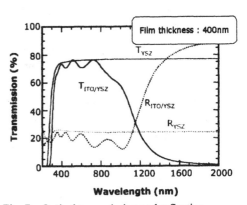

Fig. 6. Hall mobility and the carrier density of the films as a function of the SnO_2 concentration.

Fig. 7 Optical transmission and reflection spectra of ITO/YSZ and YSZ substrate.

REFERENCES

1. M. Kamei, T. Yagami, Y. Shigesato, *Appl. Phys. Lett.*, **64**, 2712 (1994).

2. N. Taga, H. Odaka, Y. Shigesato, I. Yasui, M. Kamei, T. E. Haynes, *J. Appl. Phys.*, **80**, 978 (1996).

3. E. J. Tarsa, J. H. English, J. S. Speck, *Appl. Phys. Lett.*, **62**, 2332 (1993).

4. Q. X. Jia, J. P. Zheng, H. S. Kwok, W. A. Anderson, *Thin Solid Films*, **258**, 260 (1995).

5. C. Coutal, A. Azema, J. C. Roustan, *Thin Solid Films*, **288**, 248 (1996).

6. F. O. Adurodija, H. Izumi, T. Ishihara, H. Yoshioka, K. Yamada, H. Matsui, M. Motoyama, *Thin Solid Films*, **350**, 79 (1999).

7. H. Kim, A. Pique, J. S. Horwitz, H. Mattoussi, H. Murata, Z. H. Kafafi, D. B. Chrisey, *Appl. Phys. Lett.*, **74**, 3444 (1999).

Density-of-States Effective Mass and Scattering Parameter Measurements on Transparent Conducting Oxides Using Second-Order Transport Phenomena

D. L. Young,* T. J. Coutts,* X. Li,* J. Keane,* V. I. Kaydanov,** A. S. Gilmore**
* National Renewable Energy Laboratory, 1617 Cole Blvd., Golden, CO 80401
** Colorado School of Mines, 1500 Illinois St., Golden, CO 80401

Abstract

Transparent conducting oxides (TCO) have relatively low mobilities, which limit their performance optically and electrically, and which limit the techniques that may be used to explore their band structure via the effective mass. We have used transport theory to directly measure the density-of-states effective mass and other fundamental electronic properties of TCO films. The Boltzmann transport equation may be solved to give analytic solutions to the resistivity, Hall, Seebeck, and Nernst coefficients. In turn, these may be solved simultaneously to give the density-of-states effective mass, the Fermi energy relative to either the conduction or valence band, and a scattering parameter, s, which characterizes the relaxation time dependence on the carrier energy and can serve as a signature of the dominate scattering mechanism. The little-known Nernst effect is essential for determining the scattering parameter and, thereby, the effective scattering mechanism(s). We constructed equipment to measure these four transport coefficients on the same sample over a temperature range of 30 - 350 K for thin films deposited on insulating substrates. We measured the resistivity, Hall, Seebeck, and Nernst coefficients for rf magnetron-sputtered aluminum-doped zinc oxide. We found that the effective mass for zinc oxide increases from a minimum value of $0.24m_e$ up to a value of $0.47m_e$ at a carrier density of 4.5×10^{20} cm^{-3}, indicating a nonparabolic conduction energy band. In addition, our measured density-of-states effective values are nearly equal to conductivity effective mass values estimated from the plasma frequency, denoting a single energy minimum with a nearly spherical, constant-energy surface. The measured scattering parameter, mobility vs. temperature, along with Seebeck coefficient values, characterize ionized impurity scattering in the ZnO:Al and neutral impurity scattering in the undoped material.

Introduction

Technically and commercially significant applications for transparent and conducting thin films are increasing in today's market. Some of these potentially large-volume items include: low-emissivity, selectively reflecting architectural windows; durable, antistatic coatings; transparent electrodes on flat-panel displays; solar cells; and electrochromic windows.[1] As demand increases for larger and better flat-panel displays and solar cells, current transparent and conducting materials will hinder this development.

Transparent conducting oxide (TCO) thin films represent a compromise between optical transmittance, bounded by absorbance at the optical band gap and reflectance at the plasma frequency, and electrical conductivity. As noted by Coutts et al.,[2] this compromise must pivot about maximizing the mobility. For highly degenerate, metal-like TCO films, mobility, μ, may be defined by the Drude theory,

$$\mu = q\tau/m^*, \tag{1}$$

where m* is the conductivity effective mass and τ is the relaxation time which is close to the average time between carrier-scattering elastic collisions. Obviously, either a long relaxation time or a small m* will increase the mobility. Fundamental studies of existing and novel TCO materials require knowledge of both τ and m*. Cyclotron, de Haas van Alphen, and shubnikov-de Haas resonance techniques, the traditional methods for probing the Fermi surface via the effective mass, are only applicable for high-mobility materials because of the constraint that the product μB >> 1, so that electrons complete at least one orbit before they are scattered. Effective mass values for polycrystalline, thin-film TCOs are usually inferred from optical data or assumed to be equal to single-crystal, bulk values.

In addition to the need for a direct measure of m* in TCOs, understanding the mechanism(s) for scattering of carriers in films is of great importance for improving transport properties. To better understand the limits of mobility and the dominant scattering mechanisms, a novel measurement technique, based on transport phenomena, has been applied to TCO thin-films. Transport phenomena coefficient measurements — resistivity, Hall, Seebeck, and Nernst — on thin-film TCO samples were used with solutions to the Boltzmann transport equation to make direct measurements of the density-of-states effective mass, m_d^*, and an energy-dependent scattering parameter, s. We have constructed an instrument to enable these measurements for films grown on electrically insulating substrates. Our group has applied this technique to TCO thin-film samples and has made, to our knowledge, the first direct measurements of m_d^* and s on films of cadmium stannate, zinc stannate, zinc oxide, tin oxide, and cadmium oxide. In this paper, we will discuss transport theory, our transport coefficient instrument, and data obtained from our zinc oxide study.

Theory

Traditionally, a relaxation time approximation has been used for analytic solutions to Boltzmann's transport equation in the study of charge transport in materials. This approximation assumes a mean time, τ, between transitions, induced by scattering events, from an excited to an equilibrium distribution or vice versa. In general, τ depends on energy, temperature and direction, especially for impurity ion scattering, if the constant energy surface is anisotropic. In our case the constant energy surface is almost spherical, hence we can assume an isotropic τ. The frequency of elastic collisions, $\tau^{-1}(E)$, depends on the density-of-states function, $g(E)$, and the square of the matrix element of the scattering transition from an initial energy state to a final state of the same energy, $w(E)$. For parabolic energy bands, $g(E) \propto E^{1/2}$, whereas $w(E)$ is a power function of energy $w(E) \propto E^{-s} \propto (k^2)^{-s}$, where k is the wave vector magnitude. Thus,

$$\tau^{-1}(E) = w(E)g(E) = E^{-s}E^{1/2} = E^{1/2-s}, \qquad (2)$$

which is often written $\tau \propto E^{s-1/2} \propto E^{s'}$, with s' = s - 1/2. For the parabolic band model, s' has been identified[3] for several scattering mechanisms that are shown below:

> Acoustic Phonon scattering —> s' = -1/2, s = 0
> Neutral Impurity scattering —> s' = 0, s = 1/2
> Ionized Impurity scattering —> s' = 3/2 s = 2

In the non-parabolic, isotropic energy-band case, the dependence $E(k)$ can be described by the equation (3)

$$\frac{\hbar^2 k^2}{2m_o^*} = \gamma(E) = E + \frac{E^2}{E_1} + \frac{E^3}{E_2^2} + ..., \qquad (3)$$

where E_1 and E_2 are constants for a given material. For not too strong nonparabolicity (E_1, $E_2 > E$) it can be assumed, to a first approximation, that the transition matrix element preserves the same dependence on k. as in the parabolic case. The nonparabolic density-of-states function,

$$g_{N.P.}(E) \propto \gamma^{1/2}\left(\frac{d\gamma}{dE}\right), \qquad (4)$$

leads to

$$\tau \propto \gamma^{s-1/2}\left(\frac{d\gamma}{dE}\right)^{-1} = \gamma^{s'}\left(\frac{d\gamma}{dE}\right)^{-1}. \qquad (5)$$

Clearly, the s' and s values for different scattering mechanisms in the nonparabolic case are the same as in the parabolic case.

Using equations (2) – (5), we have recently outlined[4] how transport phenomena may be used to probe the Fermi surface and reveal scattering mechanisms in degenerate semiconductors. Our work follows original efforts by Zhitinskaya et al.[5] and Kolodziejczak et al.,[6,7] who recognized that knowledge of four transport coefficients could reveal the effective density-of-states (DOS) mass at the Fermi level,

$$m_d^* = \left(\frac{3}{|R_H|q\pi} \right)^{2/3} \frac{q\hbar^2}{k_B^2 T} \left(\alpha - \frac{Q}{|R_H|\sigma} \right) \qquad (6)$$

and s, the scattering parameter,

$$s = \frac{3}{2} \frac{\left(\frac{Q}{\mu} \right)}{\left(\alpha - \frac{Q}{\mu} \right)} + \frac{1}{2} + \lambda, \qquad (7)$$

where

$$\lambda = 2\gamma \left(\frac{d\gamma}{d\varepsilon} \right)^{-2} \frac{d^2\gamma}{d\varepsilon^2} = 3 \frac{n}{m^*} \frac{dm^*}{dn}. \qquad (8)$$

Equation (6) illustrates that the DOS effective mass may be experimentally determined by measuring the conductivity (σ), Hall (R_{Hall}), Seebeck (α), and Nernst (Q) coefficients of a sample. This technique has been coined *the method of four coefficients*[8] and was originally applied to n-type PbTe single crystals.[5] Our group has applied *the method of four coefficients* to thin-film TCO samples, using a specially designed instrument to measure the four coefficients on the same sample. A detailed discussion of these transport phenomena may be found in the literature.[3,9,10]

Experimental Procedure

Measurement of four transport coefficients on thin-films has been attempted by other groups[11] and emphasizes the need for a clever film pattern to accomplish all of the measurements on the same sample. For our instrument,[12] thin-film TCO samples are deposited on electrically insulating substrates and photolithographically etched to the pattern shown in Figure 1. The high aspect ratio of the sample conforms to the specimen shape dictated by the ASTM designation F76-86[13] for van der Pauw resistivity and Hall measurements, as well as ensuring a large Nernst voltage between contacts 1 and 3 and a

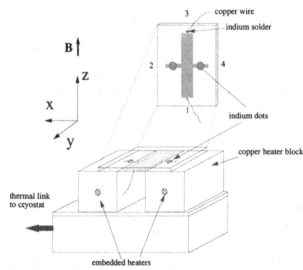

Fig. 1 Sample holder with sample placed film-side down across copper heater blocks.

large temperature gradient between contacts 2 and 4. As depicted in Figure 1, the film is placed film-side down across two copper heater blocks, with contacts 2 and 4 making ohmic contact with the heater blocks by silver paint or an indium dot. Contacts 1 and 3 are indium-soldered to fine copper wire. The heater blocks are electrically isolated from each other, and each has a copper wire attached to it to make electrical contact to the film. The heater blocks have a differential thermocouple mounted between them to measure the temperature gradient across the sample. Heater block 2 has an additional embedded thermocouple for absolute temperature measurement. The entire sample holder of Figure 1 is cooled by a closed-cycle helium cryostat for temperature-dependent measurements from 30 - 350 K. The four transport coefficients, mentioned above, may be measured by the instrument. The specific measurement sequence is outlined in reference 13.

Results

RF magnetron-sputtered ZnO and ZnO:Al thin-films grown on 7059 glass were photolithographically patterned to the shape shown in Figure 1. Film thickness was measured by profilometry. The four relevant transport phenomena coefficients (conductivity, Hall, Seebeck, and Nernst) were measured by the method and by the instrument described above, which were then used in equation (6) to calculate the DOS effective mass, m_d^*.

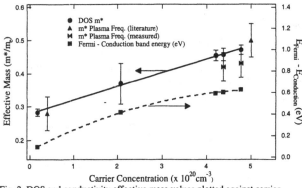

Fig. 2 DOS and conductivity effective mass values plotted against carrier concentration. Fermi energy levels above the conduction-band minimum are also plotted.

Figure 2 shows the m_d^* values plotted against the carrier density, n. We first note that m_d^* is not constant with carrier density and varies from about 0.3 to 0.47 as the carrier density changes from 2×10^{19} - 5×10^{20} cm^{-3}. This trend shows that the conduction energy band for ZnO is non-parabolic in this carrier density range and will make the energy-band term, λ, non-zero. In addition to our m_d^* values for ZnO plotted in Figure 2, conductivity effective-mass values are plotted. These values, taken from the literature[14,15] or calculated from transmission and reflection measurements, were determined by the plasma frequency. Note that the conductivity and DOS effective mass are in good agreement with each other. The relationship between conductivity and DOS effective mass values for a material with N constant energy ellipsoid of revolution surfaces is given by

$$\frac{m_d^*}{m_c^*} = N^{2/3} \frac{(2\beta + 1)}{3\beta^{2/3}}, \text{ where } \beta = \frac{m_{||}}{m_\perp}. \qquad (9)$$

With such good agreement between m_c^* and m_d^* we conclude that N must be one and that β must also be approximately one. This being the case, we may conclude that ZnO has a single-valley minimum in the conduction energy band and that the constant energy surface must be spherical to within experimental uncertainty. The single-valley minimum of ZnO is predicted by theory,[16] but the uncertainty in the plasma frequency data makes it difficult to be certain of the spherical nature of the energy surface. We note that ZnO is hexagonal and that our films are uniaxially textured with (002) orientation. Anisotrophy in the effective mass is not found in the literature, nor is it observed in the mobility above 100 K.[17]

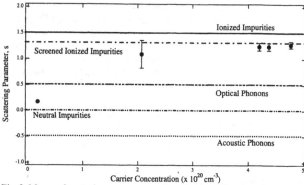

Fig. 3 Measured scattering parameter, s, plotted with predicted trends for several scattering mechanisms. The left-most sample is ZnO whereas the other four samples are ZnO:Al.

A best-fit line was drawn through the points of m_d^* vs. n, which was then used to calculate λ in equations (7) and (8). We solved for λ to determine the degree to which the band shape affects the scattering parameter, s, in equation (7). Figure 3 shows the calculated scattering parameter values for our ZnO and ZnO:Al samples, along with the predicted trends in the scattering parameter for five scattering mechanisms.

For the ZnO:Al samples, the measured scattering parameter lies near the trend expected for ionized impurity scattering (I.I.S.) with screening by free electrons. I.I.S. is predicted for these films, where aluminum is added to contribute an electron to the conduction band to dope the films n-type. The electron leaves behind an ionized aluminum atom that acts as an impurity scattering center. We will discuss below our method for accounting for the screening by free electrons.

Returning to Figure 3, we notice that for the undoped ZnO, the scattering parameter lies most closely with the neutral impurity trend. Figure 4 shows our Seebeck measurements plotted against expected Seebeck values for five scattering mechanisms. Again, ionized and neutral impurity scattering are the apparent mechanisms. Mobility vs. temperature data for both a ZnO and ZnO:Al film are shown in Figure 5. The undoped ZnO film shows a clear positive dependence of mobility on temperature, indicative of neutral impurities. As the temperature decreases, the number of neutral impurity centers will increase, decreasing the relaxation time between collisions, and thus decreasing the mobility.[10] For the

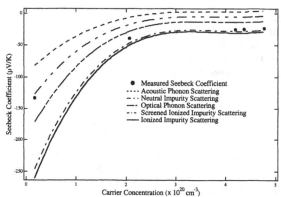

Fig. 4 Seebeck coefficient vs. carrier concentration for the same films as in Fig. 3. Shown with predicted trends for several scattering mechanisms.

doped material, the mobility is seen to be almost constant with temperature, characteristic of I.I.S. in a degenerate material.[10,18] Agreement between optical and Hall mobility data, along with temperature dependent conductivity data, rule out grain-boundary scattering as a major contributor in our ZnO films.

To extrapolate m_o^*, the effective mass at the conduction-band minimum, we use a first-order nonparabolicity approximation by using only the first two terms in equation (3). Our measured value of m_o^* is 0.24 m_e for this sample set. This value allows a calculation of the Fermi energy level relative to the conduction band.[4] Figure 2 shows the calculated Fermi energy level above the conduction band, revealing a strong degeneracy in all of the samples. Knowledge of the effective mass at the bottom of the band allows us to correct for nonparabolicity effects in the predicted mobility for films experiencing ionized impurity scattering. Following work similar to Pisarkiewicz et al.[19] we matched predicted and measured mobilities to ascertain a D.C. dielectric constant. The effective

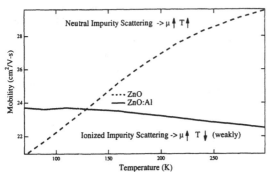

Fig. 5 mobility vs. temperature for ZnO and ZnO:Al films.

screening length, r_o, of an ionized impurity was then calculated,[10] again taking into account the nonparabolicity of the conduction band. The effective screening length adjusts the scattering parameter, s', through the following equation:

$$s' = \frac{3}{2} - \frac{d \ln \phi(k, r_o)}{d \ln E}, \quad \text{where}$$

$$\phi(k, r_o) = \ln(1 + 4k^2 r_o^2) - \frac{4k^2 r_o^2}{1 + 4k^2 r_o^2}. (10)$$

This correction is shown in Figures 3 and 4 and is in good agreement with the measured quantities.

Conclusions

We have used the *method of four coefficients* in this study to show that the conduction band in ZnO is non-parabolic, with an m_o^* value of 0.24 m_e. The dominant scattering mechanism in our ZnO thin films is neutral impurities, whereas screened ionized impurity scattering dominants the ZnO:Al films.

References

[1]H. L. Hartnagel, A. L. Dawar, A. K. Jain and C. Jagadish, *Semiconducting Transparent Thin Films* (Institute of Physics Publishing, London, 1995).

[2]T. J. Coutts, X. Wu, W. P. Mulligan and J. M. Webb, *Journal of Electronic Materials* 25 (6), 935-943 (1996).

[3]S. S. Li, *Semiconductor Physical Electronics* (Plenum, New York, 1993).

[4]D. L. Young, T. J. Coutts, V. I. Kaydanov and W. P. Mulligan, *JVST- Submitted for publication* (2000).

[5]M. K. Zhitinskaya, V. I. Kaidanov and I. A. Chernik, *Sov. Phys. - Sol. Stat.* 8 (1), 295-297 (1966).

[6]J. Kolodziejczak and S. Zukotynski, *Phys. Stat. Sol.* 5 (145), 145-158 (1964).

[7]J. Kolodziejczak and L. Sosnowski, *Acta Phys. Polon.* 21, 399 (1962).

[8]V. Kaydanov . (unpublished lecture notes), Colorado School of Mines, Golden, CO (1996).

[9]E. Putley, *The Hall Effect and Related Phenomena* (Butterworths, London, 1960).

[10]B. M. Askerov, *Electron Transport Phenomena in Semiconductors* (World Scientific, Singapore, 1994).

[11]W. Jiang, S. N. Mao, X. X. Xi, X. Jiang, J. L. Peng, T. Venkatesan, C. J. Lobb and R. L. Greene, *Phys. Rev. Lett.* 73 (9), 1291-1294 (1994).

[12]D. L. Young, T. J. Coutts and V. I. Kaydanov, *Review of Scientific Instruments* 71 (2), 462-466 (2000).

[13]"Standard Test Methods for Measuring Resistivity and Hall Coefficient and Determining Hall Mobility in Single-Crystal Semiconductors, F-76-86," in *Annual Book of ASTM Standards* (American Society of Testing and Materials, West Conshohocken, 1996).

[14]S. Brehme, F. Fenske, W. Fuhs, E. Nebauer, M. Poschenrieder, B. Selle and I. Sieber, *Thin Solid Films* (342), 167-173 (1999).

[15]K. H. Hellwege, "Landolt-Börnstein Numerical Data and Functional Relationships in Science and Technology: Semiconductors," in *Numerical Data and Functional Relationships in Science and Technology*, edited by O. Madelung, M. Schulz, and H. Weiss (Springer-Verlag Berlin-Heidelberg, Berlin, 1982), Vol. 17.

[16]S. Bloom and I. Ortenburger, *Phys. Stat. Sol. (b)* 58 (561), 561-566 (1973).

[17]P. Wagner and R. Helbig, *J. Phys. Chem. Solids* 35, 327-335 (1972).

[18]W. F. Leonard and J. T. L. Martin, *Electronic Structure and Transport Properties of Crystals*, First ed. (Robert E. Krieger Publishing Company, Malabar, FL, 1987).

[19]T. Pisarkiewicz, K. Zakrzewska and E. Leja, *Thin Solid Films* 174, 217-223 (1989).

GROWTH AND CHEMICAL SUBSTITUTION OF TRANSPARENT P-TYPE CUAlO$_2$

R.E. STAUBER**, P.A. PARILLA*, J. D. PERKINS*, D.S. GINLEY*
*University of Colorado, Boulder, CO 80304, stauber@ucsu.colorado.edu
** National Renewable Energy Lab, Golden, CO 80401

ABSTRACT

CuAlO$_2$ is one of several materials currently being investigated for application as a p-type transparent conducting oxide. In this paper we report a method of making c-axis oriented CuAlO$_2$ thin films which eliminates previously reported phase-purity and surface morphology problems. Thin film precursors of CuAlO$_2$ were deposited on YSZ (100) substrates by room-temperature radio frequency sputtering. Subsequent annealing in 10mT of oxygen at temperatures as low as 800° C resulted in phase-pure, textured CuAlO$_2$. The films were p-type and transparent with a gap of 3.5eV, but typical carrier concentrations were low (on the order of 10^{16} cm^{-3}). Oxygen anneals at 700° C in 1 atm of O$_2$ raised this to 10^{18} cm^{-3} in some samples. In order to increase the carrier concentration further, we are testing chemical substitution on the metal sublattice. Initial experiments were done with Mg as a dopant, either by direct solid-state synthesis for bulk materials or by alternating deposition of MgO and CuAlO$_2$ during the pulsed laser deposition of thin films. At Mg molar concentrations of 1% or less, the bulk material was phase-pure CuAlO$_2$ by x-ray diffraction, but was mixed-phase CuAl$_2$O$_4$ and CuO when the Mg concentration was increased to 2%. Thin films sputtered from the 1% Mg substituted target crystallized to CuAlO$_2$ after a 940° C anneal at 10T O$_2$, but were more than 30 times more resistive than the undoped films.

INTRODUCTION

Recent results have shown that CuAlO$_2$ is a p-type transparent semiconductor [1], which might have applications as a transparent conductor if the conductivity can be substantially improved. This could have considerable consequences in a wide range of devices from photovoltaics to flat-panel displays. However, thin film synthesis of CuAlO$_2$ is difficult because of the complex Cu:Al:O phase diagram. Although phase-pure bulk synthesis is relatively straight-forward, the reproducible, phase-pure growth of thin films by pulsed laser depostion (PLD) without post-deposition processing seems nearly impossible. We have used ex-situ anneals to increase the crystallinity and to reduce the concentration of the CuO and CuAl$_2$O$_4$ impurity phases usually present in our films. Our previously reported method involving crucible anneals at 1050° C over mixed CuO and Al$_2$O$_3$ powders suffered from both the persistence of impurity phases and surface morphology problems. Nonetheless, it was relatively robust, and remarkably insensitive to the way we laid down the precursor film. Since amorphous, sputtered precursor films yielded annealed films that were often better than the PLD variety, our subsequent annealing experiments have focussed on films deposited exclusively by RF sputtering at room-temperature. By changing substrates from sapphire to (100) oriented YSZ substrates, and annealing at lower temperatures and oxygen partial pressures, we can reproducibly make phase-pure c-oriented thin films up to about 2000Å in thickness.

All the p-TCO candidates under investigation today -- including $CuAlO_2$ -- still have conductivities four or five orders of magnitude lower than what has been achieved with n-type TCO's. We are attempting to reduce that gap with $CuAlO_2$, and to that end, are conducting various single-species doping experiments (with oxygen, nitrogen, magnesium, and potassium). To date, the only species we know was actually incorporated into the films (without inducing a substantial loss of phase-purity) has been Mg, but it only made the films more resistive.

EXPERIMENT

Amorphous copper-aluminum-oxide samples were grown on (100) YSZ substrates by radio-frequency sputtering of a one-inch diameter $CuAlO_2$ target. The target was mounted 3.3 cm above the substrate. After reducing the base pressure to below 10^{-4} Torr, a fixed flow rate (10.5 sccm) of N_2 gas was admitted into the chamber and the valve separating the chamber and the turbo pump was throttled to maintain a constant total pressure of 65 mT. The films were deposited at a rate of 70Å/min using an emmitted power of 50W. The film thickness was determined by means of a Sloan Dektak 3 stylus profilometer.

Ex-situ anneals were performed in a quartz tube furnace with the samples placed on an alumina sled. Rotameters and a digital flowmeter (Sierra Designs model 820) were used to monitor the ratio of the manually regulated flow rates between argon and oxygen. Using this method, a partial pressure of 10mT of oxygen was maintained in the tube during the anneal. The furnace temperature, as measured by a thermocouple inside the tube, was ramped up from 200° C to 940° C at a rate of 10 degrees per minute, then held at 940° C for 1.5 hours before gradually cooling to 200° C.

Optical measurements were performed on a Cary 5G UV-vis NIR spectrophotometer in dual beam mode with a fixed spectral width of 8nm in the near-infrared and 2nm in the UV and visible portions of the spectrum. Four-point conductivity and Hall measurements were taken on a Bio-Rad model HL5500PC using a van der Pauw electrode configuration with pressed indium contacts. Crystallinity was determined by x-ray diffraction (XRD) using Cu Kα radiation in a Scintag diffractometer.

RESULTS

Growth

In our initial work [2], $CuAlO_2$ films were grown on c-plane sapphire substrates, because Al_2O_3 is well lattice-matched to $CuAlO_2$, and has a high melting point. However, due to the persistence of the $CuAl_2O_4$ impurity phase, we suspected that at the 1050° C annealing temperatures the aluminum in the substrate was contaminating the films. Figure 1 shows a worst-case film which demonstrates the point. The top panel is the XRD spectrum for a film sputtered on a sapphire substrate then annealed at 1050°C in air, and the spectrum in the middle panel is of a film sputtered and annealed under the same conditions, except on a YSZ substrate. The film on sapphire is randomly oriented $CuAl_2O_4$, whereas the film on YSZ is phase pure (00L) oriented $CuAlO_2$, as can be seen by comparing its XRD spectrum with the expected powder pattern for $CuAlO_2$ in the bottom panel.

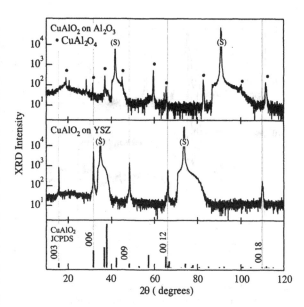

Figure 1: **Effect of substrate.** XRD patterns of films sputtered on different substrates, then annealed in air at 1050° C for 1.5 hours (top panel – sapphire substrate; middle panel – YSZ substrate). The PDF for $CuAlO_2$ is shown on a linear scale in the bottom panel

Even with the YSZ substrates, the surface of the films would sometimes be covered with mm-wide circles, which we hoped could be eliminated by reducing the processing temperature. Using existing phase diagram information [3], we created the oxygen partial pressure vs. temperature phase diagram shown in Figure 2. The shaded area is the region of stability for the $CuAlO_2$ phase. It suggests

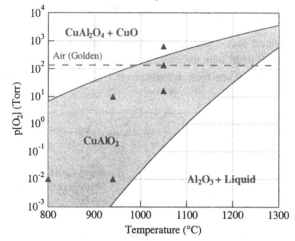

Figure 2: **Cu-Al-O Phase Diagram.** The shaded area is the region of stability for $CuAlO_2$. The triangles identify the (pO_2,T) conditions for this annealing experiment. The dashed line shows pO_2 of air at our location.

that lower temperature anneals are possible provided we lower the oxygen partial pressure (lower left hand corner of the graph). We made several samples to evaluate this, and annealed them at

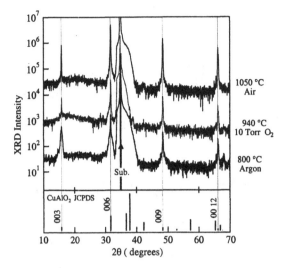

Figure 3: XRD patterns of three films sputtered on YSZ substrates then annealed at the conditions of the shaded area of Figure 2.

the conditions indicated by the triangles in Figure 2. The partial pressure of the bottom two triangles was estimated from the vendor's stated impurity level of the argon gas used during the anneal (no oxygen was intentionally introduced into the tube furnace). In every case, the films annealed at the (pO_2, T) conditions inside the shaded area came out phase-pure $CuAlO_2$ whereas those outside were mixed-phase $CuAlO_2$, CuO and $CuAl_2O_4$.

Figure 3 contains the XRD data from three of the triangles in the previous figure. It shows that phase-pure and textured films were grown down to 800° C (a 250° reduction in processing temperature) although some broadening can be observed in the diffraction peaks at the lowest temperatures. No splotches were evident in any of the films grown at or below 940° C.

In addition, the films were far more transparent than our earlier $CuAlO_2$ films which contained trace $CuAl_2O_4$ impurities [2]. With the naked eye, the substrates with 2000Å thick films on them were difficult to distinguish from blank substrates. Figure 4 shows the optical transmittance spectrum, normalized to a YSZ substrate, for one such film. The direct allowed

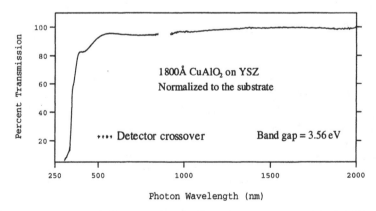

Figure 4: Optical transmittance of a film sputtered on two-sides polished YSZ then annealed at 940° C in 10T O_2 for 1.5 hours.

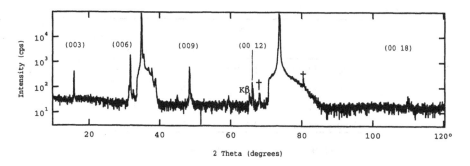

Figure 5: XRD pattern of thin film with ~2% Mg doping. The † symbols represent CuO impurities.

band-gap was 3.56eV, as estimated by a plot of α^2 vs hv.

Chemical Substitution

We attempted to increase the carrier density of the films by chemical substitution. We first tried oxygen anneals, hoping interstitial oxygen would add holes to the lattice. Anneals in flowing oxygen at 800° C or more resulted in a decomposition of the $CuAlO_2$ into $CuAl_2O_4$ and CuO, consistent with expectations from Fig. 2. However, 1.5hr 700°C anneals did not alter the phase of the films. Though the 700° C anneals appeared to affect the transport properties (one sample had $N = 10^{18}$ cm^{-3}), no consistent change in carrier density was observed. In a different series of experiments, KOH was evaporated on the surface of the films and then annealed at 700° C. The same decomposition of $CuAlO_2$ into $CuAl_2O_4$ and CuO occurred as in the oxygen annealing experiments.

When 2% magnesium was mixed in with our standard target recipe and annealed at 1050°C, $CuAl_2O_4$ and CuO were formed in addition to $CuAlO_2$. But when the concentration of Mg was reduced to 1%, phase pure material with the $CuAlO_2$ structure was formed at the same temperature. We then tried a similar doping level in the films (by sputtering from the doped target) and achieved almost phase pure films. ICP measurements confirmed that there was actually Mg in the film. Figure 5 shows the X-ray diffraction spectrum of this film. It shows c-oriented $CuAlO_2$ peaks with very small CuO lines.

Unfortunately, the Mg incorporation actually increased the resistivity from about 75 Ω-cm in the undoped films to 2500 Ω-cm. Table 1 summarizes the relevant transport measurements. It has been speculated that a potential problem with p-type doping in oxides is that the films could self-compensate by producing O vacancies [4], so we tried annealing the Mg doped films in oxygen (700° C for 1.5 hours) but this did not appear to change the resistivity.

	N cm^{-3}	ρ Ω-cm	μ cm^2/Vs
Undoped film	$1 * 10^{17}$	75	.9
O_2 doped film	$1.4 * 10^{18}$	20	.2
Mg doped film	$2.4 * 10^{16}$	2500	.1
Undoped bulk	$4 * 10^{15}$	335	5.4

Table 1: Summary of transport properties

CONCLUSIONS

Thin film synthesis of $CuAlO_2$ is problematic, but transparent, phase-pure and textured film growth is possible by post-deposition annealing of $CuAlO_x$ precursors sputtered on YSZ substrates. The undoped film carrier density of 10^{17} has not been improved with simple doping experiments. Future efforts will focus on ion implantation and various co-doping schemes [5].

ACKNOWLEDGEMENTS

The authors thank Sally Asher for SIMS and XPS measurements. This work was supported by the U.S Department of Energy under contract no. AC36-98-GO10337 and the National Photovoltaics Program.

The National Renewable Energy Laboratory assisted in meeting the publication costs of this article.

REFERENCES

1. H. Kawazoe, M. Yasukawa, H. Hyodo, et al., Nature **389**, 939 (1997).
2. R. E. Stauber, J. D. Perkins, P. A. Parilla, et al., Electrochemical and Solid State Letters **2**, 654 (1999).
3. K. T. Jacob and C. B. Alcock, Journal of the American Ceramic Society **58**, 192 (1975).
4. M. Joseph, H. Tabata, and T. Kawai, Jpn. J. Appl. Phys. **38**, L1205 (1999).
5. H. Katayama-Yoshida and T. Yamamoto, **12** (1999).

IR-TRANSPARENT ELECTRICALLY CONDUCTIVE CuAl$_x$O$_y$ DEPOSITED BY REACTIVE MAGNETRON CO-SPUTTERING

L. F. JOHNSON*, M. B. MORAN*, E. SAVRUN**, M. SARIKAYA***, R. R. KOLEGA*
*Code 4T4110D, Naval Air Warfare Center Weapons Division, China Lake, CA 93555-6001
**Sienna Technologies, Inc., Woodinville, WA 98072-6426
***Dept. of Material Science and Engineering, University of Washington, Seattle, WA

ABSTRACT

This paper reports on progress in the fabrication of IR-transparent electrically conductive copper aluminum oxide (CuAl$_x$O$_y$) by reactive magnetron co-sputtering from high-purity-Cu and -Al targets in an argon-oxygen-gas mixture. Recent equipment modifications have resulted in much better control of deposition parameters like forward and reflected power and, consequently, much better control of film composition. Applying the correct amount of power to each target and adjusting the oxygen-partial pressure have significantly reduced the growth of surface-oxide layers on the metal targets. Equipment and process improvements have eliminated the sputter-rate inconsistencies and arcing that led to lack of composition control in our earliest films. We now have much better control of film composition and are beginning to understand the relationship between the electro-optical properties and the molecular structure of the films using Fourier transform infrared (FTIR) spectroscopy, electron spectroscopy for chemical analysis (ESCA), inductively coupled plasma (ICP) emission spectroscopy, high resolution electron microscopy (HREM) and electron energy loss spectroscopy (EELS). A pair of weakly intense FTIR absorption bands at 1470 and 1395cm^{-1} is present in films that have enhanced electrical conductivity and IR transparency. Understanding the origin of these bands could speed development of CuAlO$_2$ as a wide-bandgap-conductive oxide since these bands are clearly associated with enhanced conductivity and carrier mobility. The best film to date transmits about 80% in the mid-wave IR and has a sheet resistance of 160Ω/sq.

INTRODUCTION

Recently, the first example of a transparent conductive oxide (TCO) with p-type conductivity was demonstrated [1]. Kawazoe's group used laser ablation to deposit thin films of CuAlO$_2$ exhibiting p-type conduction. This is an exciting result because the higher effective-hole mass of the p-type carriers should push the plasma resonance further into the IR. A durable, p-type TCO like CuAl$_x$O$_y$ with a tailorable bandgap and transparency in the IR could revolutionize the design and fabrication of photovoltaics and make solar energy a much more affordable alternative to fossil fuels. Kawazoe used x-ray diffraction (XRD) to show that p-type CuAlO$_2$ films deposited by laser ablation were polycrystalline and exhibited the crystalline structure of a novel class of metal oxides known as delafossites. Single crystals of delafossite metal oxides exhibit very anisotropic electrical properties. Specifically, the electrical conductivity is high in the direction perpendicular to the c-axis of the unit cell and is orders of magnitude lower in the direction parallel to the c-axis [2]. The delafossite-CuAlO$_2$-unit cell is made up of layers of Cu$^+$ cations, one atomic-dimension in thickness that are basically metallic. These layers of Cu$^+$ are bound to layers of octahedrally coordinated Al^{3+} ions by O-Cu-O dumb-bells. The sheets of Cu$^+$ metallic layers enhance electrical conductivity in the direction perpendicular to the c-axis while the oxygen atoms retard electrical conductivity in the direction parallel to the c-axis. Our sputter-deposited films are amorphous and exhibit isotropic electrical conductivity. By making the films oxygen-deficient, it should be possible to enhance the electrical conductivity without degrading the IR transparency significantly.

Mat. Res. Soc. Symp. Proc. Vol. 623 © 2000 Materials Research Society

FTIR spectra presented here will show that a pair of weakly intense absorption bands at 1470 and 1395cm^{-1} is present in CuAl$_x$O$_y$ films that have enhanced electrical conductivity and IR transparency. The fact that the frequencies of the 1470 and 1395cm^{-1} bands are about twice those of the major phonons in Cu$_2$O and Al$_2$O$_3$ is significant and indicates that this pair of bands may involve cumulated Cu⋯O⋯Al⋯O⋯Cu bonds along the c-axis. Higher-order bonding tends to enhance carrier mobility. Furthermore, higher-order bonding in a metal oxide would likely result from an oxygen deficiency. Understanding the origin of these bands could speed development of magnetron-sputter-deposited CuAl$_x$O$_y$ as a wide-bandgap-conductive oxide since these bands are clearly associated with enhanced conductivity and carrier mobility. The delafossite structure of CuAlO$_2$ to some degree mimics the structures of high-temperature-superconducting-copper-oxide compounds on an atomic scale. The pair of bands at 1470 and 1395cm^{-1} may be associated with the phonon-assisted electrical conduction and Cooper-pair phenomena that are used to explain superconductivity [3]. Hall-effect measurements also show that our CuAl$_x$O$_y$ films are p-type so lattice vibrations probably are involved in the enhanced conductivity.

Cumulated Cu⋯O⋯Al⋯O⋯Cu bonds would require p_z orbitals on O to overlap with p_z orbitals on Al and d_z^2 orbitals on Cu atoms. Using a new technique that combines conventional XRD with electron-beam diffraction, Zuo et al. were able to observe <u>directly</u> the classic textbook shape of a d_z^2 orbital in p-type Cu$_2$O [4]. The work by Zuo et al. is expected to be a first step toward understanding high-temperature-superconducting-copper-oxide compounds and may help explain the enhanced conductivity and IR transparency in our sputter-deposited CuAl$_x$O$_y$ films.

EXPERIMENTAL APPROACH

An automated-research-coating (ARC) system equipped with three magnetron guns is used to deposit CuAl$_x$O$_y$ by co-sputtering from high-purity-Al and -Cu targets in a reactive-Ar-O$_2$ mixture. The purity of each of the 2-in-diam-metal targets is at least 0.99999. The ARC system is equipped with two 600W-rf-power supplies and one 250W-dc-power supply. The water-cooled chamber is a 12-in-diam by 14-in-high stainless-steel cylinder and is configured for downward sputtering onto a rotating substrate table with a target-substrate distance of about 5.5in. By rotating the substrate underneath the guns, it is possible to obtain good film uniformity over an 8-in-diam wafer. For the CuAl$_x$O$_y$ films, the rotation speed of the substrate table was set to 10rpm. The vacuum system consists of a 250-liter-per-sec turbomolecular pump from Varian and a direct-drive oil-filled rough pump Trivac model D8B from Leybold. As the chamber is back-filled, the turbopump automatically slows down to half-speed to minimize the gas load and prevent excessive wear on the bearings. For these experiments, the O$_2$-partial pressure ranged from 0.7% to 10%. With a combined Ar-O$_2$-gas-flow rate of 27sccm and a pump speed of 125 liters per sec, the total pressure is maintained at about 14mtorr. The chamber was heated during pump-down using quartz lamps. With a thermocouple reading of 150°C, the base pressure of the chamber before starting the backfill was less than 3 x 10^{-6} torr.

For most of the deposition runs, there were two substrates: a glass microscope slide partially covered by a mask for thickness measurement and a Si wafer for FTIR spectroscopy. To allow for IR transmittance, the resistivity of the Si wafers was specified to be greater than 20Ω-cm. Step heights were measured using a Tencor P-10 contact-stylus profiler. Room temperature resistivities of films deposited onto microscope slides were measured using a four-point probe. Transmission spectra were measured using a Bio-Rad FTIR spectrophotometer. A single-beam spectrum of a coated-Si wafer was divided by a single-beam-background spectrum of an uncoated-Si wafer for all of the FTIR transmission spectra shown here. This approach was used to eliminate Si absorption bands from the spectra since only absorption from the coating was important in this study. In addition, electron spectroscopy for chemical analysis (ESCA) was performed on a limited number of samples to determine the elemental compositions of the films. FTIR, ESCA and four-point-probe measurements were obtained with films that ranged in thickness from about 1500 to 5000-Å-thick. A 500-Å-thick film was deposited onto several 3-mm-diam-carbon grids for high resolution electron microscopy (HREM) and electron energy loss spectroscopy (EELS). To provide an adequate amount of material for the inductively coupled plasma (ICP) emission measurement, a 4-in-diam Si wafer was coated with a 1.12-μm-thick film.

RESULTS AND DISCUSSION

Recent equipment modifications have resulted in much better control of deposition parameters like forward and reflected power and, consequently, much better control of film composition. The automated research coater (ARC) was originally equipped with a switch box that allowed the operator to choose different combinations of power supplies for different targets using PC-controlled software. It recently became obvious that the switch box was interfering with the measurement and control of power to each of the targets. Removing the switch box and modifying the rf-matching networks eliminated many of the deposition-control problems.

The best results to date have been obtained using dc power for the Cu target and rf power for the Al target. The black trace in figure 1 shows the transmission spectrum of a 2850-Å-thick $CuAl_xO_y$ film that has a resistivity of 0.0046Ω-cm and a sheet resistance of 160Ω/sq. Deposition parameters were 170W-dc power applied to the Cu target, 325W-rf power applied to the Al target, and 1.4%-O_2-partial pressure. The deposition rate was about 3-Å-per-sec. Prior to turning on the O_2-partial pressure and before starting the deposition of the $CuAl_xO_y$ film, approximately 19Å of Cu metal was deposited using 100% Ar and a second high-purity-Cu target. The gray trace in figure 1 is an FTIR spectrum of one of the first $CuAl_xO_y$ films deposited two years ago and is included to show that significant progress has been made. Since the initial demonstration of p-type IR-transparent $CuAl_xO_y$ coatings by magnetron-sputter deposition two years ago, we have succeeded in reducing the resistivity of the $CuAl_xO_y$ by more than a factor of 4 from 0.020 to 0.0048Ω-cm. In addition, transmission in the mid-wave IR has increased from 70% to almost 90%.

The role of the thin underlying Cu layer is not fully understood. By itself, the Cu is too thin to contribute significantly to the conductivity. This was verified when four-point-probe measurements were made on a glass slide half-coated with a 19-Å-thick-Cu layer. The 19-Å-thick-Cu-coated half was found to be as electrically insulating as the uncoated half of the glass slide. Results presented below show that the thin underlying Cu layer appears to promote the growth of the appropriate microstructure needed for enhanced conductivity and IR transparency.

Figure 2 summarizes the effect of depositing thin Cu layers of varying thickness at the beginning of the $CuAl_xO_y$ coating run. Deposition conditions for the outer layer of $CuAl_xO_y$ were the same for all three films: 170W-dc power applied to the Cu target, 280W-rf power applied to the Al target, 1.1%-O_2-partial pressure and a 15-min-long-deposition time. For the film in trace a, no thin layer of Cu was deposited at the beginning of the coating run. The resulting 3140-Å-thick $CuAl_xO_y$ film was very transparent but was only semi-conductive with a resistivity of 51Ω-cm and a sheet resistance of 1.6 x 10^6Ω/sq. For the film in trace b, approximately 13Å of Cu was deposited first followed by the outer layer of $CuAl_xO_y$. The resulting 2640-Å-thick Cu/$CuAl_xO_y$ film was more than 80% transparent from 3125 to 1042cm^{-1} and was conductive with a resistivity of 0.040Ω-cm and a sheet resistance of 1470Ω/sq. For the film in trace c, approximately 19Å of Cu metal was deposited first followed by the outer $CuAl_xO_y$. The resulting 2680-Å-thick Cu/$CuAl_xO_y$ film was more than 70% transparent from 4000 to 1961cm^{-1} and was very conductive with a resistivity of 0.0055Ω-cm and a sheet resistance of 206Ω/sq. Notice that the thickness values for the films deposited onto Cu were about 85% of the thickness value for the film deposited directly onto the substrate even though the deposition conditions and deposition times were identical for all three CuAl$_x$O$_y$ layers.

Experiments were done to determine if replacing the thin Cu layer with a thin layer of Al or tin (Sn) would also promote the growth of a CuAl$_x$O$_y$ film that has enhanced conductivity and IR transparency. Replacing the thin Cu layer with a thin layer of Al resulted in CuAl$_x$O$_y$ films that were significantly more transparent in the IR but were also more than three orders of magnitude less conductive. In other words, a thin Al layer does not promote the growth of a film that has enhanced electrical conductivity and IR transparency; the CuAl$_x$O$_y$ film that grows on top of the thin Al layer has properties very similar to a CuAl$_x$O$_y$ film that is grown on a bare substrate. A thin Sn layer also does not promote the growth of a CuAl$_x$O$_y$ film that has enhanced electrical conductivity and IR transparency.

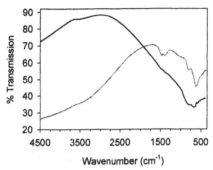

Figure 1. FTIR spectra for two CuAl$_x$O$_y$ films. The black trace is for a 2850-Å-thick film that has a resistivity of 0.0046Ω-cm and a sheet resistance of 160Ω/sq. The gray trace is for a 3800-Å-thick film that has a resistivity of 0.02Ω-cm and sheet resistance of 540Ω/sq.

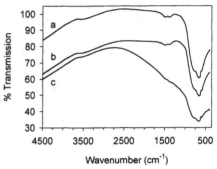

Figure 2. FTIR spectra for films having underlying Cu layers of varying thickness. Deposition parameters for all three outer CuAl$_x$O$_y$ layers were the same. Trace a is for a film with no underlying Cu layer; the 3140-Å-thick CuAl$_x$O$_y$ has a resistivity of 51Ω-cm and sheet resistance of 1.6 x 10^{6}Ω/sq. Trace b is for a film with an underlying 13-Å-thick-Cu layer; the 2640-Å-thick Cu/CuAl$_x$O$_y$ has a resistivity of 0.040Ω-cm and sheet resistance of 1470Ω/sq. Trace c is for a film with an underlying 19-Å-thick-Cu layer; the 2680-Å-thick Cu/CuAl$_x$O$_y$ has a resistivity of 0.0055Ω-cm and sheet resistance of 206Ω/sq.

The black trace in figure 3 is an FTIR spectrum of a 2893-Å-thick CuAl$_x$O$_y$ film with a resistivity of 0.0148Ω-cm and a sheet resistance of 510Ω/sq. The gray trace is an FTIR spectrum of the same film after it was annealed in a high-vacuum chamber at 600°C in O$_2$ for five hours. The peak transmission before annealing was 92.9% and after annealing was 119.3%. The annealed film was very electrically insulating with a sheet resistance of greater than 1 x 10^{9}Ω/sq.

Another difference between the as-deposited and annealed film is the absence of the pair of small absorption bands at 1470 and 1395cm^{-1} in the spectrum of the annealed film. The pair of weakly intense bands at 1470 and 1395cm^{-1} is present in spectra of films that exhibit enhanced electrical conductivity. When these bands are absent, the CuAl$_x$O$_y$ films have high values of resistivity. It is possible that the enhanced conductivity of sputter-deposited CuAl$_x$O$_y$ films could be a result of overlapping d orbitals on neighboring Cu^{1+} atoms in the plane perpendicular to the c-axis of the delafossite-unit cell. Overlapping *d* orbitals also would explain why the sputter-deposited $CuAl_xO_y$ films absorb strongly in the visible. Another possibility is that the 1470 and 1395cm^{-1} bands involve vibrational modes of the entire Cu-O-Al-O-Cu sequence along the c-axis of the delafossite-unit cell. Cuprous oxide (Cu_2O) absorbs strongly at 609cm^{-1} [5]. Randomly oriented Al_2O_3 has a strong absorption centered at about 670cm^{-1} with shoulders at 560 and 750cm^{-1} [5]. The fact that the frequencies of the 1470 and 1395cm^{-1} bands are about twice those of the major lattice vibrations in Cu_2O and Al_2O_3 is significant and indicates that these modes may involve cumulated Cu···O···Al···O···Cu bonds. Higher-order-π bonding would tend to enhance carrier mobility. Higher-order bonding in a metal oxide also would result from an oxygen vacancy. Understanding the origin of the pair of bands at 1470 and 1395cm^{-1} could accelerate the development of $CuAl_xO_y$ as a wide-bandgap-conductive oxide since the bands are clearly associated with enhanced electrical conductivity and carrier mobility.

Figure 4 is a high-resolution-ESCA spectrum for one of the most electrically conductive and IR transparent $CuAl_xO_y$ films. The spectrum has been deconvolved into three distinct peaks with the Cu $3p^1$ peak at 79.36eV contributing about 12%, the Cu $3p^3$ peak at 77.43eV contributing about 28.5% and the Al 2p peak at 74.76eV contributing about 59.5%. Although ESCA is only semi-quantitative, the high-resolution ESCA spectrum clearly shows that the film is Al rich. The more quantitative method of inductively coupled plasma (ICP) emission spectroscopy shows the Al:Cu ratio in the $CuAl_xO_y$ film is about 2:1. Even with a very non-stoichiometric composition,

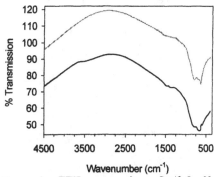

Figure 3. *FTIR spectra for a $CuAl_xO_y$ film before and after O_2 annealing. The black trace is for a 2893-Å-thick $CuAl_xO_y$ film that had a resistivity of 0.0148Ω-cm and a sheet resistance of 510Ω/sq before annealing. The gray trace is for the same film after annealing. The annealed film is very insulating.*

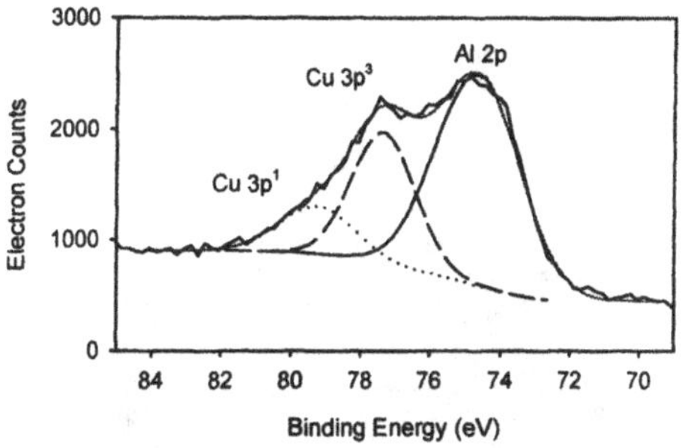

Figure 4. *High-resolution ESCA spectrum of a $CuAl_xO_y$ film showing the deconvolved peaks for Cu $3p^1$ at 79.36eV, Cu $3p^3$ at 77.43eV and Al 2p at 74.76eV.*

the $CuAl_xO_y$ film is very conductive with a resistivity of 0.0051Ω-cm, a sheet resistance of 246Ω/sq, and a peak IR transmission of 67%. Like the black trace in figure 1, the FTIR spectrum of this film (not shown here) has a broad absorption from about 2500 to 800cm^{-1} resembling the broad phonon absorption in Al_2O_3 films. This broad absorption along with resistivity measurements indicates that most of the extra Al goes into the $CuAl_xO_y$ films as oxide rather than free metal. The excess Al-O bonds make the $CuAl_xO_y$ films extremely hard and scratch resistant. However, too many Al-O bonds eventually degrade the electrical conductivity.

Atomic force microscopy (AFM) and high-resolution electron microscopy (HREM) indicate that magnetron-sputter-deposited $CuAl_xO_y$ is not a single-phase material. A second Cu-rich phase appears to be contributing to the enhanced electrical conductivity. Although not shown here, the phases and grain sizes in the AFM images are similar to those in the HREM images. The HREM images shown in figures 5 and 6 indicate that the 500-Å-thick $CuAl_xO_y$ film consists of islands of crystalline elemental-Cu particles in an amorphous Cu-Al-O matrix. The size of the Cu particles is about 40 to 50nm. Electron energy loss spectroscopy (EELS) was used to show that the dark spots in the bright field image in figure 5 are cubic-Cu crystallites and that the Cu-Al-O matrix has a ratio of Al:Cu of 1:1 with O attached to both Al and Cu. Figure 6 is simply a dark field image of the same area shown in figure 5 where the light spots are now the cubic-Cu crystallites. It is likely that the elemental-Cu particles are partly a result of diffusion of the thin layer of Cu that was deposited onto the substrate and then overcoated with about 480 Å of $CuAl_xO_y$. At first, the Al:Cu ratio determined by EELs appears to contradict the ICP and ESCA data that show the Al:Cu ratio is 2:1. This apparent contradiction is probably related to the fact that the overlying $CuAl_xO_y$ layers in the ICP and ECSA samples were much thicker. The $CuAl_xO_y$ layers in figures 5 and 6 had to be thinner to allow for transmission of electrons through the HREM samples. As the thickness of the $CuAl_xO_y$ layer increases, the ability of the thin underlying layer of Cu to promote the growth of the correct microstructure and composition for enhanced transparency and conductivity probably diminishes. It may be necessary to periodically increase and then decrease the concentration of elemental Cu during the deposition of the overlying $CuAl_xO_y$ layer to maintain the optimum composition and microstructure. Preliminary attempts to do this indicate that the amount of elemental Cu needed is very small.

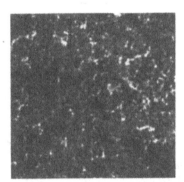

Figure 5. High resolution electron micrograph (150,000X) of a $CuAl_xO_y$ film (bright field image).

Figure 6. High resolution electron micrograph (150,000X) of a $CuAl_xO_y$ film (dark field image).

REFERENCES

1. H. Kawazoe, M. Yasukawa, H. Hyodo, M. Kurita, H. Yanagi, and H. Hosono, "P-Type Electrical Conduction in Transparent Thin Films of $CuAlO_2$," *Nature*, **389**, 939-942, (1997).
2. M. Tanaka, M. Hasegawa, T. Higuchi, T. Tsukamoto, Y. Tezuka, S. Shin, and H. Takei, *Physica B*, **245**, 157-163 (1998).
3. D. K. Finnemore, "Superconducting Materials," *Concise Encyclopedia of Solid State Physics*, ed. R. G. Lerner and G. L. Trigg (Addison-Wesley, 1983) 260-263.
4. J. M. Zuo, M. Kim, M. O'Keeffe, and J. C. H. Spence, "Direct Observation of d-Orbital Holes and Cu–Cu Bonding in Cu_2O," *Nature*, **401**, 49-52 (1999).
5. E. D. Palik, *Handbook of Optical Constants of Solids II*, (Academic Press, San Diego, 1991).

ROOM TEMPERATURE GROWTH OF INDIUM TIN OXIDE FILMS BY ULTRAVIOLET-ASSISTED PULSED LASER DEPOSITION

V. CRACIUN*, D. CRACIUN**, Z. CHEN*, J. HWANG***, R.K. SINGH*
*Materials Science & Engineering, University of Florida, Gainesville, FL 32611
**National Institute for Laser, Plasma and Radiation Physics, Bucharest, Romania
***Physics Department, University of Florida, Gainesville, FL 32611

ABSTRACT

The characteristics of indium tin oxide (ITO) films grown at room temperature on (100) Si and Corning glass substrates by an in situ ultraviolet-assisted pulsed laser deposition (UVPLD) technique have been investigated. The most important parameter, which influenced the optical and electrical properties of the grown films, was the oxygen pressure. For oxygen pressure below 1 mtorr, films were metallic, with very low optical transmittance and rather high resistivity values. The resistivity value decreased when using higher oxygen pressures while the optical transmittance increased. The optimum oxygen pressure was found to be around 10 mtorr. For higher oxygen pressures, the optical transmittance was better but a rapid degradation of the electrical conductivity was noticed. X-ray photoelectron spectroscopy investigations showed that ITO films grown at 10 mtorr oxygen are fully oxidized. All of the grown films were amorphous regardless of the oxygen pressure used.

INTRODUCTION

Indium tin oxide (ITO) thin films are widely used for optoelectronic devices as they combine a good electrical conductivity with high transparency in the visible range. There are a number of interesting applications such as anode contact in organic light-emitting diodes [1, 2] or coating of flexible polymer substrates for ultralight mobile display panels [3] where the use of a low processing temperature is very important. The use of the pulsed laser deposition (PLD) technique, which has several important advantages [4], has allowed the growth of good quality indium tin oxide (ITO) thin films at relatively low temperatures and even room temperature [5-8]. Other techniques such as synchrotron radiation ablation [9] or plasma-ion assisted evaporation [10] were also employed to deposit ITO films at room temperature. We investigated the use of an in situ ultraviolet-assisted PLD technique (UVPLD) for the growth of ITO films at room temperature. The UV source photodissociates molecular oxygen and provides ozone and atomic oxygen during the growth [11]. These more reactive gases have been shown to promote the crystalline growth at lower temperatures than those normally used during conventional PLD [12]. Moreover, UV+ozone is known to be an effective way to clean organic contaminants from the substrate [13, 14], a fact that can also improve the quality of the deposited layers.

EXPERIMENT

The PLD system employed is presented elsewhere in much more detail [15, 16] and it is only briefly described here. An excimer laser (KrF, λ=248 nm, laser fluence ~2 J/cm^2, repetition rate 5 Hz) was used to ablate ITO targets (99.99% purity). The oxygen pressure was varied from 0.1 mtorr to 50 mtorr. A vacuum compatible, low pressure Hg lamp, which allows for in-situ UV irradiation of the substrate during the laser ablation-growth process, was fitted to the PLD system. Films were deposited onto (100) Si wafers and Corning glass substrates that were placed at 10.5 cm in front of the target.

The crystalline structure of the grown films was investigated by x-ray diffraction (XRD) and transmission electron microscopy (TEM). The chemical composition was determined by x-ray photoelectron spectroscopy (XPS, Perkin Elmer 5100, Mg Kα radiation) and Auger electron spectroscopy (AES). The optical properties of films grown on Si substrates were investigated by spectroscopic ellipsometry (VASE, Woollam Co.) at 70° while those of film deposited on Corning glass were analyzed by optical spectrophotometry. The sheet resistance was measured by a four point probe method and the surface morphology by atomic force microscopy (AFM).

RESULTS

XRD and TEM investigations showed that all films deposited, regardless of the oxygen pressure used, were amorphous. Crystalline films were deposited by UVPLD at a substrate temperature of only 100 °C. The effect of the oxygen pressure on the resistivity of the grown ITO films is shown in Fig. 1. Values measured for several films, which were grown by conventional PLD for comparison reasons, are also shown in Fig. 1. As one can note, there is a continuous improvement of the electrical conductivity with the increase of the oxygen pressure up to a value of 10 mtorr. A further increase of the oxygen pressure resulted in a steep increase of the electrical resistivity. A similar dependence has been already reported for ITO films grown by PLD [7, 8]. It is worth noting that by using the UVPLD technique it was possible to obtain resistivity values below 4×10^{-4} Ω-cm.

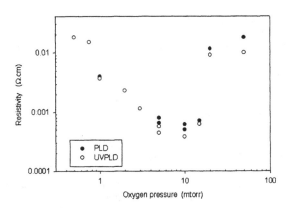

Figure 1. Resistivity values versus oxygen pressure for ITO films grown by PLD and UVPLD.

The optical transmittance of a typical ITO film grown under the optimum oxygen pressure, i. e. 10 mtorr, is displayed in Fig. 2. One can note that the film exhibited a transmission almost as good as the Corning glass substrate. The refractive index and extinction coefficient values estimated from the transmittance curve are shown in Table I. The optical transmittance of a film grown under 2 mtorr oxygen is also shown. It is obvious that the effect of oxygen pressure on the optical transmittance is very important.

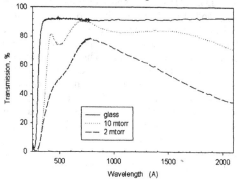

Figure 2. Optical transmittance of ITO thin films deposited by UVPLD on Corning glass.

Table I. Optical parameters of the ITO film showed in Fig. 2.

Thickness	Wavelength	Refractive index	Extinction coefficient
200 nm	422 nm	2.1	0.018
	500 nm	2.0	0.011
	730 nm	1.9	0.003
	1120 nm	1.5	0.016

The optical properties of ITO films deposited on Si were investigated by VASE. In Fig. 3 the refractive index and extinction coefficient values of films deposited by PLD (328 nm thick) and UVPLD (317 nm thick) under 10 mtorr oxygen are shown. One can note that apart from the region of high absorption (270-310 nm) the refractive index values were quite close to the reference ITO values. Also, the extinction coefficient values were very small, confirming the transparency of these films determined by spectrophotometry.

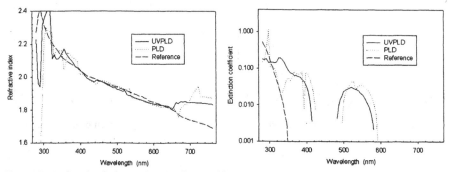

Figure 3. Refractive index and extinction coefficient values of ITO films grown under 10 mtorr.

XPS investigations were performed on samples grown on Si under 5 and 10 mtorr of oxygen by conventional PLD and UVPLD. Typical examples of survey and high resolution scans for In $3d_{5/2}$, Sn $3d_{5/2}$, and O 1s spectra acquired from a sample that was in situ sputter-cleaned for 3 min by a 2 keV Ar ion beam are displayed in Figs 4-7. The In and Sn peaks correspond to oxidized positions, In_2O_3 and SnO_2, respectively [17-19]. The oxygen peak exhibited a small shoulder on the high energy side. This shape indicates the presence of two oxidation states. The peak located at 530 eV corresponds to O bound to In and that at 531.0 eV corresponds to O bound to Sn. The O 1s spectra acquired from the as-received surface exhibited a much larger shoulder towards high energy side. This result indicated some Sn segregation in the surface region, which was confirmed by measuring the In/Sn ratio: this increased from around 10 in the surface region to ~13.3 in the bulk. Another XPS result worth mentioning is that the overall stoichiometry of the film was around $In_{0.44}Sn_{0.03}O_{0.53}$ for both PLD and UVPLD grown films. AES investigations also confirmed the growth of ITO films. It also showed that there was no lateral variation in the chemical composition of the films.

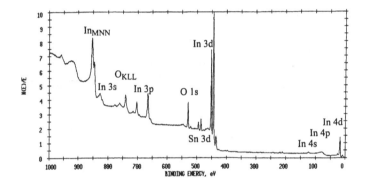

Figure 4. XPS survey spectra of an ITO film grown by UVPLD under 10 mtorr O_2; the sample was sputtered clean with 2.0 keV Ar ions for 3 min; take-off angle 45°.

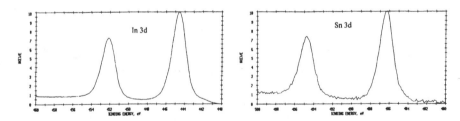

Figure 5. XPS narrow scan of the In 3d region.   Figure 6. XPS narrow scan of the Sn 3d region.

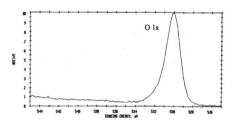

Figure 7. XPS narrow scan of the O 1s region

The surface morphology was investigated by AFM. A typical image is displayed in Fig. 8. The films were quite smooth, with an average root-mean-square (RMS) roughness of around 0.3 nm. This is a very good value, comparable to other reported results [8, 17].

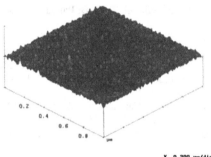

X 0.200 μm/div
Z 10.000 nm/div

Figure 8. AFM image (1 μm x 1μm) of the ITO film grown on Si.

It has been found that UVPLD grown films exhibited slightly better electrical and optical properties than PLD grown films under identical conditions. However, neither XPS nor TEM investigations could find substantial differences in the microstructure or composition between these films. UV radiation and ozone help to better clean the substrate. According to XPS results there was always a slightly lower C 1s peak due to contamination on UVPLD grown samples compared to PLD samples. The ozone and atomic oxygen formed by photodissociation during UVPLD are more reactive than molecular oxygen and can oxidize metals even at room temperature. Also, because In and Sn can react with oxygen during PLD even in gas phase [5], it is very likely they will react even more with ozone. Also, the surface mobility of adatoms is increased by UV irradiation [20]. It was found that under identical conditions the PLD grown samples were a little bit thicker than UVPLD grown ones, implying that the desorption of loosely bound adatoms was also increased during UV irradiation. All these factors could contribute to the observed better characteristics of UVPLD grown ITO films.

CONCLUSIONS

Good quality ITO films have been grown at room temperature on Si and Corning glass substrates using an in situ ultraviolet-assisted pulsed laser deposition technique. Under optimum oxygen deposition conditions, films possessing high transparency and a resistivity lower than 4×10^{-4} Ω-cm have been routinely grown.

REFERENCES

1. H. Kim, A. Pique, J. S. Horwitz, H. Mattoussi, H. Murata, Z. H. Kafafi, and D. B. Chrisey, Appl. Phys. Lett. **74**, 3444 (1999).
2. C. W. Tang and S. A. Van Slyke, Appl. Phys. Lett. **51**, 913 (1987).
3. F. Matsumoto, Asia Display'95, p. 31 (1995).
4. D. B. Chrisey and G. K. Hubler, Pulsed Laser Deposition of Thin Films, Wiley, New York, 1994.
5. R. Teghil, V. Marotta, A. Giardini Guidoni, T. M. Di Palma, C. Flamini, Appl. Surf. Sci. **138-139**, 522 (1999).
6. F. O. Adurodija, H. Izumi, T. Ishira, H. Yoshioka, K. Yamada, H. Matsui, and M. Motoyama, Thin Solid Films **350**, 79 (1999).
7. Y. Yamada, N. Suzuki, T. Makino, and T. Yoshida, J. Vac. Sci. Technol. A **18**, 83 (2000).
8. H. Kim, C. M. Gilmore, A. Pique, J. S. Horwitz, H. Mattoussi, H. Murata, Z. H. Kafafi, and D. B. Chrisey, J. Appl. Phys. **86**, 6451 (1999).
9. Y. Akagi, K. Hanamoto, H. Suzuki, T. Katoh, M. Sasaki, S. Imai, M. Tsudagawa, Y. Nakayama, and H. Miki, Jpn. J. Appl. Phys. **38**, 6846 (1999).
10. S. Laux, N. Kaiser, A. Zoller, R. Gotzelmann, H. Lauth, and H. Bernitzki, Thin Solid Films **335**,1 (1998).
11. D. L. Baulch, R. A. Cox, R. F. Hampson, Jr.,J. A. Kerr, J. Troe, and R.T. Watson, J. Phys. Chem. Ref. Data **9**, 295 (1980).
12. C. E. Otis, A. Gupta, and B. Braren, Appl. Phys. Lett. **62**, 102 (1993).
13. S. K. So, W. K. Choi, C. H. Cheng, L. M. Leung, and C. F. Kwong, Appl. Phys. A **68**, 447 (1999).
14. K. Sugiyama, H. Ishii, Y. Ouchi, and K. Seki, J. Appl. Phys. **87**, 295 (2000).
15. V. Craciun and R.K. Singh, Electrochemical and Solid-State Lett. **2**, 446 (1999).
16. V. Craciun, R.K. Singh, J. Perriere, J. Spear, and D. Craciun, Appl. Phys. A **69** [Supl.], S531 (1999).
17. S-S. Kim, S-Y. Choi, C-G. Park, H-W. Jin, Thin Solid Films **347**, 155 (1999).
18. D. Rhode, H. Kersten, C. Eggs, and R. Hippler, Thin Solid Films **305**, 164 (1997).
19. F. Zhu, C. H. A. Huan, K. Zhang, and A. T. S. Wee, Thin Solid Films **359**, 244 (2000).
20. H. Wengenmair, J. W. Gerlach, U. Preckwinkel, B. Stritzker, and B. Rauschenbach, Appl. Surf. Sci. **99**, 313 (1996).

Room temperature operation of UV-LED composed of TCO hetero p-n junction, p-SrCu$_2$O$_2$ / n-ZnO

Hiromichi Ohta[*], Ken-ichi Kawamura[*], Masahiro Orita[*], Nobuhiko Sarukura[*,**], Masahiro Hirano[*], Hideo Hosono[*,****]

[*]Hosono Transparent ElectroActive Materials, ERATO, Japan Science and Technology Corporation, KSP C-1232, 3-2-1, Sakado, Takatsu-ku, Kawasaki-shi, 213-0012, JAPAN

[**]Institute for Molecular Science, Okazaki National Research Institute, Myodaiji, Okazaki 444-8585, JAPAN

[***]Materials and Structures Laboratory, Tokyo Institute of Technology, Nagatsuta, Midori-ku, Yokohama 226-8503, JAPAN

e-mail : ohta@team.ksp.or.jp

ABSTRACT

Room temperature operation of UV LED is realized for the first time using a hetero p-n junction composed of transparent conductive oxides, p-SrCu$_2$O$_2$ and n-ZnO. Ni/ SrCu$_2$O$_2$/ZnO/ITO multi-layered film was epitaxially grown on an extremely flat YSZ (111) surface by a PLD. The grown films were processed by a conventional photolithography, followed by reactive ion etching to fabricate p-n junction diode. The resultant device exhibited rectifying I-V characteristics inherent to p-n junction whose turn-on voltage was about 1.5V. A relatively sharp electro-luminescence band centered at 382nm was generated when applying the forward bias voltage larger than the turn-on voltage of 3V. The red shift in the EL peak was noticed from that of photo-luminescence (377nm), which was most likely due to the difference in the excited state density between the emission processes. The EL band is attributed to transition in ZnO, probably to that associated with electron-hole plasma. The photo-voltage was also generated when the p-n junction was irradiated with UV light of which energy coincided with both exciton and band-to-band transitions in ZnO.

INTRODUCTION

ZnO has been known to emit visible light efficiently when excited with either electron beams or UV light. Utilizing this feature together with n-type conductivity, it has been commercialized as green phosphor for vacuum fluorescence devices and CRTs. The material is also promising for UV light emitting phosphor, as it has a direct band structure with an energy gap of 3.3eV. Additionally, binding energy of exciton in ZnO is about 60meV, sufficient enough to be thermally stabilized even at room temperature. [1]

Laser operation excited with electron beams or UV light has been reported so far. [2-5] However, no UV emission associated with p-n junction has been realized to date. Primary reason for that is its difficulty to prepare p-type conductive ZnO, although several attempts including laser doping or co-doping[6] have been made. Thus, ELs of green or UV light has been obtained only from MIS diode structure using SiO as insulator. [7]

Our approach to realizing ZnO based light emitting diode is rather unique in a sense that TCOs exhibiting p-type conductivity other than p-ZnO was employed as a partner of n-ZnO to form hetero p-n junction, in place of developing homo p-n junction of ZnO. Among several p-TCOs, [8-11] which we have developed recently, p-SrCu$_2$O$_2$ was selected in this study, primary because it can be deposited as low as 350°C. The low temperature deposition process makes it possible to minimize chemical reaction at the SrCu$_2$O$_2$ / ZnO interface. Further, SrCu$_2$O$_2$ could be hetero-epitaxially grown on ZnO, as the lattice mismatch between 5 lattices of ZnO (0001) and 6 lattices of SrCu$_2$O$_2$ (112) is less than 0.9%. That is, so called "domain matching epitaxy [12]" is expected to work in this combination. In general, epitaxial grown is possibly to result in the formation of high quality, optically active p-n junction.

EXPERIMENTAL

Diode fabrication

ITO, ZnO, SrCu$_2$O$_2$ and Ni films were successively grown on an yittria-stabilized-zirconia (YSZ) single crystal substrate by a pulsed laser deposition method, in which KrF (λ=248nm) excimer laser was used as an exciting source. N-type conductivity was realized with an introduction of excess Zn and/or oxygen defect into ZnO lattice by controlling the atmosphere during the deposition, while p-type conductivity was obtained with a dope of K$^+$ ions into Sr^{2+} site in SrCu$_2$O$_2$. The carrier density in each film is set to be 5 x 10^{17}cm^{-3}, corresponding to the depletion layer thickness of about 100nm. The deposition temperature for ZnO was raised up to 800°C, which made it possible for ZnO films to grow epitaxially on the ITO film.[13-15] This could lead to good crystal quality of ZnO. On the other hand, that for the SrCu$_2$O$_2$ film was lowered down to 350°C with a hope to form an abrupt interface between ZnO and SrCu$_2$O$_2$. Thickness of ZnO and SrCu$_2$O$_2$ was 500nm and 200nm, respectively, which was much thicker than the depletion layer.

The obtained multi-layered films were processed by a conventional photolithography technique, followed by a reactive ion etching process to fabricate a mesa structure having a

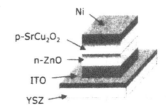

Fig. 1. Schematic illustration of fabricated p-n junction.

surface area of 500mm x 500mm. Au wire was attached to both ITO film (n-electrode) and to Ni metal surfaces (p-electrode). A schematic illustration of the diode structure is shown in Fig. 1.

Evaluation

Crystalline quality and orientation were analyzed by X-ray diffraction measurements (ATX-G, Rigaku Co.), in which out-of-plane XRD pattern (synchronous scan of 2θ with ω in the horizontal plane), in-plane XRD pattern (synchronous scan of $2\theta\chi$ with ϕ in the azimuth plane), out-of-plane locking curve (2θ fixed ω scan) and in-plane locking curve ($2\theta\chi$ fixed ϕ scan) were obtained. Lattice image near the $SrCu_2O_2$ / ZnO interface was observed by high resolution TEM (JEM-2000EX, JEOL).

I-V characteristic was evaluated by applying DC voltage to the diode. Electro-luminescence (EL) as a result of electric current injection or photo-luminescence (PL) excited with He-Cd (325nm) laser irradiation was focused into a bundle fiber through the transparent n-electrode (ITO) and guided to a spectrometer with multi-channel detector. For photo-voltage measurement, monochromized and chopped light from Xe lamp was irradiated onto the p-n junction through the transparent n-electrode and produced photo-voltage was detected by a lock-in amplifier.

RESULTS

Figure 2 shows out-of-plane (a) and in-plane (b) XRD patterns of the multi-layered film. The out-of-plane XRD pattern indicated that $SrCu_2O_2$ (112), ZnO (0001) and

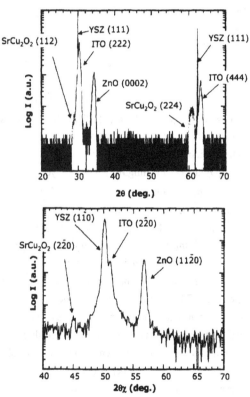

Fig. 2. HR-XRD patterns of multi-layered film. (a) Out-of-plane XRD pattern, (b) In-plane XRD pattern

ITO (111) planes were laminated vertically on YSZ (111). FWHM of the out-of-plane locking curve for ZnO and $SrCu_2O_2$ were 0.2 and 1 degree, respectively. In-plane XRD pattern together with locking curves for ITO and ZnO indicate that (111) of ITO and (0001) ZnO were hetero-epitaxially grown on YSZ (111) both in lateral and vertical direction. However, in-plane locking curve for ($1\bar{1}0$) of $SrCu_2O_2$ implied only part of this plane was laterally oriented with respect to the ZnO layer.

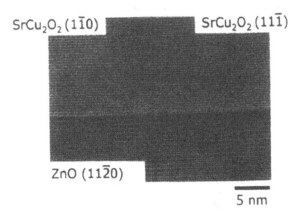

Fig. 3. Cross sectional HR-TEM image near the p-n junction region.

Figure 3 shows a cross sectional TEM image near the p-n junction, in which electron beams were injected along [$11\bar{2}0$] of ZnO. The stacking of ZnO (0001) and $SrCu_2O_2$ (112) planes in terms of deposition direction was clearly observed having a very abrupt hetero-epitaxial interface. It was also noticed there existed several domain boundaries in $SrCu_2O_2$ layer, indicating the $SrCu_2O_2$ layer were composed of several crystallographic domains. For instance, observed plane in the left region of Fig. 3 corresponds to ($1\bar{1}0$), which was expected from the domain matching epitaxy, and that in the right one is assigned as ($11\bar{1}$), which is twisted by 90 degree from ($1\bar{1}0$). Those observations are in good agreement with those by XRD, implying the interface based on the domain matching epitaxy between $SrCu_2O_2$ and ZnO was only partially realized in our specimen.

The diode exhibited rectifying I-V characteristics inherent to p-n junction, when applying DC bias voltage to the electrodes. The turn on voltage was about 1.5V, which is much smaller than that expected from band gap energy of ZnO or $SrCu_2O_2$. With an increase in the bias voltage larger than 3V in the forward direction, relatively sharp emission band centered at 382nm was observed. The emission intensity enhanced with an increase in the electric current (voltage) as shown in upper part of Fig. 4. Photo-luminescence spectrum of the same device was shown in the lower part of this figure, where excitation spectrum for photo-voltage was also shown by dotted line. PL band is attributed to transition associated with exciton in ZnO. Then, it is suggested that photo-carrier was generated directly by band-to-band excitation and indirectly excited through exciton state. Auger process among exciton may be responsible for

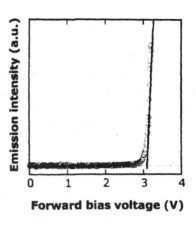

Fig. 5. LED emission intensity as a function of forward bias voltage. The turn-on voltage is about 3V.

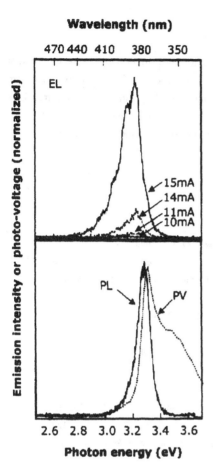

Fig. 4. UV emission spectrum of p-SrCu₂O₂ / n-ZnO p-n junction LED. (a) EL: emission due to current injection, (b) PL: photoluminescence excited with 325nm He-Cd laser light, PV: activation spectrum for photo-voltage (Photo-voltage of 0.26mV was obtained when excited with monochromated 0.2mW Xe lamp at λ = 374nm.)

the indirect excitation. Small red shift of EL band from PL was observed, which may be resulted from the excitation density difference in emission process.[5,16] EL is also due to transition in ZnO, most likely to that associated with electron-hole plasma, which were generally observed in high density excitation. Bias voltage dependence of EL intensity, summarized in Fig. 5 demonstrated the turn on voltage was about 3V, in good agreement with band gap energy of ZnO (3.3eV). From these observations, it is concluded that EL is resulted from electron injection through the hetero p-n junction and it is likely attributed to electron hole recombination in ZnO. Obtained efficiency of our LED is in the order of 10^{-3}% and this low efficiency is probably due to non-homogeneity of the junction, as suggested by XRD and TEM.

CONCLUSIONS

In summary, a UV light-emitting diode was realized at room temperature for the first time using a hetero p-n junction composed of p-SrCu$_2$O$_2$ and n-ZnO. EL is resulted from electron injection through the hetero p-n junction and it is likely attributed to electron hole recombination in ZnO. Obtained efficiency of our LED is in the order of 10^{-3}%. The low efficiency could be improved through the optimization of the device fabrication process.

REFERENCES

1. D. G. Thomas, *J. Phys. Chem. Solids*, **15**, 86 (1960).

2. F. H. Nicoll, *Appl. Phys. Lett.*, **9**, 13 (1966).

3. D. C. Reynolds, D. C. Look, B. Jogai, *Solid State Commun.*, **99**, 873 (1996).

4. D. M. Bagnall, Y. F. Chen, T. Yao, S. Koyama, M. Y. Shen, T. Goto, *Appl. Phys. Lett.*, **70**, 2230 (1997).

5. M. Kawasaki, A. Ohtomo, I. Okubo, H. Koinuma, Z. K. Tang, P. Yu, G. K. L. Wong, B. P. Zhang, Y. Segawa, *Mat. Sci. Eng.*, **B56**, 239 (1998).

6. M. Joseph, H. Tabata, T. Kawai, *Jpn. J. Appl. Phys.*, **38**, L1205 (1999).

7. B. W. Thomas, D. Walsh, *Electronics Lett.*, **9**, 362 (1973).

8. H. Kawazoe, M. Yasukawa, H. Hyodo, M. Kurita, H. Yanagi, *Nature*, **389**, 939 (1997).

9. H. Yanagi, H. Kawazoe, A. Kudo, M. Yasukawa, and H. Hosono, *to be published in J. Electroceram.*

10. A. Kudo, H. Yanagi, H. Hosono, *Appl. Phys. Lett.*, **73**, 220 (1998)

11. A. Kudo, H. Yanagi, K. Ueda, H. Hosono, H. Kawazoe, and Y. Yano, *Appl. Phys. Lett.*, **75**, 2851 (1999).

12. J. Narayan, P. Tiwari, X. Chen, J. Singh, R. Chowdhury, T. Zheleva, *Appl. Phys. Lett.*, **61**, 1290 (1992).

13. H. Ohta, H. Tanji, M. Orita, H. Hosono, H. Kawazoe, *Mat. Res. Soc. Symp. Proc. Vol. 570*, 309 (1999).

14. H. Ohta, M. Orita, M. Hirano, H. Tanji, H. Kawazoe, H. Hosono, *to be published in Appl. Phys. Lett.*,

15. C. H. Yi, I. Yasui, Y. Shigesato, *Jpn. J. Appl. Phys.*, **34**, 1638 (1995).

13. J. M. Hvam, *Phys. Status. Solidi*, **B 63**, 511 (1974).

OXYGEN IONS DIFFUSITIVES IN TIN DIOXIDE BASED THIN FILMS

N.Y. Shishkin, I.M. Zharsky, Belarus State University of Technology, Minsk 220050, ul. Sverdlov 13a, Belarus

ABSTRACT

Oxygene ions diffusitives in tin dioxide based thin films were measured with electrochemical method. The Pd content 0-3% at. thin films were investigated. The dependencies of the diffusitives vs. content of the layer were obtained.

EXPERIMENT

Ambipolar diffusion of oxygene ions in tin dioxide thin films was investigated. Electrochemical method with reversible oxygen electrode (Pt dispersed on cubic ZrO_2) and electron filters were used. The diffusion equation can be solved in the approximation of fast diffusion and dominant electronic conductivity (really, tin dioxide is an electronic semiconductor). The solution looks like:

$I = [2q(N-No)D * /L] \exp (-(D * t/4L^2),$

I – current, N – concentration of electrons after equilibrium establishment, and No – before feeding voltage, L – thickness of the film, t – time, q – charge (electron, D * - chemical diffusitive.

Thin film of tin dioxide were investigated with a content of a palladium (atomic %) 0; 0,5; 0,7; 1; 2 and 3. The films were obtained by magnetron sputtering of the pure tin target (with appropriate palladium surface content in mosaic form) and subsequent oxidation. The thickness of the films was about 40 nm. Apart from the work, temperature dependencies of conductivity and that in different atmospheres (air with CO, No_x) were measured. The method utilized for film production ensures absence of channels and through defects. That was confirmed by electron microscopy investigation.

RESULTS

Logarithm of the diffusitive dependence vs. inverse temperature has a linear character with two different slopes. The low temperature (the point of transition is 700K) region corresponds to the diffusitive and the high temperature region includes defects formation process. For the pure tin dioxide energy of activation of the diffusion is 0.3 eV in both regions. The Do is 7.6E-8 in low temperatures and 2.3E-7 square cm/s in high. Maximum value of the diffusitive was found for the film with 0.7-1% at.Pd for the highest defectivity of the material. Activation energy grows with Pd content. Thus, maximum conductivity changes with CO gas influence can be found for the 0.7-1% at. Pd films.

CONCLUSION

The possibility of the diffusitives measurement in thin film was shown. The values obtained are in good agreement with those for bulk material, being naturally higher for possible electric field existence originated from oxygene species adsorbed. The fieled accelerates oxygene ions as it is shown for thin films oxidation

Properties of a novel amorphous transparent conductive oxide, $InGaO_3(ZnO)_m$

M.Orita,[*] H.Ohta,[*] M.Hirano,[*] H.Hosono,[*] K.Morita,[**] H.Tanji,[**] H.Kawazoe[**]

[*]   Hosono Transparent ElectroActive Materials, ERATO, JST, KSP C-1232, 3-2-1
   Sakado, Kawasaki, 213-0012, JAPAN, orita@team.ksp.or.jp

[**] R&D center, HOYA corporation, 3-3-1 Musashino, Akishima, 196-8510, JAPAN.

ABSTRACT

Novel amorphous transparent conductive oxides, $InGaO_3(ZnO)_m$, where m is an integer less than four, was developed. Optical transmittance in the visible region exceeded over 80 % and the electric conductivity at 300 K was as large as 400 S/cm. Both Seebeck and Hall coefficients exhibited negative values, indicating the conduction was n-type. It was suggested that 4s orbital of Zn^{2+} played a significant role for the formation of the extended state responsible for the conduction, while In^{3+} acted as a modifier for the stabilization of amorphous state.

INTRODUCTION

Beside well known amorphous transparent semiconductors, which could be grouped into two categories (tetrahedral and chalcogenide ones), amorphous transparent conductive oxides (TCO) such as In_2O_3[1] and Cd_2SnO_4[2] have been developed, in which 5s orbital of metal ions contributes to the conduction. Based on a working hypothesis that oxides composed of metal ions with electronic configuration $(n-1)d^{10}ns^0$ $(n \geq 5)$ might become conductive, novel amorphous TCOs such as a-$AgSbO_3$[3], a-Cd_2GeO_4[4], and a-Cd_2PbO_4[5] have been found. In general, unoccupied state, or the conduction band, in oxides composed of $(n-1)d^{10}ns^0$ metal ions are mainly formed by the vacant ns orbital. When the ns orbital radii were large enough, the orbital could be delocalized around the neighboring metal ions by the overlap interaction. In addition, when the fracture of the metal ions was high enough, the delocalized state might extend all over the compound. The electron transfer through the extended state would lead to the good conductivity. It was reported that such extended state was formed by the 5s orbital of Cd ions in $xCdO$-$(1-x)GeO_2$ system when the x values were larger than 0.5.[6] The value was considered to be a threshold beyond which the delocalized state was spread all over the system. Since the 4s orbital of Ge was unlikely effective for the conductivity, the principal quantum number n should be larger than five to have sufficient overlap interactions. Thus, $3d^{10}4s^0$ cations such as Zn^{2+}, Ga^{3+}, Ge^{4+} have been excluded from constituents effective for the conductivity. Actually, no amorphous TCOs based on the $3d^{10}4s^0$ metal ions have been reported so far as far as we know.

We have reported recently that a series of $InGaO_3(ZnO)_m$ crystalline, where m value was an integer, became transparent conductors.[7] Particularly, good conductivity of 500 S/cm was obtained for the films with the m value of one.[8] Although the molar fraction of In^{3+} ions in the system remained 0.33, which was smaller than the threshold value of 0.5, the 5s orbital of In^{3+} ions formed the extended states at the bottom of the conduction band.[9] This may be

291

tentatively explained by the following consideration. The compound has layered structure composed of an alternative stacking of InO_2 and $GaZnO_2$ layers. In ions locates two dimensionally on the InO_2 layer, which makes it possible for the In 5s orbital to form the extended state within the layer, in spite of the fact that the fraction was below the threshold. In other words, it was suggested that the good conductivity took place in the layer two dimensionally. Therefore, it is of great interest as to whether the good conductivity could be obtained for the corresponding amorphous state.

EXPERIMENT

Thin film specimens were prepared by means of pulsed laser deposition. Dense sintered disks of $InGaO_3(ZnO)_m$ compositions with single phase were prepared as targets by conventional ceramic processes followed by polishing using a diamond grinder. They were mounted in an ultra high vacuum chamber of 1×10^{-6} Pa at base pressure. An SiO_2 substrate with a dimension of $10 \times 10 \times 0.5$ mm was set parallel to the target being separated by 30 mm. The substrate and the target were rotated mechanically. Oxygen gas was introduced through a mass flow meter, and the pressure of the chamber was maintained at 0.8 Pa. KrF excimer pulsed laser beam (wavelength of 248 nm, repetition of 10 Hz) was focused on the target for 20 minutes. Power density at the target surface was 4 to 6 J/cm^2 per pulse. No intentional heating of substrate was made.

Chemical compositions of the film specimens were measured by an X-ray fluorescence method (Rigaku, XRF-3080E3). Amorphous and crystal phases were analyzed by an X-ray diffraction (XRD, Rigaku RINT), electron beam diffraction and Transmission Electron Microscope (TEM). Film thickness was measured with a stylus surface profiler (Taylar-Hobson, Talystep). Transmission spectra were measured by a spectrometer in the wavelength range from 200 to 1000 nm. Electrical conductivity and Hall constants were measured by a van der Pauw method. Measurements of thermopower were made under temperature differences of 5 to 20 °C at ambient temperature.

RESULTS

Chemical compositions of the prepared films were analyzed by XRF and shown in Fig. 1, where molar ratios of Zn^{2+} ions to Ga^{3+} ions in the films were plotted versus m values of targets. A linear relationship was obtained and the slope was 1.3, in spite that contents of ZnO in films

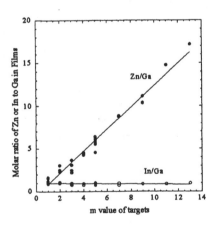

Fig. 1 Zn^{2+}/Ga^{3+} and In^{3+}/Ga^{3+} ratios in the films versus m values in the targets.

such as $ZnGa_2O_4$ tended to be much smaller than those in the targets. Evaporation of ZnO from once deposited on substrates was considered to be significant when substrates were heated up to elevated temperatures. The ratio of In^{3+}/Ga^{3+} remained almost unity.

Crystalline phases were observed in the cases of the m values larger than five. An XRD pattern for m=5 film exhibited a sharp peak at 32° characteristic of the corresponding crystalline phase (Fig. 2).[10] Only halos were seen near 22° and 34° for any films of m less than four, without the sharp peak. The halo at 22° was attributed to a SiO_2 substrate and that at 34° was due to the film. Electron diffraction measurements of m=1 specimen had only diffuse ring patterns

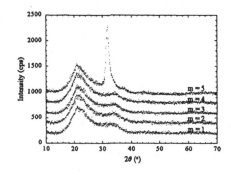

Fig. 2 XRD patterns of a-$InGaO_3(ZnO)_m$ films.

and any diffraction spot was not seen (Fig. 3a). A cross sectional TEM image of the film (Fig. 3b) showed homogeneous amorphous structure and no any sign of crystalline particle was found.

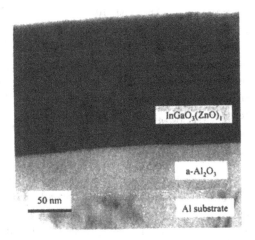

Fig 3a. (upper left) Electron diffraction pattern of a-$InGaO_3$·ZnO film for a selected area.

b. (right) TEM cross sectional view of the a-$InGaO_3$·ZnO film.

Optical transmission spectra of the amorphous films were demonstrated in Fig. 4. Each film was transparent through the visible region. Optical band gaps derived from Tauc plots of the absorption spectra increased from 2.85 eV to 3.00 eV as m value decreased (Table 1). The widening of the gap was likely due to an increase in Ga_2O_3 concentration. An averaged transmittance from 400 nm to 800 nm was over 80 %.

Table 1 summarized electric and optical properties of the films at room temperature. Seebeck coefficients of the amorphous films were ranged from −53 to −66 μK/V, which were smaller than that of crystalline ZnO (-714 μK/V) by one order of magnitude. The negative sign indicated that the films were n type conductors. Signs of Hall coefficients of a-InGaO$_3$(ZnO)$_m$ system were also negative. Sign anomaly of Hall coefficient, commonly seen in the other classes of amorphous conductors such as a-Si and a-As$_2$S$_3$, have been attributed to very short mean free path down to the order of atomic distance.[12] The mean free path of the a-InGaO$_3$(ZnO)$_m$ system derived from Hall measurements on the basis of the free electron model was around 10 nm, which was considered to be large enough to suppress the sign anomaly. The carrier density and the mobility were in the order of 10^{20}/cm^3 and 10 cm^2/Vs respectively, which were comparable to those of crystalline m=1 films.[8] It was in contrast to chalcogenide conductors, where a change in the conductivity by the phase transition from crystal to amorphous is in the order of 10^5. Arrhenius plot of the conductivity was shown in Fig. 5. Variation in the conductivity through all

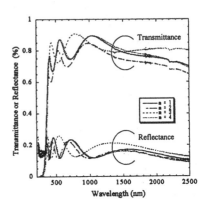

Fig. 4 Transmittance and reflectance for a-InGaO$_3$(ZnO)$_m$ films.

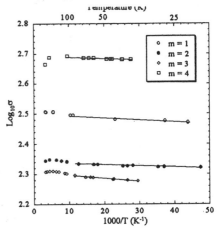

Fig. 5 Arrhenius plots on conductivity for a-InGaO$_3$(ZnO)$_m$ films.

Table 1 Optical and Electric Properties of amorphous InGaO$_3$(ZnO)$_m$ films.

Phase of Target (m value)	Thickness (nm)	Tauc gap (eV)	Averaged transmittance (%)	Seebeck Coefficient (μV/K)	Transport Properties		
					σ (S/cm)	n 10^{20}/cm^3	μ cm^2/Vs
1	251	3.00	81	-53	256	0.77	20.8
2	205	2.95	85	-62	281	1.37	12.8
3	238	2.95	81	-61	171	0.60	17.7
4	234	2.85	79	-66	408	1.22	20.9

temperature regions was about 3%. The activation energy derived from the slope was around 0.1 meV, which was much smaller than the thermal energy of room temperature (26 meV). The carrier density in the order of $10^{20}/cm^3$ may cause the degeneracy in conductivity.

It was revealed by XRD patterns that In^{3+} ions more than 17 % were required for the formation of the amorphous phases. It is known that the In^{3+} ion has wide variety of coordination numbers from 4 to 8 depending on kinds of oxides. Even in one lattice of In_2O_3,[11] for example, the bond angle is distributed from 78.4° to 113.3°. The flexibility of bond angle of O-In-O implies that In^{3+} ions act as an amorphous modifier that may release the internal stress of the O-Zn-O network.

The good conductivity of a-$InGaO_3(ZnO)_m$ films indicates the existence of the extended states. Since the fraction of In ions is smaller than 0.5, and that of Zn is larger than 0.5, it is considered that Zn 4s contributes to form the extended states. It was observed that Ge 4s was not contributive to the conductivity in the $xCdO \cdot (1-x)GeO_2$ system. It is considered that ion radii are another difference between Ge^{4+} ion and Zn^{2+} ion, which affects the overlap interaction. Slater's orbital radius of $Ge^{4+}4s$ (1.08 A) is much smaller than that of $Cd^{2+}5s$ (1.80 A), while radius of $Zn^{2+}4s$ (1.54 A) is rather larger than that of $In^{3+}5s$ (1.49 A) (Table 2). Thus the overlap integral between Zn 4s orbitals are comparable to that between the In 5s orbitals in the a-$InGaO_3(ZnO)_m$ system, and it is quite reasonable that Zn 4s orbital contributes to the formation of the extended states. Then the hypothesis is likely generalized by using the orbital radius,[13] instead of the principle quantum number.

Table 2   Radius of Slater type ns orbitals (unit:A)

n	II B	III A	IV A	V A
4	Zn^{2+}	Ga^{3+}	Ge^{4+}	As^{5+}
	1.54	1.27	1.08	0.94
5	Cd^{2+}	In^{3+}	Sn^{4+}	Sb^{5+}
	1.80	1.49	1.26	1.10
6	Hg^{2+}	Tl^{3+}	Pb^{4+}	Bi^{5+}
	1.99	1.64	1.39	1.21

CONCLUSIONS

Amorphous transparent conductive films of $InGaO_3(ZnO)_m$ of m values less than 4 were prepared by a pulsed laser deposition method. Optical band gaps were estimated to be 3 eV from Tauc plots. Averaged transmittance in the visible range was over 80 %. Dominant carrier was electron from Hall and Seebeck coefficients, and no sign anomaly of Hall coefficient was observed. Conductivity as large as 400 S/cm was obtained at the maximum. It is suggested that In^{3+} ion acts as amorphous modifier and that Zn 4s orbitals contributes to the formation of the extended electronic states. The present system was the first amorphous TCOs based on ZnO.

ACKNOWLEDGMENTS

The authors thank K.Shimizu (Keio University) for TEM observation and M.Koizumi (HOYA Co.) for help with XRF analyses.

REFERENCES

1. J.R.Bellingham, W.A.Phillips and C.J.Adkins, J.Phys.Condens.Matter 2 (1990) 6207.
2. A.J.Nozik, Phys.Rev. B 6 (1972) 453.
3. M.Yasukawa, H.Hosono, N.Ueda and H.Kawazoe, Jpn.J.Appl.Phys. 34 (1995) L281.
4. H.Hosono, N.Kikuchi, N.Ueda and H.Kawazoe, Appl.Phys.Lett. 67 (1995) 2663.
5. H.Hosono, Y.Yamashita, N.Ueda and H.Kawazoe, Appl.Phys.Lett. 68 (1996) 661.
6. N.Kikuchi, H.Hosono, H,Kawazoe, H.Oyoshi and S.Hisita, J.Am.Ceram.Soc. 80 (1997) 22.
7. M.Orita, M.Takeuchi, H.Sakai and H.Tanji, Jpn.J.Appl.Phys. 34 (1995) L1550.
8. M.Orita, H.Sakai, M.Takeuchi, Y.Yamaguchi, T.Fujimoto and I.Kojima, Trans. Mater. Res. Soc. Jpn. 20 (1996) 573.
9. M.Orita, H.Tanji, M.Mizuno, H.Adachi and I.Tanaka, Phys. Rev. B 61 (2000) 1811.
10. JCPDS card, 40-0255.
11. M.Marezio, Acta Crystallographica 1 (1967) 1948.
12. K.Shimakawa, S.Narushima, H.Hosono and H.Kawazoe, Phil.Mag.Lett. 79 (1999) 755.
13. N.Kikuchi, Doctoral thesis, Tokyo Institute of Technology (1997).

FABRICATION OF MgIn$_2$O$_4$ THIN FILMS WITH LOW RESISTIVITY ON MgO (100) SURFACE BY PLD METHOD

R. NOSHIRO*, K. UEDA*, H. HOSONO*, and H. KAWAZOE**
*Materials & Structures Laboratory, Tokyo Institute of Technology, Nagatsuta, Midori-ku, Yokohama 226-8503
**R&D Center, HOYA Corp., Musashino, Akishima 196-8510, Japan   e-mail: kawazoe@rdc.hoya.co.jp

ABSTRACT

Thin films of MgIn$_2$O$_4$ spinel, which is a recently discovered TCO material, were deposited on MgO (100) surface by PLD. The thin films were prepared under low oxygen partial pressure to enhance formation of oxygen vacancies, from which carrier electrons were generated. X-ray analyses and AFM observations suggest epitaxial growth of the grains with diameter of 100~200nm. The grains showed strong orientations both along the normal of the thin film and in plane. Epitaxial growth of the spinel was also confirmed by high-resolution transmission electron microscopic observations. The lattice image of the interface region suggests formation of structural imperfections such as dislocations, grain boundaries and amorphous phase in significant fraction. Strong optical absorption due to electron carriers was detected in near infrared region. Very large Burnstein-Moss shift was observed in ultraviolet region, and the optical band gap was estimated to be 4.3eV. DC conductivity observed was 4.5×10^{3}Scm^{-1}, which is the highest value reported for the material so far. Concentration and Hall mobility of carrier electrons were found to be 2.1×10^{21}cm^{-3} and 14 cm^{2}V^{-1}s^{-1}, respectively.

INTRODUCTION

In the flat panel industries developing a transparent electrode thin film with higher conductivity is a current issue. A promising approach would be refinement of the fabrication process conditions of ITO thin films. Actually very high conductivity of 1.4×10^{4}Scm^{-1} was reported in the laboratory made ITO thin films [1-4]. We have proposed an alternative approach to the problem, finding new and highly conductive materials having a potential of replacing ITO. Spinel lattice was selected, because it contains a linear chain of edge sharing MO$_6$ octahedra running along [110]. The linear chain is expected to play a role of highway for carrier electrons. MgIn$_2$O$_4$ [5-8], ZnGa$_2$O$_4$ [9] and CdGa$_2$O$_4$ [10] spinel oxides were found to be a new n-type TCO material. Conductivity of MgIn$_2$O$_4$ polycrystalline thin films deposited by RF sputtering on glass substrates [7,8] was changed drastically in 10 orders of magnitude from 10^{-8}Scm^{-1} to 10^{2}Scm^{-1} upon H$^+$- or Li$^+$-implantation at room temperature without post annealing. However, Hall mobility of the thin films remained around 2~3 cm^{2}V^{-1}s^{-1}, which was almost one order of magnitude smaller than that of the doped ITO thin films. The most probable cause of the low mobility was supposed to be the scattering at grain boundaries.

The purpose, therefore, of the present study is to examine whether higher conductivity of the spinel oxide is observable or not by fabricating the thin film with higher crystallinity. MgIn$_2$O$_4$ thin films are grown epitaxially on (100) surface of a single crystalline MgO substrate [11]. Electron carriers were generated from oxygen vacancies. Reported will be structures of the thin films and MgIn$_2$O$_4$/MgO interface examined by X-ray diffraction and transmission electron microscopy, optical transmission characteristics, and electrical properties.

EXPERIMENT

Thin films of MgIn$_2$O$_4$ spinel were deposited on (100) surface of single crystalline MgO substrates by laser ablation. MgO substrate was mechanically polished and annealed under oxygen at 1200°C for 10h prior to the deposition. The lattice mismatch between MgO and MgIn$_2$O$_4$ spinel is 5.2%. Deposition conditions of the thin films are summarized in Table I. An emission from a KrF excimer laser with wavelength of 248nm was used for excitation. The thin films were prepared under low oxygen partial pressure in order to generate carrier electrons from oxygen vacancies.

Surface structure or texture of the as-deposited films was examined by using an atomic force microscope (AFM). Orientation of $MgIn_2O_4$ crystals in the films was identified by X-ray powder diffraction and pole figure measurements. Cross-sectional images of $MgIn_2O_4/MgO$ interfacial region were obtained by a high-resolution transmission electron microscope (HR-TEM). Concentration profile of the constituent cations, Mg^{2+} and In^{3+}, were estimated by secondary ion mass spectrometry (SIMS) and Rutherford backscattering spectroscopy (RBS).

Table I Deposition conditions of $MgIn_2O_4$ thin films by laser ablation

Target	Sintered disk of $MgIn_2O_4$
Substrate	MgO (100)
Substrate temp	450 °C
Laser power	6.5 J cm^{-2} pulse^{-1}
Frequency	20 Hz
O_2 gas pressure	2.3×10^{-1} Pa

Optical transmission spectra of the thin films were measured in the region from ultra violet (UV) to near infrared (NIR) at room temperature by using a conventional spectrophotometer. Electrical conductivity and Hall voltage of the thin films were measured at room temperature in van der Pauw electrode configuration. Ohmic contact was provided by a sputtered Au electrode.

RESULTS AND DISCUSSION

Structure Analysis

AFM images of the MgO substrate subjected to the annealing is shown in Fig.1. Step and terrace structure is clearly seen in the surface of MgO substrate, although sharp spikes or peaks

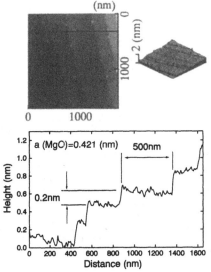

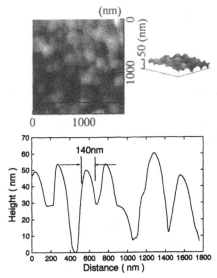

Fig.1   AFM image of MgO substrate subjected to the annealing under oxygen at 1200°C for 10h (upper left). Upper right is a bird-eye view. Distribution of the height along the line in the upper left was depicted in the lower figure. Typical step height was around 0.2nm, which was equivalent to a half of the lattice constant of MgO crystal.

Fig.2   AFM images of $MgIn_2O_4$ thin film. Upper right is a bird-eye view. Distribution of the height along the line in the upper left was depicted in the lower figure.

298

are aligned at the edges of the terrace. Height of the steps was about 0.2nm which was equivalent to a half of the lattice constant, the height of a single atomic layer. Width of the terraces ranged from about 100nm to 500nm and the average roughness was estimated to be 8×10^{-2}nm.

AFM image of the surface of the spinel thin films is shown in Fig.2. Pillar like structure was clearly noticed. The diameter of grains ranged from about 100 to 200 nm. Because of the columnar structure the average surface roughness was found to be 13nm, which was far bigger than that of the MgO substrate.

Thickness of the thin films was 600~1200nm. Depth profile of Mg/In ratio in the thin film estimated by SIMS using 15.1 keV O^{2+} beam was almost flat from the surface to the substrate, but the ratio was slightly lower than the stoichiometric value of 0.5. Since sputtering rate of In is higher than that of Mg, atomic ratio of Mg/In would probably be measured as smaller than 0.5. On the contrary the profile of Mg/In ratio estimated by RBS is not flat in the thin film: almost stoichiometric in the top half and smaller than 0.5 in the bottom half. At the present moment we have no conclusive answer to the discrepancy.

Only the diffraction peak of (004) was observed in the out-of-plane (2θ-θ scan)(Fig. 3 (a)). X-ray diffraction pattern indicates strong orientation of (004) plane in parallel with the surface of the substrate. The lattice constant estimated from the diffraction line indicated expansion of 0.1% along the normal of the thin film. In the measurements of the in-plane X-ray diffraction pattern (ϕ-scan) (Fig. 3(b)), in which the detector was fixed at the diffraction angle of (440) and equivalents, clear four peaks were observed. The angle difference of the neighboring peaks was 90 degree. This is the indication of the fact that each crystal grain has similar orientation also in plane. FWHM of the rocking curve of this diffraction line was found to be 0.68 degree.

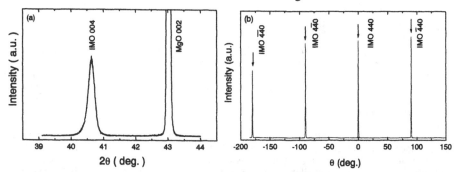

Fig.3 X-ray diffraction patterns of $MgIn_2O_4$ thin film deposited on MgO (100) surface. Deposition conditions of the thin film is given in Table I. (a) is the pattern obtained by 2θ-θ scan, and (b) is that for (440) and the equivalents obtained by ϕ-scan, respectively. FWHM of the diffraction line for (440) was found to be 0.68 degree by the rocking curve measurements.

Cross-sectional HR-TEM photos of the thin film (a) and interface region (b) are displayed in Fig. 4. It is noted in the image (a) that the surface of the thin film is not flat and grain size is 100~200nm in consistent with AFM observations. Texture of the thin film seems to be inhomogeneous. Density of dislocations and grain boundaries is high in the region close to the substrate. In the image (b), firstly noted is that the thin film on the whole is epitaxially grown on MgO (100) surface in harmony with the results of X-ray analysis. However some unexpected observations were obtained. The interface of the substrate and the spinel thin film is not atomically flat or clear, although the substrate having flat step and edge structure was used in the fabrication. This may originate from the chemical reaction of the substrate with the In species in the film. Hesse et al. observed formation and the epitaxial growth of $MgIn_2O_4$ spinel at the interface between MgO single crystal and In_2O_3 powder [11]. More surprisingly, a lot of structural imperfections are involved in the thin film: dislocations, grain boundaries and considerable fraction of amorphous phase. The structural imperfections affect X-ray diffraction only on the width of the diffraction lines as a first approximation.

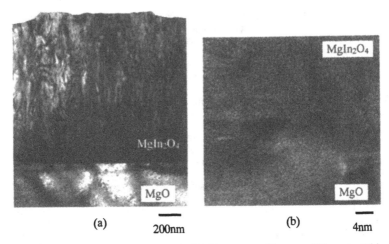

(a) 200nm   (b) 4nm

Fig.4 Cross-sectional TEM images of MgIn$_2$O$_4$/MgO structure. Texture of the spinel thin film (a) and lattice image of the region near the interface (b).

Optical and Electrical Properties

Optical transmission spectrum of the MgIn$_2$O$_4$ spinel is displayed in Fig. 5. Firstly noted in the figure is the strong absorption and reflection in NIR region, and the weak transmission loss is extending to visible range. In harmony with the intense loss in NIR region, a very large Burnstein-Moss shift is seen in the ultraviolet region. All these observations suggest that quite high concentration and probably small effective mass of carrier electrons were generated in the spinel thin film. The inset shows the plot $(\alpha h\nu)^2$-$h\nu$ for estimation of the direct allowed band gap. This amounts to 4.3eV.

Electrical properties of the MgIn$_2$O$_4$ thin film was measured with the Van der Pauw method at room temperature and results are summarized in Table II. Conductivity of this film was found to be 4.5×10^3 Scm^{-1}, this being the highest conductivity reported for MgIn$_2$O$_4$ spinel and about a half of ITO thin films commercially available. The high conductivity observed results mainly from very high electron density of 2.1×10^{21}cm^{-3} and partially from relatively large Hall mobility of 14 cm^{2}V^{-1}s^{-1}.

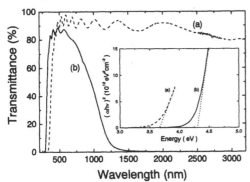

Fig. 5

Optical transmission spectra of MgIn$_2$O$_4$ spinel thin films deposited on MgO (100) surface by PLD. Spectrum (a) is obtained for the thin film prepared under high oxygen partial pressure, and this film is an almost electrical insulator because of low concentration of the oxygen vacancies. Spectrum (b) is obtained for the conductive film deposited under low partial pressure of oxygen. The inset shows the plot of $(\alpha h\nu)^2$-$h\nu$ for estimating direct allowed band gap.

The mobility is also the highest value we have obtained so far.

Although the highest conductivity was observed for $MgIn_2O_4$ spinel, this is not considered to be an intrinsic property. It was evidenced that concentration of electron carriers could be increased to about $2\times10^{21}cm^{-3}$, which is the highest value to avoid generation of the optical absorption in visible range. The problem, therefore, to be solved to obtain higher conductivity is how we can realize higher Hall mobility of the thin film. A possible approach would be further improvement of crystal quality of the thin films. As discussed in the previous subsection, the thin film obtained in the present study includes a lot of structural imperfections such as amorphous region, dislocations and grain boundaries. The maximum mobility hitherto observed for the polycrystalline thin films obtained by RF sputtering was~2 $cm^2V^{-1}s^{-1}$. Then an improvement in crystal quality would increase the Hall mobility. Selection of the substrate material with smaller lattice mismatch and the refinement of the deposition conditions are thought to be most effective factors.

Table II Electrical properties of $MgIn_2O_4$ thin film deposited on MgO (100) surface. Deposition conditions are given in Table I.

Conductivity	4.5×10^3 Scm^{-1}
Electron concentration	2.1×10^{21} cm^{-3}
Hall mobility	14 cm^{2}V^{-1}s^{-1}

CONCLUSIONS

$MgIn_2O_4$ spinel thin films were deposited on (100) surface of a single crystalline MgO substrate by PLD. The electron carriers were introduced from oxygen vacancies. In order to increase concentration of the vacancy, low oxygen partial pressure was employed in the deposition. Following results were obtained by structure analyses and measurements of optical and electrical properties of the thin films.

X-ray analyses and AFM observations suggest epitaxial growth of the grains with diameter of 100~200nm. The grains showed strong orientations both along the normal of the thin film and in plane. Epitaxial growth of the spinel was also confirmed by high-resolution transmission electron microscopic observations. The lattice image of the interface region suggests formation of structural imperfections such as dislocations, grain boundaries and amorphous phase in significant fraction.

Strong optical absorption due to electron carriers was detected in near infrared region. Very large Burnstein-Moss shift was observed in ultraviolet region, and the optical band gap was estimated to be 4.3eV. DC conductivity observed was 4.5×10^3Scm^{-1}, which is the highest value reported for the material so far. Concentration and Hall mobility of carrier electrons were found to be $2.1\times10^{21}cm^{-3}$ and 14 cm^{2}V^{-1}s^{-1}, respectively. Further improvement in the quality of the thin film was discussed to be a key factor to obtain the thin film with higher mobility.

ACKNOWLEDGMENTS

This work was in part supported by NEDO. The authors thank M. Yuuki of HOYA Corp. for TEM observation.

REFERENCES

1. P. Nath, R. F. Bunshah, Thin Solid Films, **69**, 63 (1980).
2. S. Bay, R. Banerjee, N. Basu, A. K. Batabyal, A. K. Barna, J. Appl. Phys.,**54**, 3497 (1983).
3. I. A. Rauf, J. Matter. Sci. Lett., 12, 1902 (1993).
4. H. Ohta, M. Orita, M. Hirano, H. Tanji, H. Kawazoe, H. Hosono, Appl. Phys. Lett., **76**, 2740 (2000).
5. N. Ueda, T. Omata, N. Hikuma, K. Ueda, H. Mizoguchi, T. Hashimoto, and H. Kawazoe, Appl. Phys. Lett. **61**, 1954 (1992)
6. H. Un'no, N. Hikuma, T. Omata, N. Ueda, T. Hashimoto, H. Kawazoe, Jpn. J. Appl Phys. **32**, L1250 (1993)
7. H. Kawazoe, N.Ueda, H. Un'no, I. Omata, H. Hosono, H. Tanoue, J. Appl. Phys. **76**, 7935 (1994)

8. H. Hosono, H. Un'no, N. Ueda, H. Kawazoe, N. Matsunami, H. Tanoue, Nucl. Instr. Meth. Phys. Res. B**106**, 517 (1995)
9. T. Omata, N. Ueda, K. Ueda, H. Kawazoe, Appl. Phys. Lett. **64**, 1077 (1994)
10. T. Omata, N. Ueda, N Hikuma, K. Ueda, H. Mizoguchi, T. Hashimoto, H. Kawazoe, Appl. Phys. Lett. **62**, 499 (1993)
11. D. Hesse, H. Sieber, P. Werner, R. Hillebrand, J. Heydenreich, Z. Phys. Chem. **187**, 161 (1994)

Film Deposition Methods

CHARACTERISTICS OF ULTRAVIOLET-ASSISTED PULSED LASER DEPOSITED THIN OXIDE FILMS

V. CRACIUN[1], R.K. SINGH
Department of Materials Science & Engineering, University of Florida, Gainesville, FL 32611

ABSTRACT

The properties of thin oxide films such as Y_2O_3, ZnO and $Ba_{0.5}Sr_{0.5}TiO_3$ grown using an in situ ultraviolet (UV)-assisted pulsed laser deposition (UVPLD) technique have been studied. With respect to films grown by conventional PLD under similar conditions but without UV illumination, the UVPLD grown films exhibited better structural and optical and electrical properties, especially for lower substrate temperatures. They also exhibited a better stoichiometry and contained less physisorbed oxygen than the conventional PLD grown layers. These improvements can be traced to several factors. Firstly, deep UV photons and ozone ensure a better in situ cleaning of the substrate prior to the deposition. Secondly, the presence of more reactive gaseous species like ozone and atomic oxygen formed by photodissociation of molecular O_2 promotes the growth of more oxygenated films. Thirdly, absorption of UV photons by adatoms could result in an increased of their surface mobility. All these factors have a beneficial effect upon crystalline growth, especially for moderate substrate temperatures. For optimised growth conditions, the crystalline quality and properties of ultraviolet-assisted pulsed laser deposited films was similar to that of films grown using conventional PLD at substrate temperatures of at least 200 °C higher.

INTRODUCTION

Among various techniques for growing thin films, pulsed laser deposition (PLD) hasemerged as one of the most promising due to several important advantages such as stoichiometric transfer, abrupt interfaces, layer by layer growth a.s.o. [1, 2]. One particular advantage that helped PLD to gain its wide spread acceptance is the use of a relatively low substrate temperature during the growth process. However, for many advanced technology applications, a further reduction of the process temperatures is highly desirable. A lower deposition temperature will prevent or at least limit harmful film and/or ambient gas-substrate interaction [3, 4], unwanted substrate interdiffusion processes, and re-evaporation of volatile components [5, 6]. It is also well known from rapid thermal processing practice that substrate temperature non-uniformity and reproducibility is considerably reduced for processing temperatures below 500 °C [7]. With the exception of very few materials, most high quality PLD grown materials, such as high temperature superconductors, ferroelectrics or piezoelectrics, which helped established its reputation, require substrate temperatures in excess of 650 °C. If one wants to lower the processing temperature without adversely affecting the crystalline quality, stoichiometry, and properties, then a non-thermal source of energy and/or a more reactive gaseous atmosphere should be provided during the growth process.

---

[1] Permanent address: Institute of Atomic Physics, Bucharest, Romania

Ion beam assisted PLD has been successfully demonstrated for the growth of several materials even at room temperature [8-10]. However, it is rather expensive and complicated to be easily implemented, especially for reactive PLD, which uses gas pressures of around tens to hundreds of mbar. The use of a more reactive gaseous atmosphere such as N_2O, NO_2, or ozone has been shown to improve the general properties of the grown layers and allow for a reduction of the substrate temperature [11-13]. Photon-assisted PLD also showed great promise [14-16]. This is a process where either a part of the incoming laser pulse used for ablation or a second laser pulse is used to irradiate the substrate during the growth of the film. The growth process which takes place on the substrate is delayed by at least several μs with respect to the ablation process initiated by the incoming laser beam because of the necessary time for the ablated atoms and ions to travel from the target to the substrate. The best results using this method have been obtained when using a second laser, which was fired after a certain optimum delay time depending on the target-substrate distance employed in that particular experiment. There are two drawbacks when considering this two lasers technique. First of all it is rather expensive and complex. Secondly, the pulsed laser beam, which directly irradiates the substrate, can induce an appreciable heating of its outermost layer [17]. This is so even for modest fluence values of several tens of mJ/cm^2, thereby precluding its application to sensitive substrate materials such as plastics. One has also to consider the case of the deposition of transparent thin films, which do not absorb very well the laser beam. Moreover, optical interference effects when the growing film is very thin and possess a refractive index different from that of the substrate can further complicate this process [18].

The energy of adatoms is increased by absorption of photons. This can also increase their surface mobility, which will have a beneficial effect upon the crystalline quality of the grown layers, as suggested during UV–assisted ion beam growth of TiN films [19]. A variant of the laser-assisted PLD, which replaces the second laser with an inexpensive UV source, namely a low pressure Hg lamp, is presented here. The short wavelength UV radiation (λ=185 nm) emitted by such lamps can dissociate molecular oxygen and form ozone and atomic oxygen, which were also shown to help the crystallinity of the grown films. In comparison with the laser-assisted PLD technique, the UV source can be used during the cooling stage as well. It has been shown that an UV-assisted anneal in oxygen of PLD grown oxide films has dramatically improved their crystallinity and properties [20-22]. We have investigated the microstructure and properties of Y_2O_3, $Ba_{0.5}Sr_{0.5}TiO_3$ (BST), and ZnO (results for indium tin oxide (ITO) will be presented in Symposium U) films grown by the UVPLD technique at moderate temperatures. The results are compared with those obtained from films grown using conventional PLD under similar conditions.

EXPERIMENT

An excimer laser (KrF, λ=248 nm) emitting 25 ns long pulses was used to ablate the targets (99.99% purity at least). The laser fluence used was varied from 1.5 to 2.5 J/cm^2 and the repetition rate was set at 5 Hz. The oxygen pressure was optimized for each studied material by monitoring the FWHM of the main x-ray diffraction peak [23-26]. It has been generally found that for normal deposition conditions (without UV irradiation) the optimum pressure was slightly higher than that required during UVPLD. The lower optimum oxygen pressure required by UVPLD is an indication of the higher reactivity of ozone and atomic oxygen species created by

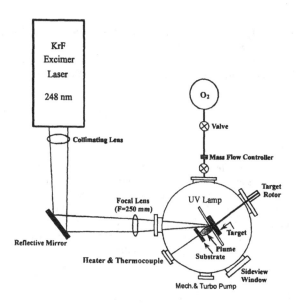

Figure 1. PLD set-up: the position of the Hg lamp is indicated.

UV radiation-induced photo-dissociation of molecular oxygen [27]. The films were grown on (100) Si, Corning glass, and (001) sapphire substrates that were cleaned in acetone, then sonicated in methanol, dipped in 1% HF and then blown dry with N_2. The substrates were located at ~10 - 8 cm in front of the target. Several films were also deposited on very thin Si_3N_4 membranes (~10 nm thick) for transmission electron microscopy (TEM) investigations. The deposition temperature was measured with a thermocouple attached to the substrate holder and checked with an infrared pyrometer. A vacuum compatible, low pressure Hg lamp having a fused silica envelope, which allows more than 85% of the emitted 184.9 nm radiation (around 6 % of the 25 W output) to be transmitted, was added to the PLD system (see Figure 1). It was used for in situ UV irradiation during the laser ablation-growth process and the cooling stage. The lamp was situated at 6 - 7 cm in front of the substrate, just below the target. It was switched on before the start of the deposition process to clean the substrate of any organic contaminants [28, 29].

The crystalline structure of the grown layers was investigated by x-ray diffraction (XRD, Philips 3720). For pole figures, ϕ scans, and rocking curve measurements an X'Pert MRD Philips system was used. The chemical composition and bonding were investigated by x-ray photoelectron spectroscopy (XPS, Perkin Elmer 5100, Mg Kα radiation), and the optical properties by variable angle spectroscopic ellipsometry (VASE, Woollam Co.). The microstructure of the films grown on Si_3N_4 membranes was investigated by transmission electron microscopy (Philips 420C at 120 kV) while the sheet resistance was measured by a four point probe method. Rutherford backscattering spectrometry (RBS) in random and channeling mode was used to assess the stoichiometry, thickness, and epitaxial quality of the grown layers or determine more accurately the oxygen content. The capacitance-voltage characteristics for MOS capacitors were measured with a Hg probe at 100 kHz using a HP4275A LCR meter. The current-voltage (I-V) characteristics were determined by depositing Au contacts onto the films and measuring the leakage current of the formed structures with a semiconductor parameter analyzer.

RESULTS

Cristallinity

The most impressive effect of the UV irradiation upon crystallinity has been seen for Y_2O_3 thin films. In Fig. 2 one can note that highly textured Y_2O_3 films can be grown at much lower temperature by UVPLD than by conventional PLD. For BST and ZnO the differences were subtler. For example, both UVPLD and PLD grown ZnO films were highly textured, showing only (002) and (004) reflections. However, when comparing, for films grown at the same substrate temperature, the FWHM of the (002) rocking curve (see Fig. 3) one can clearly see that UVPLD grown film exhibited better crystallinity.

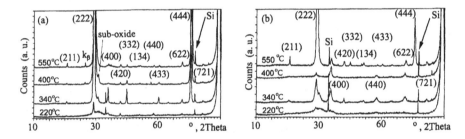

Fig. 2. XRD diffraction patterns of Y_2O_3 thin films grown by (a) UVPLD and (b) PLD

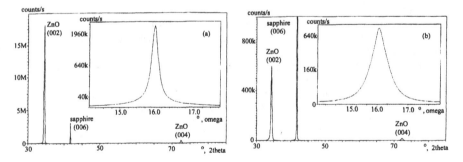

Fig. 3. XRD patterns of ZnO thin films grown at 550 °C by (a) UVPLD and (b) PLD; the insets show the ZnO (002) rocking curve

The evolution of crystallinity of layers grown at a fixed substrate temperature of 400 °C under the optimum oxygen pressure for various deposition times (i.e. thicknesses) was studied for yttria. It was found that several crystalline orientations were present after only one minute deposition time. With increasing deposition time and film thickness, the ratio of the intensities of the (222) peak to other peaks increases significantly, thus suggesting that there is a typical case of the survival of the fastest growth mode.

TEM investigation confirmed the better crystallinity of UVPLD grown films. In Fig. 4 typical bright-field TEM micrographs of Y_2O_3 films grown on membranes are displayed. The crystallites of the UVPLD grown film are more homogeneous and the film appears denser, exhibiting larger grains, ~20 nm versus ~10 nm for the PLD grown film. The accompanying selected area electron diffraction patterns reflected this difference in size between the two samples. The pattern of the UV grown sample is much spottier than that of the non-UV sample. Similar results regarding the microstructure of the grown films have been obtained for BST and ZnO films [30]. These XRD and TEM results indicated that during UVPLD the surface mobility of adatoms was significantly increased, promoting the growth of high crystalline quality layers at relatively moderate temperatures.

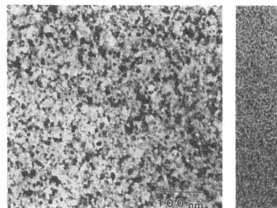

Figure 4. Bright-field TEM micrographs of Y_2O_3 thin films grown at 400 °C by UVPLD and PLD.

Stoichiometry

The XPS investigations conducted revealed interesting details regarding the chemical bonding of atoms in the grown layers. In Fig. 5 the region of the O 1s photoelectron peak acquired from Y_2O_3 films grown by UVPLD and PLD at different substrate temperatures are shown. The samples were sputter cleaned with Ar ions (5 min, 4 kV) until no C contamination was found on the surface. The acquired O 1s peak could be generally fitted using two peaks centered at around 529 eV and 531 eV, which correspond to oxygen atoms (O^{2-}) bound into the chemical compound , ie. Y_2O_3, (or BST and ZnO for these materials), and physisorbed oxygen (or a hydroxide), respectively [31, 32].

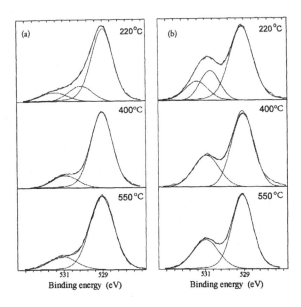

Figure 5. XPS high resolution spectra of the O 1s region acquired from Y_2O_3 thin films grown by (a) UVPLD and (b) PLD at different temperatures.

One can clearly see from Fig. 5 that the amount of physisorbed oxygen was much smaller in the UVPLD samples than that observed in the PLD samples. The films grown at 220 °C showed a third oxygen peak centered at 530.2 eV, which very likely corresponds to a sub-stoichiometric yttrium oxide phase, in agreement with the results for the Y 3d peak, which also showed two components [33].

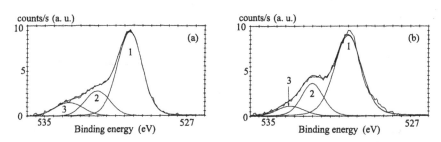

Figure 6. XPS spectra of as-received BST films grown at 650 °C by (a) UVPLD and (b) PLD: three oxygen peaks are visible. Labels 1, 2, and 3 correspond to oxygen atoms bound in BST, a surface stressed layer, and physisorbed, respectively.

A somehow different situation was encountered when analyzing the surface composition of BST films. As one can note from Fig. 6, where the O 1s photoelectron spectra acquired from as-received samples are displayed, there are 3 oxygen peaks present. Besides the usual BST perovskite phase, an additional phase was found, most likely caused by the presence of mechanical stresses and/or oxygen vacancies [25]. The presence of this new phase will negatively affect the electrical characteristics of the grown structure. One can also note that the amount of oxygen atoms bound to BST phase were several percent greater in the UVPLD grown sample.

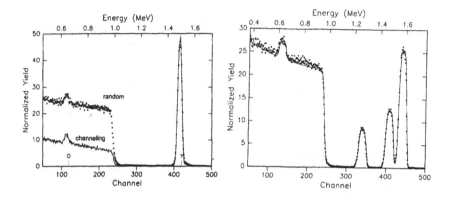

Figure 7. RBS spectra of a thin Y_2O_3 film   Figure 8. RBS spectra of a thin BST film

XPS analysis generally showed that UVPLD grown films contained less physisorbed oxygen and possessed a higher amount of oxygen atoms bound stoichiometrically into the chemical compound. This higher oxygen content was confirmed by the RBS results. Typical examples of RBS investigations for thin films are displayed in Figs. 7 and 8. According to the RUMP simulation program [34], the stoichiometry of the UVPLD deposited films was $Y_2O_{2.98\pm0.03}$ and $Ba_{0.5\pm0.02}Sr_{0.5\pm0.02}Ti_{1.0}O_{3+\delta}$ (δ around 0.1 and 0.2-0.3 for UVPLD and PLD), respectively.

Optical properties

The measurement of the optical properties by VASE was conducted in the 280-760 nm wavelength range. It should be mentioned that films grown at lower substrate temperatures, which were not very well crystallized and contained a significant amount of a suboxide phase, did not exhibit a smooth variation of Δ and ψ, the ellipsometric parameters. It was not always possible to accurately fit the measured data with a model and extract meaningful thickness and refractive index values for these grown layers.

A typical example of the optical quality of the grown layers is shown in Fig. 9, where the refractive index values obtained by fitting the recorded spectroscopic ellipsometry data from

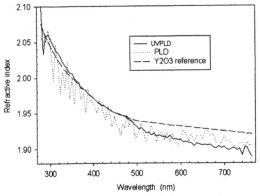

Figure 9. Refractive index versus wavelength of Y_2O_3 thin films deposited at 500 °C.

samples grown at 500 °C are displayed. For comparison reason, the refractive index of a reference standard Y_2O_3 layer [35] is also shown. The UVPLD grown sample exhibited refractive index values closer to those of the reference Y_2O_3 layer than the PLD grown sample. However, there was not a significant difference between the refractive index values of samples grown at higher substrate temperatures, both being quite close to the reference n values, as one can see in Fig. 9b. It is also worth mentioning that the extinction coefficient values estimated by the fitting program were always negligible, indicating that the films are highly transparent. Similar good optical results have been obtained for BST [25], ZnO [26] and ITO thin films.

Electrical properties

Figure 10 shows the capacitance versus voltage plots for MOS capacitors fabricated from BST films while in Fig 11 a typical example of leakage current values are displayed.

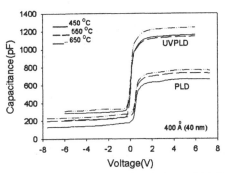

Fig. 10. C-V characteristics of 40 nm thick BST films deposited at various temperatures.

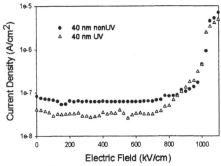

Fig. 11. Leakage current characteristics of BST films deposited at 650°C.

One can note that the UVPLD grown films showed much higher accumulation values than the films grown by PLD. At a field of 100 kV/cm 40 nm thick UVPLD-grown films deposited at 650°C by UVPLD exhibited a leakage current of 6×10^{-8} A/cm^2, which was nearly a factor of 1.5 better than that measured for PLD deposited films. Such values are among the best reported so far for BST films. Although the leakage current mechanism involved in this case is not clear, we expect it to be a Schottky conduction mechanism [36], as the current is almost constant until 750 kV/cm beyond which it shows a steep increase.

Other important electrical parameters such as flat band voltage (V_{fb}), density of states (D_{it}), and equivalent oxide thickness (T_{eq}) extracted from the measured C-V characteristics of the films are summarized in Table I. The equivalent oxide thickness values obtained for the best UVPLD films were around 10 Å, which is comparable to the best values reported [37]. Similarly the interface trap density data listed in Table I show that UVPLD grown BST films exhibited much lower values. The values listed in Table I suggest that UV irradiation during the ablation growth process significantly helped to improve MOS capacitor characteristics.

TABLE I. The electrical parameters of the BST films

Temp. (°C)	Thickness (nm)	T_{eq} (nm)		V_{fb} (V)		D_{it} (eV^{-1}cm^{-2})	
		UVPLD	PLD	UVPLD	PLD	UVPLD	PLD
450	20	0.6	1.68	+1.01	+1.12	7.1×10^{11}	2.1×10^{12}
450	30	1.10	1.75	+1.02	+1.09	7.8×10^{11}	2.9×10^{12}
450	40	1.06	1.86	+1.05	+1.10	8.1×10^{11}	4.3×10^{12}
550	20	1.00	1.57	+1.02	+1.15	6.6×10^{11}	2.0×10^{12}
550	40	1.08	1.69	+1.02	+1.15	7.9×10^{11}	3.4×10^{12}
650	20	0.98	1.47	+0.80	+1.15	5.6×10^{11}	1.8×10^{12}
650	30	1.01	1.56	+1.00	+1.10	6.9×10^{11}	2.2×10^{12}
650	40	0.99	1.62	+1.02	+1.12	7.2×10^{11}	2.3×10^{12}

One of the major problem of the grown of high k dielectric constant is the formation of the interfacial layer as one can see in Fig. 12. This is not a UVPLD problem *per se*. Depth profiling XPS investigations have shown that an interfacial SiO$_2$ layer always forms during the deposition process and not prior to it. However, further investigations are underway to minimize its thickness and, if would not be possible to complete eliminate it, at least to improve its quality.

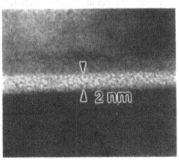

Figure 12. High-resolution cross-section TEM of a BST/Si structure; the presence of a 2 nm thick SiO$_2$ interfacial layer is clearly visible

DISCUSSIONS

The results presented here clearly showed that the microstructure and stoichiometry of the grown films has been influenced by the UV irradiation during the ablation-deposition process. A significant improvement of the optical and electrical properties of the grown films has been seen. There are several factors that could explain these findings. Firstly, deep UV radiation and ozone help to better clean the substrate and the surface of the film of carbon and other organic contaminants. According to XPS results there was always a slightly lower C 1s peak due to contamination on UVPLD grown samples. Secondly, the ozone and atomic oxygen formed by photodissociation are more reactive than molecular oxygen and can oxidize metals even at room temperature. This can help the chemical oxidation reactions, both on the substrate and, perhaps, even in gas phase. Thirdly, the surface mobility of adatoms is increased by UV irradiation. Although we don't have yet a direct proof of this, an indirect evidence was found. It was noted that under identical deposition conditions, PLD grown samples were always a little bit thicker than UVPLD grown ones. For example, for Y_2O_3 thin films, the growth rate at 400 °C for UVPLD and PLD was estimated to be approximately 0.21 and 0.27 nm/pulse, respectively. This implies that desorption of loosely bound adatoms was also increased during UV irradiation. Unless these adatoms can find a suitable low energy position within the growing lattice during their residence time, they will have a higher probability of being desorbed. Without UV irradiation, the chances for adatoms to be trapped and then buried as physisorbed atoms or in unfavorable positions, which would result in the formation of growth defects or sub-stoichiometric phases, may be higher. All these factors could contribute to the observed better characteristics of UVPLD grown thin oxide films.

CONCLUSIONS

The structure and properties of several oxide thin films grown by an in situ ultraviolet – assisted PLD have been studied. With respect to conventional PLD technique, the optimum oxygen pressure required to grow high quality films was found to be somewhat lower, a consequence of the more reactive gaseous species created by photodissociation of molecular O_2. In situ UV irradiation, which exposes each deposited atomic layer to the action of energetic photons and reactive gaseous species during laser ablation-deposition in oxygen atmosphere, resulted in the growth of highly textured layers at moderate substrate temperatures. These layers also exhibited very good optical and electrical properties and a better overall stoichiometry, containing less physisorbed oxygen. The quality and properties of ultraviolet-assisted pulsed laser deposited layers was similar to that of layers grown using conventional PLD at substrate temperatures of at least 200 °C higher.

ACKNOWLEDGEMENTS:

We would like to thank J. Perriere for RBS measurements, J. Spear for XRD measurements, J. M. Howard for CV and IV measurements, and N. D. Bassim for TEM analysis. Part of this research was funded by a grant from DOE #99020291

REFERENCES

1. R. K. Singh and J. Narayan, Phys. Rev. **B 43**, 8843 (1990).
2. R. K. Singh and D.Kumar, Mat. Sci. Engr. Reports **R22**, 113 (1998).
3. J. Park, C. M. Rouleau, and D. H. Lowndes, MRS Symp. Proc. Vol. 526, p. 27 (1998).
4. R. A. McKee, F. J. Walker, J. R. Conner, Appl. Phys. Lett. **59**, 782 (1991).
5. W. Prusseit, S. Corsepius, M. Zwerger, Physica **C201**, 249 (1992).
6. J. S. Horwitz, K. S. Grabowski, D. B. Chrisey, and R. E. Leuchtner, Appl. Phys. Lett. **59**, 1565 (1991).
7. R. Nagabhusman and R. K. Singh, in Transient Thermal processing of Materials, D. N. Ravindrai and R. K. Singh editors, TMS publications, Warrendale PA, 1997.
8. N. S. Gluck, H. Sankur, and W. J. Gunning, J. Vac. Sci. Tecnol. **A7**, 2983 (1989).
9. G. Reisse, B. Keiper, S. Weissmantel, H. Johansen, R. Scolz, and T. Martini, Thin Solid Films **241**, 119 (1994).
10. B. Holzapfel, V. Betz, D. Schlafer, H.-D. Bauer, and L. Schultz, IEEE Tran. Appl. Supercond. **9**, 1051 (1999).
11. G. Koren, A. Gupta, and R. J. Baseman, Appl. Phys. Lett. **54**, 1920 (1989).
12. C. E. Otis, A. Gupta, and B. Braren, Appl. Phys. Lett. **62**, 102 (1993).
13. H. Tabata, T. Kawai, and S. Kawai, Appl. Phys. Lett. **58**, 1443 (1991).
14. S. Otsubo, T. Minamikawa, Y. Yonezawa, T. Maeda, A. Morimoto, and T. Shimizu, Jap. J. Appl. Phys. **28**, 2211 (1989).
15. A. Morimoto, S. Mizukami, T. Shimizu, Mater. Res. Soc. Symp. Proc. 275, 371 (1992).
16. J. S. Horwitz, D. B. Chrisey, K. S. Grabowski, C. A. Carosella, P. Lubitz, and C. Edmondson, Mater. Res. Soc. Symp. Proc. 285, 391 (1993).
17. R. C. Estler, N. S. Nogar, R. E. Muenchausen, R. C. DyeC. FlammeJ. A. Martin, A. R. Garcia, and S. Foltyn, Mat. Lett. **9**, 342 (1990).
18. C. Boulmer-Leborgne, A. L. Thomann, P. Andreazza, C. Andreazza-Vignolle, J. Hermann, V. Craciun, P. Echegut, D. Craciun, Appl. Surf. Sci. **125**, 137 (1998).
19. H. Wengenmair, J. W. Gerlach, U. Preckwinkel, B. Stritzker, and B. Rauschenbach, Appl. Surf. Sci. **99**, 313 (1996).
20. V. Craciun, D. Craciun, and I. W. Boyd, Electronics Letters **34**, 1527 (1998)
21. V. Craciun, I. W. Boyd, D. Craciun, P. Andreazza, J. Perriere, Appl. Surf. Sci. **138-139**, 587 (1999).
22. V. Craciun, I. W. Boyd, D. Craciun, P. Andreazza, and J. Perriere, J. Appl. Phys. **85**, 8410-8414 (1999).
23. K. G. Cho, D. Kumar, S. L. Jones, D. G. Lee, P. H. Holloway, and R. K. Singh, J. Electrochem. Soc. **145**, 3456 (1998).
24. V. Craciun and R.K. Singh, Electrochemical Solid-State Letters **2**, 446 (1999).
25. V. Craciun and R. K. Singh, Appl. Phys. Lett., **76**, 1932 (2000).
26. V. Craciun, R.K. Singh, J. Perriere, J. Spear, and D. Craciun, J. Electrochem. Soc. **147**, 1077 (2000).
27. D. L. Baulch, R. A. Cox, R. F. Hampson, Jr.,J. A. Kerr, J. Troe, and R.T. Watson, J. Phys. Chem. Ref. Data **9**, 295 (1980).
28. S. J. Pearton, F. Ren, C. R. Abernathy, W. S. Hobson, and H. S. Luftman, Appl. Phys. **58**, 1416 (1991).

29. D. W. Moon, A. Kurokawa, S. Ichimura, H. W. Lee, and I. C. Jeon, J. Vac. Sci. Technol. A**17**, 150 (1999).

30. V. Craciun, J. Perriere, N. Bassim, R.K. Singh, D. Craciun, and J. Spear, Appl. Phys. A **69** S531, Suppl. S Dec. (1999).

31. G. M. Ingo and G. Marletta, Nucl. Instr. Meth. **B 116**, 440 (1996).

32. J. P. Duraud, F. Jollet, N. Thromat, M. Gautier, P. Maire, C. le Gressus, and E. Dartyge, J. Am. Ceram. Soc. **73**, 2467 (1990).

33. V. Craciun, E. S. Lambers, N. D. Bassim, R.K. Singh, and D. Craciun, J. Mater. Res. **15**, 488 (2000).

34. L. R. Doolittle, Nucl. Instr. Methods Phys. Res. B **15**, 227 (1986).

35. Handbook of optical constants of solids II, pp. 1090-1092, Ed. by D. Palik, Boston, Academic Press 1991.

36. C. S. Hwang, B. T. Lee, C. S. Kang, J. W. Kim, k. H. Lee, H. J. Cho, H. Horii, W. D. Kim, S. I. Lee, Y. B. Roh, and M. Y. Lee, J. Appl. Phys. **83**, 3703 (1998).

37. R. A. McKee, F. J. Walker, and M. F. Chisholm, Phys. Rev. Lett. **81**, 3014 (1998).

DEVELOPMENT AND IMPLEMENTATION OF NEW VOLATILE Cd AND Zn PRECURSORS FOR THE GROWTH OF TRANSPARENT CONDUCTING OXIDE THIN FILMS VIA MOCVD

J.R. BABCOCK, A. WANG, N.L. EDLEMAN, D.D. BENSON, A.W. METZ, M.V. METZ, T.J. MARKS

Department of Chemistry and the Materials Research Center, Northwestern University, Evanston, IL 60208-3113, tjmarks@casbah.acns.nwu.edu

ABSTRACT

For the growth of thin zinc group metal oxide films [i.e. cadmium oxide (CdO), cadmium stannate (Cd_2SnO_4), and zinc oxide (ZnO)] via metal-organic chemical vapor deposition (MOCVD), volatile Cd and Zn precursor families are needed. Starting with Cd, β-ketoiminates of varying substitution were prepared to elucidate structure-property relationships. The nature of the ligand substituents strongly influences the melting point (liquid precursors are desired). Unlike conventional Cd β-diketonates, these complexes are monomeric as determined by x-ray crystallography. Despite these advantageous characteristics, attempts to grow CdO thin films in a cold-wall MOCVD reactor using two such derivatives were not successful. This class of Cd complex appears to decompose thermally with time-- a likely cause of the poor performance. Therefore, a new series of more thermally stable Cd precursors was sought. Using the chelating diamine N,N,N',N'-tetramethylethylenediamine (TMEDA), monomeric β-diketonates were prepared. The molecular structure of $Cd(hfa)_2(TMEDA)$ (hfa = 1,1,1,5,5,5-hexafluoropentane-2,4-dionate) confirms the monomeric structural assignment. This series of Cd complexes is appreciably more volatile and sublime more cleanly than the aforementioned β-ketoiminates, as determined by vacuum thermogravimetric analysis (TGA). In addition to this advantageous characteristic, these complexes are easily prepared under ambient laboratory conditions from commercially available starting materials in a single step. Following the protocol established for Cd, a volatile series of Zn precursors was also prepared. For the Zn series, the melting point was effectively tuned through variation of both the β-diketonate and diamine ligands. The use of the Cd and Zn β-diketonate precursors in the successful growth of CdO and ZnO thin films, respectively, by MOCVD is also presented.

INTRODUCTION

The efficient growth of transparent conducting oxide (TCO) thin films is of tremendous technological importance [1]. Despite the widespread use of the current TCO of choice, tin-doped indium oxide (ITO), characteristics of this material are not optimum for certain applications. Incompatibility of tin with other components of some devices as well as the relatively narrow transparency window and low conductivity (~1000-5000 S/cm) of ITO illustrate two of these drawbacks. To address these and other limitations associated with ITO, new TCO materials have been sought possessing conductivities and optical transparencies rivaling or exceeding those of ITO. Zn is an important metal component in some new TCO materials. Our group has grown thin films of Zn-In-O and Sn-doped Zn-In-O using simple metal β-diketonate precursors [2]. These MOCVD-derived films exhibit electrical properties superior to ITO. Recently, thin films of another promising new TCO, cadmium stannate (Cd_2SnO_4 or CTO), were deposited using rf sputtering by Coutts, et al [3]. Initial analysis of the electronic properties of CTO thin films reveals that the carrier mobility is higher than that of ITO by a

317

factor of 2-4. CTO is of particular interest as a compatible TCO layer in CdSe and CdTe photovoltaic devices, the properties of which are highly dependent on the nature of the TCO [4]. A potential complimentary route to this physical vapor deposition (PVD) approach to CTO films is MOCVD. As with other chemical vapor deposition techniques, MOCVD offers several attractions including scalability, conformal coverage, growth at higher pressures, and growth at lower growth temperatures. Cd-containing oxide thin films have not yet been grown using MOCVD, likely due to the lack of suitable Cd precursors [5].

Useful MOCVD precursors must be sufficiently volatile to efficiently deliver the metal-containing species to the substrate, and sufficiently thermally stable to survive several heat/cool cycles during the film growth process. Dimethylcadmium (CdMe$_2$) meets these criteria and, despite its high toxicity, is frequently employed in the thin film growth of cadmium chalcogenide films [6]. However, attempts to grow CTO using CdMe$_2$ and tetramethyltin were unsuccessful [7]. A review of the literature on Cd^{2+} compounds containing ligands typically employed for oxide precursors, such as β-diketonates and alkoxides, fails to reveal any viable candidates for Cd CVD. Most Cd^{2+} compounds containing such mono- or bidentate ligands are not discreet monomeric complexes, a common problem of small charge-to-radius ratio metal ions. For example, the solid state structure of bis(2,4-pentanedionato)cadmium (II) [Cd(acac)$_2$] shows that this complex is polymeric [8]. Our group has previously addressed this issue using tridentate β-ketoiminate ligands with lanthanide (La^{3+}, Ce^{3+}, Nd^{3+}, Er^{3+}) and alkaline earth (Ca^{2+}, Sr^{2+}, Ba^{2+}) precursors or a combination of β-diketonate and neutral polydentate amines with Mg^{2+} [9]. A similar approach seemed logical for Cd^{2+} and Zn^{2+} [10]. While Zn(dpm)$_2$ (dpm = 2,2,6,6-tetramethyl-3,5-heptanedionato) has been successfully employed as a precursor for MOCVD [2, 11] and is monomeric [12], it is a relatively high-melting solid (132-134°C) and remains a solid under typical film growth conditions. The details for the synthesis and characterization of improved Zn and Cd precursors are presented in this communication along with our preliminary film growth studies using these complexes.

EXPERIMENTAL

Precursors

Bis(β-ketoiminato)cadmium (II) complexes 2a-c. Representative procedure for bis(4-N-2-ethoxyethylimino-2-pentanonato)cadmium (II) (2b): In a N$_2$-filled glovebox, a 3-neck, 500 mL round bottom flask fitted with a glass stopper, rubber septum, and reflux condenser is charged with Cd[N(TMS)$_2$]$_2$ (24 mmol) [13]. The vessel is then removed from the glovebox, interfaced to a Schlenk line, and 200 mL toluene (dried over and distilled from molten Na) is added via syringe. 4-N-2-Ethoxyethylimino-2-pentanone (1b; 50 mmol; 2.1 equiv.) is injected and the solution stirred at 65°C for 2 h. The volatiles are next removed *in vacuo*, and 9.3 g (20 mmol; 86% yield) of pure 2b was obtained as a pale yellow solid via vacuum sublimation (100°C / 10^{-4} Torr).

For bis(4-N-2-methoxyethylimino-2-pentanonato)cadmium (II) (2a). Mp 148-152°C; ^{1}H NMR (δ, C$_6$D$_6$): 1.56 [s, 6 H, C(N)CH$_3$], 2.05 [s, 6 H, C(O)CH$_3$], 3.04 (t, 4 H, OCH$_2$CH$_2$N), 3.09 (s, 6 H, OCH$_3$), 3.28 (t, 4 H, OCH$_2$CH$_2$N), 4.80 [s, 2 H, C(O)CHC(N)]; Anal. Calcd. for C$_{16}$H$_{28}$N$_2$O$_4$Cd: C, 45.24; H, 6.64; N, 6.59. Found: C, 45.28; H, 6.67; N, 6.71.

For 2b. Mp 98-100°C; ^{1}H NMR (δ, C$_6$D$_6$) 1.06 (t, 6 H, OCH$_2$CH$_3$), 1.57 [s, 6 H, C(N)CH$_3$], 2.07 [s, 6 H, C(O)CH$_3$], 3.10 (t, 4 H, OCH$_2$CH$_2$N), 3.38 (quart, 4 H, OCH$_2$CH$_3$), 3.45 (t, 4 H, OCH$_2$CH$_2$N), 4.82 [s, 2 H, C(O)CHC(N)]; Anal. Calcd. for C$_{18}$H$_{32}$N$_2$O$_4$Cd: C, 47.74; H, 7.12; N, 6.19. Found: C, 47.75; H, 7.17; N, 6.30.

For bis(2,2-dimethyl-5-N-2-ethoxyethylimino-3-hexanonato)cadmium (II) (2c). Mp 146-149°C; ^{1}H NMR (δ, C$_6$D$_6$): 1.05 (t, 6 H, OCH$_2$CH$_3$), 1.34 [s, 18 H, C(O)C(CH$_3$)$_3$], 1.60 [s, 6 H, C(N)CH$_3$], 3.13 (t, 4 H, OCH$_2$CH$_2$N), 3.63 (quart, 4 H, OCH$_2$CH$_3$), 3.50 (t, 4 H, OCH$_2$CH$_2$N), 5.01 [s, 2 H, C(O)CHC(N)]; Anal. Calcd. for C$_{24}$H$_{44}$N$_2$O$_4$Cd: C, 53.68; H, 8.26; N, 5.22. Found: C, 53.82; H, 8.35; N, 5.23.

Bis(β-diketonato)(TMEDA)cadmium (II) complexes 3a-d. Representative procedure for bis(2,2-dimethyl-6,6,7,7,8,8,8-heptafluoro-3,5-octanedionato)(N,N,N',N'-tetramethylethylenediamine)cadmium (II), Cd(fod)$_2$(TMEDA) (3d): A 3-neck, 500 ml round bottom flask equipped with a mechanical stirrer is charged with Cd(NO$_3$)$_2$·4H$_2$O (99.999% metals purity; 31 mmol), 100 mL deionized H$_2$O, and TMEDA (31 mmol; 1.0 equiv.). The resultant solution is then treated with a suspension prepared by adding 2,2-dimethyl-6,6,7,7,8,8,8-heptafluoro-3,5-octanedione (62 mmol; 2.0 equiv.) to 120 mL of a 0.5 M solution of NaOH in ethanol/water (3:1). A pale yellow oil immediately forms, and the resultant mixture is stirred for 30 min. The crude mixture is then extracted with Et$_2$O (3 x 200 ml), the combined Et$_2$O fractions dried over anhydrous Na$_2$SO$_4$, and 22.5 g (27 mmol; 88% yield) of pure **3d** is obtained as a pale yellow viscous oil via vacuum distillation (118°C / 0.1 Torr).

For bis(2,2,6,6-tetramethyl-3,5-heptanedionato)(N,N,N',N'-tetramethylethylenediamine)cadmium (II), Cd(dpm)$_2$(TMEDA) (3a). Mp 154-156°C; ^{1}H NMR (δ, C$_6$D$_6$): 1.30 [s, 36 H, C(CH$_3$)$_3$], 1.98 (t, 4 H, NCH$_2$CH$_2$N), 2.26 [t, 12 H, N(CH$_3$)$_2$], 5.72 [s, 2 H, C(O)CHC(O)]; Anal. Calcd. for C$_{28}$H$_{54}$O$_4$N$_2$Cd: C, 56.51; H, 9.14; N, 4.71. Found: C, 56.75; H, 9.52; N, 4.97.

For bis(1,1,1-trifluoro-2,4-pentanedionato)(N,N,N',N'-tetramethylethylenediamine)cadmium (II), Cd(tfa)$_2$(TMEDA) (3b). Mp 121-124°C; ^{1}H NMR (δ, C$_6$D$_6$): 1.70 (s, 6 H, CH$_3$), 1.75 (t, 4 H, NCH$_2$CH$_2$N), 2.03 [t, 12 H, N(CH$_3$)$_2$], 5.56 [s, 2 H, C(O)CHC(O)]; Anal. Calcd. for C$_{16}$H$_{24}$F$_6$N$_2$O$_4$Cd: C, 35.93; H, 4.52; N, 5.24. Found: C, 35.87; H, 4.47; N, 5.32.

For bis(1,1,1,5,5,5-hexafluoro-2,4-pentanedionato) (N,N,N',N'-tetramethylethylenediamine) cadmium (II), Cd(hfa)$_2$(TMEDA) (3c). Mp 88-91°C; ^{1}H NMR (δ, C$_6$D$_6$): 1.52 (t, 4 H, NCH$_2$CH$_2$N), 1.83 [t, 12 H, N(CH$_3$)$_2$], 6.18 [s, 2 H, C(O)CHC(O)]; Anal. Calcd. for C$_{16}$H$_{18}$F$_{12}$N$_2$O$_4$Cd: C, 29.90; H, 2.82; N, 4.36. Found: C, 29.78; H, 2.76; N, 4.43.

For 3d. ^{1}H NMR (δ, C$_6$D$_6$): 1.07 [s, 18 H, C(CH$_3$)$_3$], 1.83 (t, 4 H, NCH$_2$CH$_2$N), 2.07 [t, 12 H, N(CH$_3$)$_2$], 6.00 [s, 2 H, C(O)CHC(O)]; Anal. Calcd. for C$_{26}$H$_{36}$F$_{14}$N$_2$O$_4$Cd: C, 38.13; H, 4.43; N, 3.42. Found: C, 37.81; H, 4.29; N, 3.44.

Bis(β-diketonato)(ethylenediamine)zinc (II) complexes 4a-e. Representative procedure for bis(2,2,6,6-tetramethyl-3,5-heptanedionato) (N,N'-diethylethylenediamine) zinc (II), Zn(dpm)$_2$(N,N'DEA) (4c): A 3-neck, 500 ml round bottom flask equipped with a mechanical stirrer is charged with Zn(NO$_3$)$_2$·6H$_2$O (99.999% metals purity; 30 mmol), 100 mL deionized H$_2$O, and N,N'-diethylethylenediamine (30 mmol; 1.0 equiv.). The resultant solution is then treated with a suspension prepared by adding Hdpm (61 mmol; 2.0 equiv.) to 120 mL of a 0.5 M solution of NaOH in ethanol/water (3:1). A heavy colorless precipitate immediately forms, and the resultant suspension is stirred for 30 min. The solid is collected on a glass frit, dried over P$_4$O$_{10}$ *in vacuo* for 16 h, and 12.3 g (22 mmol; 74% yield) of pure **4c** is obtained as a crystalline yellow solid via vacuum sublimation (80°C / 10^{-4} Torr).

For bis(2,2,6,6-tetramethyl-3,5-heptanedionato)(N,N,N',N'-tetramethylethylenediamine)zinc (II), Zn(dpm)$_2$(TMEDA) (4a). Mp 196-202°C; ^{1}H NMR (δ. C$_6$D$_6$): 1.28 [s, 36 H, C(CH$_3$)$_3$], 2.11 (s, 4 H, NCH$_2$CH$_2$N), 2.24 [s, 12 H, N(CH$_3$)$_2$], 5.67 [s, 2 H, C(O)CHC(O)]; Anal. Calcd. for C$_{28}$H$_{54}$N$_2$O$_4$Zn: C, 61.36; H, 9.93; N, 5.11. Found: C, 61.26; H, 9.85; N, 5.27.

For bis(2,2,6,6-tetramethyl-3,5-heptanedionato)(N,N,N'-trimethylethylenediamine)zinc (II), $Zn(dpm)_2(TriMEDA)$ *(4b).* Mp 111-113°C; ^1H NMR (δ, C_6D_6): 1.30 [s, 36 H, $C(CH_3)_3$], 2.06 [br s, 2 H, $(CH_3)_2NCH_2CH_2N(CH_3)H$], 2.14 [br s, 2 H, $(CH_3)_2NCH_2CH_2N(CH_3)H$], 2.20 [s, 9 H, $(CH_3)_2NCH_2CH_2N(CH_3)H$], 5.70 [s, 2 H, $C(O)CHC(O)$]; Anal. Calcd. for $C_{27}H_{52}N_2O_4Zn$: C, 60.72; H, 9.81; N, 5.24. Found: C, 60.58; H, 9.90; N, 5.33.

For 4c. Mp 93-96°C; ^1H NMR (δ, C_6D_6): 0.94 (t, 6 H, $NHCH_2CH_3$), 1.06 (s, 2 H, $NHCH_2CH_3$), 1.29 [s, 36 H, $C(CH_3)_3$], 2.30 (s, 4 H, NCH_2CH_2N), 2.61 (quart, 4 H, $NHCH_2CH_3$), 5.73 [s, 2 H, $C(O)CHC(O)$]; Anal. Calcd. for $C_{28}H_{54}N_2O_4Zn$: C, 61.36; H, 9.93; N, 5.11. Found: C, 61.14; H, 9.74; N, 4.88.

For bis(2,2,6,6-tetramethyl-3,5-heptanedionato) (N,N-diethyl-ethylenediamine) zinc (II), $Zn(dpm)_2(N,NDEA)$ *(4d).* Mp 88-92°C; ^1H NMR (δ, C_6D_6): 0.89 [t, 6 H, $N(CH_2CH_3)_2$], 1.07 (s, 2 H, NH_2), 1.30 [s, 36 H, $C(CH_3)_3$], 2.14 (s, 2 H, $Et_2NCH_2CH_2NH_2$), 2.28 (s, 2 H, $Et_2NCH_2CH_2NH_2$), 2.55 [quart, 4 H, $N(CH_2CH_3)_2$], 5.78 [s, 2 H, $C(O)CHC(O)$]; Anal. Calcd. for $C_{28}H_{54}N_2O_4Zn$: C, 61.36; H, 9.93; N, 5.11. Found: C, 61.36; H, 9.84; N, 4.90.

For bis(2,2,7-trimethyl-3,5-octanedionato) (N,N-diethyl-ethylenediamine) zinc (II), $Zn(tmod)_2(N,NDEA)$ *(4e).* ^1H NMR (δ, C_6D_6): 0.95 [d, 12 H, $CH_2CH(CH_3)_2$], 1.03 (t, 6 H, $NHCH_2CH_3$), 1.19 [s, 18 H, $C(CH_3)_3$], 2.10 [d, 4 H, $CH_2CH(CH_3)_2$], 2.14 [sep, 2 H, $CH_2CH(CH_3)_2$], 2.48 (s, 4 H, NCH_2CH_2N), 2.70 (quart, 4 H, $NHCH_2CH_3$), 5.53 [s, 2 H, $C(O)CHC(O)$]; Anal. Calcd. for $C_{28}H_{54}N_2O_4Zn$: C, 61.36; H, 9.93; N, 5.11. Found: C, 61.34; H, 9.52; N, 2.40—satisfactory N analysis was not obtained.

Film Growth

Thin films of both CdO and ZnO were grown in the previously described [14] cold-wall MOCVD reactor. For CdO, the precursor reservoir containing $Cd(hfa)_2(TMEDA)$ was maintained at 68-80°C with an Ar carrier flow rate of 45-80 sccm. The precursor-rich carrier gas was mixed with O_2 (flowing at 100 sccm through a water reservoir) immediately upstream of the susceptor (growth temperature = 350-380°C for glass substrates) at a total working pressure of 2.6 Torr. For ZnO, the precursor reservoir containing $Zn(dpm)_2(NN'DEA)$ was maintained at 85°C with an Ar carrier flow rate of 50-60 sccm. The precursor-rich carrier gas was mixed with O_2 (flowing at 150 sccm) immediately upstream of the susceptor (growth temperature = 475-500°C for Corning 7059 glass substrates) at a total working pressure of 3.0 Torr. The commercial substrates used were washed with acetone and immediately placed into the reactor. Growth parameter dependence of deposited film structure and crystallinity was investigated by x-ray diffraction scans (XRD). A film growth rate of ~31 Å/min was estimated for CdO using the total growth time and film thickness as measured by a Tencor P-10 profilometer.

RESULTS

Precursor Synthesis, Characterization

Polydentate β-ketoiminate ligands (**1a-c**) have previously proven extremely effective in saturating the coordination sphere of small charge-to-radius ratio metals since they occupy three coordination sites and bear only a (-1) charge. Because of this, divalent metal ions such as Cd^{2+} can achieve preferred octahedral coordination geometry via binding two β-ketoiminate ligands. This stable hexacoordinate configuration should prevent oligomerization and thereby maximize volatility and thermal stability. These ligands also possess three variable organic substituents, two in the ketoiminate framework and one in the ether lariat, which allows tuning of melting point and volatility characteristics. Under rigorously anaerobic conditions, the synthesis of bis(β-ketoiminato)cadmium (II) complexes **2a-c** is readily accomplished via the protonolysis reaction between bis[bis(trimethylsilyl)amido]cadmium(II) and reagents **1a-c** outlined in Equation 1. The key to this synthetic route is that the hexamethyldisilazane byproduct can be easily removed *in*

$$Cd[N(SiMe_3)_2]_2 + 2 \quad \xrightarrow[RT, 3 h]{toluene} \quad + 2 HN(SiMe_3)_2 \quad (1)$$

1a, R=R'=R''=Me

1b, R=R'=Me, R''=Et

1c, R='Bu, R'=Me, R''=Et

2a-c

vacuo leaving essentially pure products. Further purification is accomplished using vacuum sublimation. The pale-yellow crystalline products decompose slowly when exposed to air, and therefore are best handled in, and stored under, an inert atmosphere.

A second approach to achieving coordinative saturation for metal ions such as Cd^{2+} and Zn^{2+} is to use neutral ancillary ligands to supplement the anionic ligands. This technique has proven extremely effective with Ba^{2+}, and several groups have prepared and used Ba MOCVD precursors possessing both β-diketonate and Lewis basic polyether ancillary ligands [15]. Recently we extended the scope of ancillary ligation to include polydentate amines for preparation of Mg^{2+} precursors useful in the MOCVD growth of high-quality MgO thin films [9c]. Following the protocol for Mg^{2+}, potential Cd^{2+} and Zn^{2+} precursors were also prepared using the simple aqueous route shown in Equation 2 under ambient laboratory conditions. The crude products are insoluble in the $H_2O/EtOH$ reaction solvent, a characteristic that facilitates

$$M(NO_3)_2 \cdot xH_2O + 2 \quad + \quad \xrightarrow[RT, 1 h]{NaOH \; EtOH, H_2O} \quad (2)$$

M=Cd: x=4

M=Zn: x=6

M = Cd: 3a-d

M = Zn: 4a-e

straightforward isolation via filtration. The resultant air-stable complexes are purified using vacuum sublimation, after drying over P_4O_{10} to remove any residual water. This facile one-step preparation is completed using only commercially available, air-stable reagents.

An important physical characteristic of useful MOCVD precursors is the melting point. The ideal precursor melting point falls within a relatively small range. If the melting point is too low, the complexes are sticky, waxy solids or heavy oils that are difficult to handle and transfer to precursor reservoirs. On the other hand, high-melting solids are typically less volatile and remain in the solid state under typical film growth conditions. Solids also tend to sinter, and the resulting surface area depletion causes diminished volatility and variable precursor delivery rate. The present precursors were engineered with these considerations in mind, and as shown in Table I, the melting points of these complexes are highly dependent on the nature of the organic substituents. From these data, **2b**, **3c**, and **4c** melt closest to the ideal range (ca. 50-100°C) for use as MOCVD precursors.

Table I. Melting points of several potential Cd and Zn MOCVD precursors.

complex	R^1	R^2	R^3	R^4	R^5	R^6	MP (°C)
2a	Me	Me	Me	-	-	-	148-152
2b	Me	Me	Et	-	-	-	98-100
2c	^{t}Bu	Me	Et	-	-	-	146-149
3a	^{t}Bu	^{t}Bu	Me	Me	Me	Me	154-156
3b	CH_3	CF_3	Me	Me	Me	Me	121-124
3c	CF_3	CF_3	Me	Me	Me	Me	88-91
3d	^{t}Bu	$CF_2CF_2CF_3$	Me	Me	Me	Me	Liq.
4a	^{t}Bu	^{t}Bu	Me	Me	Me	Me	196-202
4b	^{t}Bu	^{t}Bu	Me	Me	Me	H	111-113
4c	^{t}Bu	^{t}Bu	Et	H	Et	H	93-96
4d	^{t}Bu	^{t}Bu	Et	Et	H	H	88-92
4e	^{t}Bu	$CH_2CH(CH_3)_2$	Et	H	Et	H	Liq.

To confirm our goal of preparing volatile, monomeric complexes, x-ray crystallographic analyses were performed on single-crystal samples (grown from pentane solutions at –30°C) for one example from each precursor class. The results of this investigation are shown in Figure 1 [16]. Both ligand configurations clearly satisfy the Cd^{2+} and Zn^{2+} coordinative requirements and lead to discreet monomeric complexes. In **2b**, both β-ketoiminate ligands bind in a tridentate fashion. The Cd-O_{ether} bond is considerably longer [2.614(9) Å] than the Cd-$O_{ketonate}$ bond [2.220(8) Å], and the average Cd-N bond length is 2.224(10) Å. The bond angles typify a severely distorted octahedral geometry, with the greatest departure from 90° in the O2A-Cd-N1A angle of 69.8(3)° and the O1A-Cd-O1B angle of 110.00(8)°. The average Cd-$O_{ketonate}$ bond in **3c** [2.351(3) Å] is longer than that of **2b**. As would be expected, the average Cd-N bond in **3c** is longer than that of **2b** as a consequence of the dative nature of the diamine ligation. The shorter average metal-ligand bond distances in the Zn complex **4a** [Zn-O: 2.044(2) Å; Zn-N: 2.245(3) Å] relative to **3c** can be attributed to periodic trends. When compared to **2b**, a lesser degree of distortion is observed in the ligand bite angles of **3c** and **4a**. The average O-Cd-O angle in **3c** is 80.17(8)°, and the N2C-Cd-N1C angle is 78.80(10)°. The corresponding parameter in complex **4a** is even closer to 90°, with an average O-Zn-O angle of 87.04(8)° and a N1C-Zn-N2C angle of 81.07(11)°. Despite the fact that **3c** and **4a** were prepared via an aqueous route, no coordinated water is observed, a characteristic supported by thermal studies (*vide infra*). This contrasts the structure and thermal behavior of the glyme adducts of Zn(hfa)$_2$ [17].

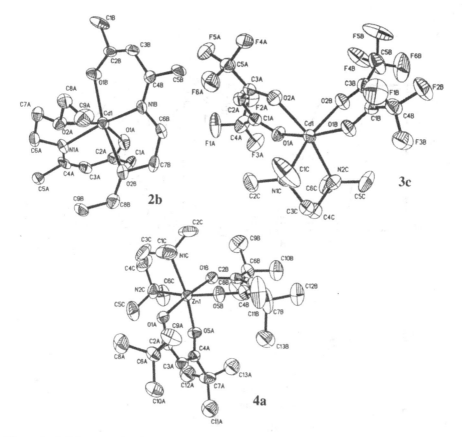

Figure 1. Solid state molecular structures of MOCVD precursors **2b**, **3c**, and **4a**.

All CVD precursors were analyzed using reduced pressure thermogravimetric analysis (TGA) to compare volatility characteristics [18]. Figure 2A shows the weight loss data as a function of temperature for Cd complexes **2a**, **3a**, **3c**, and **3d**. From these data and the linear thermal activation plots of sublimation rate vs. 1/T in Figure 2B, complex **3c** cleanly volatilizes with no decomposition (weight retention) at the lowest temperatures. In general, the β-diketonate/diamine complexes appear to be more volatile than β-ketoiminates. The difference in volatility characteristics between our new Zn precursors and Zn(dpm)₂ is less distinct (see Figures 2C and 2D). The diamine coordinated Zn complexes equal or exceed the volatility of Zn(dpm)₂ while melting, in some cases, at a much lower temperatures (*vide supra*). The one exception is precursor **4e**, which appears to decompose upon volatilization as evidenced by the ~10% residue.

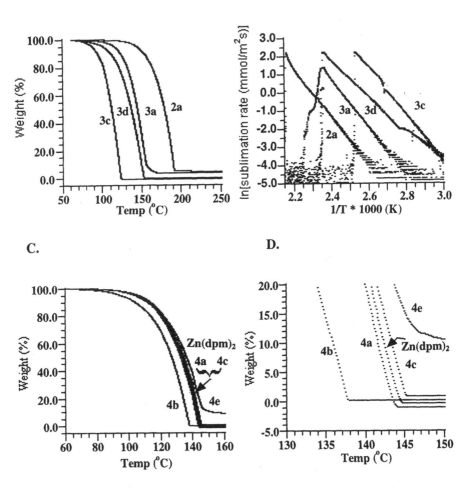

Figure 2. Reduced pressure TGA comparison of volatility characteristics of potential Cd MOCVD precursors. The weight loss comparison (A) and subsequent thermal activation plot (B) were recorded at a ramp rate of 1.5°C/min and at 5.00 (±0.05) Torr N_2 pressure. The weight loss comparison of some Zn complexes (C) with the region of interest expanded for clarity (D) is also shown.

Film Growth Studies

Phase-pure CdO thin films were grown on glass substrates at a growth rate of ~3.1 nm/min using **3c** as the precursor and O_2/H_2O as the oxidant. X-ray diffraction analysis of the as-deposited films reveals only the (00l) reflection of cubic CdO, and the absence of the other observable orientations, suggesting a high degree of texturing (see Figure 3A). Similarly, highly textured ZnO films, as evidenced by XRD data in Figure 3B, have also been grown on Corning 7059 glass using **4c** as the volatile precursor and O_2 as oxidant. Preliminary transport measurements as described elsewhere [2] have also been carried out for CdO films. At 298K, the conductivity of an as-grown CdO film is ~3000 S/cm with n-type transport (Figure 4A). This value is significantly higher than that of CdO thin films grown by PVD methods, such as pyrolysis (~ 600 S/cm) [19] or magnetron reactive sputtering (~ 200 S/cm) [20]. The corresponding room temperature carrier concentration (Figure 4B) and mobility (Figure 4C) values are 4.4×10^{20} cm^{-3} and 45 $cm^2/V \cdot s$, respectively. The yellowish CdO films show a high optical transmittance (~85%) in the visible and near-infrared region, as shown in Figure 4D. However, the intrinsic low energy gap (~2.6 eV as estimated from the optical transmission spectrum) results in considerable optical absorption at higher wavelengths, which can be shifted by doping [21].

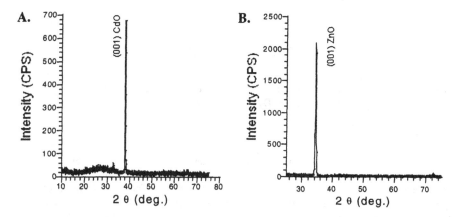

Figure 3. X-ray diffraction θ-2θ of an MOCVD-derived film of CdO deposited on glass at 450°C (**A**) and ZnO deposited on glass at 475°C (**B**).

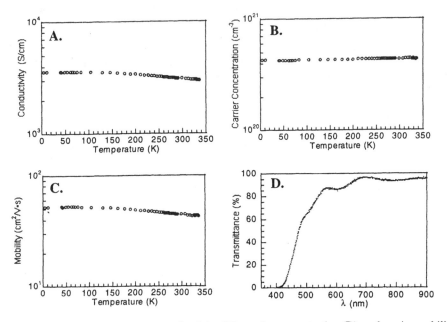

Figure 4. Variable-temperature conductivity (**A**), carrier concentration (**B**), and carrier mobility (**C**), along with the optical transmittance (**D**) of an MOCVD-derived CdO thin film.

CONCLUSIONS

This communication reports the synthesis and characterization of two new classes of monomeric, highly volatile cadmium and one class of zinc MOCVD precursors. Solid state x-ray structural analysis on one example from each class confirms the monomeric character of these complexes. The melting points of these new precursors are easily tunable via variations in organic substituents on the ligands. One Cd complex was utilized in preliminary film growth studies to deposit CdO on glass at temperatures as low as 350°C. Similarly, high-quality ZnO films were deposited on glass at 475°C. Preliminary transport and optical properties of the CdO films reveal that it is an excellent TCO material as deposited, and efforts to enhance these properties through the introduction of various dopants is currently underway.

ACKNOWLEDGEMENTS

This work was supported by the National Science Foundation MRSEC Program through the Northwestern Materials Research Center and in part by the Department of Energy through the National Renewable Energy Laboratory (subcontract no. AAD-9-18668-05). We thank Ms. M.A. Lane and Prof. C.R. Kannewurf for preliminary transport measurements.

REFERENCES

1. (a) Z. M. Jarzebski, Phys. Status Solidi A **71**, p. 13 (1982). (b) C. G. Granquist, Appl. Phys. A, **52**, p. 83 (1991).

2. A. Wang, N. L. Edleman, J. R. Babcock, T. J. Marks, M. A. Lane, P. W. Brazis, C. R. Kannewurf, MRS Symp. Series, in press.

3. X. Wu, T. J. Coutts, and W. P. Mulligan, J. Vac. Sci. Technol. A **15**, p. 1,057 (1997).

4. S. N. Alamri and A. W. Brinkman, J. Phys. D: Appl. Phys. **33**, p. L1 (2000).

5. Thin films of CdO have been deposited using spray pyrolysis, see: R. Ferroard and J. A. Rodriguez, Thin Solid Films **347**, p. 295 (1999).

6. H. Dumont, S. Fujita, and S. Fujita, J. Cryst. Growth **145**, p. 570 (1994); W. S. Kuhn, D. Angermeier, R. Druilhe, W. Gebhardt, and R. Triboulet, J. Cryst. Growth **183**, p. 535 (1998); S. Stolyarova, N. Amir, and Y. Nemirovsky, J. Cryst. Growth **184/185**, p. 144 (1998); M. Niraula, T. Aoki, Y. Nakanishi, and Y. Hatanaka, J. Appl. Phys. **83**, p. 2,656 (1998).

7. T. J. Coutts, private communication.

8. E. N. Maslen, T. M. Grearey, C. L. Raston, and A. H. White, J. Chem. Soc., Dalton Trans., p. 400 (1975).

9. (a) Lanthanide precursors: J. A. Belot, A. C. Wang, R. J. McNeeley, L. Liable-Sands, A. L. Rheingold, and T. J. Marks, Adv. Mater. (Chem. Vap. Dep.) **5**, p. 65 (1999). (b) Heavier alkaline earth precursors: D. L. Schulz, B. J. Hinds, D. A. Neumayer, C. L. Stern, and T. J. Marks, Chem. Mater. **5,** p. 1,605 (1993). (c) Mg^{2+} precursor: J. R. Babcock, D. D. Benson, A. Wang, N. L. Edleman, J. A. Belot, M. V. Metz, and T. J. Marks, Adv. Mater. (Chem. Vap. Dep.), in press.

10. Zn^{2+} and Cd^{2+} β-diketonates have been complexed with diamines, but not for MOCVD applications. See for example: P. O. Dunstan, J. Chem. Eng. Data **44**, p. 243 (1999); L. Bustos, J. H. Green, J. L. Hencher, M. A. Khan, and D. G. Tuck, Can. J. Chem. **61**, p. 214 (1983); R. B. Bindy and M. Goodgame, Inorg. Chim. Acta **36**, p. 281 (1978).

11. V. P. Ovsiannikov, G. V. Lashkarev, Y. A. Mazurenko, and M. E. Bugaeva, Proc. Electrochem. Soc. **97**, p. 1,020 (1997).

12. F. A. Cotton and J. S. Wood, Inorg. Chem. **3**, p. 245 (1964).

13. H. Bürger, W. Sawodny, U. Wannagat, J. Organomet. Chem. **3**, p. 113 (1965).

14. B. J. Hinds, R. J. McNeely, D. L. Studebaker, T. J. Marks, T. P Hogan, J. L. Schindler, C R. Kannewurf, X. F. Zhang, and D. Miller, J. Mater. Res. **12**, p. 1,214 (1997).

15. (a) R. Gardiner, D. W. Brown, P. S. Kirlin and A. L. Rheingold, Chem. Mat. **3**, p. 1,053 (1991). (b) S. H. Shamlian, M. L. Hitchman, S. L. Cook and B. C. Richards, J. Mater. Chem. **4**, p. 81 (1994). (c) G. Malandrino, I. L. Fragala, D. A. Neumayer, C. L. Stern, B. J. Hinds and T. J. Marks, J. Mater. Chem. **4**, p. 1,061 (1994). (d) T. J. Marks, J. A. Belot, C. J. Reedy, R. J. McNeely, D. B. Studebaker, D. A. Neumayer and C. L. Stern, J. Alloy. Compd. **251**, p. 243 (1997).

16. Complete crystallographic data for **2b**, **3c**, and **4a** has been deposited at the Cambridge Structure Database.

17. A. Gulino, F. Castelli, P. Dapporto, P. Rossi, and I. Fragala, Chem. Mater. **12**, p. 548 (2000).

18. B. J. Hinds, R. J. McNeely, D. B. Studebaker, T. J. Marks, T. P. Hogan, J. L. Schindler, C. R. Kannewurf, X. F. Zhang, D. J. Miller, J. Mater. Res. **12**, p. 1,214 (1997).

19. L. C. S. Murthy and K. S. R. K. Rao, Bull. Mat. Sci. **22**, p. 953 (1999).

20. T. K. Subramanyam, B. R. Krishna, S. Uthanna, B. S. Naidu, P. J. Reddy, Vacuum **48**, p. 565 (1997).

21. A. Wang, J.R. Babcock, N.L. Edleman, S. Yang, A.J. Freeman, M.A. Lane, C.R. Kannewurf, T.J. Marks, submitted for publication.

EXPLORATION OF NEW PROPERTIES OF OXIDES
BY THE GROWTH CONTROL USING PULSED LASER EPITAXY

Hideomi Koinuma[1)2)3)], Takashi Koida[1)], Daisuke Komiyama[1)], Mikk Lippmaa[3)4)], Akira Ohtomo[4)], Masashi Kawasaki[3)4)]

[1)]Frontier Collaborative Research Center, Tokyo Institute of Technology, 4259 Nagatsuta, Midori, Yokohama 226-8503, Japan

[2)]CREST, Japan Science and Technology, Japan

[3)]National Institute for Research in Inorganic Materials, 1-1 Namiki, Tsukuba, 305-0044, Japan

[4)]Department of Innovative and Engineered Materials, Tokyo Institute of Technology, 4259 Nagatsuta, Midori, Yokohama 226-8503, Japan

ABSTRACT

Metal oxides exhibit a wide variety of properties originating mainly from strong electron correlation. Electronic properties of oxides had been utilized mostly in the bulk form till recently and people frequently observed significant changes in such properties when these materials were converted into thin films. For the electronic application in recent years, hybridization and integration of materials in thin film forms are becoming more and more important. In view of our recent studies on high-Tc superconducting junctions and ZnO light emitting devices, this paper is devoted to stimulate the exploration of oxides as new innovative electronic materials by discussing the control of epitaxial thin film growth.

1. INTRODUCTION

Electrofunctional oxides are generally composed of more elements including volatile oxygen than currently used semiconductors to give versatile crystal phases which can be interchanged by external energies. Because of this structural complexity, the well developed thin film technology in semiconductors cannot be simply applied to oxides. When we try to fully extract the potential ability of oxides in thin films and devices, we must pay a lot of

Mat. Res. Soc. Symp. Proc. Vol. 623 © 2000 Materials Research Society

attention to structural sensitivity in the fabrication method and process parameters.

Since our first success in the fabrication of high-Tc superconducting $La_{2-x}Sr_xCuO_{4-\delta}$ thin films by an off-axis sputtering method[1], we have been elucidating key factors for high quality epitaxial thin film growth of complex oxides[2]. In this paper, these key factors are discussed in relation to their significance for device application and exploration of new structures and properties of oxide materials.

2. FUNCTIONAL PRORERTIES OF OXIDES

Electronic properties of semiconductors are almost exclusively dominated by charge carriers, whereas those of metal oxides depends on spin and quantized energy states of phonons and photons as well. Correlation among charge, orbital, spin, phonon and photon can produce a wide variety of properties in metal oxides and are put together in Table1. Recent topics include intrinsic Josephson effect in high Tc cuprates[3], colossal magnetoresistance (CMR) in $(La,Sr)MnO_3$[4], quantum paraelectricity in $SrTiO_3$ [5], and excitonic UV lasing at room temperature in ZnO[6]. Atomically controlled epitaxial film growth is important for exerting these properties in plane films and multilayer devices, and also for constructing artificially designed lattices and superlattices that may exhibit novel physical properties.

Table 1 Functional properties of oxides

OUT \ IN	Charge	Spin	Phonon	Photon
Charge e, h, ion couper pair	conductivity superconductivity	electromagnetism high power magnet	dielectlicity piezoelectricity	light emitting devices Kerr effect Josephson effect
Spin	CMR Hall effect	para-, ferro- magnetism Meissner effect	magnetostriction phonon maser	Faraday effect Kerr effect NMR, ESR
Phonon	piezo-, pyro- electricity Seebeck effect superconductivity	magnetic phase transition	ultrasonic reflection thermal conduction	black radiation
Photon	photonconductivity photovoltaic effect	magnet-optic disk	photorefractive effect photoacaustic effect	photo luminescence optical fiber nonlinear optics laser

3. GROWTH CONTROL OF OXIDE FILMS

We have been developing comprehensive technologies for atomically regulated lattices of oxides through the studies on high-Tc superconducting(HTSC) thin films(eg. $YBa_2Cu_3O_{7-\delta}$(YBCO)) and Josephson tunnel junction for many years[7]. Fig. 1 illustrates junction structure and key points to realize the junction formation. In spite of apparent simple trilayer junction structure, it was necessary to deposit the following eight layers for us to observe a junction characteristics: $SrTiO_3$ substrate / SrO(1ML) / BaO(1ML) / YBCO buffer layer(10nm) / a-YBCO transient layer(40nm) / a-YBCO highly crystalline bottom layer(300nm) / $PrGaO_3$ insulating layer(1.2-3.4nm) / a-YBCO buffer layer(5nm) / a-YBCO(100nm). Substantially important requirements are high crystal quality, orientation control, atomic scale flatness of surface and interfaces of these layers, and atomic scale precision in the insulating(I) layer thickness.

3-1. Film deposition system: laser MBE

Laser molecular beam epitaxy(LMBE) system was designed as a most powerful tool to compromise most of the key factors and verify atomically controlled epitaxial growth of oxides[8]. One of our specially designed high temperature LMBE systems is illustrated in Fig.2[9]. The system consists of ultrahigh vacuum chamber, a pulsed KrF excimer laser with

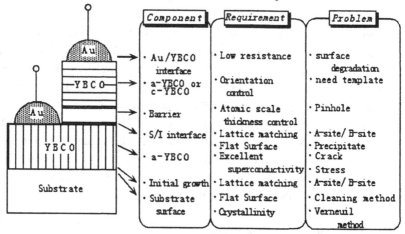

Fig. 1 Schematic cross sectional view of a trilayer junction and problems and requirements to fabricate the junction.

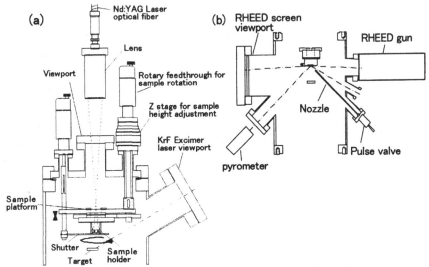

Fig. 2 Schematic drawing of one of our laser MBE chamber.
Nd-YAG laser is irradiated to the back of the sample holder to heat the substrate.(a)
The temperature is monitored by pyrometer.(b)

optics, a 300W continuous wave Nd:YAG laser as a heat source of the substrates, RHEED(reflection high-energy electron diffraction), and CAICISS(coaxial impact collision ion scattering spectroscopy). Substrate temperature can be heated up to 1400℃ in oxygen atmosphere. A focused pulsed KrF excimer laser beam is impinged onto a sintered target of the material to be deposited. Films with same composition of the targets were deposited under an atmosphere of O_2. Monitoring of RHEED pattern and intensity of the specular beam spot is performed, which enables the fabrication of atomically regulated epitaxial films. The determination of the surface terminating layer is done by CAICISS measurement. The crystal structure of films is characterized by X-ray diffractometer and the surface morphology is by AFM and SEM. High resolution transmission electron microscopy (HR-TEM) analysis is also employed to examine the lattice and interface structure.

3-2. Atomic layer control of substrate surface

Not only physical (lattice parameter, thermal expansivity, surface morphology) matching, but also chemical (surface terminating atomic layer) affinity is primarily important factor to make a heterojunction well defined at the interface. In the preparation of HTSC tunnel(SIS) junction, both of the superconducting(S) and insulating(I) layer thickness must be regulated on an atomic scale in view of the short coherence length of HTSC($\zeta_{ab} \sim 2$nm in the ab-plane and $\zeta_c \sim 0.3$nm along the c-axis). In order to achieve this important requirement, both physical and chemical factors of a substrate should be taken into account. The commercially available single–crystal SrTiO$_3$(STO) wafers are prepared by so-called mechanochemical polishing and have a small corrugation of 0.2 to 0.8nm as smooth as that of Si wafer. However, this surface is not sufficiently smooth for using atomically regulated epitaxiy. This problem was solved by developing the atomic layer lift-off of STO substrates to make the surface composed of atomically flat terraces and 0.4nm(unit cell) steps by the buffered HF-NH$_4$F(BHF) solution [10]. The method harnesses the chemical difference of basic A-site layer(SrO) and weakly acidic B-site layer(TiO$_2$) and the BHF treated surface was determined by CAICISS to be exclusively terminated with TiO$_2$ atomic layer. Other single crystal substrate materials(LaAlO$_3$, NdGaO$_3$, Al$_2$O$_3$ etc) can also be made atomically flat by a similar procedure or just annealing under appropriate conditions[11], contributing much to improve the epitaxial quality of thin films to be deposited on them.

3-3. Dimension controlled epitaxy of SrTiO$_3$ films

STO thin films were deposited on the BHF treated STO(100) substrate in 10^{-6} Torr of oxygen in the high temperature LMBE chamber. At a substrate temperatures below 900℃, the STO film growth proceeded in layer-by-layer mode, as clearly indicated by the oscillatory nature of the specular RHEED spot intensity in Fig.3(a). This mode of film growth enables digital control of film thickness, thus being useful for I layer deposition in SIS tunnel junction.

Usually the crystalinity and physical properties of epitaxial thin films are inferior to those of bulk single crystal. We would like to demonstrate a solution for this problem by the dimension control in epitaxy. In general, crystal quality of the films fabricated by step-flow mode was found to be better than that of layer-by–layer mode. By increasing the deposition temperature of STO to 1200℃, RHEED intensity behavior changed as depicted in Fig.3(b). After every excimer laser pulse, the RHEED intensity dropped sharply and recovered. The intensity recovery time was strongly dependent on temperature and laser pulse frequency. This behavior corresponds to a momentary increase of surface roughness by adatoms adhesion (nucleation) and their subsequent dissociation and diffusion to the step edges to recover atomically flat terraces. Thus, the growth mode is convertible between the layer-by-layer and the step-flow by such parameters as temperature, laser power and frequency, and terrace width of the substrate[12].

A 320nm thick layer-by-layer film at 800℃ and a 380nm thick step-flow film at 1200℃ were grown on 0.5wt% Nb-doped (100) STO substrates. Platinum contacts (0.5mm φ) were used as top electrodes to measure the dielectric property. The ε_r value(1050 at 40K) and dielectric loss factor(0.015 at 4K and <10kHz) of the layer-by-layer film are comparable to the best results reported on SrTiO$_3$ films so far (Fig.4(a)). The step-flow film exhibited a higher maximum ε_r value of 4880 at 50K(Fig.4(b)). The maximum ε_r value further jumped to 12700 at 4.2K(inset) when the Pt electrode was biased at +0.8V to compensate the asymmetric work function of the Nb-doped STO and Pt electrodes[13]. This ε_r value is as high as the value for unstrained single crystal. Typical tanδ value of the step-flow film was around 0.05 at 4K.

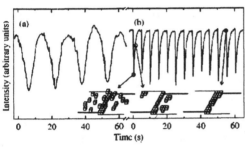

Fig.3 RHEED specular spot intensity during homoepitaxy of SrTiO$_3$ (a) at 800℃ and 2Hz, and (b) 1200℃ and 0.5Hz.

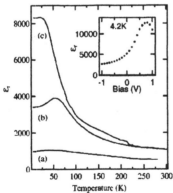

Fig. 4 Temperature dependence of the dielectric constant ε_r of a film grown at 800°C in the layer-by-layer mode(a). Dielectric constant of a step-flow film grown at 1200°C, measured during cooling (b) and during heating at a +0.8V dc bias(c). The bias dependence of ε_r at 4.2K is shown in the inset. The measurement frequency was 500kHz.

3-4. High Tc superconducting films

For the deposition of HTSC films, we must pay attention to some other factors. These are controlling of oxygen defficiency, effect of top most atomic layer of substrate and orientation control of layered crystal lattice. Since the first point has been discussed frequently and can be estimated by the Ellingham diagram[14], we focus on the latter two problems.

(1) Atomic layer effect

Such a cubic crystal as $A^{2+}B^{4+}O_3$ perovskite consists of alternating stacking of AO and BO_2 atomic layers along the a, b, and c-axis. To study the effect of substrate terminating layer on the initial growth of YBCO thin films on STO, we used TiO_2 and SrO terminated substrates. The latter substrate was obtained by the deposition of 1 monolayer SrO on the BHF treated substrate. Fig.5 shows the AFM images for YBCO half monolayer deposited on the both substrates [7]. On the SrO terminated substrate, only one unit cell roughness on step and terrace substrate was observed, while on the TiO_2 terminated substrate, higher roughness of many precipitates having rectangular shape was observed. These results suggested us that the formation of precipitates during the initial growth of YBCO thin films depended on the chemical affinity between YBCO precursors and the surface atomic layer of STO substrate. When laser ablated YBCO precursors(CuO, BaO, YO_x, ...) reach TiO_2 terminated surface, dominant CuO(B-site) fragments tend to aggregate rather than to adhere homogeneously to TiO_2 layer, while they can adhere homogeneously to SrO surface due to the high affinity between CuO and SrO layers. Hence, it is better to use A-site terminated perovskite substrate for getting YBCO thin films with better surface morphology.

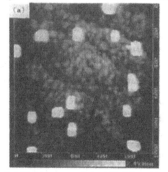 

Fig. 5 AFM images of the YBCO half monolayer on (a) TiO_2 and (b) SrO surfaces. The scanning area was $1 \times 1\mu m^2$.

2) Orientation control

In view of the anisotropy in the structure and coherence length of HTSC, it is desirable to grow films with their c-axis oriented parallel to the substrate surface for the purpose of fabricating SIS junction. The orientation of YBCO films is known to be controllable by deposition temperature. In case of using STO(100) substrate, a-YBCO was fabricated below 600℃, while c-YBCO was dominant above 700℃ [15,16]. a-YBCO film fabricated at this low temperature, however, exhibited poor superconducting properties because of its low crystal quality.

High quality a-YBCO film was obtained by the following procedure[17]. 1) One monolayer of SrO on BHF treated STO substrate, 2) One monolayer of BaO to reduce the lattice mismatch between STO substrate and subsequent YBCO, 3) 10nm a-YBCO seed layer at 580 ℃ to force the c-axis parallel to the substrate surface, 4) 40nm a-YBCO at temperatures increasing from 580 to 735℃ at a rate of 10℃/min to gradually improving the crystallinity of YBCO with keeping a-axis orientation, and 5) YBCO film at 735℃. By inserting the temperature gradient layer, this film showed high crystalinity, a-axis orientation(>99%), and Tc exceeding 90K.

4. APPLICATION OF GROWTH CONTROL TO NANO STRUCTURED FILMS

The advanced growth technology described above can be extended to the exploration of nano structured oxide films and their quantum properties. With a focus on dimension controlled epitaxy, some examples are given below.

4-1. Perovskite superlattices

The clear observation of RHEED oscillation makes it possible for us to fabricate oxide superlattices which may exhibit novel electronic or optical properties. Since one period of RHEED oscillation corresponds to exactly one unit cell deposition in layer-by-layer growth mode, atomically designed superlattices can be fabricated. We have fabricated $ABO_3/AB'O_3$ and $ABO_3/A'BO_3$ superlattices, where A and A' are Sr, Ba, and (La_{1-x},Sr_x) while B and B' are Ti, V, Ru, and Mn, by LMBE under the observation of RHEED intensity for digital control of each layer thickness.

4-2. HTSC wire

NdBa$_2$Cu$_3$O$_{7-\delta}$(NBCO) has the highest Tc among the 123 family of HTSC and higher decomposition temperature than YBCO. Fig.6 shows the AFM image of a half unit cell thick NBCO film deposited by increasing the growth temperature to 850°C in 200mTorr oxygen on a BHF treated STO(100) substrate. It is clearly seen that the film growth occurred preferentially along the steps, forming wires along the step lines by the step-flow growth[18].

4-3. ZnO nano dot film

In the case of fabrication of heteroepitaxial films, the problem of lattice mismatch is serious. Usually, this problem is solved by inserting buffer layers as was seen in GaN film growth on sapphire substrate[19]. However, there is a case that such lattice mismatch can be utilized favorably to manifest a new property which is hard to be observed in bulk oxide.

ZnO films in wurtzite was grown by LMBE on lattice mismatched(18.3%) sapphire(0001) substrates by a spiral growth mode to show the morphology of hexagonally shaped nanocrystals assembled closely each other as shown in Fig.7. The hexagonal

Fig. 6 AFM iamge of wire structure of
NdBa$_2$Cu$_3$ O$_{7-\delta}$.

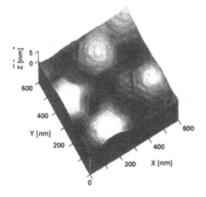

Fig. 7 AFM image of a ZnO nano crystal
film with thickness of 200nm.

nanocrystals form columns with grain boundaries running parallelly from the interface between ZnO and sapphire to the film surface and the average spacing is almost the same as the nanocrystal size seen in AFM image. The formation of the nanocrystals was explained by the higher order epitaxy[20]. From this hexagonal nano-dot ZnO film, we observed optically pumped UV(390nm) lasing at room temperature[6]. This new ZnO film laser attracts our interest because not only of its short wavelength but also of its low threshold energy. This unique and promising stimulated emission is presumed to originate from the collision of exciton-exciton confined in the hexagonal nano structure and the Fabry-Perot cavity naturally occurring in the faceted plane(Fig.8). Studies are in progress on the bandgap engineering by ZnO/Zn-Mg-O superlattices to control the laser wavelength as well as on the p-type doping in ZnO to fabricate current injected lasing.

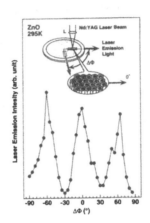

Fig. 8 Spontanious emission intensity generated by exciton-exciton collision process as a function of the angle between pumping stripe and in-plane crystal axis. 0° corresponds the ZnO [10$\bar{1}$0] direction. The inset depicts the experimental configuration. Lasing occurs at every 60° in the directions normal to the parallel edges of hexagons.

CONCLUSIONS

Through years of work, we have been elucidating and partly solving the problems of epitaxial thin film growth of oxides. Key issues are atomic scale finishing of substrate, the conquest or utilization of lattice and thermal expansion mismatch, and the nucleation and growth control which influences the orientation and dimension of growing thin films. Careful consideration is required on physics, chemistry, and electronic states of oxide materials as well as on the thermodynamics and kinetics of growth process so that we could understand and solve the problems and realize not only the currently interested Josephson tunnel junction, high dielectric oxide film growth on silicon, and ultraviolet laser emission but also the expansion to a new field of oxide based electronics.

REFERENCES

1. H. Koinuma, M. Kawasaki, M. Funabashi, T. Hasegawa, K. Kishio, K. Kitazawa, K. Fueki, and S. Nagata, J. Appl. Phys. **62**, p.1524 (1987)

2. H. Koinuma, MRS Bull., **19**(9), p.21 (1994).

3. R. Kleiner, F. Steinmeyer, G. Kukel and P. Muller, Phys. Rev. Lett. **68**, p.2394 (1992).

4. K. Tamasaku, Y. Nakamura and S. Uchida, Phys. Rev. Lett. **68**, p.1455 (1992).

5. Y. Inaguma, J.-H. Sohn, I.-S. Kim, M. Ito and T. Nakamura, J. Phys. Soc. Jpn. **61**, p.3831 (1992).

6. Z.K. Tang, P. Yu, G.K.L. Wong, M. Kawasaki, A. Ohtomo, H. Koinuma and Y. Segawa, Solid State Commun. **103**, 459 (1997)

7. H. Koinuma, M. Kawasaki, S. Ohashi, M. Lippmaa, N. Nakagawa, M. Iwasaki and X.G. Qui, Proc. of SPIE. **3481**, p153 (1998).

8. K. Koinuma, N. Kanda, J. Nishino, A. Ohtomo, H. Kubota, M. Kawasaki and M. Yoshimoto, Appl. Surf. Sci. **109/110**, p.514 (1997)

9. S. Ohashi, M. Lippmaa, N. Nakagawa, H. Nagasawa, H. Koinuma, and M. Kawasaki, Rev. Sci. Instru. **70**, p.178 (1999)

10. M. Kawasaki, K. Takahashi, T. Maeda, R. Tsuchiya, M. Shinohara, O. Ishiyama, T. Yonezawa, M. Yoshimoto and H. Koinuma, Science **266**, p.1540 (1994).

11. M. Yoshimoto, T. Maeda, T. Ohnishi, H. Koinuma, O. Ishiyama, M. Shinohara, M. Kubo, R. Miura and A. Miyamoto, Appl. Phys. Lett. **67**, p.2615 (1995)

12. M. Lippmaa, N. Nakagawa, M. Kawasaki, S. Ohashi and H. Koinuma, Appl. Phys. Lett. **76** p.2439 (2000)

13. M. Lippmaa, N. Nakagawa, M.Kawasaki, S. Ohashi, Y. Inaguma, M. Itoh and H. Koinuma, Appl. Phys. Lett. **74**, p.3543 (1999).

14. H.J.T. Ellingham, J. Soc. Chem. Ind. **63**, p.125 (1944)

15. Inam, C. Rogers, R. Ramesh, K. Remchning, L. Farrow, D. Hart, T. Venkatesan and B. Wilkens, Appl. Phys. Lett. **57** p.2482 (1990)

16. K. Fujito, M. Kawasaki, J.P. Gong, R. Tsuchiya, Y. Yoshimoto and H. Koinuma, Trans. Mater. Res. Soc. Jpn. **19A**, p.541 (1991)

17. H. Koinuma, K. Fujito, R. Tsuchiya, andM. Kawasaki, Physica C **235-240** p.731 (1994)

18. H. Koinuma, N. Kanda, J. Nishino, A. Ohtomo, H. Kubota, M. Kawasaki, M. Yoshimoto, Appl. Surf. Sci. **109** p.514 (1997)

19. H. Amano, M. Kito, K. Hiramatsu, I. Akasaki, Jpn. J. Appl. Phys. **28**, L2112 (1989)

20. M. Kawasaki, A. Ohtomo, I. Ohkubo, H. Koinuma, Z.K. Tang, P. Yu, G.K.L. Wong, B.P. Zhang, Y. Segawa, Materials Science and Engineering B56, p.239 (1998).

Preparation of PZT-YBCO heterostructure on YSZ coated Si by KrF laser ablation

Kenji Ebihara[1], Fumiaki Mitsugi[1], Tomoaki Ikegami[1], and Jagdish Narayan[2]

[1]Department of Electrical Engineering and Graduate School of Science and Technology, Kumamoto University, Kurokami 2-39-1, Kumamoto 860-8555, Japan
[2]Department of Materials and Science, North Carolina State University, Raleigh, NC, USA

ABSTRACT

KrF excimer laser ablation technique is used to fabricate the ferroelectric $Pb(Zr_xTi_{1-x}O_3)$(PZT) capacitor on Si(100) substrate. The superconducting $YBa_2Cu_3O_{7-x}$(YBCO) and the colossal magnetoresistive $La_{0.8}Sr_{0.2}MnO_3$(LSMO) thin films are used as bottom electrodes for the ferroelectric capacitive structure. The YBCO and LSMO films were studied to understand the interface problems with perovskite oxide films. The fabricated PZT/YBCO/YSZ/Si capacitor shows the ferrolectric properties of the remanent polarization of $20\mu C/cm^2$ and the coercive field of 40 kV/cm. Post-thermal annealing improves the ferroelctric properties and the results are comparable with that of the PZT/YBCO/MgO(100) structure. The leakage current of the PZT capacitor is discussed.

INTRODUCTION

Ferroelectric nonvolatile random access memories(FeRAM) have attracted much attention from the viewpoint of their high-speed and low voltage operation, low power consumption, long-term endurance and large-scale integration. The lead zirconate titanate($Pb(Zr,Ti)O_3$:PZT) thin films have been one of the promising candidates for the FeRAM application because of their high remanent polarization and low coercive field. The direct integration of ferroelectric films into the silicon technology without damaging the underlying semiconductor devices has been hampered due to degrading of the ferroelectric properties [1]. In fabricating the capacitor stack consisting of ferroelectrics and electrodes, the choice of bottom electrode materials is very important to avoid forming an insulating oxide at the PZT-electrode interfaces. We have studied the PZT / perovskite oxide $YBa_2Cu_3O_{7-x}$(YBCO) heterostructures for the ferroelectric–superconducting devices[2-4]. The integration of these structures into high integrated silicon-based devices requires the use of buffer layer material between the bottom electrodes and the silicon substrate. YSZ buffer layer is very effective for preparing PZT/YBCO/YSZ/Si(100) structure having high-quality ferroelectric properties for a nonvolatile memory device.

In this study, we report the fabrication of the ferroelectric PZT/YBCO heterostructures on YSZ-coated silicon substrate. Recently, colossal magnetoresistance (CMR) in manganese perovskite has attracted considerable attention due to its physical properties and potential applications. The CMR materials are closely latticed matched to the perovskite ferroelectrics such as PZT. We also report the preparation of $La_{1-x}Sr_xMnO_3$(LSMO) film needed for PZT/LSMO structure, in which LSMO is used as a bottom electrode. KrF excimer laser ablation technique was used to prepare the thin films and the heterostructures.

EXPERIMENT

The ferroelectric PZT , superconducting YBCO , buffer YSZ and colossal magnetoresistance LSMO films were deposited by KrF excimer laser ablation method. Details of the experimental system are given elsewhere[4]. The ambient oxygen gas was fed into the stainless steel chamber (ϕ=280 mm) evacuated to a base pressure of 10^{-7} Torr. Lambda Physik LPX305icc KrF excimer laser beam (λ=248 nm, pulse duration=25 ns, maximum output=850 mJ/pulse) was directed on the rotating stoichiometric ceramic targets of $PbZr_{0.52}Ti_{0.48}O_3$, $YBa_2Cu_3O_{7-x}$ and $(ZrO_2)_{0.97}$ $(Y_2O_3)_{0.03}$. The target was ablated and the resulting plasma plume was directed on the Si(100) substrate placed in front of the target. Figure 1 shows a schematic of the PZT / YBCO / YSZ heterostructures and the crystal structure.   The distance between the targets and Si substrate was 40 mm. All films were deposited at a laser repetition rate of 5 Hz. Deposition of YSZ on the Si (100) substrate was carried out in an oxygen pressure of 1 mTorr at 775°C and laser energy density of 3 J/cm². The second YBCO layer was deposited at 710°C in an oxygen pressure of 200 mTorr. Finally, PZT was grown at 550°C in an oxygen pressure of 100 mTorr. The laser fluence for YBCO and PZT films was 2 J/cm². After deposition whole heterostructures were kept in an oxygen pressure of 600 Torr at 400°C for an hour and cooled down to the room temperature. The LSMO thin film was deposited on MgO substrate using a sintered $La_{0.8}Sr_{0.2}MnO_3$ target.

The ferroelectric characteristics of the PZT film such as *P-E* hysteresis loop and switching fatigue property were measured by Sawyer-Tower circuit. LCR meter (NF Electronic Instruments LCZ2345) was used to measure the dielectric constant and tanδ. The whole structures were examined using x-ray diffractometer (Rigaku RINT2000/PC) with CuKα radiation. The surface morphology of the film was observed by atomic force microscopy (Seiko Instruments Inc. SPI3800N). The film thickness was estimated by cross-sectional scanning electron microscopy (JEOL JSM-T200).

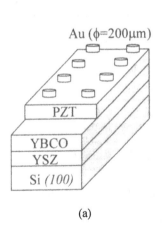

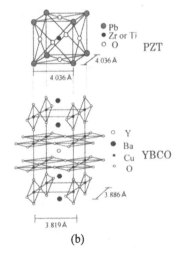

Figure 1. The schematic diagram of the Au/PZT/YBCO/YSZ heterostructures (a) and crystal structure of the PZT / YBCO (b).

RESULTS AND DISCUSSION

Figure 2 shows the *P-E* hysteresis loop for the ferroelectric PZT films of the capacitive structures which were deposited on the silicon substrate and MgO(100) substrate. The film thickness for PZT and YBCO was 400nm. The cylindrical Au top electrode of a diameter 200μm was deposited on the PZT surface using thermal evaporation method. The Au/PZT/YBCO/YSZ/Si(100)   sample has remanent polarization Pr of 20μC/cm² and coercive field of 40 kV/cm, while the sample deposited on the MgO substrate shows Pr of 33μC/cm² and Ec of 40 kV/cm, respectively. The triangular voltage V_{p-p} of 30V at 1kHz was applied to measure the *P-E* hysteresis loops.

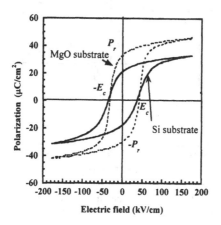

Figure 2. The substrate dependence of the *P-E* hysteresis loop.

Post-thermal annealing treatment (O₂:600Torr, 400°C, 1 hour) of the buffer YSZ and the bottom YBCO electrode was done to improve the ferroelectric quality of the PZT/YBCO heterostructure. Figure 3 shows the *P-E* hysteresis curves before and after the post-thermal annealing for each layer. It is observed that the post-annealing procedure gives the high-quality PZT/YBCO structure. After post-annealing, the remanent polarization of 34μC/cm² and the coercive field of 40 kV/cm were achieved and are comparable with the properties of the MgO substrate sample. Figure 4 shows the XRD patterns of the PZT/YBCO/YSZ/Si(100) heterostructures with (a) and without the post-annealing(b). PZT(52/48) thin film shows (00*l*) orientation. The diffraction intensity of YBCO(00*l*),YSZ(00*l*) and PZT(003) remarkably increased when the as-deposited YBCO and YSZ layers were post-annealed. The intensity of PZT(001) and (002) also increased. However, a marked change can not be observed due to intensity saturation in post-annealed samples.

Figure 5 shows the operating frequency dependence of the relative dielectric constant ε_r and the loss tangent *tanδ* for the heterostructures with and without the post-annealing. It is shown that the dielectric constant decreases gradually with increase of operating frequency. The gradual increase of tanδ is observed in the high frequency region. This is related to the increase of the resistive current component in the high frequency. Figure 6 shows the leakage current versus electric field characteristics of PZT/YBCO/YSZ/Si(100) heterostructures with and without post-annealing. Although at low electric field the thermal annealing treatment contributes to decrease in the leakage current, in the high electric field region no significant deference is observed. On the other hand, the PZT/YBCO/MgO structure showed the leakage current of the order from 10^{-9} (at 20kV/cm) to 10^{-7} A/cm^2 (at 160kV/cm).

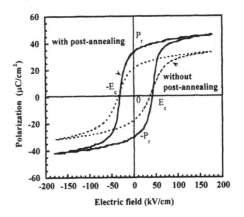

Figure 3. The post-annealing effect on the *P-E* hysteresis property of the Au/PZT/YBCO/YSZ/Si.

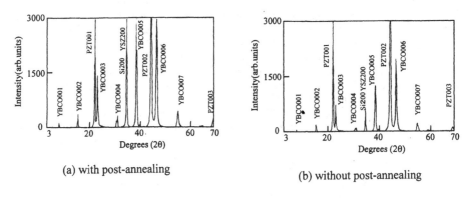

(a) with post-annealing

(b) without post-annealing

Figure 4. The XRD (θ-2θ) spectra of the PZT/YBCO/YSZ/Si heterostructures.

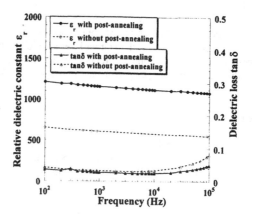

Figure 5. The post-annealing effect on the
dielectric constant and loss tanδ of the
PZT/YBCO/YSZ/Si.

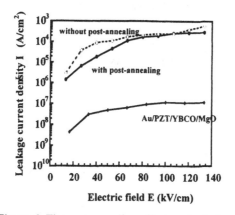

Figure 6. The post-annealing effect on the leakage
current density of the Au/PZT/YBCO/YSZ/Si.

We also fabricated the PZT/LSMO heterostructures on the MgO(100) substrate. The LSMO film used as a bottom electrode was deposited at the substrate temperature of 800°C under the ambient O_2 pressure of 100mTorr with laser fluence of $2J/cm^2$. The P-E hysteresis loop has Pr of 22 $\mu C/cm^2$ and coercive field of 50 kV/cm at $10V_{p-p}$ (1kHz). This heterostructure has a potential of a promising future device which is affected by electric field and magnetic field.

CONCLUSION

We have studied the ferroelectric and crystalline properties of the PZT/YBCO and the PZT/LSMO structures fabricated by KrF excimer laser ablation technique. High-quality PZT films were deposited on the YBCO and LSMO bottom electrodes. YSZ coated silicon substrate is useful to obtain the c-axis oriented perovskite electrode such as YBCO film.

ACKNOWLEDGEMENT

The authors would like to thank Kohei Mimura, Graduate Student of Kumamoto University for his technical support to this work. This work has been supported by a Grant-in-Aid for Science Research(No. 10045046,1998-2000) by the Ministry of Education, Science, Sports and Culture of Japan.

REFERENCES

1. M.Alexe, St.Senz, A.Pignolet, J.F.Scott, D.Hesse, and U.Goself; Materials Research Society Symposium Proceeding."Ferroelectric Thin FilmsVI",Vol.493,517(1998).
2.H.Kurogi, Y.Yamagata, K.Ebihara, N.Inoue; Surface and Coatings Technology, Vol.100/101, 424(1998).
3.A.M.Grishin, M.Yamazato, Y.Yamagata, and K.Ebihara ; Appl.Phys.Lett.Vol.72,620(1998).
4.K.Ebihara,F.Mitsugi, M.Yamazato, T.Ikegami, J.Narayan ; Materials Research Society Symposium Proceeding" Materials for Smart Systems III"(2000,in press).

PULSED LASER DEPOSITION OF EPITAXIAL SrVO₃ FILMS ON (100)LaAlO₃ AND (100)Si

P.W. YIP and K.H. WONG
Department of Applied Physics, The Hong Kong Polytechnic University, Hung Hom,
Kowloon, Hong Kong, People's Republic of China

ABSTRACT

Thin films of SrVO₃ have been grown on (100)LaAlO₃ and TiN buffered (100)Si substrates by pulsed laser deposition. The films were deposited in temperature range of 450°C - 750°C and under ambient oxygen pressure between 10^{-6} and 10^{-2} Torr. Their structural properties were characterized using a four-circle x-ray diffractometer. High quality SrVO₃ films were obtained at growth temperatures above 500°C without post annealing. Heteroepitaxial relationship of $<100>_{SrVO_3} \parallel <100>_{LaAlO_3}$ and $<100>_{SrVO_3} \parallel <100>_{TiN} <100>_{Si}$ were observed for films deposited at ≥ 550°C. X-ray photoelectron spectroscopic studies of the films suggest that the vanadium is mainly tetravalent and pentavalent. Charge transport measurements show that the films vary from semiconducting to highly conducting for different growth conditions. Resistivity of a few micro-ohm cm was recorded for some of the epitaxial SrVO₃ films.

INTRODUCTION

Currently, there is great interest in fabricating highly conducting oxide films for use as electrodes in applications such as ferroelectric capacitor and solid-oxide fuel cell. Commonly used conducting oxides include $YBa_2Cu_3O_{7-x}$, $La_{1-x}Sr_xCoO_3$, and doped lanthanum manganates. $SrVO_3$ (SVO) has a perovskite cubic structure with a lattice constant of 3.84 Å. It is known to have very low electrical resistivity and is therefore a suitable material for conducting oxide electrodes. Indeed it has been reported that high quality SVO films can be fabricated on oxide single crystal substrates and resistivities of the order of $\mu\Omega$-cm have been achieved [1]. SVO films directly grown on Si substrates have also been attempted. Only poor crystalline structure and high resistivity films, however, have been obtained [1,2]. SVO films grown on Si through thin buffer layers of Sr [3] and yttria-stabilized zirconia (YSZ) [1] have been reported. High quality epitaxial films have been demonstrated at substrate temperature of 650°C. This relatively high deposition temperature, however, is incompatible with the processing temperature of about 500°C commonly used in Si fabrication technology.

In this report, we present the results of fabricating SVO thin films on LaAlO₃ (LAO) and TiN buffered Si by pulsed laser deposition (PLD) method. Very high quality epitaxial SVO films grown on LAO substrates have been obtained and resistivities of 3 $\mu\Omega$-cm and 19 $\mu\Omega$-cm were recorded at 77 K and 300 K respectively. High quality heteroepitaxial SVO/TiN/Si structures have also been fabricated and a low processing temperature of 550°C has been demonstrated.

Mat. Res. Soc. Symp. Proc. Vol. 623 © 2000 Materials Research Society

EXPERIMENT

In this work the SVO films were grown on (100)LAO and (100)Si substrates by the usual pulsed laser deposition (PLD) method. A KrF excimer laser ($\lambda = 248$ nm) with a repetition rate of 10 Hz was used. The laser fluence was kept at 2 and 3 J/cm^2 for the TiN and SVO film depositions respectively. The target and substrate distance was maintained at 4.5 cm. A rotating multi-target holder incorporated into our chamber was used for in-situ deposition of TiN and SVO heterostructures. The SVO targets were prepared by the standard solid state reaction using high purity $SrCO_3$ and V_2O_5 powders. They were initially weighted and mixed according to the stoichiometric composition ratio. They were then ground into fine powders and sintered at 850°C for 5 hours in air. Afterwards they were pressed into circular pellets of 22 mm diameter and 5 mm thick. Finally the pellets were sintered at 950°C for 5 hours in air again. The TiN targets were bought from a commercial supplier with a stated purity of 99.5 %.

The substrates were ultrasonically degreased with acetone and thoroughly rinsed with deionized water. Si substrates were additionally etched by 10 % HF solution for 10 minutes to rid of the oxide layer. All substrates measuring 5×10 mm^2 were adhered to the faceplate of the ohmic heater with high temperature silver paste. A thermal couple was embedded in the faceplate and placed directly underneath the substrate for accurate monitoring of the film deposition temperatures. Before the ablation, the chamber was evacuated to a base pressure of 2×10^{-7} Torr by a cryo-pump. SVO films were deposited under ambient oxygen pressure between 10^{-6} and 10^{-2} Torr and in temperature range of 450°C - 750°C. TiN films were grown at the base pressure with a fixed substrate temperature of 550°C. No additional post-annealing was made. The structural properties of the as-deposited SVO/LAO and SVO/TiN/Si were characterized by a four-circle X-ray diffractometer using CuK_α radiation. The resistivity-temperature (R-T) curves of the SVO films were measured by a standard four-point probe technique over a temperature range from 77 K to room temperature. X-ray photoelectron spectroscopic (XPS) study of the SVO films was carried out to identify the valency state of Vanadium.

RESULTS

SrVO₃/LAO

SVO was grown on LAO substrate at deposition temperature range of 450°C - 750°C. The ambient oxygen pressure was kept between 10^{-6} and 10^{-2} Torr. Except those deposited at high vacuum of 10^{-6} Torr, all SVO films show poor crystallinity and high resistivity. So, in subsequent studies, SVO films were grown at 10^{-6} Torr ambient pressure. Fig.1 shows the X-ray θ - 2θ diffraction pattern for SVO/LAO at substrate temperature of 650°C. It is seen that only strong (h00) reflections were recorded. The X-ray ω-scan rocking curves of the (200)SVO reflection shows a full width at half maximum (FWHM) of about 0.9°. It indicates that the film is highly oriented. The X-ray 360° ϕ-scan on (220)SVO and (220)LAO planes as depicted in Fig.2 shows the characteristic four-fold symmetry. This suggests that the SVO film is cube-on-cube epitaxially grown on LAO.

The resistivity-temperature (R-T) curves of the SVO film fabricated at 700°C is shown in Fig.3. Apparently, the SVO film is of good metallic properties. Resistivity at room temperature of the as-grown films is 19 $\mu\Omega$-cm. It decreases with temperature to 3 $\mu\Omega$-cm at 77 K.

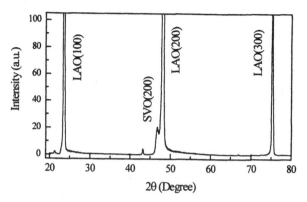

Fig.1   X-ray θ-2θ diffraction pattern of SVO/LAO with the SVO layer grown
under high vacuum and deposition temperature of 650°C.

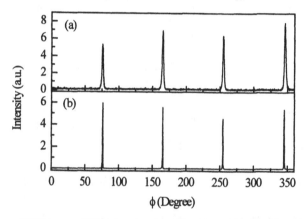

Fig. 2   360° φ-scan diffraction pattern of (a) (220)SVO, (b) (220)LAO.

SrVO₃/TiN/Si

Fig.4 shows the X-ray θ - 2θ diffraction patterns for the SVO/TiN/Si heterostructures
deposited at substrate temperatures of 750°C (a), 700°C (b), 650°C (c), 600°C (d), 550°C (e).
It is seen that highly oriented SVO films with strong (h00) reflections were prepared. The
crystallinity improves at higher deposition temperatures. The X-ray ω-scan rocking curves on
the (200) reflection of the SVO and TiN films deposited at 550°C give 1.7° and 1.6° at FWHM
respectively. Fig.5 shows the X-ray 360° φ-scan on the (220)SVO, (220)TiN and (220)Si
diffraction. Both the TiN and SVO films were deposited at 550°C. The characteristic four-fold
symmetry is observed and cube-on-cube epitaxial growth is suggested. The low temperature
processing of the SVO/TiN/Si heterostructure is important for developing integrated devices on
Si. If SVO is directly grown on Si, there is strong reaction between them. Previous study by

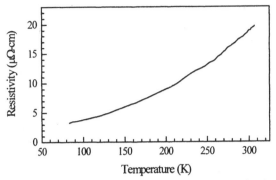

Fig.3   Resistivity against temperature of SVO on LAO. The deposition conditions
for SVO was at 700 °C and under high vacuum.

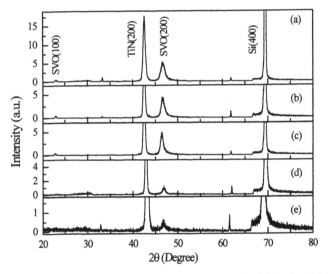

Fig. 4   X-ray θ-2θ diffraction pattern for SVO/TiN/Si with both of the SVO and
TiN layers grown under high vacuum and with deposition temperature of
TiN at 550 °C and that of SVO at (a) 750 °C   (b) 700 °C   (c) 650 °C   (d)
600 °C   (e) 550 °C

XPS on SVO grown on Si shows that the VO_2 is reduced by silicon to metallic vanadium [3].
In our present work, however, the TiN buffer layer not only provide a lattice matched site for
subsequent SVO film growth, but a good chemical diffusion barrier. Fig.6 shows the X-ray
photoelectron core level spectrum of V 2p and O 1s of as-grown film deposited at 550°C under
high vacuum. In a perovskite SVO structure, the vanadium should be tetravalent. The binding
energy of V^{4+} and V^{5+} are 516.3 and 517.6 ev respectively. Our XPS results suggest that our
SVO films are composed of both of V^{4+} and V^{5+}. The measured ratio of the V/Sr by XPS is
1.30/1.48, which is close to the nominal composition of 1/1.

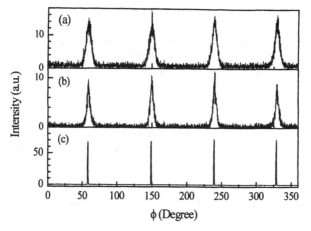

Fig.5   360° φ scan diffraction pattern of (a) (220)SVO, (b) (220)TiN, and (c) (220)Si.

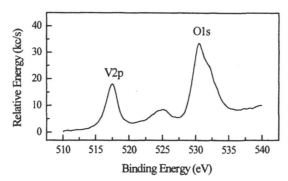

Fig.6   X-ray photoelectron core level spectrum of V 2p and O 1s of as-grown film deposited at 550°C under high vacuum.

CONCLUSIONS

We have prepared conductive films of SVO on LAO and TiN buffered Si by pulsed laser deposition method. Very high quality epitaxial SVO films grown on LAO substrates have been obtained and resistivities of 3 μΩ-cm and 19 μΩ-cm were recorded at 77 K and 300 K respectively. High quality heteroepitaxial SVO/TiN/Si structures have also been fabricated and a low processing temperature of 550°C has been demonstrated.

ACKNOWLEDGEMENTS

This work was supported by a Research Grant of the Hong Kong Polytechnic University under the Code No.G-V724. One of us (P.W. YIP) is grateful for the award of research studentship from the Hong Kong Polytechnic University.

REFERENCES

1. D.L. Ritums, N.J. Wu, X. Chen, D. Liu and A. Ignatiev, "Conducting and interfacial properties of epitaxial SVO films", Space Technology and Applications International Forum (1998), pp672-677.

2. Hirotoshi Nagata, Tadashi Tsukahara, Mamoru Yoshimoto and Hideomi Koinuma, "Laser molecular beam epitaxy of single-crystal $SrVO_{3-x}$ films", Thin Solid Films, 208 (1992), pp264-268.

3. B.K. Moon, E. Tokumitsu, H. Ishiwara, "Formation of high-dielectric oxide films on $SrVO_{3-x}Si$ substrates", Materials Science and Engineering B41 (1996), pp157-160.

Growth of ZnO/MgZnO Superlattice on Sapphire

J.F. Muth, C.W. Teng, A.K. Sharma[1], A. Kvit[1], R.M. Kolbas, J. Narayan[1]
Department of Electrical and Computer Engineering, North Carolina State University,
Raleigh, NC 27695
[1]Department of Materials Science and Engineering, North Carolina State University,
Raleigh, NC 27695-7916.

ABSTRACT

The optical and structural properties of ZnO/ MgZnO superlattices were
investigated by transmission electron microscope, transmission measurement and
photoluminescence. The uncoupled wells ranged in thickness from ~30 Å to 75 Å.
Modulation of the Mg content was observed in Z-contrast TEM indicating the alloy
composition was periodic. The density of stacking faults in the superlattice was extremely
high, however the photoluminescence in the narrowest well case was blue shifted, and
substantially brighter than comparable bulk layers of ZnO and MgZnO indicating that the
emission was enhanced. Excitonic features were observed in the optical absorption
spectra and also revealed that diffusion of Mg from the barrier layers into the well was
occurring.

INTRODUCTION

A great deal of research has been performed in wide band gap semiconductors
resulting in the commercialization of group III nitride blue lasers, blue and green light
emitting diodes and ultraviolet photodetectors for use in display optical data storage and
solar-blind detection applications.[1] As an alternative to the GaN material system ZnO
alloys are of substantial interest. There are many similarities between GaN and ZnO,
they are both wurtzite and have similar band gaps, both exhibit strong excitonic emission.
However the exciton binding energy is nearly 3 times as large in ZnO (~60 meV) which
makes excitonic effects even more pronounced.[2] As yet, p-type doping of ZnO is not
technologically feasible although some reports indicate that nitrogen may act as an
acceptor.[3]

We have recently been focusing on the growth of MgZnO alloys to investigate the
potential of bandgap engineering for the ZnO material system. While the equilibrium
solubility of Mg in ZnO is ~2 percent through pulsed laser deposition we have been able
to achieve metastable alloys with Mg concentrations of up to 36 percent.[4] The absorption
and photoluminesence spectra indicated that the exciton persists despite alloy broadening
at room temperature. These alloys have been shown to be thermally stable for
temperatures less than 700 °C, indicating that formation of stable heterojunction
interfaces should be practical.[5] A superlattice structure comprised of ZnO and $Mg_{0.2}Zn_{0.8}O$
has also been demonstrated by Ohtomo et al., indicating that ZnO alloy based quantum
structures should be feasible[6].

In this work we report on the growth of a MgZnO superlattice by PLD. The
superlattice was characterized by high-resolution transmission electron microscopy,

transmission measurements and photoluminescence. In optical transmission measurements, the excitonic features of the absorption were enhanced and slightly blue shifted. The photoluminescence from the sample was very bright and blue shifted from the corresponding ZnO band edge value. While several samples were examined optically, only one sample has been analyzed by TEM. In transmission electron microscopy, the z-contrast technique indicated that the Mg content was modulated according to the expected period of the superlattice. High-resolution transmission microscopy revealed numerous horizontal stacking faults. The interface between the MgZnO barriers and ZnO wells was poorly defined. The well thickness was also wider than expected, lessening the confinement which complicates the analysis. However, we found the study illuminating since it provides insight into the broadening mechanisms and growth issues that are expected to be important in the growth of quantum wells and heterostructures in this material system.

EXPERIMENTAL DETAILS

The MgZnO superlattice and bulk films in this study were deposited by pulsed laser deposition on c-plane double-side polished sapphire. Before deposition the sapphire was cleaned in an ultrasonic bath using acetone and methanol. The vacuum system was evacuated to ~5x 10^{-8} Torr and the substrate temperature was maintained at 650 °C during deposition. The low temperature was intended to minimize Mg diffusion. During growth a pulsed KrF excimer laser (λ=248 nm, pulse width=25 ns, and repetition rate=10 Hz) with laser energy densities in the range of 3-4 J/cm^2 was used to ablate MgZnO and ZnO sintered targets. The composition of the MgZnO target was the same as that used to grow $Mg_{36}Z_{64}O$ bulk layers. A 1000 Å buffer layer of ZnO was deposited first to promote a smooth growth mode. Then 10 alternating layers of ZnO, ranging from ~30-75 Å in thickness and MgZnO (~120 Å), were deposited. The thickness of the barrier layers was chosen to ensure that the wells would be uncoupled. The thickness of the well was controlled by deposition time, with the rate determined from the growth of thicker films.

RESULTS AND DISCUSSION

TEM of the ZnO/sapphire and MgZnO/sapphire interfaces showed that the first 100 Å of ZnO or MgZnO had a very high density of defects. High resolution TEM also indicated the formation of a spinel phase at the interface boundary in the case of the ZnO/sapphire interface. After the first 100 Å the defect density was greatly reduced and improved as the film thickness was increased. Threading dislocations were the dominant defect in MgZnO bulk layers as shown in Figure 1. Threading dislocations were observed to propagate from the interface to the surface. Some stacking faults were also observed in the bulk MgZnO layer.

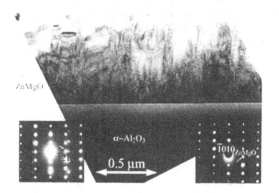

Figure 1. TEM of bulk MgZnO alloy. Threading dislocations dominate, while some stacking faults are visible in the upper left corner. The insets show the epitaxial relationship between the sapphire and MgZnO layer is maintained.

The superlattice structure is dominated by stalking faults as shown in Figure 2. Additionally, a higher density of stacking faults is found in the ZnO layer. The density of stacking faults made it very difficult to detect interfaces between ZnO wells and MgZnO barriers. However using Z-contrast technique it was found that the brightness was modulated periodically with the periodicity expected, indicating that the Mg concentration was not interdiffusing to the point of giving a homogeneous alloy in the superlattice region. The interfaces of the one sample examined were certainly not clearly defined, which make independent confirmation of the well thickness by TEM impractical.

The absorption spectra for three superlattices, ZnO, and $Mg_{36}Zn_{74}O$ at room temperature are shown in Figure 3. The relative absorbance is scaled such that the "A" exciton peak is the same for each sample. In the case of the 34 % Mg sample the thickness of the sample was such that the above band gap absorption was approximately the same as the other samples. In the ZnO sample the "A" and "B" excitons are very apparent, with the "A" exciton being very pronounced. As the deposition time of the ZnO well is decreased from 18 to 10 seconds (~75 to 30 Å in well thickness) the barrier layer composition decreases from ~29% to ~19 percent. The line width of the exciton resonance in the barrier layer also decreased as the well width was increased, indicating that alloy broadening was increasing with the width of the wells.

The "A" exciton resonance of the ZnO well layers appears to be more sensitive to the broadening mechanisms than the "B" exciton. With the increased well widths the "B" exciton became much more pronounced at room temperature. The "C" exciton, while visible, but diminished in strength to the A and B excitons, in the ZnO bulk sample at 77 K was not apparent in any of the well layers comprising the superlattice at room temperature or 77 K. The absorption spectra consistently shifts to higher energies as the well width is decreased. The formation of low percentage MgZnO alloys would also be expected to blue shift the absorption spectra. Relatively little is known about this material system, and while the amount of strain and magnitude of piezoelectric effects are expected to be less than that of the GaN/AlGaN system they should not be discounted. The blue shift of the narrowest well width is approximately 3 times that of what can be conservatively estimated for the strain effect, and the number of defects present in the film should also relax the thin film.

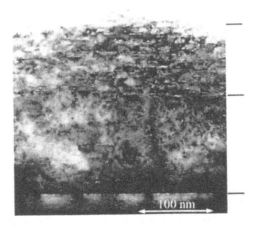

Figure 2. TEM image showing a large number of stacking faults in the superlattice region. A higher than normal number of stalking faults is also apparent in the ZnO buffer layer.

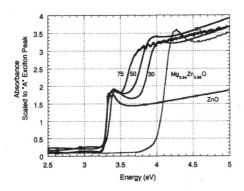

Figure 3. Absorbance of superlattice films, bulk MgZnO and bulk ZnO film scaled to "A" exciton absorbance. Note that the barrier layer composition decrease as the well deposition time increases. Curves labeled with their estimated well thicknesses in Å. The thickest well is very similar to bulk ZnO. The 50 and 30 Å wells are blue shifted.

The photoluminescence (PL) spectra of ZnO and ZnO/MgZnO superlattice films, at room temperature, excited by the 270 line of an Ar^+ ion laser is shown in Figure 4. The luminescence of bulk ZnO is usually very bright. The relative intensity of the PL of the superlattice films was substantially brighter, but was not quantitatively measured. The PL of bulk MgZnO and ZnO films is also usually red shifted with respect to the optical absorption edge. In the superlattices comprised of 30 Å and 50 Å wells the PL was blue shifted. In the PL of a superlattice of wells intended to be ~12 Å thick, only the barrier layer luminescence, and the ZnO emission from the buffer layer are observed. In this sample, we believe no enhancement of emission was obtained. In the superlattice of 50 Å wells, the buffer layer is visible as a shoulder on the long wavelength side. Enhanced emission at about 362 nm and the barrier layer PL is apparent at ~340 nm. The 30 Å well, with emission at ~360 nm, has a line width comparable to that of the bulk ZnO film. The buffer layer emission was not apparent. To explain this PL, we conjecture that while clearly defined interfaces are not visible in TEM, and numerous stacking faults are present, confinement effects are still present to enhance the emission. A counter argument is that the emission is from lower concentrations of MgZnO formed by diffusion of the Mg from the barriers into the well regions. However, the TEM indicates that the alloy formed in this manner is certainly not homogeneous. The Z contrast TEM indicates that spatial variation is on the order of sizes where quantum effects should start to have an influence. In this case the wells are not simple square wells, but confinement effects could still provide enhancement.

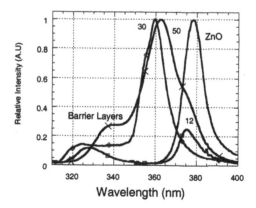

Figure 4. Intensity of PL spectra for MgZnO superlattices. The photoluminescence of the superlattices and a bulk ZnO layer are scaled to 1 for clarity. The actually efficiency of the 30 Å and 50 Å superlattices was actually significantly brighter than that of the bulk ZnO. The PL of the superlattice of ~12 Å wells was diminished in comparison and is shown proportionally.

In conclusion, the optical and structural properties of ZnO/ MgZnO superlattices were investigated by transmission electron microscope, transmission measurement and photoluminescence. The uncoupled wells ranged in thickness from ~30 Å to 75 Å. Modulation of the Mg content was observed in Z-contrast TEM indicating the alloy composition was periodic. The density of stacking faults in the superlattice was extremely high, however the photoluminescence in the narrowest well case was blue shifted, and substantially brighter than comparable bulk layers of ZnO and MgZnO indicating that the emission was enhanced. Excitonic features were observed in the optical absorption spectra and also revealed that diffusion of Mg from the barrier layers into the well was occurring.

REFERENCES

1. S.J. Pearton, J.C. Zolper, R.J. Shul, and F.Ren, J. Appl. Phys **86**, 1 (1999)
2. J.F. Muth, R.M. Kolbas, A.K. Sharms, S. Oktyabrsky, and J. Narayan J. Appl. Phys. **85**, 7884, (1999)
3. J. Mathew, H. Tabata, T. Kawai, Jpn. J. App. Phys. Part 2 **38**, L1205 (1999)
4. C.W. Teng, J.F. Muth, Ü. Ögür, M.J. Bergmann, H.O. Everitt, A.K. Sharma, C.Jin, and J. Narayan, Appl. Phys. Lett. **76**, 979 (2000)
5. A. Ohtomo, R. Shiroki, I. Ohkubo, H. Koinuma, M. Kawasaki, Appl. Phys. Lett. **75**, 4088 (1999)
6. A. Ohtomo, M. Kawasaki, I. Ohkubo, H. Koinuma, Appl. Phys. Lett. **75**, 980 (1999)

SINGLE QUANTUM WELL HETEROSTRUCTURES OF MgZnO/ZnO/MgZnO ON C-PLANE SAPPHIRE

S. Choopun, D. M. Chalk, W. Yang, R. D. Vispute, S. B. Ogale, R. P. Sharma, and T. Venkatesan
CSR, Department of Physics, University of Maryland, College Park, Maryland 20742

ABSTRACT

The single quantum well heterostructures of MgZnO/ZnO/MgZnO were grown on c-plane sapphire substrate by pulsed laser deposition. The well width was varied from 10 nm to 40 nm by controlling the deposition rate via number of laser pulsed on ZnO target. Using photoluminescence spectroscopy, we have observed a blue shift with respect to a thick ZnO reference sample when the well width was decreased. These results were fitted with calculations based on the simple square well model using the appropriate electron and holes effective masses. The quantized-energy and band offset as a function of well width, growth conditions, interface roughness, and possible quantum size effects on the quantum wells are discussed.

INTRODUCTION

Over the past several years, gallium nitride (GaN) has been the topic of intense research due to its potential for application in several opto-electronic devices. Indeed, GaN blue light emitting diodes (LEDs) and laser diodes (LDs) have been realized using many different designs including double heterostructures, single quantum wells (SQW), and multiple quantum wells (MQW) [1]. However, due to considerably high dislocation densities in the GaN films and a short lifetime of the laser diodes, a search for new and stable wide band gap materials is necessary. Recently, zinc oxide (ZnO) has drawn interest as a potential alternative to GaN for use in various UV and blue light emitting devices [2,3]. ZnO is a direct band gap semiconductor that crystallizes in the wurtzitic, hexagonal close packed (HCP) structure. It has an energy gap of 3.3eV at room temperature; the band gap can be narrowed or widened by alloying the ZnO films with either CdO or MgO, respectively. Moreover, optically pumped lasing in ZnO films has been successfully demonstrated at room temperature [4,5]. Quantum well structures result in enhanced light emission efficiency due to the carrier confinement effect in the active layer sandwiched between larger band gap barrier layers.

ZnO/$Mg_{0.2}Zn0_{0.8}$O superlattice structures have been studied using the photoluminescence (PL) and photoluminescence excitation (PLE) spectroscopy [6]. A blue shift of the PL and PLE spectra has been observed indicating a quantum-size effect. However, the superlattice structures are complicated for fitting calculation. SQW structures are desirable for fundamental studies and calculation.

Here, we report for the first time, MgZnO/ZnO/MgZnO single quantum well heterostructures grown on c-plane sapphire substrate by pulsed laser deposition (PLD). PL and PLE results showed the evidence of quantum confinement effects. The results are compared with predictions based on the simple square well model.

EXPERIMENT

A KrF pulsed excimer laser was used for ablation of the ZnO and $Mg_{0.1}Zn_{0.9}$O targets. The heterostructures were deposited onto c-plane (0001) sapphire substrates at 750°C in an oxygen background pressure of 1 X 10^{-4} Torr [2,3]. The SQW heterostructures consisted of a narrow band gap thin layer of ZnO (3.3 eV) sandwiched epitaxially between two wider band gap Mg_xZn_{1-x}O (3.6 eV) layers. The Mg content of the targets used in this study was fixed at 0.1 based on the quality of the films from XRD and RBS/ion channeling results. The Mg content of Mg_xZn_{1-x}O films grown by PLD has been previously reported [7,8] to be 2.5 times larger than that of the targets due to differences in the vapor pressure of ZnO and MgO at high temperature. The thickness of ZnO layer was varied from 10, 20 to 40 nm, where as thickness of the bottom MgZnO layer was 200 nm for all experiments, and the top layer was fixed at 100 nm. The film

Mat. Res. Soc. Symp. Proc. Vol. 623 © 2000 Materials Research Society

thickness was controlled by varying the deposition time calibrated from the average growth rate per pulse of the single film layer on sapphire.

Photoluminescence (PL) measurements were performed with both CW and pulsed lasers at room temperature and at liquid nitrogen temperature (77 K). A He-Cd laser (325 nm) laser and N-gas (337 nm) were used as the CW and pulsed excitation source, respectively. The output signal was collected from the edge-emitting configuration.

RESULTS AND DISCUSSION

The 77 K CW PL spectra of the SQW samples with 10, 20 and 40 nm quantum well thickness are shown in Fig 1. For reference sample, PL from 500-nm thick ZnO film on sapphire is also shown in Fig. 1. This reference sample has a peak position at 369 nm corresponding to 3.36 eV. This emission peak is attributed to radiative recombination of bound excitons. The spectrum also shows the two-shoulder peak at a higher wavelength position that is attributed to donor-acceptor pair transitions at 374 nm with phonon replicas at 384 nm. The peak position at around 340 nm represents the emission from MgZnO barrier layer. The variation of the peak position from 339 to 342 nm could be caused by variation of Mg content in the layer due to small interdiffusion during the growth process. The more the Mg content the lower the wavelength emission (higher energy). The weak peaks between 340 nm and 370 nm belong to the emission from the ZnO active layer in the SQWs. These peaks can be clearly observed in the pulsed PL

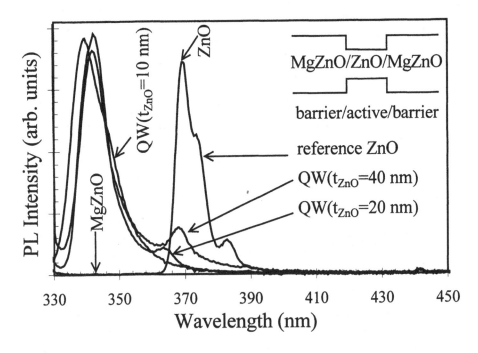

Fig. 1 CW PL spectra of the SQW samples with 10, 20 and 40 nm quantum well (ZnO) thickness at 77 K. The PL of 500 nm thick ZnO films is also shown as a reference sample.

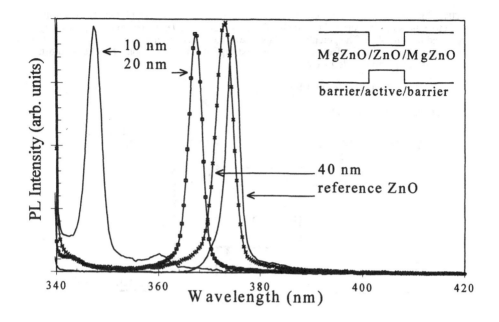

Fig. 2 The pulsed PL spectra of the SQW sample with 10, 20 and 40 nm quantum well (ZnO) thickness at 77 K. The PL of 500 nm thick ZnO films is also shown for reference.

spectra in Fig. 2. The peak position for a SQW sample decreases as the well thickness decreases with respect to a thick ZnO sample. We believe that this blue shift of the peak position relative to ZnO films is due to a quantum size effect.

The peak position and corresponding energy from CW PL is summarized in Table I. These data were fitted with the calculation based on the simple square well model.

Table I Summary of the peak position and corresponding energy from CW PL spectra.

L_{exp} (nm)	Peak position		Peak position		$\Delta E_c + \Delta E_v$ (eV)
	λ (nm)	Eg (eV)	λ (nm)	Eg (eV)	
10	339.63	3.6510	346.90	3.5745	0.2175
20	342.16	3.6240	363.45	3.4117	0.0548
40	341.80	3.6279	368.13	3.3684	0.0114
500	369.38	3.3570			

Our model assumes a finite square well potential for both conduction and valence bands, and that transitions occur only between the first quantum state of a bound electron and a bound hole. The calculation parameters are defined as shown in Fig. 3. E_{PL}, E_g(MgZnO) and E_g(ZnO) are obtained from the continuous PL spectra as shown in Table I. L_{exp} is the thickness of the active ZnO layer. ($\Delta E_C + \Delta E_V$) is derived from the difference of E_{PL} and $E_g(ZnO)$. The value of effective mass of electron (0.24) and hole (0.59) in ZnO has been taken from the previously reported literature[9]. The potential well (V) can be obtained from the band offset. However, in our knowledge the band offset between ZnO and MgZnO is not available. So, the band offset is treated as one parameter defined as

$$V = C[E_g(MgZnO)-E_g(ZnO)] \qquad (1)$$

where C has a value from 0 to 1.

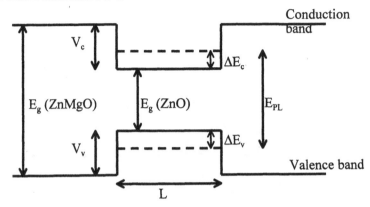

Fig. 3 Schematic diagram of the calculation parameters.

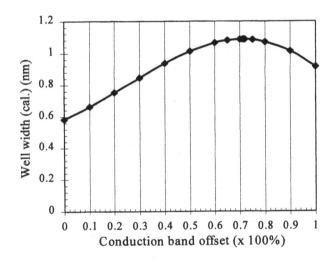

Fig 4 The calculated well width as a function of conduction band offset for 10 nm thick quantum well sample.

By assuming the well widths of the conduction and valence bands to be equal, the well width was calculated as a function of conduction band offset as plotted in Fig. 4 for 10 nm thick quantum well sample. The calculated well width varied between 0.58-1.08 nm as the band offset increased from 0 to1. Surprisingly, the well width varied in the narrow range as a function of band offset in each SQW sample. For comparison, the experimental and calculated well width are summarized in Table II. The calculated values are about one order of magnitude less than that of experimental values. A possible explanation for this difference is that Mg content in MgZnO barrier layer diffused into the well layer making the well layer thinner than expected. It has been reported [10] that the thermodynamic solubility limit of Mg in MgZnO films is actually 15 mole%. By comparing the band gap value to the previous report [7], the actual Mg content in our case should be around 19 mole. This value exceeds the thermodynamic solubility limit and the diffusion should occur during the growth process at 750 °C. To avoid the diffusion, a barrier layer with Mg content less than 15 mole% would be ideal for this experiment. The other possible explanations are the over-simplification of the model, interface roughness of ZnO films, band bending, strain effect or quantum confined stark effect.

Table II Summary of the experimental and calculated well width.

L_{exp} (nm)	L_{cal} (nm)
10	0.58-1.08
20	2.39-4.44
40	6.49-12.07

CONCLUSIONS

We have successfully grown the single quantum well heterostructures based on ZnO and Mg ZnO alloys. A quantum size effect has been demonstrated from a blue shift of PL peak position with respect to a thick ZnO film on sapphire. By using the simple square well model, the well width as a function of band offset has been calculated. The calculated well width does not agree well with calibrated thickness of ZnO active layer presumably due to the interdiffusion of Mg.

ACKNOWLEDGEMENTS

This work was supported by NSF-UMD MRSEC seed Grant No. DMR 9632521. The authors gratefully acknowledge the support of NSF through a contract supervised by Dr. Khosla. The authors would like to thank Dr. A. Iliadis for discussions. One of the author, S. Choopun was supported by DPST scholarship, Thai government.

REFERENCES

[1] S. Nakamura and G. Fasol, (The Blue Laser Diode, Springer, Berlin, 1997).
[2] S. Choopun, R. D. Vispute, W. Noch, A. Balsamo, R. P. Sharma, T. Venkatesan, A. Illiadis and D. C. Look, Appl. Phys. Lett. 75, 3947 (1999).
[3] R. D. Vispute, S. Choopun, Y. H. Li, D. M. Chalk, S. B. Ogale, R. P. Sharma, T. Venkatesan and A. Iliadis, MRS proceeding 1999-2000 (in press).
[4] M. Kawasaki, A. Ohtomo, I. Ohkubo, H. Koinuma, Z. K. Tang, P. Yu, G.K.L. Wang, B. P. Zhang, and Y. Segawa, Mater. Sci. Eng B56, 239 (1998).
[5] D. M. Bagnall, Y. F. Chen, Z. Zhu, T. Yao, S. Koyama, M. Y. Shen, and T. Goto, Appl. Phys. Lett. 70, 2230 (1997).
[6] A. Ohtomo, M. Kawasaki, I. Ohkubo, H. Koinuma, T. Yasuda and Y. Segawa, Appl. Phys. Lett. 75, 980 (1999).

[7] A. Ohtomo, M. Kawasaki, T. Koida, K. Masubuchi, H. Koinuma, Y. Sakurai, Y. Yoshida, T. Yasuda, and Y. Segawa, Appl. Phys. Lett. **72**, 2466 (1998).

[8] A. K. Sharma, J. Narayan, J. F. Muth, C. W. Teng, C. Jin, A. Kvit, R. M. Kolbas, and O. W. Holland, Appl. Phys. Lett. **75**, 3327 (1999).

[9] L. I. Berger, (*Semiconductor Materials*, CRC press, Florida, 1997), p.184.

[10] A. Ohtomo, R. Shiroki, I. Ohkubo, H. Koinuma, and M. Kawasaki, Appl. Phys. Lett. **75**, 4088 (1999).

EVALUATION OF LSCO ELECTRODES FOR SENSOR PROTECTION DEVICES

R. W. Schwartz and M. T. Sebastian
The Gilbert C. Robinson Department of Ceramic and Materials Engineering
Clemson University
Clemson, SC

M. V. Raymond
DigitalDNA™ Laboratories, Motorola Inc.
Austin, TX

ABSTRACT

We have evaluated lanthanum strontium cobalt oxide ($La_{0.50}Sr_{0.50}CoO_x$; LSCO 50/50) as a candidate "transparent" electrode for use in an electrostatic shutter-based infrared sensor protection device. The device requires that the electrode be transparent (80% transmission) and have moderate sheet resistance (300 – 500 Ω/sq.). To meet these needs, the effects of post-deposition annealing on the resistivity and optical absorption characteristics of sputter deposited LSCO thin films were studied. The as-deposited films were characterized by an absorption coefficient of ~ 12,500 cm^{-1} and resistivities of ~ 0.08 to 0.5 Ω-cm. With annealing at 800°C, the resistivity decreased to 350 $\mu\Omega$-cm, while the absorption coefficient increased to ~ 155,000 cm^{-1}. By using a post-deposition annealing step at 800°C and controlling film thickness, it appears that a standard LSCO 50/50 material may possess the requisite conductivity and optical transmission properties for this sensor protection device.

INTRODUCTION

"Optical" information, whether obtained by electronic sensors or the vision of an individual soldier or aircraft pilot, is becoming of greater importance in a range of battlefield management scenarios. Because of the increased reliance on optical information, protection of sensor systems is, therefore, also becoming ever more important. At the same time, the threat of damage to these sensors by antagonistic forces is also increasing. While the U.S. and other governments have agreed to prohibit the use of weapons that are designed to cause blinding, the use of tunable lasers by terrorist organizations still poses a significant threat [1]. Thus, devices that provide optical limiting and serve to protect sensor systems are of great importance to the U.S. Military because of the increasing threat of optical "warfare" and the critical need for battlefield intelligence information that is obtained by human sources and electronic sensors.

One sensor protection approach is an electrostatic shutter [2] that can be operated at frequencies approaching 100 kHz. A schematic of this shutter, which is currently under development at MCNC [3], is illustrated in Fig. 1. The device is being designed for utilization with sensors that operate in the 3 –5 and 8 – 12 μm bands. In the "on" state, the shutter is open (curled) and the underlying sensor device is utilized to gather information. When an optical threat is sensed, through an associated photodiode or similar device, a signal is sent to close the shutter, thus protecting the sensor. The shutter is actuated via an electrostatic mechanism.

For the operation of the device, stringent material requirements are placed on the shutter itself as well as the supporting substrate, electrode, and insulating layers. In general, the substrate, electrode and insulator should be as transparent as possible to limit the reduction in the sensitivity of the device, but the electrode must also possess an adequate conductivity to close

365

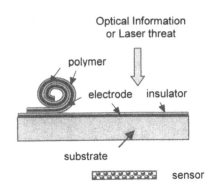

Optical Information
or Laser threat

polymer

electrode   insulator

substrate

sensor

Fig. 1. Electrostatic shutter for the protection of optical sensor systems [2,3].

the shutter electrostatically. Target goals for the transparency and resistance of the electrode are 80% transmission and a sheet resistance of 300 to 500 Ω/sq.

The focus of the present research is to evaluate lanthanum strontium cobalt oxide (LSCO) for use as the lower electrode. While a variety of materials may be considered for this layer, the moderate resistance and high transparency demanded place fairly stringent requirements on material performance. Since LSCO demonstrates a "free carrier" conduction mechanism, it is expected that increases in the conductivity of the material will be accompanied by a decrease in transparency. For the design of new materials, the optimal approach is to achieve high conductivity through enhancement of the carrier mobility, rather than an increase in carrier concentration. This method improves conductivity without degrading the transmission characteristics of the material as extensively. However, due to previous experience with LSCO, and preliminary results that indicated that resistivity was strongly dependent on post-deposition annealing conditions [4], we have evaluated the suitability of LSCO for this application.

In this paper, we report on the effects of post deposition annealing on the resistivity and IR transmission characteristics of sputter deposited LSCO. A composition of ($La_{0.50}Sr_{0.50}CoO_x$) was utilized since it was previously reported to have the lowest resistivity [5-8]. We have attempted to meet the device requirements of 80% transmission and 300 Ω/sq. sheet resistance, which may be considered as engineering parameters, through control of intrinsic material properties (extinction coefficient and resistivity) and tailoring of film thickness. In the present study, these properties are manipulated through control of crystalline quality via post-deposition annealing at different temperatures. To maximize the transparency of the electrode layer, it may be anticipated that thin layers will be required. Therefore, we also discuss our preliminary results on the effects of film thickness on the resistivity of the LSCO layers.

EXPERIMENTAL

The LSCO films were deposited by rf-magnetron sputtering with two different systems: (i) a Unifilm Technology, Inc. PVD-300 with a 3" diameter target that utilizes substrate rotation and scanning (lateral motion of the substrate under the source) to provide a high degree of film uniformity [4]; and (ii) a fairly standard Kurt J. Lesker system with a 3" sputter gun. Films were deposited onto (100) $LaAlO_3$, BaF_2, and (100) MgO, and thickness was varied from 15 to 150 nm through control of deposition time. Both systems were evacuated to a background pressure in the range of 10^{-7} torr prior to deposition. Table I provides a summary of the deposition conditions utilized. Following deposition, the films were annealed in air at temperatures ranging from 300 to 850°C for times of either 30 or 60 minutes.

Table I. Experimental conditions for the sputter deposition of LSCO 50/50 thin films.

Deposition Parameter	Unifilm Technology, Inc. PVD-300 [4]	Lesker Sputtering System
O_2:Ar sputter gas ratio	0:100 – 50:50 (sccm)	0:100 – 50:50 (sccm)
Deposition Pressure	10 mtorr	10 – 40 mtorr
RF power density	1.5 w/cm^2	4.4 w/cm^2
Target to substrate distance	3.5 cm	3 – 5 cm
Substrate scanning mode	97% uniformity; 4" diameter	NA
Deposition temperature	Ambient	Ambient

Resistivities of the films were characterized using a standard 4-point probe resistance measurement technique. Infrared transmission properties were studied with a Nicolet Magna550 spectrometer and subtracting the substrate transmission characteristics as the background. Thicknesses of the films were determined either by cross-sectional SEM method or by using the SEM results to calculate the sputtering rate, and then specifying a particular sputter deposition time. Absorption coefficients of the films were estimated using a simple Beer's Law approximation; reflection losses were neglected. While the estimate is not as accurate as actual measurements of refractive index and absorption coefficient, it at least gives a general feeling regarding the absorbtivity of the films. Film composition was not investigated in this study, but previous studies have shown that the composition of the films may vary during deposition [4].

RESULTS AND DISCUSSION

Fig. 2 shows the effects of post-deposition annealing on the resistivity of the LSCO films. The films prepared previously by Raymond [4] were approximately 150 nm in thickness while those shown for the current study were 50 nm, except for the 850°C film that is 145 nm thick. In the as-deposited state, the resistivity of the film deposited by Raymond [4] is approximately 0.3 Ω-cm. Similar resistivity values (0.08 to 0.5 Ω-cm) were obtained for the as-deposited films of the present study prepared with a different sputtering system. Atomic force microscopy investigations suggest that the as-deposited films are at least partially crystallized, although we have not verified this with x-ray diffraction at this time. With annealing at 800 – 850°C, the resistivity of both sets of films decreased by about 3 orders of magnitude to ~ 350 – 750 µΩ-cm. The lower value compares favorably with the best reported values for this material [5-8].

The resistivity of the films also varied systematically as a function of post-deposition annealing temperature. For the films prepared by Raymond, annealing at lower temperatures appears to initially increase the resistivity (note the measured value for the 400°C anneal), followed by a strong decrease in resistivity with annealing temperature. In contrast, while the results of the present study also suggest a strong relationship between annealing temperature and resistivity, no increase in resistivity is observed for lower annealing temperatures, at least for the 50 nm films. Further, the dependence of resistivity on annealing temperature does not appear to be as strong as for the samples prepared earlier by Raymond [4]. The results do indicate, however, that for both thicknesses (50 and 150 nm), it is possible to "tune" the resistivity of the films by the choice of the post-deposition annealing temperature.

Since conductivity is related to the concentration of free carriers, the variation in resistivity with annealing suggests that the optical properties of the films in the infrared region (which are also dependent on carrier concentration) should also strongly depend on the post-deposition annealing conditions. This relationship between transmission and annealing temperature is shown in Fig. 3. It may be seen that the optical transparency decreases concomitantly with the resistivity, shown previously in Fig. 2. The values reported in this figure are for a wavelength of

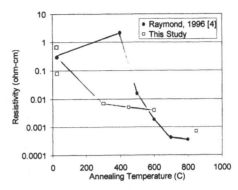

Fig. 2. Resistivity of LSCO electrodes as a function of post-deposition annealing conditions.

8.3 μm, which is just above the cut-off of the LaAlO₃ substrate. The percent transmission for the unannealed 150 nm film is ~ 82% while the corresponding film annealed at 800°C displays a transmission of only ~ 9%. This suggests that as the post- deposition annealing temperature is increased, the concentration of free carriers is increased, which promotes more intense absorption. The transmission results for the as-deposited film explain our interest in this material as a "transparent oxide conductor." In the unannealed state, the transmission is high but the resistivity is still relatively low (< 1 Ω-cm).

We have also calculated the effective absorption coefficients of the films as a function of annealing temperature using Beer's law and the results are shown in Fig. 4. In these calculations we have neglected reflection losses. As expected from the measured transmission characteristics of the films, an increase in annealing temperature causes a significant increase in the absorption coefficient of the film. Values of the absorption coefficient range from ~ 12,500 cm⁻¹ for the as-deposited film to ~ 155,000 cm⁻¹ for the film annealed at 800°C. While these values are only rough estimates they suggest that the optical absorption of the materials is not as strong a function of annealing temperature as resistivity. While the resistivity varied by nearly three orders of magnitude over the annealing temperature range studied, the absorption coefficients varied by slightly more than one order of magnitude. We also note that by studying the

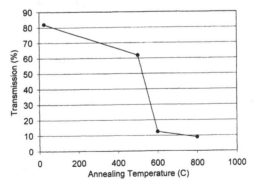

Fig. 3. Transmission of 150 nm thick LSCO 50/50 films at 8.3 μm as a function of post-deposition annealing temperature.

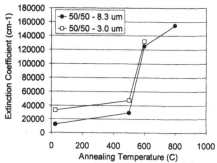

Fig. 4. Calculated absorption coefficients of LSCO 50/50 thin films as a function of annealing temperature.

transmission characteristics of films with different thicknesses, an estimate of the reproducibly of the calculated absorption coefficients is possible. Calculated absorption coefficients were within 5% for films that varied in thickness from 50 to 150 nm; i.e., similar absorption coefficients were obtained for the 50 nm films prepared on BaF_2 and MgO as for the 150 nm films prepared on $LaAlO_3$. These results imply that for the annealing temperatures investigated, no chemical interaction occurred between the BaF_2 or MgO substrates and the LSCO films.

Based on these results we also speculate that there is a well-defined relationship between resistivity and absorption coefficient in LSCO, although we have not studied this relationship in detail. The measured absorption coefficients at 3.0 μm are apparently higher but we attribute this difference to an increase in reflection losses at this wavelength which are not accounted for in our simple Beer's law calculation. The increase in reflection is expected due to the higher refractive index at shorter wavelength.

The absorption coefficient and resistivity data were then used to determine the suitability of the LSCO films annealed under different conditions for use in the electrostatic shutter application. The results of these calculations are presented in Table II, in terms of the film maximum thickness that may be used while still retaining 80% transmission, and the minimum film thickness required to obtain a sheet resistance of 300 or 500 Ω/sq. It may be seen in the table that the best opportunity to use LSCO for this application occurs for the highest annealing temperature where crystalline perfection would be expected to be greatest. For LSCO annealed at 800°C, to obtain a sheet resistance of 500 Ω/sq., the film needs to be at least 5 nm thick, while to obtain the desired transparency, the film must be thinner than 14 nm.

Hence, following this preliminary study, LSCO still appears to be a viable candidate electrode for this application. However, this evaluation neglects the fact that as film thickness is decreased, it may no longer be possible to retain bulk material property values. To begin to investigate the effect of thickness on resistivity, films with thicknesses ranging from 15 to 145 nm were prepared, subjected to different annealing conditions, and characterized. The results of

Table II. Opportunities for engineering the resistance and transparency properties of LCSO through the control of annealing conditions.

Post-Deposition Annealing Temp. (°C)	Maximum Thickness for 80% Trans. (nm)	Minimum Thickness for 500 Ω/sq. (nm)	Minimum Thickness for 300 Ω/sq. (nm)
As-Deposited	180	3400	5670
500	74	210	350
600	16	26	43
800	14	5	9

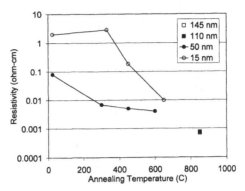

Fig. 5. Effect of film thickness of LSCO 50/50 on the measured resistivity as a function of annealing temperature.

these preliminary studies are shown in Fig. 5. The resistivity of the films seems to be more dependent on thickness at the lower annealing temperatures, while at annealing temperatures above 650°C, this dependence seems less pronounced. However, further studies are still required for films 8 to 15 nm in thickness that are annealed at 800 to 850°C for more conclusive evidence regarding this observation.

CONCLUSIONS

We have studied the effects of post-deposition annealing on the resistivity and infrared transparency of LSCO materials. Higher annealing temperatures resulted in a three order of magnitude decrease in resistivity from ~ 0.3 Ω-cm (as-deposited) to 350 $\mu\Omega$-cm (800°C). Concomitant with the decrease in resistivity was an increase in the optical absorption throughout the infrared spectral region. The calculated absorption coefficients increased by approximately one order of magnitude for the same change in annealing conditions. Despite this increase in absorption, LSCO appears to be a viable electrode for this device, primarily due to the moderate sheet resistance that is required for this application.

ACKNOWLEDGMENTS

This work was supported by the U.S. Defense Advanced Research Projects Agency (DARPA) under contract DAAD19-99-1-003. A portion of this work was also carried out at Sandia National Laboratories and was supported by the U.S. Department of Energy under contract DE-AC-04-94ALAL85000.

REFERENCES

1. J. S. Shirk, Optics & Photonics News, April (2000), p. 19.
2. C. G. Kalt, U.S. Patent 3,989,357 (1975).
3. S. Goodwin-Johansson and G. E. McGuire, private communication (2000).
4. M. V. Raymond, et al., Mat. Res. Soc. Symp. Proc., **433**, 145 (1996).
5. G. H. Jonker and J. H. Van Santen, Physica, **15**, 120 (1953).
6. H. Ohbayashi, et al., Jpn. J. Appl. Phys., **13**, 1 (1974).
7. J. T. Cheung, et al., Appl. Phys. Lett., **62**, 2045 (1993).
8. S. Madhukar, et al., J. Appl. Phys., **81**, 3543 (1997).

VOLATILE, FLUORINE-FREE β-KETOIMINATE PRECURSORS FOR MOCVD GROWTH OF LANTHANIDE OXIDE THIN FILMS

N.L. EDLEMAN, J.A. BELOT, J.R. BABCOCK, A.W. METZ, M.V. METZ, C.L. STERN, T.J. MARKS*
Department of Chemistry and the Materials Research Center, Northwestern University, 2145 Sheridan Rd., Evanston, IL 60208-3113, *tjmarks@casbah.acns.nwu.edu

ABSTRACT

Lanthanide oxide thin films are of increasing scientific and technological interest to the materials science community. A new class of fluorine-free, volatile, low-melting lanthanide precursors for the metal-organic chemical vapor deposition (MOCVD) of these films has been developed. Initial results from a full synthetic study of these lanthanide-organic complexes are detailed.

INTRODUCTION

Lanthanide-containing films are of great current interest to the materials science community [1]. Applications include device buffer layers (e.g., CeO_2) [2-4], superconductors (e.g., $LnBa_2Cu_3O_{7-\delta}$) [5], doped phosphors [6-8], and solid-state fuel cells [9]. MOCVD offers many attractions for oxide film growth, including lower equipment cost, easier scale-up, conformal deposition on a variety of complex substrates, lower growth temperatures, and faster growth rates [10]. However, the success of MOCVD depends on the availability of volatile, thermally stable precursors which exhibit constant vapor pressure and the ability to selectively form the desired film composition at the substrate surface[11,12].

The design and realization of new molecular MOCVD precursors offers a considerable synthetic challenge. For many applications, the metal ion of interest is of large size and therefore has many coordination sites which require passivation. Large ion size is often further complicated by a low ionic charge, demanding the use of multidentate anionic and/or donating ligands in order to ensure a monomeric metal complex [13,14]. In addition, the use of metal-carbon bonded ligand frameworks (e.g., a substituted cyclopentadienyl ring [15]) or fluorinated substituents [9,16] often leads to impurity incorporation in films and compromises desired film properties.

No commercially-available lanthanide precursor satisfies all of the above criteria. A common type employs non-fluorinated β-diketonate ligands **1**. Metal-organic complexes based on **1** suffer falling volatility over film growth times due to oligomerization, thermal decomposition, and sintering [9, 16-18]. In addition, these sources frequently require undesirably high growth temperatures [16, 19, 20]. The temperature of $Ln(\mathbf{1})_{3 \text{ or } 4}$ precursors often must be ramped in order to provide a constant precursor vapor pressure in the reactor [21]. β-diketonate precursor volatility and thermal stability have been improved by the use of fluorinated substituents **2**, but formation of fluoride phases in the resulting films is a problem, and F^- species corrode certain (i.e., metal) substrates and reactor components [22].

The goal of our earlier research was to design a multidentate, non-fluorinated ligand system that would address the general precursor considerations discussed above *and* could successfully be applied to the entire lanthanide series, with trivalent ionic radii ranging from 1.17-1.00 Å (La^{3+}-Lu^{3+}). In addition, a precursor that is a *liquid* at typical reactor operating temperatures would avoid the sintering problem commonly encountered with the

371

O   O
  ‖ ‖
R — R

1

acac: R = Me
dpm: R = ^{t}Bu

O   O
  ‖ ‖
R_1 — R_2

2

hfa: $R_1 = R_2 = CF_3$
fod: $R_1 = Me$, $R_2 = CF_2CF_2CF_3$

R_3
|
O
|
O   N
  ‖ ‖
R_1 — R_2

3

3a: $R_1 = {}^tBu$, $R_2 = R_3 = Me$
3b: $R_1 = {}^tBu$, $R_2 = Et$, $R_3 = Me$
3c: $R_1 = R_2 = Me$, $R_3 = Et$
3d: $R_1 = {}^tBu$, $R_2 = Et$, $R_3 = Et$
3e: $R_1 = {}^tBu$, $R_2 = Me$, $R_3 = {}^iPr$

abovementioned Ln(**1**)$_{3 \text{ or } 4}$ and Ln(**2**)$_{3 \text{ or } 4}$ sources. Therefore, the synthesis and characterization of low-melting, non-fluorinated lanthanide complexes bearing homoleptic multidentate β-ketoiminate ligands **3** has been a central goal [23]. This ligand was originally explored by our laboratory for coordinative saturation of alkaline earth ions [24]. The MOCVD growth of high-quality, epitaxial CeO$_2$ films using one of the new cerium precursors was carried out, demonstrating the advantages of this new precursor class [25]. This film growth was achieved at lower substrate temperatures than possible with other sources. The aim of the present study was to elaborate and investigate the effect on precursor melting point and volatility characteristics of various alkyl substitution at the R_1, R_2, and R_3 sites of ligand **3**. The new complexes are directly compared to those previously communicated [23], Ln(**3a**)$_3$, where Ln = Ce^{3+} (**4**), Nd^{3+} (**5**), and Er^{3+} (**6**).

EXPERIMENTAL

Tris[2,2-dimethyl-5-N-(2-methoxyethyl-imino)-3-heptanato] lanthanide complexes, Ln(3b)$_3$, Ln = Ce (7), Er (8): In a N$_2$-filled glovebox, a 2-neck, 100-mL round bottom flask fitted with a rubber septum and reflux condenser is charged with Ln[N(SiMe$_3$)$_2$]$_3$ [26] (2.4 mmol). The reaction vessel is then removed from the glovebox, interfaced to a Schlenk line, and 24 mL xylenes (dried and distilled from molten Na) added via a syringe. The reaction flask is then immersed in a pre-heated oil bath (~60°C) and the reaction mixture is stirred until homogeneous. Immediately, ligand **3b** (7.9 mmol, 3.3 eq.; distilled and stored over molecular sieves in an N$_2$-filled storage tube) is injected into the reaction flask, and the solution refluxed overnight (14 h). The volatiles are then removed in vacuo (1 Torr), and the resulting waxy solid is recrystallized from pentane at -30°C to afford a crystalline product.

Complex **7**: m.p. 99-101°C; ^{1}H NMR (C$_6$D$_6$): δ -10.3, -8.2, -2.0, 1.2, 2.1, 5.1, 12.9; Elemental analysis for C$_{36}$H$_{66}$O$_6$N$_3$Ce (%): Calcd: C, 55.65; H, 8.56; N, 5.41; Found: C, 55.57; H, 8.74; N, 5.52.

Complex **8**: m.p. 81-82°C; ^{1}H NMR (C$_6$D$_6$): δ -81.0, -49.4, -16.7, -2.7, 1.8, 35.6; Elemental analysis for C$_{36}$H$_{66}$O$_6$N$_3$Er (%): Calcd: C, 53.77; H, 8.27; N, 5.23; Found: C, 53.52; H, 8.46; N, 5.27.

Tris[4-N-(2-ethoxyethyl-imino)-3-pentanato] lanthanide complexes, Ce(3c)$_3$, 9: In a N$_2$-filled glovebox, a 2-neck, 100-mL round bottom flask fitted with a rubber septum and reflux condenser is charged with 1.93g Ce[N(SiMe$_3$)$_2$]$_3$ (3.4 mmol). The reaction flask is then removed from the glovebox, interfaced to a Schlenk line, and 34 mL xylenes (dried and distilled from molten Na) added via syringe. The reaction flask is then immersed in a pre-heated oil bath (~60°C) and the solution stirred until homogeneous. Immediately, ligand **3c** (11.3 mmol, 3.3 eq.; distilled and stored over molecular sieves in an N_2-filled storage tube) is injected and the solution refluxed overnight (14 h). The volatiles and excess ligand reagent are then removed in vacuo (1 Torr) to yield the spectroscopically pure, somewhat waxy complex **9**: m.p. 76-78°C; ^{1}H NMR (C_6D_6): δ -11.0, -7.8, -1.5, 1.6, 7.2, 12.3.

RESULTS

The Ln(**3**)$_3$ complexes are synthesized in a single-step amine elimination reaction by refluxing the corresponding lanthanide tris[bis(trimethylsilyl)amides] with an excess of free ligand in high-boiling xylenes overnight (Eq. 1). The desired species are readily isolated, since the only reaction byproduct is the volatile amine $HN(SiMe_3)_2$. The products are purified by recrystallization, sublimation, or by removing any excess ligand in vacuo to yield homoleptic, monomeric, colorful lanthanide complexes [Ce(**3**)$_3$, gold; Nd(**3**)$_3$, blue-green; Er(**3**)$_3$, pink]. The complexes are stable under an inert atmosphere. Melting points are found in the 60-100°C range, affording liquid, constant surface area, thermally stable precursors for MOCVD.

The various alkyl substituents were deliberately selected such that lanthanide precursor physical properties could be assessed with respect to substitution effects at a particular molecular site or sites. As listed in Table I, several melting point trends are noteworthy. First, the melting point decreases with ionic radius across the lanthanide series for a given ligand system (i.e., m.p. **4**, 95-98°C; **5**, 78-81°C; **6**, 65-68°C). Second, introduction of larger alkyl groups (e.g., substitution of ethyl for methyl) at the R_2 site results in an increase in melting point. (i.e., m.p. **6**, 65-68°C; **8**, 81-82°C). Additionally, the introduction of a ligand with only small substituents (i.e., **3c**) leads to a noticeable decrease in precursor melting point, as well as room temperature consistency more like an wax than a solid, as shown with compound **9**. Initial preparation of Er(**3d**)$_3$ and Ce(**3e**)$_3$ complexes reveals that more sterically encumbered substituents at the R_3 site lead to lower melting, more wax-like products than **8** or **4**, respectively.

Table I. Synthetic yield, melting point, and coordination number of new lanthanide MOCVD precursors

Complex	Yield (%)	m.p.(°C)	Coordination #
7, Ce(**3b**)3	49	99-101	N/A
4, Ce(**3a**)3	85	95-98	8
9, Ce(**3c**)3	60	76-78	N/A
5, Nd(**3a**)3	85	78-81	8
8, Er(**3b**)3	44	81-82	7
6, Er(**3a**)3	85	65-68	7

Single-crystal x-ray diffraction studies reveal a plausible explanation for the observed melting point characteristics. Generally, crystal structures of the 8-coordinate Ce^{3+} and Nd^{3+} complexes [i.e. **4**, **5**, and Nd(**3b**)$_3$] reveal that two of the three ligands are coordinated to the metal through all three possible donor sites; however, the third ligand is bidentate, with a non-coordinating ether moiety (Figure 1). Conversely, the solid-state structures of the Er^{3+} complexes **6** and **8** are 7-coordinate, with two dangling ether groups (Figure 1). This difference is a manifestation of the lanthanide contraction, with the smaller Er^{3+} ion being electrostatically satisfied at a lower coordination number. The crystal structures demonstrate that the present ligand is capable of completely saturating the Ln^{3+} environment, regardless of metal ion radius. It can be speculated that increased steric bulk at the R_3 site leads to lower precursor melting points due to decreased symmetry of the complex. Similarly, a greater R_1 to R_2 size difference (i.e., R_1 = ^tBu, R_2 = Me versus R_1 = ^tBu, R_2 = Et) also lowers the symmetry of the complex, giving a lower-melting precursor.

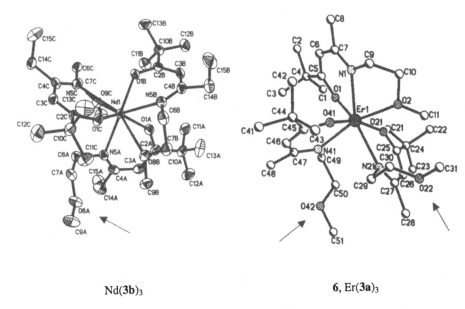

$$Nd(3b)_3 \qquad\qquad 6, Er(3a)_3$$

Figure 1. X-ray Crystal Structures of complexes Nd(**3b**)$_3$ and **6**; arrows indicate non-coordinated R_3 ether groups

The present lanthanide precursors were further characterized by vacuum thermogravimetric analysis (TGA) [27]. An Arrhenius plot (sublimation rate versus inverse temperature, Figure 2) for compounds **4** and **7** and commercially available Ce(dpm)$_4$ [28] allows direct comparison of the volatility characteristics of these precursors. Complexes **4** and **7** clearly have substantially higher vaporization rates and at lower temperatures than Ce(dpm)$_4$, a characteristic preferred for optimum film growth. In addition, the sublimation rates of the β-ketoiminate precursors increase sharply upon melting, reaffirming the advantages of liquid precursors. It can also be seen that the more highly substituted complex **7** has lower volatility than **4**.

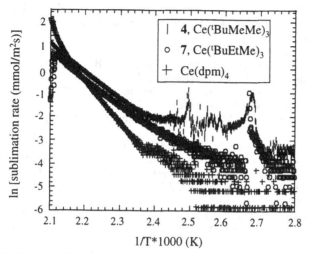

Figure 2. Vacuum TGA data for new and commercial cerium precursors

CONCLUSIONS

A new series of fluorine-free lanthanide MOCVD precursors has been prepared which exhibit improved volatility and lower melting points than commercial sources. The β-ketoiminate ligand is a versatile, multidentate system, producing saturated, monomeric lanthanide complexes proven to be excellent MOCVD precursors. Substitution of various alkyl moieties at the keto, imino, and ether sites of the β-ketoiminate ligand significantly affects the melting point and volatility characteristics of the complexes, facilitating the design of customized lanthanide precursors.

ACKNOWLEDGMENTS

This work was supported by the National Science Foundation (CHE-9807042). NLE wishes to acknowledge the support of a graduate fellowship through the Northwestern Materials Research Center (NSF-MRSEC, DMR-9632472).

REFERENCES

1. T.S. Lewkebandara, C.H. Winter, Chemtracts- Inorg. Chem. p. 271 (1994).
2. M. Suzuki, T. Ami, Mater. Sci. Eng. B **41**, p. 166 (1996).
3. T. Inoue, Y. Yamamoto, M. Satoh, A. Ide, S. Katsumata, Thin Solid Films **282**, p. 24 (1996).
4. X.D. Wu, R.C. Dye, R.E. Muenchausen, S.R. Foltyn, M. Maley, A.D. Rollett, A.R. Garcia, N.S. Nogar, Appl. Phys. Lett. **58**, p. 2165 (1991).
5. J.L. Macmanus-Driscoll, Adv. Mater. **9**, p.457 (1997).
6. X.Z. Wang, B.W. Wessels, Appl. Phys. Lett. **67**, p. 518 (1995).
7. P.D. Rack, A. Naman, P.H. Holloway, S.S. Sun, R.T. Tuenge, MRS Bull. **21**, p. 49 (1996).
8. C.R. Ronda, T. Justel, H. Nikol, J. Alloys Compd., **277**, p. 669 (1998).
9. J. McAleese, J.C. Plakatouras, B.C.H. Steele, Thin Solid Films **280**, p. 152 (1996).

10. T. Kodas, M. Hampden-Smith, *The Chemistry Of Metal CVD*, VCH, Weinheim, 1994.
11. K.-H. Dahmen, T. Gerfin, Progress in Crystal Growth and Characterization of Materials **27**, p.117 (1993).
12. T.J. Marks, Pure and Appl. Chem. **67**, p. 313 (1995).
13. J.A. Belot, D.A. Neumayer, C.J. Reedy, D.B. Studebaker, B.J. Hinds, C.L. Stern, T.J. Marks, Chem. Mater. (Chem. Vap. Dep.) **9**, p. 1638 (1997).
14. J.R. Babcock, D.D. Benson, A. Wang, N.L. Edleman, J.A. Belot, M.V. Metz, T.J. Marks, Adv. Mater. (Chem. Vap. Dep.), in press.
15. J. Nukeaw, J. Tanagisawa, N. Matsubara, Y. Fujiwara, Y. Takeda, Appl. Phys. Lett. **70**, p. 84 (1997).
16. M. Becht, T. Gerfin, K.H. Dahmen, Chem. Mater. **5**, p.137 (1993).
17. I.E. Graboy, N.V. Markov, V.V. Maleev, A.R. Kaul, S.N. Polyakov, V.L. Svetchnikov, H.W. Zandbergen, K.H. Dahmen, J. Alloys Compd. **251**, p. 318 (1997).
18. R. Hiskes, S.A. DiCarolis, R.D. Jacowitz, Z. Lu, R.S. Fiegelson, R.K. Ronte, J.L. Young, J. Cryst. Growth **128**, p.781 (1993).
19. M. Becht, F. Wang, J.G. Wen, T. Morishita, J. Cryst. Growth **170**, p. 799 (1997).
20. Z. Lu, R. Hiskes, S.A. DiCarolis, A. Nel, R.K. Ronte, R.S. Fiegelson, J. Cryst. Growth **156**, p. 227 (1995).
21. S. Liang, C.S. Chern, Z.Q. Shi, P. Lu, Y. Lu, B.H. Kear, J. Cryst. Growth **151**, p.359 (1995).
22. D. Chadwick, J. McAleese, K. Senkiw, B.C.H. Steele, Appl. Surf. Sci. **99**, p. 417 (1996).
23. J.A. Belot, A. Wang, R.J. McNeely, L. Liable-Sands, A.L. Rheingold, T.J. Marks, Adv. Mater. (Chem. Vap. Dep.) **5**, p. 65 (1999).
24. D.L. Schulz, B.J. Hinds, D.A. Neumayer, C.L. Stern, T.J. Marks, Chem. Mater. **5**, p. 1605 (1993).
25. A. Wang, J.A. Belot, T.J. Marks, P.R. Markworth, R.P.H. Chang, M.P. Chudzik, C.R. Kannewurf, Physica C **320**, p. 154 (1999).
26. D.C. Bradley, J.S. Ghorta, F.A. Hart, J. Chem. Soc, Dalton Trans., p. 1021 (1973).
27. The applicability of this analysis is explained in B.J. Hinds, R.J. McNeely, D.B. Studebaker, T.J. Marks, T.P. Hogan, J.L. Schindler, C.R. Kannewurf, X.F. Zhang, D.J. Miller, J. Mater. Res. **12**, p. 1214 (1997).
28. Purchased from Strem Chemicals, Inc., Newburyport, MA 01950-4098.

PREPARATION AND CHARACTERIZATION OF EPITAXIAL Bi₂WO₆ THIN FILMS PREPARED BY A SOL-GEL PROCESS

Keishi Nishio*, Chikako Kudo*, Tsutomu Nagahama*, Takaaki Manabe**, Iwao Yamaguchi**, Yuichi Watanabe* and Toshio Tsuchiya*
*Dept. of Mater. Sci. and Tech., Science University of Tokyo, Noda Chiba, 278-8510, JAPAN
k-nishio@rs.noda.sut.ac.jp
**National Institute of Materials and Chemical Research, Tsukuba, Ibaraki, 305-8565, JAPAN

ABSTRACT

We have succeeded in an epitaxial growth of bismuth tungstate (Bi_2WO_6, BWO) thin films, one of the bismuth layer-structure ferroelectrics (BLSF), on $SrTiO_3$ (001) single crystal substrates by a spin coating process. BLSF are known to be one of the promising materials for ferroelectric random access memory (FeRAM) devices. A homogeneous coating solution was prepared with tungsten hexachloride and bismuth 2-ethylhexanoate as raw materials, and 2-(2-methoxyethoxy) ethanol and formamide as solvents. The as-coated thin films were sintered at temperatures from 500 to 800°C for 1h in air. BWO crystallized at temperatures above 500°C. Any crystal phase was not observed in the thin films except for ($00l$) phases of BWO in the XRD patterns. It was confirmed that the thin film was growing epitaxially by measurement of XRD pole figure. The crystallographic relationship of the film and substrate was BWO(001)//STO(001), BWO[110]//STO[100].

INTRODUCTION

Bismuth layer structure ferroelectrics (BLSF) in the system $(Bi_2O_2)^{2+}(A_{n-1}B_nO_{3n-1})^{2-}$ (n=1 to 5) have a structure in which the $(Bi_2O_2)^{2+}$ layers and quasi-perovskite units are piled up along the crystallographic c-axis [1-6]. Bi_2WO_6 (BWO) is the simplest representative of the BLSF family (n=1). It is polar at low temperature but undergoes a nonferroelectric reconstructive phase transition near 950 to 960°C [7,8]. In a ferroelectric thin film, the degradation of ferroelectric properties by the size effect becomes a problem. The size effect is caused by small crystal grains. The crystal grains become small when the film thickness is decreasing. It also has been reported even in $BaTiO_3$, $PbTiO_3$, and $Pb(Zr_xTi_{1-x})O_3$ thin films [9,10]. It may be possible to solve the problem by producing these ferroelectrics as epitaxial films. Preparation of epitaxial Bi_2WO_6 thin films has been reported by pulsed laser deposition (PLD) [7]. The BWO thin film shows epitaxial growth along the [100] direction on a [100] oriented Nb-doped $SrTiO_3$ substrate.

The sol-gel process has received much interest due to a potential of precise control of chemical stoichiometry, homogeneity, low temperature processing and decreasing costs [11-15]. Preparation of epitaxial thin films by sol-gel deposition has also been reported [16]. In this study, an epitaxially grown film of BWO was prepared by sol-gel deposition, and the evaluation of its properties was carried out.

At present, DRAMs are used as memories for computers. DRAMs lose information, when the power of the computer is cut. A ferroelectric memory has a working speed which is equivalent to that of the DRAM, and it does not lose information, since the ferroelectric property is utilized. The write is excellent with 10^{12} to 10^{13} cycles without degradation.

Ferroelectric memories are expected as a memory for computer in the next generation [7,17]. The ferrolelctric properties of the bismuth layer structure ferroelectrics (BLFS) are very promising for the application to the ferroelectric random access memory (FeRAM) devices. The coercive field of BLSF is lower than that in ferroelectric simple perovskite oxides, e.g., $Pb(Zr_xTi_{1-x})O_3$, which enables one to operate the FeRAM devices under lower applied voltage.

EXPERIMENTAL

Generally, the following materials are used as starting materials for the sol-gel method: metal salts, organic salts and metal-organic complexes, metal alkoxides, etc. However, a metal alkoxide is very sensitive to atmospheric moisture [18]. When multicomponent ceramics are prepared, it is difficult to control the hydrolysis rates of the raw materials. In this study, we used metal salts and organic solvents which formed a complex with metal ions. $SrTiO_3$ (001) (STO) single crystals were used as substrates. The lattice constants of the single crystal are $a=b=c=0.3905nm$, whereas those of BWO are $a=0.547$, $b=0.544$ and $c=1.640nm$:It should be noted that the lattice constants a and b of BWO are about as large as $\sqrt{2}$ times those in STO.

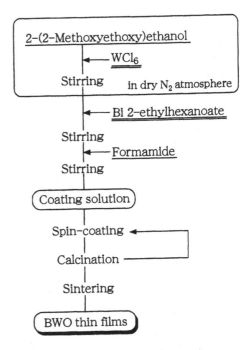

Figure 1 Preparation procedure of Bi_2WO_6 thin films.

BWO thin films were prepared on STO substrate by spin coating, followed by appropriate heat treatment. Figure 1 represents the preparation procedure of the Bi_2WO_6 (BWO) thin films. Tungsten hexachloride and bismuth 2-ethylhexanoate were used as raw materials. In order to reduce the hydrolysis rate of the metal complex ions and the capillary pressure of pores in the film, we used 2-(2-methoxyethoxy) ethanol and formamide [16,19,20]. Tungsten hexachloride was dissolved in 2-(2-methoxyethoxy) ethanol in dry nitrogen atmosphere. Subsequently bismuth 2-ethylhexanoate and formamide were added to the solution. The coating solution thus prepared was kept at 0°C in order to suppress the hydrolysis.

The calcination temperature of 450°C, had been decided by preliminary thermogravimetric and differential thermal analysis (TG-DTA) measurements using a Rigaku TG-DTA8078 system : At this temperature, the organic products constituted were fully removed by combustion. The spin coating and calcination were repeated 5 times. The calcined thin films were sintered at a temperature from 500°C to 800°C for 1h in air. The X-ray diffraction (XRD) patterns were obtained using a Rigaku diffractometer model CN4148 with a thin film attachment for identification of crystal phase and evaluation of the crystallinity of prepared thin film with Cukα(1.542 Å) radiation and a graphite monochromator. X-ray pole figure investigations were carried out using a MAC Science model M03X-HF system with CuKα (1.542 Å) radiation. In the measurement of the pole figure, the 113 diffraction (main peak of polycrystal BWO) which did not overlap with the diffraction peak of the substrate was used. The surface observation of the prepared thin films was carried out using atomic force microscopy (AFM SII SPA300 system).

RESULTS AND DISCCUSION

An exothermic peak with weight loss that seemed to correspond to combustion of the organic solvents which remain in the powder was observed up to 500°C in the TG-DTA spectrum. Another exothermic peak which was not accompanied with weight loss at 950°C was observed, which agrees with the orthorhombic phase transition temperature of BWO to its tetragonal phase [7,8].

Figure 2 shows XRD patterns (measured by theta-2theta method) for thin films heat-treated at various temperatures. As seen in fig. 2, clear diffraction peaks besides the intense ones due to the STO substrate are observed for the films heat-treated at temperatures above 600°C. For the film heat-treated at 500°C, only a weak peak could be observed at 32.7° beside the intense peaks due to the substrate. Thus, we recognized that crystallization took place effectively for films heat-treated at a temperature above 600°C. All diffraction peaks labeled by ● could be assigned to *00l* diffraction of BWO. The results indicated that the *c*-axis of BWO is oriented perpendicular to the surface of the STO substrate. In

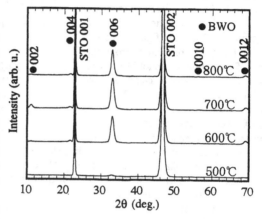

Figure 2 XRD θ-2θ scans of thin films prepared on STO single crystal substrate, heat-treated at various temperatures for 1h.

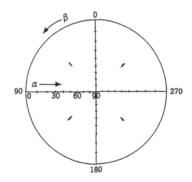

Figure 3 Pole figure of BWO thin film heat-treated at 600°C for 1h.

order to clarify the exact nature of the c-axis orientation of the thin films, a pole figure measurement was carried out. Figure 3 shows a result of this pole figure measurement in which the diffraction of the 113 plane of BWO is used. As seen in fig.3 only four spots were observed clearly : The result implies an epitaxial growth of the BWO thin film on the STO substrate As the spots of a (100) oriented (a-axis-oriented) epitaxial film would appear almost in the same positions of a pole figure recorded with the (113) reflection, from this pole figure alone it cannot be decided, whether an a-axis- or c-axis-oriented film has grown. Therefore a beta scan using the (115) reflection was performed, because such a scan permits to distinguish between a- and

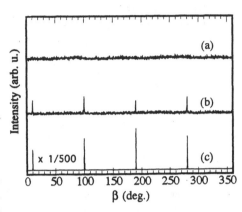

Figure 4 XRD beta scans of BWO thin film on STO heat-treated at 600°C for 1h
(a) a-axis of BWO(115) diffraction, (b) c-axis of BWO (115) diffraction and (c) STO (101) diffraction.

c-axis orientation of the film. The result is shown in figures 4. The diffraction peaks appeared at every 90° at alpha=49.5° (diffraction by 115 of c-axis oriented BWO) in fig.4 (a), whereas no peaks were observed at alpha=27.3° (diffraction by 115 of a-axis oriented BWO) in fig.4 (b). We therefor recognized that the prepared BWO thin film was a perfect c-axis oriented film. Figure 4 (c) shows a result of the beta scan for STO substrate using the (101) diffraction. As seen in fig.4 (c), the diffraction peaks appeared at every 90° as well as those appeared in fig.4 (b). From this results, the epitaxial relationship between BWO and STO can be represented by BWO(001)//STO(001) and BWO[110]//STO[100]. Figure 5 shows the configuration relationship between the BWO film and STO. The BWO unit cell has been placed for the position in which it is rotated by 45° with respect to the STO unit cell. The

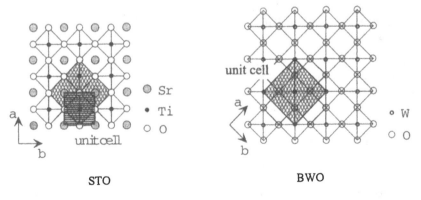

Figure 4 Schematic projection of Bi_2WO_6 and $SrTiO_3$

380

lattice constants of STO (cubic) and BWO (orthorhombic) are $a=b=c=0.3905$ and $a=0.547$, $b=0.544$, $c=1.64$ respectively. The mismatches of a and b are $Ma=1.2\%$ and $Mb=1.6\%$ calculated from $\sqrt{2}$ times the lattice constant of STO and from the lattice constant of BWO.

The surfaces of prepared BWO thin films observed by using AFM are shown in figure 6. The particle size of the thin film prepared at 500°C was less than 0.1 μm. Large aggregates were observed only in the film surface prepared at 600°C. The large aggregates were formed of small grains. The aggregates were not observed in the film prepared at 700 and 800°C. With the increase in heat treatment temperature, the grain size slightly increased. The average grain size of the film prepared at 800°C was about 0.2μm.

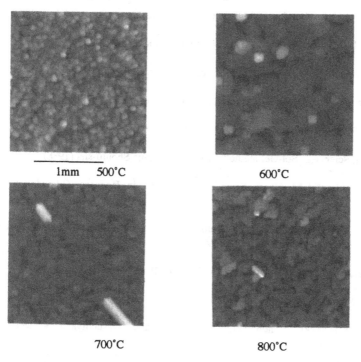

   1mm   500°C   600°C

      700°C   800°C

Figure 6 AFM images of the surface of BWO thin films on STO substrates heat-treated at various temperatures for 1h.

CONCLUSIONS

We have succeeded in the preparation of epitaxial thin films of bismuth tungstate on SrTiO$_3$ (001) single crystal substrates by a spin coating process. The crystallographic relationship of the film and substrate was BWO(001)//STO(001), BWO[110]//STO[100]. This relationship was to be expected from the fact that the a and b lattice constants of BWO are nearly equal to $\sqrt{2}$ times that of STO.

REFERENCES

1. D. A. Payne and Jahar L. Mukherjee, Appl. Phys. Lett., **29**, p. 748 (1976)

2. M. T. Montero, P. Millan, A. Castro, Mater. Res. Bull., **33**, p.1103 (1998)

3. E. C. Subba Rao, J. Am. Ceram. Soc., **45**, p.166 (1962)

4. B. Aurivillius and P. H. Fang, Phys. Rev., **126**, p. 893 (1962)

5. V. K. Yanovskii and V. I. Voronkova, Phys. Stat. Sol. (a), **93**, p. 57 (1986)

6. S. Kojima, J. Phys., Condensed Matter, **10**, p. L327 (1998)

7. M. Hamada, H. Tbata, T. Kawai, Thin Solid Films, **306**, p. 6 (1997)

8. R. E. Newnham, R. W. Wolfe, J. F. Dorrian, Mat. Res. Bull., **6**, p. 1029 (1971)

9. S. Yamamichi, T. Sakuma, K. Takemura, Y. Miyasaka, Jpn J. App. Phys., **30**, p. 1 (1991)

10. K. Abe, S. Komatu, Jpn. Appl. Phys., **32**, p. 1157 (1993)

11. S. Sakka, Science of Sol-Gel method, in Japanese, Agne Shoufu-sha (1988).

12. H. Yanagida and M. Nagai, Science of Ceramics, in Japanese, Gihoudou Shyuppan (1993).

13. S. Hirano, T. Yogo, K. Kikuta,H. Urahata, Y. Isobe, T. Morishita, K. Ogiso and Y. Ito, Mat. Res. Soc. Symp. Proc., **271**, p. 331 (1992)

14. Yu-Fu Kuo and Tseung-Yuen Tseng, J. Mat. Sci., **31**, p. 6361 (1996)

15. Yihu, J. Mat. Sci., **31**, p. 4255 (1996)

16. K. Nishio, N. Seki, J. Thongrueng, Y. Watanabe and T. Tsuchiya, J. Sol-Gel Sci. and Tech., **16**, p. 37 (1999)

17. R. E. Newnham, R. W. Wolfe, J. F. Dorrian, Mat. Res. Bull., **6**, p. 1029 (1971)

18. H. Dislich, J. Non-Cryst. Solids, **57**, p. 371 (1983)

19. T. Nakano, M. Kumagai and T. Funahashi, (Proceedings of Fall Meeting The Ceramic Society of Japan, 1990) p. 242

20. S. Wallace and L. L. Hench, (Mat. Res. Soc. Symp. Proc., **32**, 1984) p. 47

PROPERTIES OF AMORPHOUS AND CRYSTALLINE Ta$_2$O$_5$ THIN FILMS PREPARED BY METALORGANIC SOLUTION DEPOSITION TECHNIQUE FOR INTEGRATED ELECTRONIC DEVICES

P.C. JOSHI, M.W. COLE
US Army Research Laboratory, WMRD, Aberdeen Proving Ground, MD 21005, joshipc@aol.com

ABSTRACT

We report on the properties of Ta$_2$O$_5$ thin films prepared by the metalorganic solution deposition (MOSD) technique on Pt-coated Si, n$^+$-Si, and poly-Si substrates. The effects of post-deposition annealing temperature on the structural, electrical, and optical properties were analyzed. The electrical measurements were conducted on MIM and MIS capacitors. The dielectric constant of amorphous Ta$_2$O$_5$ thin films was in the range 29.2-29.5 up to 600 °C, while crystalline thin films, annealed in the temperature range 650-750 °C, exhibited enhanced dielectric constant in the range 45.6-51.7. The dielectric loss factor did not show any appreciable dependence on the annealing temperature and was in the range 0.006-0.009. The films exhibited high resistivities of the order of 10^{12}-10^{15} Ω-cm at an applied electric field of 1 MV/cm in the annealing temperature range of 500-750 °C. The temperature coefficient of capacitance was in the range 52-114 ppm/°C for films annealed in the temperature range 500-750 °C. The bias stability of capacitance, measured at an applied electric field of 1 MV/cm, was better than 1.41 % for Ta$_2$O$_5$ films annealed up to 750 °C.

INTRODUCTION

Ta$_2$O$_5$ thin films have potential for numerous microelectronic applications such as gate dielectric of metal-insulator-semiconductor devices, optical waveguides, electroluminescent display devices, and surface acoustic wave (SAW) devices [1-5]. Tantalum oxide thin films are attractive for scaled down capacitor in ultra large scale integrated (ULSI) circuits because of their high dielectric constant, low dielectric loss, low leakage current, low defect density, and good temperature and bias stability. The high dielectric constant and low dielectric loss materials are also attractive for microwave applications [4]. For successful integration into microelectronic devices, extremely reliable Ta$_2$O$_5$ thin films are desired. The properties of Ta$_2$O$_5$ thin films have been reported to be strongly dependent on the fabrication method, nature of substrate and electrode material, and post-deposition annealing treatment. The technical literature shows a wide variation in the reported structural, dielectric, and insulating properties of amorphous and crystalline Ta$_2$O$_5$ thin films. An understanding of process-structure-property correlation is important to understand and compare various thin film studies reported in the literature and exploit Ta$_2$O$_5$ thin films for devices. In this paper, we report on the systematic study of structural, optical, dielectric, and insulating properties of amorphous and crystalline Ta$_2$O$_5$ thin films fabricated by MOSD technique. Detailed studies were conducted on MIM and MIS capacitors to analyze the influence of film/substrate interface on the electrical properties.

EXPERIMENT

Ta$_2$O$_5$ thin films were fabricated by metalorganic solution deposition technique using tantalum ethoxide as precursor. 2-methoxyethanol was used as solvent. In the experiment, tantalum ethoxide was initially dissolved in 2-methoxyethanol. Finally, acetic acid was added to

383

prevent rapid hydrolysis and precipitation of the metal oxide. The Ta_2O_5 thin films were deposited on platinized silicon, n^+-Si, and poly-Si substrates by spin coating using a photoresist spinner. Prior to film deposition, the silicon substrates were cleaned by a room temperature technique entitled spin etching to remove the native silicon oxide and make the substrate surface hydrogen terminated [6]. After spinning on to various substrates, films were kept on a hot plate (~350 °C) in air for 10 min to remove solvents and other organic. The post-deposition annealing of the films was carried out in a furnace at various temperatures in an oxygen atmosphere. The structure of the films was analyzed by glancing angle x-ray diffraction using CuK_α radiation at 40 kV. The surface morphology of the films was examined by Digital Instrument's Dimension 3000 atomic force microscope using tapping mode with amplitude modulation. The optical properties were measured by variable angle spectroscopic ellipsometry. Rutherford backscattering spectroscopy (RBS) was employed to determine film composition and interfacial properties. Elemental distribution within the bulk of the Ta_2O_5 film and Ta_2O_5/Si interface was determined by Auger electron spectroscopy (AES) using a Perkin-Elmer PHI660 scanning Auger microscope. The electrical measurements were conducted on films in MIM and MIS configurations. The platinum electrodes were sputter deposited through a shadow mask on the top surface of the films to form MIM and MIS capacitors. The dielectric properties were measured with HP4194A impedance/gain-phase analyzer. The insulating properties of the Ta_2O_5 thin films were analyzed through the measurement of leakage current versus voltage (I-V) characteristics using HP4140B semiconductor test system.

RESULTS

Figure 1 shows the XRD profiles of the films deposited on Pt-coated Si wafers. The as-pyrolyzed films were amorphous in nature and the post-deposition annealing of the films was carried out in the temperature range 500-750 °C. The Ta_2O_5 thin films were amorphous up to 600 °C and a well-crystallized orthorhombic phase was obtained at an annealing temperature of 650 °C with no evidence of secondary phases. The peak intensity and sharpness were found to increase with increase in temperature in the range 650-750 °C, indicating enhanced crystallinity and an increase in grain size with increasing annealing temperature. The presence of an intense diffraction peak corresponding to (200) plane and the relatively weaker diffraction from (001) and (002) planes implied that the deposited Ta_2O_5 thin films possessed a strong preferential orientation. Similar trends were observed in the XRD patterns of Ta_2O_5 thin films deposited on silicon substrates (n^+ and poly) by the MOSD technique indicating preferential orientation. The lattice constants a, b, and c, calculated using (001), (200), and (1,11,1) peaks in the XRD pattern, were 6.225, 39.8521, and 3.8609 Å, respectively. These values are in good agreement with those reported for bulk Ta_2O_5; suggesting that that the films were crystallized in the orthorhombic phase [7]. The surface morphology of the films, as shown in Fig. 2, was smooth with no cracks and/or defects. The films exhibited a dense microstructure and fine grain size. There was no appreciable effect of the annealing temperature on the microstructure of amorphous Ta_2O_5 thin films, while crystalline films showed an increase in grain size with

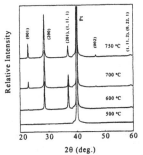

FIG. 1. XRD patterns of Ta_2O_5 thin films annealed at various temperatures for 30 min.

increasing annealing temperature, which is consistent with the XRD studies. The average surface roughness was found to increase with increase in annealing temperature with average surface roughness lower than 4 nm at 750 °C. Figure 3 shows a RBS spectrum for Ta_2O_5 thin film deposited on n^+-Si substrate together with a simulated curve. The sharp edges of the peaks and close correspondence between the simulated and experimental curves indicated stable film formation with a sharp film/substrate interface and smooth surface features. The calculation of the film stoichiometry gave an oxygen/tantalum elemental ratio equal to 2.7 which was possibly due to absorbed moisture or contamination. The RBS spectra for the as-pyrolyzed and 700 °C annealed sample, as shown in Fig. 3(b), indicated negligible diffusion of Ta into Si substrate. The characteristic depth in the substrate was found to be 33 nm by fitting an exponential profile to the experimental profile. The AES depth profile, as shown in Fig. 4, showed nearly constant Ta and O Auger intensities throughout the bulk of the Ta_2O_5 film and a rapid drop off at the Ta_2O_5/Si interface with little interdiffusion of the constituent elements. The AES surface profile did not show any Si diffusion to the surface or impurities, which contributed to the excellent dielectric and leakage current properties.

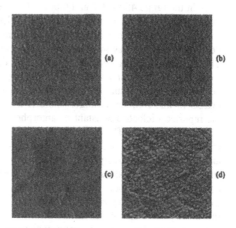

FIG. 2. AFM micrographs of Ta_2O_5 thin films annealed at (a) 500 °C, (b) 600 °C, (c) 650 °C, and (d) 750 °C. (Scan area 1×1 um²)

The small signal dielectric measurements were conducted on MIM and MIS capacitors by applying an *ac* signal of 10 mV amplitude. The effects of applied frequency, film thickness, and post-deposition annealing temperature on the dielectric properties were also analyzed. Table I shows the effect of the post-deposition annealing temperature on the dielectric properties of Ta_2O_5 thin films. There was no appreciable effect of the annealing temperature on the dielectric properties as long as the films were in amorphous state. The dielectric constant and the dissipation factor for amorphous Ta_2O_5 thin films were in the range 29.2-29.5 and 0.006-0.009, respectively, up to an annealing temperature of 600 °C. The dielectric constant was found to increase to 45.6 at an annealing temperature of 650 °C as the films were well crystallized. The dielectric constant of crystalline Ta_2O_5 thin films was found to increase with increase in annealing temperatures while the loss factor did not show any appreciable dependence on the annealing temperature. The dielectric constant and the dissipation factors

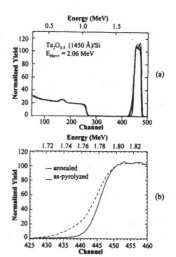

FIG. 3. RBS spectra of Ta_2O_5/n^+-Si structure

were in the range 45.6-51.7 and 0.006-0.009, respectively, for films annealed in the temperature range 650-750 °C. The increase in the dielectric constant of the crystalline thin films with annealing temperature may be attributed to improved crystallinity and increase in grain sizes and density of the films as indicated by XRD and AFM studies.

The permittivity showed no appreciable dispersion with frequency up to about 1 MHz indicating that the values were not masked by any surface layer effects or electrode barrier effects in this frequency range. There is a wide variation in the reported dielectric constant of amorphous and crystalline Ta_2O_5 thin films [8]. The dielectric constant of amorphous and crystalline thin films has been reported in the range (14-31) and (24-60), respectively. The dielectric properties of thin films strongly depend on the fabrication method, nature of substrate and electrode material, post-deposition annealing treatment, crystallographic orientation, microstructure, thickness of samples, and film-substrate interface characteristics. So it is very important to compare the thin film structure, lattice constant, and morphology with bulk to establish structure-property correlation and understand the reported properties of Ta_2O_5 thin films fabricated by various techniques. The dielectric constant of β-Ta_2O_5 is expected to show a maximum along a-axis due to anisotropy of the crystal structure. The Clausius-Mosotti relation shows that the dielectric permittivity strongly depends on the molar volume, which in turn depends on the lattice constants [9]. Thin film lattice constant values are strongly influenced by the nature of substrate, as one surface of the film is adhered to the substrate, as opposed to bulk material where all the surfaces are free. The observed high dielectric constant for the present crystalline Ta_2O_5 thin films was attributed to strong a-axis orientation and difference between the thin film and bulk lattice constant values.

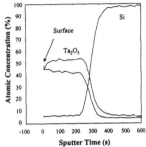

FIG. 4. The AES depth profile of Ta_2O_5/n^+-Si structure

The effects of the substrate on the dielectric properties were analyzed by measurement of dielectric properties on MIM and MIS capacitors. The MIS capacitor is the most commonly used active component for silicon ULSI circuits. The device technology has been hindered by the very high density of states at the interface of Si and insulator. A poor interface between the dielectric film and silicon leads to high leakage currents, higher frequency dispersion, and high defect trapped charges. Several attempts have been made to improve the effective dielectric constant and leakage current characteristics of Ta_2O_5 based MIS capacitors by optimizing the fabrication techniques and/or post-deposition annealing treatment. In the present case, the dielectric properties of Ta_2O_5 thin films deposited on n^+-Si and poly-Si substrates were compared with those deposited on Pt-coated Si substrates to analyze the effects of silicon oxidation on the

Table I. Effects of post-deposition annealing temperature on the dielectric and optical properties of Ta_2O_5 thin films

T_A (°C)	ε_r (at 100 kHz)	tan δ	n (at 630 nm)	TCK (ppm/°C)
500	29.2	0.007	1.98	+ 51
600	29.5	0.009	2.01	+ 52
650	45.6	0.009	2.05	+ 66
700	50.4	0.007	2.06	+ 77
750	51.7	0.008	2.08	+114

Table II. Effect of substrate on the dielectric properties of Ta_2O_5 based MIM and MIS capacitors

Substrate	T_{Anneal} = 600 °C		T_{Anneal} = 750 °C	
	ε_r	tan δ	ε_r	tan δ
n^+-Si	28.5	0.018	43.7	0.018
poly-Si	29.0	0.008	45.7	0.017
Pt-coated Si	29.5	0.008	51.7	0.008

effective dielectric properties. The dielectric constant of Ta_2O_5 thin films was found to be lower on n^+-Si and poly-Si substrates, as shown in Table II, as compared to films deposited on Pt-coated Si substrates due to the growth of an interfacial oxide layer on silicon substrates. There was no significant difference in the dielectric constant of amorphous Ta_2O_5 thin films, annealed at 600 °C, deposited on various substrates indicating the absence of any appreciable oxide thickness at film-substrate interface. The films annealed at 750 °C showed a much lower dielectric constant on silicon substrates as compared to films deposited on Pt-coated Si substrates indicating the presence of an appreciable oxide growth at the interface. The thickness of the oxide layer was calculated by comparing the dielectric constant of MIM and MIS capacitors by assuming the formation of a uniform oxide layer between the film and substrate with no interdiffusion. The thickness of the silicon oxide layer was calculated from the relation, $1/C_T = 1/C_f + 1/C_{SiO_2}$, by considering the total capacitance (C_T) of the MIS capacitor to be a series combination of Ta_2O_5 film capacitance (C_f) and SiO_2 layer capacitance (C_{SiO_2}). For the present Ta_2O_5 films the thickness of the silicon oxide layer was found to less than 2.5 nm on both n^+-Si and poly-Si substrates indicating that the combination of spin etching and MOSD technique was effective in minimizing the film-substrate interface reaction and providing good interfacial properties. Annealing temperature effects up to 1000 °C have been analyzed for Ta_2O_5 thin films in an attempt to improve their electrical properties. Annealing at high temperatures severely degrades the film/Si interface and requires close control of processing parameters to improve the electrical properties. Low temperature processing is necessary to minimize interfacial reactions and integrate Ta_2O_5 thin films into semiconductor devices.

The temperature and bias stability of the dielectric properties of Ta_2O_5 thin films were also analyzed to establish their reliability for integrated electronic applications. The dielectric constant and the dissipation factor were relatively unchanged with measurement temperature in the range 25-125 °C indicating good temperature stability of the present Ta_2O_5 thin films. Table I shows the temperature coefficient of capacitance of Ta_2O_5 thin films, measured in the temperature range 25-125 °C, as a function of annealing temperature. The films showed a low TCK of +52 ppm/ °C up to an annealing temperature of 600 °C. The TCK for the crystalline films, as shown in Table I, was found to be higher than that of amorphous films. Crystalline Ta_2O_5 thin films annealed at 750 °C exhibited a low TCK of +114 ppm/°C establishing good reliability of present Ta_2O_5 thin films for integrated capacitor applications. The C-V measurements were conducted on MIM capacitors to analyze the bias stability by applying a small ac signal of 10 mV amplitude and of 100 kHz frequency across the capacitor, while the dc electric field was swept from a negative bias to positive bias. There was no appreciable change in the capacitance of amorphous Ta_2O_5 thin films up to an applied electric field of 1 MV/cm indicating good bias stability. The change in the film capacitance was found to be 1.41 % for crystalline Ta_2O_5 thin films, annealed at 750 °C, at an applied electric field of 1 MV/cm. The loss factor also showed good bias stability and was less than 1% up to 1 MV/cm. The low temperature coefficient of capacitance and good bias stability of the present Ta_2O_5 films indicate completeness of oxidation and low defect concentration in amorphous and crystalline Ta_2O_5 thin prepared by MOSD technique.

The optical properties of Ta_2O_5 thin films were determined by ellipsometry. The experimentally determined psi and delta values were transformed into refractive index by assuming an optical model consisting of single homogeneous nonabsorbing film on Pt-coated-Si substrates. The refractive index and the band-gap values were found to be 2.08 (at 630 nm) and 5.11 eV, respectively, for a 0.15-μm-thick film annealed at 750 °C. The effects of post-deposition annealing temperature on the refractive index were also analyzed. The refractive index value was found to increase with the increase in annealing temperature, as shown in Table

I, which may be attributed to a change in the microstructure of the films from amorphous to crystalline, and an increase in crystallinity and density of the films.

The leakage current characteristics of the Ta_2O_5 thin films were measured using MIM capacitors by applying *dc* voltages with a step delay time of 30 s. The leakage current density showed a strong dependence on the post-deposition annealing temperature and was considerably higher for crystalline thin films as compared to amorphous thin films. The resistivity of amorphous thin films was found to be of the order of 10^{15} Ω-cm while that of crystalline thin films was of the order of 10^{12} Ω-cm at an applied electric field of 1 MV/cm. The high resistivity observed in the present films shows the completeness of phase formation and oxidation of Ta_2O_5 thin films prepared by MOSD technique.

CONCLUSIONS

Amorphous and crystalline Ta_2O_5 thin films exhibiting good structural, dielectric, and insulating properties were successfully deposited on Pt-coated Si, n^+-Si, and poly-Si substrates by MOSD technique. The dielectric constant of amorphous thin films, annealed up to 600 °C, was in the range 29.2-29.5, however, crystalline thin films exhibited significantly enhanced dielectric constant in the range 45.6-51.7 for films annealed in the temperature range 650-750 °C. The loss factor did not show any appreciable dependence on the annealing temperature and was in the range 0.006-0.009 for films annealed in the temperature range 500-750 °C. The films deposited on n^+-Si and poly-Si substrates exhibited good-film/silicon interfacial characteristics. The growth of the oxide layer at the interface was calculated to be lower than 2.5 nm at an annealing temperature of 750 °C. Ta_2O_5 thin films exhibited high resistivity, measured at an applied electric field of 1 MV/cm, in the range 10^{12}-10^{15} Ω-cm for amorphous and crystalline thin films. The temperature coefficient of capacitance was in the range 52-114 ppm/°C for films annealed in the temperature range 500-750 °C. The high resistivity, low temperature coefficient of capacitance, and good bias stability of dielectric properties establish the reliability of Ta_2O_5 thin films for microelectronic applications.

REFERENCES

1. J. Baliga, Semiconductor International **22**, 79 (1999).

2. M. Anthony, S. Summerfelt, and C. Teng, Texas Instr. Tech. J. **12**, 30 (1995).

3. P. Balk, Adv. Mater. **7**, 703 (1995).

4. F. S. Barnes, J. Price, A. Hermann, Z. Zhang, H.-D. Wu, D. Galt, and A. Naziripour, Integr. Ferroelectr. **8**, 171 (1995).

5. Y. Nakagawa and T. Okada, J. Appl. Phys. **68**, 556 (1990).

6. D. B. Fenner, D. K. Biegelsen, and R. D. Bringans, J. Appl. Phys. **66**, 419 (1989).

7. R. S. Roth, J. L. Waring, and H. S. Parker, J. Solid State Chem. **2**, 445 (1970).

8. P. C. Joshi and M. W. Cole, J. Appl. Phys. **86**, 871 (1999).

9. R. D. Shannon, J. Appl. Phys. **73**, 348 (1993).

The seeding effect of lanthanum nickel oxide ceramic/ceramic nanocomposite thin films prepared by the MOD method

Yu ZHANG, Qifa ZHOU, Helen Lai-wa CHAN, Chung-loong CHOY
Department of Applied Physics and Materials Research Center, The Hong Kong Polytechnic University, Hunghom, Kowloon, Hong Kong China

ABSTRCT

Lanthanum nickel oxide (LNO) is a conducting ceramic which has potential to be used as electrodes in multilayer ceramic actuators. Thick LNO films have been formed by incorporating nanosized LNO powder (annealed at 700°C, with diameter around 100 nm) into a LNO solution prepared by a metal-organic decomposed (MOD) method. Three different weight percents, 2%, 4%, and 10% of LNO powder have been added. The structural variation of the ceramic/ceramic composite film with annealing temperature was studied by X-ray diffraction and differential thermal analysis. The crystallization temperature of the film is found to decrease from ~590°C to ~510°C due to the seeding effect introduced by the nano-powder.
Key words: seeding effect, metallic organic decomposition (MOD), Lanthanum Nickel Oxide

INTRODUCTION

Nanosized particles are of interest in fundamental as well as applied research because many material- properties change drastically when the crystallite size reaches the nano/submicrometer range [1]. Since nano-scaled particles have relatively larger surface areas, they tend to agglomerate to minimize the total surface energy. Agglomeration adversely affects their properties [2,3], for example: different reaction temperature, faster grain growth speed, closer and more uniform film quality, etc.

Recently, it has been reported that the formation of PZT thin film may be controlled by seeding with PZT powder [1]. Applying the seeding effect in order to enhance the transformation kinetics and to control the development of a desired phase has been successfully investigated.

Lanthanum nickel oxide (LNO) is a conductive ceramic material that can be used in fabricating interleaving electrodes in PZT multilayer devices[2,3,4]. In this work, new investigation of using seeding effect to reduce the annealing temperature of LNO films is presented. Various weight percentage of nano-sized lanthanum nickel oxide (LNO) powder fabricated by MOD method were added into a 0.5M LNO solution [4,6] as seeds. Then the LNO film was spin-coated onto a $Pt/Ti/SiO_2/Si$ multilayer substrate. Using rapid thermal annealing process (RTA) and by changing the annealing temperature and annealing time, samples with different properties were obtained.

EXPERIMENTS

Lanthanum nickel oxide (LNO) powder has been produced by using a metal-organic decomposed (MOD) method [4,5] and annealed at 700°C for 60 minutes. The crystallization process of LNO was studied by X-ray diffraction (XRD, Philips x'pert XRD system), and differential thermal analysis (DTA, Perkin-Elmer 1700).

The LNO metal-organic decomposed solution was produced by combining the $La(NO_3)_3$ and $Ni(NO_3)_3$ solutions in a molar ratio of 1:1. The concentration of the final solution was about 0.5 M. Polyvinyl alcohol (PVA) with a weight percentage of about 5% was added as a binder. A small amount of glycerin was also added to increase the viscosity of the solution [2,3,5]. The chemical reaction equation is shown below:

$$La(NO_3)_3 + Ni(NO_3)_3 + 3H_2O = LaNiO_3 + 6HNO_3$$

LNO seeds were added into the pure LNO solution with different seeding amounts, including 2%, 4%, and 10% weight percent. After stirring and powder dispersion in an ultrasonic bath, a uniform suspension was formed.

The composite LNO solution was then spin-coated onto the Pt/Ti/SiO$_2$/Si substrate. The spinning speed was 100 rpm for the first 6 seconds then 3500 rpm for the next 30 seconds. The spin-coated film was heat-treated by the rapid thermal annealing process (RTA) in order to evaporating the water and organic solvent inside the film. The heat-treat procedure is 150°C for 10 min, then 350°C for 15 min. This whole process was repeated three times in order to get the final composite LNO thin film with a thickness of ~400nm. The sample was then annealed for 30 minutes at various temperatures, ranging from 550°C to 700°C. The crystallization of the film was monitored by X-ray diffraction [5,6] and the crystallite size in the LNO film was estimated through the width of the diffraction peaks.

The Scanning electron microscope (SEM) was used to study the morphology of the LNO films, and to observe the grain growth. The DTA method was also used to study the change in the phase transition temperature in the LNO film with and without seeding in order to reveal its effect.

RESULTS AND DISCUSSIONS

The XRD patterns in Fig. 1 show that the LNO powder annealed below 450°C has amorphous structure. When the annealing temperature increased to 700°C, crystalline peaks of LNO become clear.

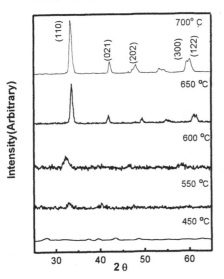

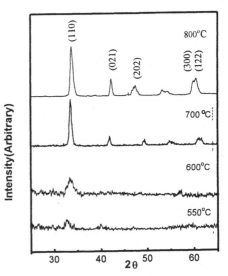

Fig.1   XRD patterns for LNO powder annealed at different temperatures.

Fig.2   XRD patterns for pure LNO films (without seeds) annealed at different temperature.

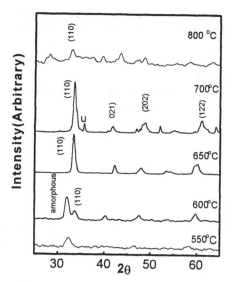

Fig.3 XRD patterns for LNO nano- composite films annealed at different temperature

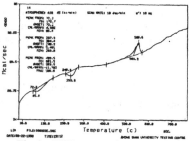

Fig.4 DTA curve for pure LNO metal-organic decomposed film.

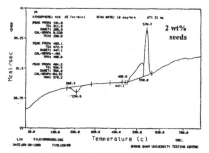

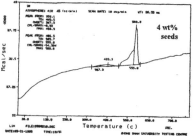

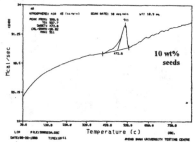

Fig.5 The composite LNO films' DTA with the different seeding amounts.

Fig. 3 is the XRD patterns of LNO nanocomposite films made by the metal organic decomposition (MOD) method after they were annealed at different temperatures. This pattern shows that the nano-composite LNO film has ~80°C lower crystallization temperature compare to the pure LNO film (Fig. 2). This is consistent with the thermal data. Fig. 4 is the DTA graph of the pure LNO film. Fig. 5 shows the DTA graphs of the composite LNO films with different amount of LNO powder (including 2%, 4%, 10% by weight) added as seeds. Compare these patterns, clear differences in the DTA curves can be observed. In the pure LNO film, DTA curve shows a prominent peak at around 587°C. In the LNO composite films, peaks are observed at ~578°C for the film with 2 wt% LNO powder, and as the amount of powder increases to 4 wt% and 10 wt%, the peak shifts to ~570°C and ~511°C, respectively, as shown in Fig.5. This indicates that the crystallization temperature is progressively lowered as the amount of seeding increases.

Table 1 shows the crystallite size of the LNO nanocomposite films annealed at different temperature, which are calculated from the full width at half maximum (FWHM) of the (110) XRD peaks using the Scherrer's equation[7]:

$$\tau = \frac{K\lambda}{\beta_r \cos\theta}$$

Where β_r is the line broadening due to the effect of small crystallite, τ is the mean crystallite dimension, K is a constant (=0.8), θ is set close to the Bragg angle where show X-ray diffraction peaks.

Annealing temperature(°C)	550	600	650	700
Crystallite size(nm)	16.2	18.4	19.5	26.4

Fig. 6 shows the surface morphology of composite films heat-treated for 30 minutes at 450°C and 600°C, respectively. For the film annealed at 450°C, it is easily seen that the powder is uniformly dispersed in an amorphous matrix. After annealing at 600°C, the matrix has crystallized and the powder disappeared into the crystalline grains of the matrix. By comparing Fig.6(b) and Fig. 7, we can see that the nanocomposite LNO film is denser, and it has better uniformity compare to the LNO film without seeding and heat-treated at 600°C.

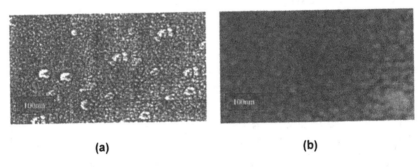

(a)   (b)

Fig. 6 Surface morphology of composite films.
(a)   sintering at 450°C; (b) sintered at 600°C

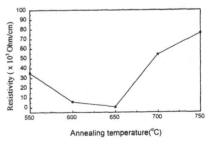

Annealing temperature(°C)

Fig. 7 Surface morphology of pure
LNO film annealed at 600°C

Fig. 8 Conductivity of the LNO composite
as a function of annealing temperature.

Fig. 8 shows the conductivity of the LNO composite film as a function of the annealing temperature. From this graph we can see that when annealed at 600°C~650°C, the film has the highest conductivity, and as the annealing temperature increases, the conductivity decreases rapidly, which maybe caused by the decomposition of the LNO as can be seen from the XRD pattern in Fig. 3[3,4,8,9].

CONCLUSION

Lanthanum nickel oxide (LNO) ceramic/ceramic nanocomposite thin films with a thickness of ~400 nm were fabricated by MOD method with various amounts of LNO powder added as seeds to induce nucleation of the crystalline phase. The seeding effect decreases the crystallization temperature from 590°C to 510°C.

Experimental results show that the gains in LNO composite film grow faster and are more densely packed. Film uniformity is also improved due to the seeding effect and the crystallization of a pure perovskite phase can be achieved at lower temperature. The conductivity of the LNO composite thin film annealed at 550°C, 600°C, 650°C, 700°C, and 750°C has been measured at room temperature. The LNO composite film annealed at 600~650°C has the highest conductivity.

ACKNOWLEDGEMENT

This work was supported by the Center for Smart Materials and the Postdoctoral Fellowship Scheme of The Hong Kong Polytechnic University.

REFERENCE

[1]. C.R.Cho, David.A.Payne. Appl. Phys. Lett. 71(20) (1997) 3013
[2] Q.F. Zhou, H.L.W. Chan and C.L.Choy, Thin Solid Films.
[3]. Q.F.Zhou, H.L.Chan and C.L.Choy. J. Mater. Process. Tech. 63 (1997) 281.
[4]. Y.Zhang, Q.F.Zhou, H.L.W.Chan and C.L.Choy. submitted to Thin Solid films..
[5]. Pramod K.Sharma, M.H.Jilavi, R.Nab, H.Schmidt. Journal of Materials Research Letters. 17(1998)823
[6]. A.Li, C.Z.Ge, P.Lu. Appl. Phys. Lett. 68(10) (1996) 1347.
[7]. A.Li, C.Z.Ge, P.Lu, D.Wu, S.B.Xiong. Appl. Phys. Lett. 70(12) (1997) 1616.
[8]. A.Wu, P.M.Vilarinho, I.M.MirandaSalvado and J.L.Baptista. Material Research Bulletin. 33(1998)59
[9]. H.B. Sharma and A.Mansingh, Ferroelectric Lett.22, 75(1997)

PREPARATION AND CHARACTERIZATION OF LiCoO$_2$ AND LiMg$_{0.05}$Co$_{0.95}$O$_2$ THIN FILMS ON POROUS Ni/NiO CATHODES FOR MCFC BY COMPLEX SOL-GEL PROCESS (CSGP)

W. ŁADA*, A. DEPTUŁA*, B. SARTOWSKA*, T. OLCZAK*, A.G. CHMIELEWSKI*, M. CAREWSKA**, S. SCACCIA**, E. SIMONETTI**, L. GIORGI**, A. MORENO**
*Institute of Nuclear Chemistry and Technology, 03-195 Warsaw, Poland, adept@orange.ichtj.waw.pl
** ENEA, TEA-CCAT, Casaccia Casaccia, AD 00100 Roma, Italy, moreno@casaccia.enea.it

ABSTRACT

The major disadvantage of Ni/NiO cathodes for a Molten Carbonate Fuel Cells (MCFC) application is dissolution of NiO in K/Li electrolyte that significantly decreases the cell lifetime. Thin films of LiCoO$_2$ or LiMg$_{0.05}$Co$_{0.95}$O$_2$ were prepared on a cathode body in order to protect them against dissolution. For preparation of starting sols the Complex Sol-Gel Process (CSGP) has been applied. These sols have been prepared by adding of LiOH to aq. acetates solution of Co^{2+}(Mg^{2+}) with ascorbic acid and then by alkalizing them with aqueous ammonia to pH=8. The cathode plates of various dimensions (to several hundreds cm^2) have been dipped in these sols and withdrawn at rate a 1.7 cm/s. Commercially sintered Ni plates were always initially oxidized by heating at various temperatures. Their microstructure and mechanical properties as a function of temperature were observed. Heat treatment should be carried out under the dead load of the ceramic plates in order to avoid their waving. The best non-folded plates were obtained by treating them for 1h at 550°C. The covered substrates were calcined for 1h at 650°C, using low heating rate1°C/min. The presence of LiCoO$_2$ in a deposited coating has been proved by EDS patterns. The resultant film thicknesses were measured by scanning electron microscopy (SEM) on the fractured cross-sections; they ranged from 0.5 to 2μm and depended on sol concentration and viscosity. A 350 hundred hours test in molten carbonates, proved that the cathode bodies covered with LiCoO$_2$ are completely prevented from dissolution of Ni in a molten K/Li electrolyte. Dissolution of LiCoO$_2$ coating was not observed as well. After treatment in a molten electrolyte SEM observations did not show any changes in microstructures and morphology of the covered cathodes.

INTRODUCTION

Fuel cells are commonly recognized as the most promising power generation systems [1,2]. However, according to the opinion of "The Economist" (November 1998) the energy from fuel cells is actually several times more expensive than the energy from conventional power generation systems. The main perspective for a substantial cost reduction is the elaboration of cheaper components of fuel cells and the improvement of their quality.

One of the most important type of a fuel cell, highly efficient and environmentally clean, are the Molten Carbonate Fuel Cells (MCFC) composed generally from Ni anode and NiO cathode and operated at 600-700°C in the presence of a corrosive liquid Li/K carbonate [1,3]. The major disadvantage of this type of a cathode is dissolution of NiO in K/Li electrolyte, which decreases significantly the cell lifetime [1,4,5]. LiCoO$_2$ cathodes show less solubility [1,6] however, they are far more expensive. S. T. Kuk et al. [7] proposed to cover NiO cathodes with LiCoO$_2$ layer. The acetate Li and Co solutions mixed with PVA were deposited on NiO plates by sol-gel dipping technique. LiCoO$_2$ thin films, were also fabricated by: spray-coating process [8], electrostatic spray pyrolysis [9,10], electrostatic deposition technique [11], r.f. sputtering [10,12], laser ablation deposition [13] and pulsed laser deposition [14]. These materials however were not used in MCFC.

The aim of the presented work was the preparation of LiCoO$_2$ and LiMg$_{0.05}$Co$_{0.95}$O$_2$ thin films on Ni/NiO in order to protect a cathode body against dissolution. A sol-gel dipping technique was selected. According to C.J. Brinker et al. [15,16] this technique is less expensive than the other known processes like chemical vapor deposition, evaporation, sputtering and laser ablation.

Mat. Res. Soc. Symp. Proc. Vol. 623 © 2000 Materials Research Society

EXPERIMENTAL

The Li-Co^{2+}- (Mg^{2+}) starting sols were prepared by a new type [17,18] of a sol-gel process, namely the Complex Sol-Gel Process (CSGP). In this process a very strong complexing agent, having high reduction power-ascorbic acid, is added to the aqueous solution acetates of cations. CSGP was successfully employed for preparation of powders and thin films of LiNi$_{0.5}$Co$_{0.5}$O$_2$ on solid metallic substrates [19]. These sols have been prepared by adding 4M LiOH to 1M aq. acetates solution of Co^{2+} (Mg^{2+}) with ascorbic acid (0.5M) then by alkalizing them with aqueous ammonia to pH=8. The sols were then diluted with water and ethanol. A positive effect of this second solvent was observed in our former work concerning the preparation of SnO$_2$ thin films [20]. This observation was also confirmed in our later work [18] as well as by other authors [21,22].

The coatings were prepared by a dipping technique [18,21] employing the motor-driven dip-coating unit with the immersion time 60 sec and a withdrawal rate 1.7 cm/sec. The porous sintered Ni plates 21.1x 29.7x 0.05cm (produced by FABRICAZIONI NUCLEARI, Italy) oxidized by heating in an ambient atmosphere for 1h at 550°C, were used as the substrates. The covered substrates were maintained for 1h at RT and soaked for 72h at 200°C, then for 1h at 400°C, and calcined for 4h at 650°C, using low heating and cooling rate 1°C/min.

The coatings were characterized by:
- X-ray diffraction (XRD), Co K$_\alpha$ (Positional Sensitive Detector, Ital Structure),
- SEM using a scanning electron microscope (Zeiss DSM 942). The samples were coated with a thin layer (~20nm) of Au. The morphology of a surface (S) was observed. The film thicknesses were measured on fractured cross-sections (CS),
- porosity measurements, using mercury porosimeter type Autopore 9220, Micrometics, USA.

The chemical stability of a LiCoO$_2$-Ni/NiO cathode in molten carbonates was determined by an immersion test. The mixture used was composed of 62 mol% Li$_2$CO$_3$ and 38 mol% K$_2$CO$_3$. The carbonate (35 g) was contained in a pure alumina crucible, over which an air/CO$_2$ 70/30 gas mixture was passed at 50 ml/min flow rate. A LiCoO$_2$-Ni cathode (0.2g, 2 mm in diameter) was accommodated on the bottom of a crucible. The immersion test performed for 350 h at 650°C. About 1 g of the melt has been periodically extracted from a crucible using an alumina pipette and transferred to a gold plate. The cooled melts were dissolved in nitric acid and analyzed for the content of Ni and Co by AAS [23]. After the immersion test, the cathode was washed in diluted acetic acid and dried for taking the scanning electron micrographs (SEM) and energy-dispersive X-ray (EDAX) analysis.

RESULTS AND DISCUSSION

In general a NiO cathode is formed in MCFC by in situ oxidation of porous nickel during the cell start-up [1]. Because in our study the cathode should be covered by a Li spinel layer before introducing it into the cell, preliminary oxidation of Ni sheets is necessary. The thermal analysis of Ni sheets is shown in Fig.1. In dynamic conditions applied during thermal analysis (10°C/min) the oxidation of porous Ni material starts at 600°C (Fig.1). In order to analyze this process in stationary conditions the samples of the substrate have been fired under various conditions. The results are presented in Table I. In this table SEM micrographs and simple mechanical tests of sheet fracture resistance are included.

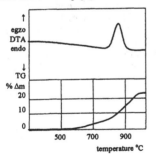

Fig.1. Thermal analysis of sintered Ni sheets.

TABLE I. Oxidation of Ni porous sheets, SEM micrographs and a fracture resistance of plates related to non-treated plates (100%)

Heating conditions⇒	Non-treated	1h, 550°C	20h, 700°C
S ——— 2.5μm			
CS ——— 1.2μm			
NiO content	0	6.7 %	81,4 %
Thickness of NiO layer (t)	0	~0.3 μm	Metallic Ni not observed
mechanical resistance	100%	50%	totally fractured

Very elastic and flat, non-treated sheets, became rigid and brittle after thermal treatment. A decrease of mechanical resistance is connected with the formation of NiO on the cathodes. Moreover they become vowed and folded. These negative features can be avoided by loading the sheets between ceramic plates before thermal treatment. In this case, a slightly higher NiO content was observed. The sheets after heating were perfectly flat.

The results of microporosity are shown in Table II. Ni plates are very porous with a large total pore area and about 15% of closed pores. It is evident that during heating a total pore area and skeletal density decrease but it appears that the closed pores volume remains on a similar level. The $LiCoO_2$ layer strongly protects Ni against oxidation and fully limits the decrease of a total pore volume.

TABLE II. The results of microporosity of porous Ni plates, treated also in an ambient atmosphere and covered with $LiCoO_2$.

No of samples and treatment procedures⇒	Ni	Ni/NiO 20h, 700°C	Ni/NiO covered with $LiCoO_2$, 1h, 800°C
TOTAL INTRUSION VOLUME	0.3339 ml/g	0.1606 ml/g	0.1784 ml/g
TOTAL PORE AREA	18.250 m²/g	0.067 m²/g	5.301 m²/g
MEDIAN PORE DIAMETER (VOL.)	12.4681 μm	12.9861 μm	11.9271 μm
MEDIAN PORE DIAMETER (AREA)	0.0045 μm	5.6988 μm	0.0039 μm
AVERAGE PORE DIAMETER	0.0732 μm	9.6229 μm	0.1346 μm
BULK DENSITY	2.1330 g/ml	2.9336 g/ml	2.8479 g/ml
APPARENT (SKELETAL) DENSITY	7.4092 g/ml	5.5484 g/ml	5.7897 g/ml

Note. Densities of: metallic Ni=8.90 g/cm³, NiO=6.72 g/cm³

TABLE III. SEM micrographs of Ni/NiO cathodes covered with $LiMg_{0.05}Co_{0.95}O_2$ fired for 4h at 650°C

Type of sol $\Rightarrow$	Parent sol (PS) Viscosity(η)= 35.4 cSt Concentration of Σ Me 46.8 g/l	1 volume of PS 1 volume of Ethanol η =14 cSt	1 volume PS 3 H_2O 4 volume Ethanol η= 2.6 cSt
S ___ 2.5μm			
CS ___ 1.2μm			
	t=1.78μm	t=1.14μm	t=0.78μm

Thin films of $LiCoO_2$ and $LiMg_{0.05}Co_{0.95}O_2$ from various sols were deposited on Ni/NiO plates oxidized for 1h at 550°C, (see Table I). The SEM micrographs of $LiMg_{0.05}Co_{0.95}O_2$ films are presented in Table III. It is evident that the best films were obtained from the 1PS:1Ethanol. The film thicknesses increase with sol concentration and viscosity.

The scanning electron micrographs and EDAX spectra of the covered electrode before and after a 350h immersion test are shown in Fig.2 and 3, respectively. The results of the analyzed melts indicate that the $Ni/NiO/LiCoO_2$ cathode is stable during the test because no trace of cobalt and nickel was detected by chemical analysis. In contrast, the separately prepared $LiCoO_2$ powder exhibited small solubility of several ppm of Co. As can be seen no changes in the morphological features of the electrodes, as well as in the mean size of particles, have been noticed. The EDAX analysis confirmed the presence of cobalt in the specimen before and after the test.

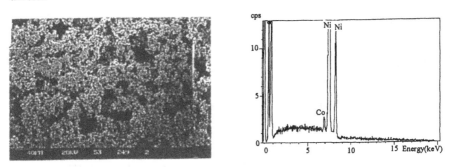

Fig.2. SEM micrograph and EDAX spectrum of $Ni/NiO/LiCoO_2$ cathode before treatment.

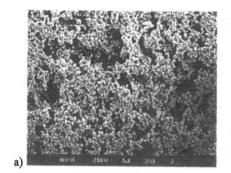 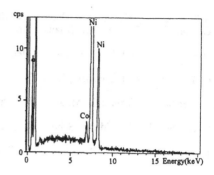

a)

Fig.3. SEM micrograph and EDAX spectrum of Ni/NiO/LiCoO$_2$ cathode after 350 h treatment at 650°C in (Li$_{0,62}$K$_{0,38}$)$_2$ CO$_3$ and pCO$_2$ = 0,3 atm, pO$_2$ = 0.14 atm.

CONCLUSIONS

1. Sol-gel process for covering Ni/NiO cathodes of dimensions 21.1x29.7x0.05cm with LiCoO$_2$ and LiMg$_{0.05}$Co$_{0.95}$O$_2$ for MCFC has been elaborated,
2. Film thicknesses varied from 0.6 to 2 µm increasing on sol concentration and viscosity,
3. Dilution of the parent sols with ethanol strongly improved the coating quality,
4. Covering of a Ni/NiO cathode body with LiCoO$_2$ was very effective as a protection against dissolution in a MCFC carbonate electrolyte,
5. Dissolution of LiCoO$_2$ coating was not observed,
6. Heat treatment should be carried out under the dead load of the ceramic plates in order to avoid their waving.

ACKNOWLEDGMENT

The authors would like to thank A. Jarzębski and A. Lachowski (Institute of Chemical Engineering, Polish Academy of Science, Poland) for porosity measurements.

REFERENCES

1. C.Yuh, R. Johnsen, M. Farooque, H. Maru, J. Power Sources 56, 1(1995).

2. K. Joon, J. Power Sources, 71, 12 (1998).

3. A.L.Dicks, J. Power Sources, 71, 111 (1998).

4. N.T. Fukui, H. Okawa, T. Tsunoka, J. Power Sources 71, 239 (1998).

5. C. Belhomme, M. Cassir, C. Tessier, E. Berthoumieux, Electrochemical and Solid State Letters, 3, 216 (2000).

6. A. Lundblad, B. Bergman, Fabrication and characterization of LiCoO$_2$ cathode material for MCFC, New Materials for Fuel Cell and Modern Battery Systems II, Editors, O. Savadogo and P.R. Roberge, ISBN 2-553-00624-1, Montreal, Canada, 1997, p86.

7. S.T. Kuk, H.J. Kwon, H.S. Chun, K.K. Kim, Properties of LiCoO$_2$-coated NiO MCFC Cathode, Fuel Cell Seminar, Orlando, Florida, USA, 1996, Proceedings, p. 367.

8. H.S.W. Chang, T.J. Lee, S.C. Lin, J.H. Jeng, J. Power Sources, 54, 403 (1995).

9. C.H. Chen, A. A. J. Buysman, E. M. Kelder, J. Schoonman, Solid State Ionics, 80, 1(1995).

10. P. Fragnaud, T. Brouse, D. M. Schleich, J. Power Sources 63, 187 (1996).

11. C. H. Chen, E. M. Kelder, J. Schoonman, J. Mat. Sci. 31, 5437 (1996).

12. J. Lee, S. Lee, H. Baik, H. Lee, S. Wang, S.M. Lee, Electrochemical and Solid State Letters, 2, 512 (1999).

13. M. Antaya, K. Cearns, J.S. Preston, J.N. Reimers, J.R. Dahn, J. Appl. Phys, 76, 2799 (1994).

14. K.A. Striebel, C.Z. Deng, S.J. Wen, E.J. Cairns, J. Electrochem. Soc. 143, 1821 (1996).

15. C.J. Brinker, G.C. Frye, A.J. Hurd, C.S. Asley, Thin Solid Films, 201, 97 (1991).

16. C.J. Brinker, G.W. Scherer, Sol-Gel Science, The Physics and Chemistry of Sol-Gel Processing ,1990 (Academic Press, San Diego).

17. A. Deptuła, W. Łada, T. Olczak, M. Lanagan, S.E. Dorris, K.C. Goretta, R.B. Poeppel, Method for preparing of high temperature superconductors, Polish Patent No 172618, June 2 1997. Claimed and valid since October, 29,1993.

18. A. Deptuła, W. Łada, T. Olczak, K.C. Goretta, A. Di Bartolomeo, S. Casadio, Sol-gel Process for preparation of YBa$_2$Cu$_4$O$_8$ from acidic acetate/ammonia/ascorbic acid systems, Mat. Res. Bull, 32, 319 (1997).

19. A.Deptuła, W.Łada T.Olczak, F.Croce, F.Ronci, A.Ciancia, L.Giorgi, A.Brignocchi, A.Di Bartolomeo, Synthesis and Preliminary Electrochemical Characterization of LiNi$_{0.5}$Co$_{0.5}$O$_2$ Powders Obtained by the Complex Sol Gel Process (CSGP), MRS Fall'97 Meeting, December 1-5,1997 Boston, USA, Proceedings, Volume 496, Symposium Y: Materials for Electrochemical Energy Storage and Conversion II- Batteries, Capacitors and Fuel Cells, Pittsburg PA, USA,1998. p. 237 – 242.

20. W.Łada, A.Deptuła, T.Olczak, W.Torbicz, D.Pijanowska, A.Di Bartolomeo, Preparation of Thin SnO$_2$ Layers by Inorganic Sol-gel Process, J. of Sol-Gel Science and Technology, 2, 551(1994).

21. X.Orignac, H.C.Vasconcelos, X.M.Du, R.M.Almeida, J.of Sol-Gel Science and Technology, 8, 243 (1997).

22. N.Özer, Thin Solid Films, 304, 310(1997).

23. S. Scaccia, Talanta, 49, 467(1999).

EFFECTS OF THE HEAT-TREATMENT CONDITIONS ON MICROSTRUCTURES OF YbBa$_2$Cu$_3$O$_{7-\delta}$ SUPERCONDUCTING FILMS FORMED BY THE DIPPING-PYROLYSIS PROCESS

J.SHIBATA*, K.YAMAGIWA**, I.HIRABAYASHI**, T.HIRAYAMA*, Y.IKUHARA***
*Japan Fine Ceramics Center, 2-4-1 Mutsuno, Atsuta-ku, Nagoya 456-8587, Japan
**Superconductivity Research Laboratory, 2-4-1 Mutsuno, Atsuta-ku, Nagoya 456-8587, Japan
***Engineering Research Institute, School of Engineering, The university of Tokyo, 2-11-16 Yayoi, Bunkyo-ku, Tokyo 113-8656, Japan

ABSTRACT

YbBa$_2$Cu$_3$O$_{7-\delta}$(Yb123) films were formed on SrTiO$_3$(STO)(001) and LaAlO$_3$(LAO)(001) substrates by the dipping-pyrolysis process. Using transmission electron microscopy, we investigated effects of the heat-treatment conditions in the processes of the dipping-pyrolysis method on microstructures of these films. As a result, we found that the high heating rates at the initial and final heat-treatments are necessary for achieving the epitaxial growth of the superconducting films.

INTRODUCTION

Since the breakthrough attributed to Bednorz and Muller in 1986 on their discovery of the superconducting oxide with the high critical temperature(T$_C$)[1], many studies have been performed to produce these high T$_C$ superconductors. The dipping-pyrolysis is a promising method for producing the superconducting films at a low cost. This method has the advantage of fabricating the films on substrates with any shape and any size as well as the advantage of enabling the chemical composition of the films to be controlled. McIntyre et al. reported that they succeeded in producing YBa$_2$Cu$_3$O$_{7-\delta}$(Y123) films with zero-field critical current densities(J$_C$) higher than 5×10^6A/cm^2 at 77K on LAO substrates[2]. The dipping-pyrolysis method contains two heat-treatment steps: the first step is the thermal decomposition of the metal organic compounds on the substrate to form an amorphous precursor film; the second step is the crystallization of the precursor film to form the superconducting final film. During these two heat-treatments, it is important to control the nucleation and growth of crystals for obtaining the good superconducting properties. In the present paper, we report the effects of the conditions at the initial and final heat-treatments on microstructures of Yb123 films.

EXPERIMENT

The homogeneous solution with a molar ratio of Yb:Ba:Cu=1:2:3 and metal concentration of 0.4mol/l was prepared by dissolving metal naphthenates in toluene. STO(001) and LAO(001) substrates were spin-coated with this solution. These spin-coated STO substrates were heat-treated at 698K in air with the different heating and cooling rates, as shown in Fig.1: (a)in the fast process, the substrate was directly inserted into the furnace kept at 698K, hold for 20min, and then rapidly removed from the furnace to room temperature; (b)in the slow process, the substrate was gradually heated at the heating rate of 0.5K/min, kept for 20min at 698K, and finally cooled down at the cooling rate of 2K/min. we abbreviate the precursor film prepared by the fast process as PF(F) and the precursor film by the slow process as PF(S). PF(F) and PF(S) were heated up to 998K in an Ar gas flow at the heating rate of 3K/min, and kept for 2hours to form the superconducting final films, which are called FF(F) and FF(S) respectively.

401

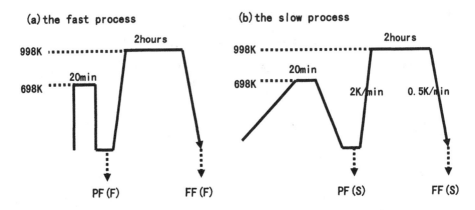

Fig.1 Schematic diagram of heating schedule for preparing Yb123 precursor and final films on STO(001) substrates: (a) the fast process; (b) the slow process.

On the other hand, the LAO(001) samples were heat-treated at 698K by the fast process, and then heated at 1023K under the different conditions of the heating rate and holding time. Table 1 lists the final heat-treatment conditions for preparing Yb123 films on LAO substrates. FF(A), FF(B) and FF(C) were produced by heating the precursor films up to 1023K at 20K/min, 3K/min and 0.5K/min respectively, and holding them for 10hours. FF(D) was fabricated by heating the precursor film at 3K/min and holding it for 2hours at 1023K. FF(E) was produced by heating the precursor film at 0.5K/min to 1023K, and immediately cooling it to room temperature. The cooling rate was 3K/min, when all the films were prepared. The thickness of these films was about 500nm.

Crystal phases and orientations in all of the precursor and final films were characterized by X-ray diffraction patterns(XRD) using Cu-K α radiation.

Cross-sectional specimens for TEM observations were prepared by "sandwiches-gluing", slicing, grinding, dimpling and then ion thinning under a cold stage (liquid N_2). Transmission electron microscope used in this work was JEM-2010 operating at 200kV with the point to point resolution of 0.194nm.

Table. 1 Final heat-treatment conditions for preparing Yb123 final films on LAO(001) substrates.

Sample Name	Heating Rate (K/min)	Holding Time at 1023K (hours)	Cooling Rate (K/min)
FF(A)	20	10	3
FF(B)	3	10	3
FF(C)	0.5	10	3
FF(D)	3	2	3
FF(E)	0.5	0	3

RESULTS

Effects of the initial heat-treatment conditions

In the XRD patterns, all the precursor films on STO substrates showed amorphous-like humps. However, TEM studies revealed the different microstructures of the precursor films prepared by heating under the various conditions. Figures 3(a) and 3(b) show cross-sectional transmission electron micrographs and electron diffraction patterns of PF(F) and PF(S), respectively. These figures were taken along the [100]STO direction. As shown in Fig.3(a), PF(F) is amorphous. In contrast, PF(S) is polycrystalline. The sizes of crystals generated in the PF(S) are about 10nm. From the electron diffraction pattern of PF(S), we calculated lattice spacings of these crystals. Most of them agreed with those for Yb123. The spacings of 0.125nm and 0.106nm, which did not agree with those for Yb123, corresponded to (222) and (131) of CuO.

Figures 4(a) and 4(b) show the cross-sectional electron micrographs of FF(F) and FF(S) respectively, observed from the [100]STO direction. As shown in Fig.4(a), FF(F) has grown into a c-axis oriented Yb123 film. This film is defective with many stacking faults, which is consistent with the streaks along the c*-axis in the selected-area diffraction pattern shown at the top left of Fig.4(a). On the other hand, FF(S) is a polycrystalline film containing Yb123 and other crystals such as Yb_2O_3. In this film, randomly oriented Yb123 crystals were seen

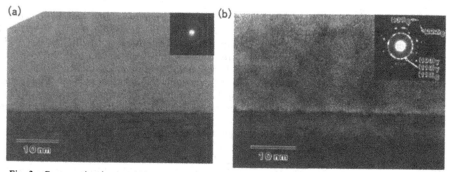

Fig. 3   Cross-sectional transmission electron micrographs and electron diffraction patterns of (a) PF(F) and (b) PF(S), observed from the [100]STO direction. PF(F) is amorphous, in contrast, PF(S) is polycrystalline. In the diffraction pattern of PF(S), 'C' and 'Y' indicates CuO and Yb123, respectively.

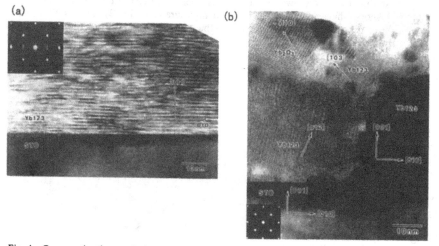

Fig. 4   Cross-sectional transmission electron micrographs of (a) FF(F) and (b) FF(S), observed from the [100]STO direction. In the FF(S), randomly oriented Yb123 crystals are generated.

and these crystals were grown over 200nm in size.

From these results of the TEM observation, we suggest the growth mechanism of the superconducting films as follows: in the case of the slow process, the metal organic compounds on the STO substrates are decomposed at different temperatures during heating gradually up to 698K (which was confirmed by differential thermal analysis(DTA)[3]). This leads to the random nucleation of nonsuperconducting crystals such as CuO and Yb_2O_3 as well as Yb123 crystals. When this polycrystalline precursor film is heated at 998K, the growth of those random nuclei in the film dominates the epitaxial growth of the Yb123 film.

As for superconducting properties, FF(F) showed a sharp resistive transition around T_C, however, FF(S) exhibited the broad transition. In conclusion, the rapid heating rate at the initial heat-treatment is necessary for achieving the epitaxial growth and the sharp resistive transition of the Yb123 film[4].

Effects of the final heat-treatment conditions

All of the precursor films on LAO substrates were amorphous as same were the precursor films prepared on the STO substrates by the fast process. FF(A), FF(B), FF(C), FF(D) and FF(E) were prepared by heating these amorphous precursor films at 1023K. In the XRD patterns of these final films, a significant difference was not seen, except that Yb124 peaks were visible in the patterns of FF(B), FF(C), FF(D) and FF(E): the intensity of (00l)LAO and (00l)Yb123 peaks was very strong, and nonsuperconducting phases such as Yb_2O_3 and $BaCuO_2$ were not detected in all the patterns[5]. Therefore, we investigated the effects of the final heat-treatment conditions on the microstructures of Yb123 films by transmission electron microscopy.

Figure 5 shows a typical cross-sectional electron micrograph of FF(A), observed from the [100]LAO direction. An electron diffraction pattern at the top right in this micrograph was obtained from the selected-area which included both film and substrate. It is noted that the film is a c-axis oriented Yb123 film with a thickness of 100nm. This film seems to have the good crystallinity, because the spots along c*-axis are clear in the electron diffraction pattern. In this figure, the surface of the film was removed by ion thinning. However, we confirmed that polycrystalline film had grown over the c-axis oriented Yb123 film in other areas.

Figure 6 shows the cross-sectional electron micrograph of FF(B). In this film, all the surface of the LAO substrate was covered with the c-axis oriented Yb123 film as well as in FF(A). However, a-axis oriented Yb123 crystals generated in the middle of the film were also visible in other regions of FF(B). In the case of FF(C), an amorphous layer is seen in the vicinity of the interface between the film and the substrate, as shown in Fig.7. An a-axis oriented Yb123 and a nonsuperconducting CuO crystal are also visible. Figure 8 shows the typical cross-sectional electron micrograph of FF(D). This film is polycrystalline, containing randomly oriented Yb123 and non-

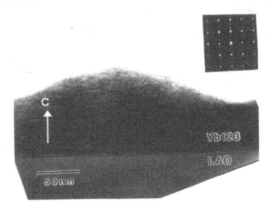

Fig: 5 Cross-sectional transmission electron micrograph of FF(A), observed from the [100]LAO direction. This film is a c-axis oriented Yb123 film with a thickness of 100nm.

superconducting crystals. The c-axis oriented Yb123 crystal is seen in the shape of an island in this film. Figures 9(a) and 9(b) show typical cross-sectional micrographs of FF(E). Many nonsuperconducting phases such as $BaCuO_2$ and $Yb_2Cu_2O_5$ were generated in this film.

On the basis of these results of the TEM observation, we discuss the effects of the heating rate and holding time at the final heat-treatment on the microstructures of the Yb123 films. Focusing our discussion on the effects of the heating rate, we can compare the microstructure of FF(C) with those of FF(A) and FF(B). In FF(A) and FF(B), the entire surface of the substrate was covered with the c-axis oriented Yb123 film. On the other hand, in FF(C) amorphous regions and nonsuperconducting crystals were seen. As a result of this, it is found that the rapid heating rate at the final heat-treatment is necessary for the epitaxial growth of the Yb123 film. Furthermore, when we compare the microstructure of FF(D) with that of FF(B), we found that the long holding time at the final heat-treatment is effective for the grain growth of the c-axis oriented Yb123 crystals. However, in the film where many nonsuperconducting crystals are generated such as in FF(E), the epitaxial growth of the Yb123 film is suppressed. This is the reason why FF(C) was not grown into the c-axis oriented Yb123 film.

When the amorphous precursor film is heated rapidly at the final heat-treatment, the heterogeneous nucleation and epitaxial growth of the Yb123 film is progressed from the

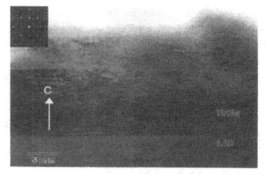

Fig. 6   Cross-sectional transmission electron micrograph of FF(B). In this film, the entire surface of the substrate was covered with the c-axis oriented Yb123 film as well as in FF(A).

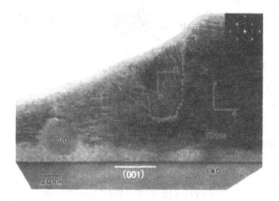

Fig. 7   Cross-sectional transmission electron micrograph of FF(C). In this film, an amorphous layer is seen in the vicinity of the interface between the substrate and the film.

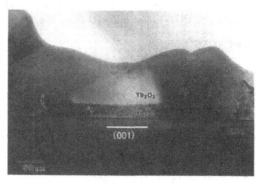

Fig. 8   Typical cross-sectional transmission electron micrograph of FF(D). The c-axis oriented Yb123 crystal is seen in the shape of an island, and Yb_2O_3 is also visible.

surface of the substrate. In contrast, the slow heating facilitates the random nucleation and growth of crystals.

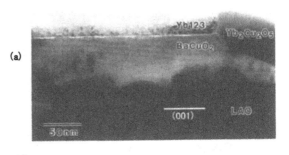
(a)

(b)
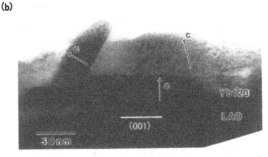

CONCLUSIONS

We observed cross sections of Yb123 precursor films and final films by transmission electron microscopy, and investigated the effects of the heat-treatment conditions in the processes of the dipping-pyrolysis method on microstructures of the films. Amorphous precursor films were prepared by heating rapidly the spin-coated substrates at 698K in air. The c-axis oriented Yb123 films were produced by the rapid heating of these amorphous precursor films over 973K. In addition, the c-axis oriented Yb123 film was able to be fabricated by the long heat-treatment of the polycrystalline precursor film including a little nonsuperconducting crystal.

Fig. 9 Typical cross-sectional transmission electron micrographs of FF(E): (a) nonsuperconducting crystals such as $BaCuO_2$ and $Yb_2Cu_2O_5$ are visible; (b) ramdomly oriented Yb123 crystals are generated in this film.

In conclusion, the high heating rates at the initial and final heat-treatments are necessary for the epitaxial growth of the Yb123 film, and the long holding time at the final heat-treatment is effective for the grain growth of the c-axis oriented Yb123 crystals.

ACKNOWLEDGMENTS

This work was supported by the New Energy and Industrial Technology Development Organization(NEDO).

REFERENCES

1. J.B.Bednorz and K.A.Muller, Z.Phys., p189, B64(1986).

2. P.C.McIntyre, M.J.Cima, J.A.Smith Jr., R.B.Hallock, M.P.Siegal and J.M.Phillips, J.Appl.phys., p.1868, 71(1992).

3. K.Yamagiwa and I.Hirabayashi, Physica C, p.12, 304(1998).

4. J.Shibata, K.Yamagiwa, I.Hirabayashi, X.L.Ma, J.Yuan, T.Hirayama and Y.Ikuhara, Jpn.J.Appl.Phys., p.5050, 38(1999).

5. J.Shibata, K.Yamagiwa, I.Hirabayashi, T.Hirayama and Y.Ikuhara, Jpn.J.Appl.Phys., in press.

INFLUENCE OF LOW TEMPERATURE-GROWN GaAs ON LATERAL THERMAL OXIDATION OF Al$_{0.98}$Ga$_{0.02}$As .

J. C. FERRER[1], Z. LILIENTAL-WEBER[1], H. REESE[2]. Y.J. CHIU[2], E. HU[2]

[1] Materials Science Division, Lawrence Berkeley National Laboratory, Berkeley, CA
[2] Department of Electrical and Computer Engineering, University of California, Santa Barbara, CA

ABSTRACT

The lateral thermal oxidation process of Al$_{0.98}$Ga$_{0.02}$As layers has been studied by transmission electron microscopy. Growing a low-temperature GaAs layer below the Al$_{0.98}$Ga$_{0.02}$As has been shown to result in better quality of the oxide/GaAs interfaces compared to reference samples. While the later have As precipitation above and below the oxide layer and roughness and voids at the oxide/GaAs interface, the structures with low-temperature have less As precipitation and develop interfaces without voids. These results are explained in terms of the diffusion of the As toward the low temperature layer. The effect of the addition of a SiO$_2$ cap layer is also discussed.

INTRODUCTION

Lateral oxidation of Al$_x$Ga$_{1-x}$As layers is a very attractive technology for the fabrication of isolating oxide layers in optoelectronic devices because of their stability, high resistivity and near planar topology. They have been used in forming self-aligned dielectric layers in the fabrication of semiconductor laser diodes and on vertical cavity surface emitting laser (VCSEL) applications due to the excellent carrier confinement provided by the oxidized layer. These methodcan also be used attention in metal-oxide-semiconductor (MOS) devices. The high quality of the oxide is attributed to the formation of stable AlO(OH) and Al$_2$O$_3$ compounds [1]. However some problems related to the excess As created during the process, and weakness of the oxide interfaces, due to structural changes in the Al$_x$Ga$_{1-x}$As layers, remain unsolved [2,3].

The influence of parameters, such as temperature, layer thickness or composition, on the oxidation process has been the subject of recent studies. In this work we study the structural changes resulting from the inclusion of a low-temperature (LT) GaAs layer. The effects of the presence of a LT-GaAs on the oxidation rates was reported previously [4] indicating a higher oxidation rate for samples including LT-GaAs layers. The influence of the incorporation of a SiO$_2$ capping layer on the quality of the oxide layer is also discussed.

EXPERIMENTAL

Samples were grown by molecular beam epitaxy (MBE) on a (100) semi-insulating GaAs substrate. Two similar structures (shown in Fig.1) were grown to be oxidized. The only difference between the two types of sample is that in one the central 300 nm thick layer is standard GaAs grown at 580°C whereas in the other sample it is low temperature GaAs, grown at 210°C. The low temperature GaAs was annealed at 600°C for two minutes in-situ and received further annealing during growth of the subsequent layers at 590°C: 100 nm of n-GaAs(Si:10^{18} cm^{-3}), 30 nm of Al$_{0.98}$Ga$_{0.02}$As, and a capping layer of 35 nm GaAs. In addition thin layers (0.5-10nm) of AlAs were grown on either side of the LT. The reference and the LT samples were processed simultaneously. Mesas were formed in the top of the samples by patterning and

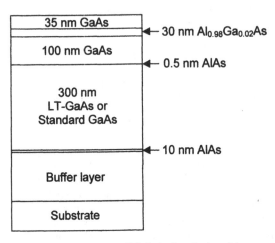

```
┌─────────────────────┐
│     35 nm GaAs      │ ◄── 30 nm Al₀.₉₈Ga₀.₀₂As
├─────────────────────┤
│     100 nm GaAs     │
│                     │ ◄── 0.5 nm AlAs
├─────────────────────┤
│       300 nm        │
│    LT-GaAs or       │
│   Standard GaAs     │
├─────────────────────┤
│                     │ ◄── 10 nm AlAs
│    Buffer layer     │
│                     │
├─────────────────────┤
│     Substrate       │
│                     │
└─────────────────────┘
```

Fig. 1. Sample structure. LT GaAs is substituted by standard GaAs in reference sample.

etching in $H_3PO_4:H_2O_2:H_2O$ (3:3:100). The etch extended past the AlGaAs but not the LT GaAs, exposing only the AlGaAs and part of the n-GaAs sidewalls to the oxidizing ambient. One more sample with LT-GaAs include a PECVD SiO_2 cap. Oxidation was carried out in a water vapor with a N_2 carrier gas, which was bubbled through water heated to 80°C and flowed over the sample, placed in a furnace held at a constant temperature of 450°C. Samples were prepared for transmission electron microscopy (TEM) observation by conventional mechanical polishing and Ar ion milling in a cooled stage until perforation. Topcon 002B and JEOL ARM microscopes were used for these studies.

RESULTS

Annealing of the LT-GaAs layers during the subsequent growth causes the formation of As precipitates that can be recognized as dark spots in Fig. 2. Previous investigations [5,6] show that LT-GaAs as grown is non-stoichiometric, containing excess As in amounts up to 1.5%. The annealing leads to a decrease of the concentration of As_{Ga} antisite defects with simultaneous formation of hexagonal As The average size of the precipitates prior to oxidation is about 4.3 nm, consistent with our previous studies [6]. A slight increase in precipitate average size, about 2%, is detected after oxidation.

After oxidation of samples with a single GaAs capping layer, both standard and LT-GaAs samples developed a homogeneous oxide layer. However in the case of the sample with standard GaAs (Fig. 3.a), As precipitates were formed above and below the original $Al_{0.98}Ga_{0.02}As$ layer as a product of the oxidation. Furthermore, the interface between the oxide and the surrounding GaAs was rough and degraded by the presence of voids that may cause delamination.

It has been proposed [1] that the products of the oxidation reaction of AlAs are mainly Al oxides and hydroxides, and AsH_3. AsH_3 is a material which was assumed to escape to the surface:

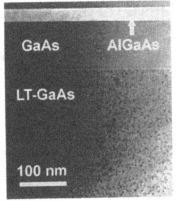

Fig. 2. Top region of the sample containing LT-GaAs. Note the As precipitates in LT layer as dark spots.

$$2AlAs + 3H_2O \rightarrow Al_2O_3 + AsH_3 \tag{1}$$

$$AlAs + 2H_2O \rightarrow AlO(OH) + AsH_3 \tag{2}$$

$$AlAs + 3H_2O \rightarrow Al(OH)_3 + AsH_3 \tag{3}$$

However our results show that a significant amount of As remains in the sample after oxidation suggesting that either arsine decomposes [7]:

$$2AsH_3 \rightarrow 2As + 3H_2 \tag{4}$$

Or that direct formation of As and H_2 takes place by substitution of reaction (4) in (1), (2) and (3).

Conversely, the sample that includes a 300 nm LT GaAs layer, instead of standard GaAs, developed higher quality oxide/GaAs interfaces (Fig. 3.b). Arsenic precipitates in the vicinity of the oxide layer are only occasionally found in this case, and the interfaces were smoother, with no voids along the interfaces. Another interesting feature is the faster oxidation rate of the sample with the LT GaAs layer (21 μm in 10 min.) compared to the standard sample (10 μm in the same time) [4].

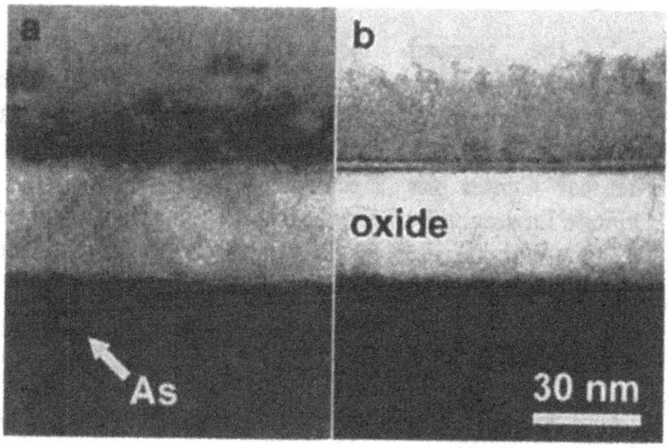

Fig. 3. Oxidized $Al_{0.98}Ga_{0.02}As$ layer in standard sample (a) and in sample with LT-GaAs (b). Arsenic segregated from the oxide and voids are found at the interfaces in (a). Interface quality is greatly improved in (b).

It is not yet clear what is the reason for reduced As precipitation near the oxidized layer when an underlying layer of annealed LT-GaAs is present. One possible explanation is that the presence of As precipitates in the annealed low-temperature layer acts as a sink for excess As so that near the oxidized layer the excess As concentration never reaches the critical value for precipitate nucleation. Another factor that could play a role is introduction of some excess Ga vacancies during annealing at of the low temperature GaAs into the layers above the low temperature layer which could facilitate As diffusion away from the oxide layer. Migration of excess As away from the oxidation front to the As precipitates in the low temperature layer is also consistent with the observed small increase in size of the As precipitates after the oxidation treatment.

Finally, we present the results for the sample that includes a top SiO_2 capping layer. The micrographs (Fig. 4) show again sharp interfaces and clean from As, like in the case that no capping layer is included.

Future work involves the capping of the sample with Si_3N_4 which is known to be non-permeable to As, acting as a barrier to outdiffusion. This will allow us to assess whether the arsenic accumulations diffuse mostly towards the LT-layer or leave the sample through the surface.

CONCLUSIONS

The influence of a low-temperature-grown GaAs layer, on the oxidation behavior of an $Al_{0.98}Ga_{0.02}As$ layer, has been investigated by TEM observations. Results show an improvement of the quality of the oxide/GaAs interfaces when a LT GaAs layer is included.

The exact reason for this improvement is not yet clear but it appears that the presence of As precipitates in the annealed low temperature layer may be acting as a sink for As thus reducing its build up near the oxidation front.

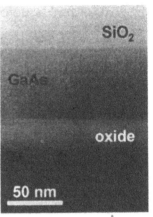

Fig. 4. Sample including a SiO₂ capping layers and LT GaAs. It exhibits clean, sharp oxide/GaAs interfaces.

ACKNOWLEDGMENTS

The research in Lawrence Berkeley National Laboratory was supported by AFOSR-ISSA-99-0012, and in UC Santa Barbara by AFOSR F49620-95-1-0394, both under the sponsorship of Dr. G. Witt. The use of the facility of the National Center for Electron Microscopy at the Ernest Orlando Lawrence Berkeley Laboratory is greatly appreciated. Dr. W. Walukiewicz is also gratefully acknowledged for helpful discussions about vacancy diffusion.

REFERENCES

1. A. R. Sugg, N. Holonyak Jr., J. E. Baker, F. A. Kish, and J. M. Dallesasse, Appl. Phys. Lett. **58**, p. 1199 (1991)

2. S. Guha, F. Agahi, B. Pezeshki, J.A. Kash, D. W. Kisker and N. A. Bojarczuk, Appl. Phys. Lett. **68**, p. 906 (1996)

3. Z. Liliental-Weber, S. Ruvimov, W. Swider, J. Washburn, M. Li, G. S. Li, C. Chang-Hasnain and E. R. Weber, SPIE proceeding series **3006**, p. 15 (1997)

4. H. Reese, Y. J. Chiu and E. Hu, Appl. Phys. Lett. **73**, p. 2624 (1998)

5. M. Kaminska, E. R. Weber, Z. Liliental-Weber, R. Leon and Z. Rek, J. Vac. Sci. Technol. **B 7**, p. 710 (1989)

6. Z. Liliental-Weber, X. W. Lin and J. Washburn, Appl. Phys. Lett. **66**, p. 2086 (1995)

7. C. I. H. Ashby, M. M. Bridges, A. A. Allerman, B. E. Hammons, Appl. Phys. Lett. **75**, p. 73 (1999)

Fabrication of $Bi_2Sr_2CaCu_2O_8$ Superconductor Thick Films on Cu Substrates

Sang-Chul Han, Tae-Hyun Sung, Young-Hee Han, Jun-Seong Lee, and Sang-Joon Kim
103-16 Munji-dong, Yusong-gu, Taejon, Korea, Power System Laboratory, Korea Electric Power
Research Institute

ABSTRACT

Well oriented $Bi_2Sr_2CaCu_2O_8$(Bi2212) superconductor thick films were formed successfully
on Cu tapes by liquid reaction between Cu-free precursors and Cu tapes.   Cu-free Bi-Sr-Ca-O
powder mixtures were screen-printed on Cu tapes and heat-treated at 850-870°C for several
minutes in air. Cu-free precursors were composed of Bi_xSrCaO_y(x=1.2-2). In order to obtain the
optimum heat-treatment condition, we studied on effects of the precursor composition, heat-
treatment temperature and time, the screen-printing thickness, and the heat-treatment atmosphere
on the superconducting properties of Bi2212 films. Microstructures and phases of films were
analyzed by XRD and optical microscopy. The electric properties of superconducting films were
examined by the four probe method. At heat-treatment temperature, the specimens were in a
partially molten state by liquid reaction between CuO in the oxidized copper tape and the
precursors. The non-superconducting phases in the molten state are mixtures of Bi-free phase
and Cu-free phases.

INTRODUCTION

An impressive progress has been made in large scale application of HTS (high temperature
superconducting) wire technology in last few years and close to a commercial performance level
[1,2]. The U.S. Department of Energy announced an award of a project to Pirelli to install and
demonstrate the HTS power cable of 100 MW scale in Detroit Edison's network by the end of
year 2000 [3].

Soon after HTS materials were discovered, scientists have dramatically improved operating
temperatures. But in manufacturing ceramic-based HTS wire they have difficulties in making
strong, flexible wires out of materials that are as brittle as blown glass. So researchers began
experimenting with many novel wire-making methods to obtain HTS wires of high critical
current density, sufficient strength and flexibility to be handled. By 1995, two manufacturing
techniques showed great potential for producing high-performance HTS wires. They are the
silver-sheathed BSCCO powder-in-tube (PIT) method [4] and the YBCO coated-conductor
method [5]. Up to date, the PIT method has been the most useful method because of its
applicability to high-production manufacturing. The coated conductor method is not likely to
perfect after 10 years because of the difficulty in the fabrication of long wires. In spite of the

411

relatively impressive performance of PIT wires, there remains a serious high cost problem for the technology to make broad commercial impact, competing directly with copper wires in power applications. The high cost is attributed to the complex process for PIT wire manufacturing and the price of the silver sheath.

The object of this study is to develop a process to fabricate $Bi_2Sr_2CaCu_2O_x$ (Bi2212) tapes at a reduced cost by using less expensive material, Cu, and by simplifying the manufacturing step. In order to utilize the excess oxidation of the copper substrate during annealing, copper-free precursors, xBi_2O_3-$SrCO_3$-$CaCO_3$ mixtures (x=1.2~2), were placed on copper tapes by the screen-printing method and heat-treated. We carried out experiments to optimize the preparation condition, such as the composition of the precursor powder, the screen-printing thickness, heat-treatment temperature and time, and the heat-treatment atmosphere.

EXPERIMENT

Bi_2O_3 (4N), $CaCO_3$ (4N), and $SrCO_3$ (3N) powders were mixed in the molecular ratio of Bi_2-$_xSrCaO_z$ (x=0, 0.5, 0.7, and 0.8) and wet-milled in a planetary pot for 2 hours. An organic formulation consisting of solvent (butyl carbitol + terpineol), binder (ethyl cellulose) and dispersant (triolein) was then added and the mixture was homogenized in an agate mortal to form a paste. The paste was printed on copper plates (10 mm in width, 10 mm in length and 0.3 mm in thickness) through a 150 mesh silk screen. The thickness of one time printing layer was about 20~25 μm. The screen-printed films were dried at 80 °C and heat-treated at temperatures between 820 °C and 880 °C for 30 sec to 120 sec in different atmospheres. The heat-treatment was carried out in a gold crucible for rapid heating. Phase identification was performed by scanning electron microscopy, x-ray diffraction (XRD) and energy-dispersive x-ray (EDX) analysis. The temperature dependence of resistance was measured by a conventional four-probe method.

RESULTS

Figure 1 shows the resulting phase colony when a powder mixture of Bi : Sr : Ca = 2 : 1 : 1 (211 precursor) in molecular ratio was screen-printed once on copper tapes and the printed thick films were preheated at 820°C for 60 sec and then heat-treated at 850-870°C for 15 - 80 sec in air. The reason for pre-heating at 820°C is to remove the organic materials. Bi-rich composition of the precursor powder is to make rapidly well-oriented superconducting thick films by using the partial melting between the copper substrate and the printed layer. The non-superconducting phases in the molten state was mixtures of Bi-free phase and Cu-free phases, as shown in Figure 2. Figure 2 shows XRD pattern of the thick film prepared by screen-printing the Bi_2SrCaO_x precursor on a Cu plate and heat-treating at 870°C for 25 sec in air.

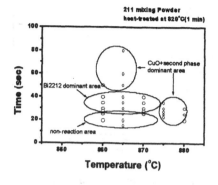

Figure 1. The phase colony for heat-treatment temperature and time of Cu/Bi₂SrCaOₓ powder.

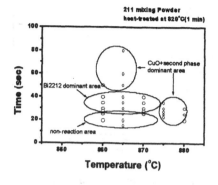

Figure 1. The phase colony for heat-treatment temperature and time of Cu/Bi_2SrCaO_x powder.

Figure 2. XRD pattern of the thick film prepared by screen-printing the Bi_2SrCaO_x precursor on a Cu plate and heat-treating at $870°C$ for 25 sec in air.

On the basis of the above results, the optimum condition of the heat-treatment and the characteristic analysis result in case of 211 precursor are as follows. Figure 3 shows the SEM image of a Bi2212 thick film, after screen-printing the 211 precursor on a copper plate, heat-treated at 820°C for 1 min, 870°C for 50 sec and 830°C for 3 min in air consecutively. The annealing at 830°C is to transform the residue liquid to Bi2212. Figure 4 shows the XRD pattern of the same Bi2212 thick film as for Figure 3. CuO or Cu₂O peak was not detected, as shown in Figure 4.

Figure 3. SEM image of the Bi2212 thick film prepared by screen-printing the Bi_2SrCaO_x precursor on a Cu plate and heat-treating at 820 ℃ for 1 min → at 870 ℃ for 50 sec → at 830 ℃ for 3 min in air.

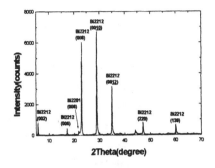

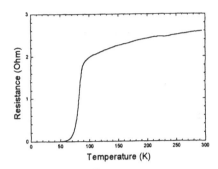

Figure 4. *XRD pattern of the same Bi2212 thick film as for Figure 3.*

Figure 5. *Resistance vs. temperature graph of the same Bi2212 thick film as for Figure 3.*

The main phase was the Bi2212 superconducting phase and the film was aligned to surface normal c-axis. In spite of the annealing at 830°C, a small amount of $Bi_2Sr_2CuO_x$ (Bi2201 : quenched liquid) still remained in thick films. Figure 5 shows a plot of resistance versus temperature of the specimen of Figure 3. Zero-resistance was obtained at 58 K, which is lower than the optimal for Bi2212. It indicates that a liquid phase is present at the grain boundary of Bi2212. Figure 6 shows the XRD patterns of the samples, after screen-printing the 211 precursor once, twice, and thrice respectively on copper plate and heat-treating at 870°C for 36 sec. In the one time screen-printed sample, the dominant phase was the Bi2212 and had its c-axis well-aligned perpendicular to the surface. In the samples screen-printed two and three times, Bi2212 phase was not detected and a small amount of Bi2201 and a large amount of CuO were detected.

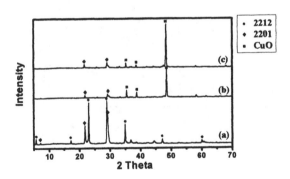

Figure 6. *XRD patterns for the specimen of Cu/Bi_2SrCaO_x calcination powder heat-treated at 870 °C for 36 sec in air : (a) 1 time printing, (b) 2 times printing, (c) 3 times printing*

414

Figure 7 shows a plot of the resistance versus temperature of the thick film prepared by screen-printing the $Bi_{1.5}SrCaO_x$ precursor on copper and annealing at 845 ℃ for 50 sec in air. Zero-resistance was obtained at 74 K. Figure 8 shows a plot of the resistance versus temperature of the thick film prepared by screen-printing the $Bi_{1.3}SrCaO_x$ precursor on copper and annealing at 855 ℃ for 50 sec in air. Zero-resistance temperature was 78 K, which is higher than that of the thick film that used the $Bi_{1.5}SrCaO_x$ precursor. The lower Bi composition of precursor powder resulted in the higher critical temperature (T_c). But the lower Bi composition generally cause the weaker bonding strength between copper and thick films and the poorer c-axis alignment.

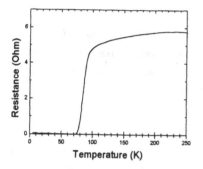

Figure 7. Electric resistance vs. temperature graph of $Cu/Bi_{1.5}SrCaO_x$ mixture powder heat-treated at 845 ℃ for 50 sec in air.

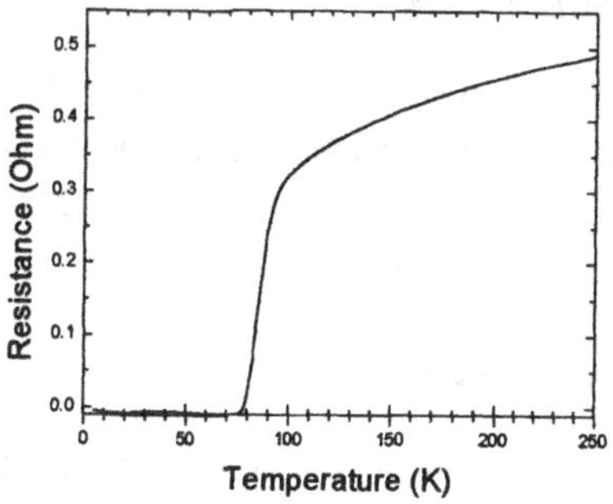

Figure 8. Electric resistance vs. temperature graph of $Cu/Bi_{1.3}SrCaO_x$ mixture powder heat-treated at 855 ℃ for 50 sec in air.

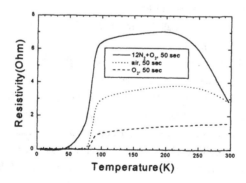

Figure 9. Electric resistance vs. temperature graph of $Cu/Bi_{1.3}SrCaO_x$ mixture powder heat-treated at 870 ℃ for 50 sec in $12N_2+O_2$(solid), air(dot), and O_2(dash), respectively.

Figure 9 shows a plot of the resistance versus temperature of the thick film prepared by screen-printing the $Bi_{1.3}SrCaO_x$ precursor on copper and annealing at 870 °C for 50 sec in $12N_2+O_2$ (solid), air (dot), and O_2 (dash), respectively. In the previous reports on the effect of the O_2 partial pressure [6,7], it was concluded that an increase of the O_2 partial pressure decreases the partial melting temperature of the Bi2212 phase. Also in this study, the lower O_2 partial pressure in annealing atmosphere caused the large amount of the liquid phase to remain in the heat-treated thick films as shown in Figure 8. Electric resistance versus temperature graph of the thick film annealed in N_2 showed an amorphous property.

CONCLUSIONS

Well oriented $Bi_2Sr_2CaCu_2O_8$(Bi2212) superconductor thick films were fabricated successfully on inexpensive copper tapes by liquid reaction between Cu-free precursors and Cu tapes. The use of Bi in excess of the Bi2212 stoichiometry was crucial for the formation of a superconducting Bi-Sr-Ca-Cu-O layer under the present heat treatment condition.

REFERENCES

1. A. P. Malozemoff, W. Carter, S. Fleshler, L. Fritzemeier, Q. Li, L. Masur, P. Miles, D. Parker, R. Parrella, E. Podtburg, G. N. Riley, M. Rupich, J. Scudiere, W. Zhang, "HTS wire at commercial performance levels", IEEE Transactions on Applied Superconductivity , 9, 2469 (1999).
2. T. Shibata, M. Watanabe, C. Suzawa, S. Isojima, J. Fujikami, K. Sato, H. Ishii, S. Honjo, Y. Iwata, "Development of high temperature superconducting power cable prototype system", IEEE Transactions on Power Delivery , 14, 182 (1999).
3. Gerry George, "Detroit Edison to Install Superconducting Cable," Transmission & Distribution World, 51, 40 (1999).
4. Y. Yamada, in Applied Physics series 6 : Bismuth-Based High-Temperature Superconductors, edited by H. Maeda and K. Togano, Marcel Dekker, NY, 289 (1996)
5. A. Goyal, D. P. Norton, J. D. Budai, M. Paranthaman, E. D. Specht, D. M. Kroeger, D. K. Christen, Q. He, B. Saffian, F. A. List, D. F. Lee, P. M. Martin, C. E. Klabunde, E. Hartfield, and V. K. Sikka, "High critical current density superconducting tapes by epitaxial deposition of $YBa_2Cu_3O_x$ thick films on biaxially textured metals", Applied Physics Letters, 69, 43 (1996)
6. J. L. MacManus-Driscoll, J. C. Bravman, R. J. Savoy, G. Gorman, and R. B. Beyers, J. Am. Ceram. Soc., 77, 2305 (1994).
7. W. Zhang and E. E. Hellstrom, Physica C, 218, 141 (1993).

GROWTH MORPHOLOGY AND ELECTRONIC STRUCTURE OF ULTRA-THIN TaO$_x$ FILMS ON Ag(100)

M. M. Howard*, C. A. Ventrice, Jr.*, H. Geisler**, D. A. Hite*** and P. T. Sprunger***
*Department of Physics, University of New Orleans, New Orleans, LA 70148
**Department of Chemistry, Xavier University, New Orleans, LA 70125
***CAMD/Department of Physics, Louisiana State University, Baton Rouge, LA 70806

ABSTRACT

A study of the growth morphology and electronic structure of TaO$_x$ films on the Ag(100) substrate has been performed to determine the properties of ultra-thin TaO$_x$ films without the influence of a mixed interfacial oxide (i.e., a disordered SiO$_2$/TaO$_x$ interface for growth on Si). The TaO$_x$ films were grown by thermal evaporation of Ta in an oxygen atmosphere of 1 x 10^{-6} Torr. Growth on a Ag(100) surface held at room temperature results in an amorphous TaO$_x$ overlayer, as determined by low energy electron diffraction. The onset of ordering of these films occurs for a post-anneal at ~500 °C. A diffraction pattern that corresponds to a multi-domain overlayer structure is observed for anneals at ~550 °C. Deposition of Ta without oxygen results in the formation of Ta islands. These results indicate that there is a very weak adsorbate-substrate interaction. Photoemission measurements of the TaO$_x$ films show the formation of a band gap with the valence band maximum residing at 3.5 eV below the Fermi level. Core level shifts of ~3.5 eV are observed for the Ta with no indication of metallic Ta at the surface.

INTRODUCTION

As device sizes in integrated circuits continue to shrink, high dielectric constant metal-oxides such as tantalum oxide are beginning to replace SiO$_2$ for use in the capacitive elements and field effect transistors of these circuits.[1-6] For a 1 Gbyte DRAM, it is estimated that feature sizes of 0.1 μm and dielectric thicknesses of < 5 nm are necessary. Within this thickness regime, direct tunneling through SiO$_2$ can occur, resulting in excessive dissipation in the circuit. The capacitance of a parallel plate capacitor is given by $c = k\varepsilon_0 A/d$, where k is the dielectric constant, ε_0 is the permittivity of free space, A is the lateral area of the capacitor, and d is the oxide thickness. Therefore, the same amount of charge can be stored with a lower operating voltage by increasing the dielectric constant of the oxide. By lowering the operating voltage, tunneling through the oxide is dramatically reduced. In addition, the lateral size of the capacitive elements can be reduced while maintaining the same capacitance by increasing the dielectric constant of the oxide, resulting in a much higher device density. Tantalum oxide films have shown the most promise for replacing SiO$_2$ films for these applications because of the compatibility with silicon process technology. The dielectric constant of tantalum pentoxide (Ta$_2$O$_5$), the most stable form of tantalum oxide, is $k \sim 25$ as compared to $k \sim 4$ for SiO$_2$.

One of the key issues in using tantalum oxide for silicon based integrated circuits is the leakage current density of the TaO$_x$ films. This leakage depends on the crystalline structure of the films and the nature of defects within the films.[3,7] The most common technique for growing TaO$_x$ films is by chemical vapor deposition (CVD). This results in conformal films over complex geometries; however, CVD grown films have a rather large defect density, primarily oxygen deficiencies and carbon impurities. Films deposited by sputter deposition of TaO$_x$ typically have a lower density of impurities but are non-conformal and have a high number of structural defects. The best control of crystalline quality is usually achieved by growing films

417

using molecular beam epitaxy (MBE) techniques (e.g., thermal evaporation of tantalum in an oxygen atmosphere). The main drawback of this technique is the formation of an interfacial SiO_2 layer.[6] This can be reduced by depositing tantalum without oxygen, followed by oxidation of the tantalum film.[5] However, this results in a film with a much higher number of structural defects. In fact, for all of these growth techniques, there will be some degree of SiO_2 formation at the interface.

Although tantalum oxide shows promise for replacing SiO_2 for many thin film applications, the basic electronic properties of the various crystallographic forms of tantalum oxide are not well understood. This is primarily due to the fact that high quality bulk crystals of tantalum oxide are not commercially available. In addition, the insulating nature of tantalum oxide would make it difficult to study bulk crystals using either electron or ion spectroscopies. One technique that has been used to study the properties of insulating metal oxides is to grow the metal oxide as a thin film on a noble-metal, single-crystal substrate.[8] The noble-metal substrate can act as a template to induce order in the metal oxide overlayer, without forming a mixed interfacial oxide. If the metal oxide film is thin enough (less than ~5 nm), any excess charge induced during ion or electron spectroscopies will tunnel into the metallic substrate. In this study, the growth morphology and electronic properties of tantalum oxide films on a Ag(100) substrate have been examined. Since silver does not oxidize at the oxygen pressures and temperatures used for the growth of the TaO_x films, the effect of the ordered Ag(100) substrate on the crystal structure of the TaO_x films can be determined directly.

There are three stable crystal forms for tantalum oxide: TaO, TaO_2, and Ta_2O_5. The most stable form of tantalum oxide is generally considered to be Ta_2O_5. It crystallizes in the vanadium oxide (V_2O_5) structure. This forms an orthorhombic lattice with lattice parameters of $a_0 = 6.2$ Å, $b_0 = 3.7$ Å, $c_0 = 3.9$ Å.[9] The crystal structure of TaO_2 is the rutile structure. This forms a tetragonal lattice with lattice parameters of $a_0 = 4.7$ Å and $c_0 = 3.1$ Å.[10] The crystal structure of TaO is the rocksalt structure, which is a face centered cubic (fcc) lattice with a two atom basis. Its lattice parameter is $a_0 = 4.4$ Å.[10]. Either TaO or TaO_2 can be formed when growing tantalum oxide under oxygen deficient conditions. The primary goal of this study is to examine the properties of Ta_2O_5. This proves somewhat problematic when trying to find a lattice matched noble-metal substrate since there are no noble metals with an orthorhombic symmetry. Since silver is a fcc crystal with a lattice parameter of $a_0 = 4.1$ Å, the closest match of Ta_2O_5 with Ag(100) would be with the b-c plane of the Ta_2O_5 that is rotated by 45°. This would result in a tensile strain of the overlayer of ~7%.

EXPERIMENT

These experiments were performed at the CAMD synchrotron in Baton Rouge, Louisiana. The growth and characterization of the films were performed in an ultra-high vacuum (UHV) chamber with a base pressure of 2×10^{-10} Torr. The TaO_x films were grown by thermal evaporation of Ta in an oxygen atmosphere of 1×10^{-6} Torr. The Ta was evaporated from a 0.5 mm Ta filament connected to two high current feedthrus and shielded with Ta foil. A shutter connected to a rotary motion feedthru was used to control deposition times. After outgassing of the Ta source, the base pressure of the system remained in the high 10^{-10} Torr range during deposition of the Ta. The deposition rates of the Ta were monitored with a quartz crystal microbalance. It is estimated that the film thicknesses are accurate to ±20%. The dosing rate for the Ta was kept at ~2 Å/min.

The clean Ag(100) surface was prepared by sputtering the crystal with Ne at an energy of 1 keV, followed by annealing for ~15 min at ~500 °C in UHV. The crystal was mounted on an X,Y,Z manipulator that allows two rotational axes for the crystal. The structure of the clean

Ag(100) surface and the Ta and TaO_x films was monitored by low energy electron diffraction (LEED). The electronic structure of the films was monitored by angle-resolved ultra-violet photoelectron spectroscopy (UPS). The UHV system incorporates a hemispherical analyzer mounted on a two-axis goniometer. The measurements were performed at the plane grating monochromator (PGM) beamline. The overall energy resolution of the system is estimated to be ~0.5 eV for the measurements presented in this study.

RESULTS

The LEED pattern of the clean Ag(100) surface exhibited sharp LEED spots with a square symmetry, which indicates that the surface is well ordered and not reconstructed. Growth of TaO_x on a Ag(100) substrate held at room temperature (RT) results in an attenuation of the Ag(100) LEED spots and an increase of the diffuse background. After ~5 Å of Ta has been deposited in an oxygen atmosphere, all of the LEED spots disappear, leaving only a diffuse background. This indicates that the TaO_x forms a uniform disordered overlayer for growth on the Ag(100) substrate held at RT. In order to induce order in the TaO_x films, the films were annealed at various temperatures while monitoring the LEED. The onset of order was observed at ~500 °C. At this temperature, broad LEED spots appear. At ~550°C, a sharp 12 spot LEED pattern was observed. However, there still remained a large amount of diffuse background. This indicates the formation of a multi-domain overlayer structure with a great deal of disorder between crystallites. The highest temperature that the sample holder could achieve was ~550 °C. Annealing to higher temperatures would presumably reduce the diffuse background further by

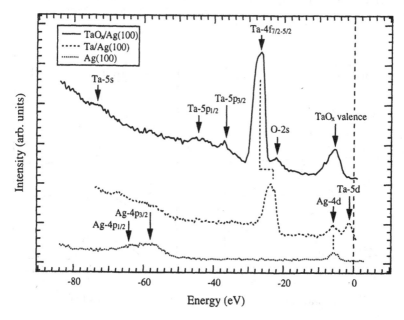

FIG. 1. Normal emission UPS spectra of the clean Ag(100) surface, ~10 Å Ta deposited on the Ag(100) surface at RT, and ~10 Å Ta deposited in an oxygen atmosphere of 1×10^{-6} Torr at RT. All spectra were taken with a photon energy of 140 eV.

coalescing the crystallites. At this time the crystal structure of the crystallites from the LEED pattern observed after the anneal to ~550 °C has not been determined.

Growth of Ta without a background pressure of oxygen resulted in an increase in the diffuse background of the LEED pattern, but the original square symmetry of the LEED remained through several monolayers of Ta deposition. In order to better understand these LEED results, overview UPS spectra were taken at a photon energy of 140 eV for Ag(100), Ta/Ag(100), and TaO_x/Ag(100) as shown in Fig. 1. For the clean Ag(100) spectrum, the emission from the filled Ag-4d band is observed from 4eV to 8 eV below the Fermi level, and a broad feature between 54 eV and 66 eV that corresponds to the Ag-4p emission is also observed. The spectrum for a 10 Å Ta film shows emission features of the Ta-5d that cross the Fermi level and the 4f emission centered at ~24 eV below the Fermi level. In addition to the Ta features, the Ag-4d emission can clearly be seen. Since 10 Å of Ta corresponds to ~4 ML, the observation of Ag features indicates that the Ta is forming islands on the surface, which leaves bare regions of Ag. Otherwise, no Ag features should be observed since the mean free path for electrons with these kinetic energies should be less than 10 Å. Another possible explanation for the observation of Ag emission features is that the Ta is alloying with the Ag. However, Ag and Ta do not form bulk alloys.[11] The spectrum of TaO_x grown by deposition of 10 Å of Ta in an oxygen atmosphere of 1 x 10^{-6} Torr at RT shows no Ag emission features. This provides evidence that the TaO_x is forming a uniform, disordered overlayer at RT. Apparently, the presence of oxygen at the surface is sufficient to reduce the mobility of the Ta at RT so that islands do not form. No emission is observed at the Fermi level, which indicates that there is no unoxidized Ta. In addition, there is a ~3.5 eV shift in the Ta core emissions. The full width at half maximum of the Ta-4f emission is approximately the same before and after oxidation. Therefore, the most probable stoichiometry of the overlayer is a fully oxidized Ta_2O_5. Otherwise, significant

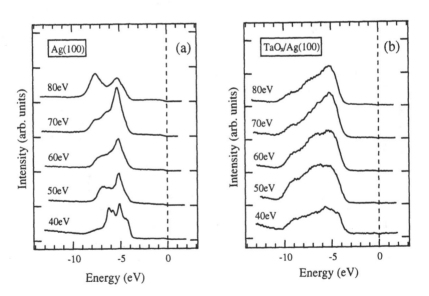

FIG 2. Normal emission UPS spectra taken at photon energies ranging from 40 eV to 80 eV for (a) the clean Ag(100) surface and (b) ~10 Å of Ta deposited in an oxygen atmosphere of 1 x 10^{-6} Torr at RT followed by an anneal to ~550 °C.

broadening of the Ta-4f emission would occur due to a mixed oxidation state of the Ta.

In order to determine the electronic structure of the films, UPS spectra were taken at normal emission as a function of incident photon energy as shown in Fig. 2. Since $k_\parallel$ is zero (normal emission), the bulk dispersion can be determined. As seen in Fig. 2(a), the emission from the 4d band of the Ag exhibits quite a bit of structure between 4 eV and 8 eV. These states disperse as a function of increasing photon energy, which is expected for a well ordered three-dimensional crystal. UPS spectra for a TaO_x film that was grown by deposition of 10 Å of Ta in an oxygen atmosphere at RT followed by an anneal at ~550 °C is show in Fig. 2(b). The onset of the valence band emission occurs at 3.5 eV below the Fermi level. Previous measurements have determined that the band gap of Ta_2O_5 is ~5 eV.[12,13] Therefore, the conduction band minimum is estimated to be ~1.5 eV above the Fermi level. Very little structure is resolved within the valence band emission, and these states are not observed to disperse. These results are expected for a multi-domain overlayer structure with considerable disorder.

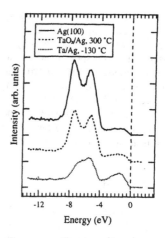

FIG 3. Normal emission UPS spectra taken at a photon energy of 80 eV for the clean Ag(100) surface, ~10 Å Ta deposited in an oxygen atmosphere at ~300 °C, and ~15 Å Ta deposited without oxygen at -130 °C.

Since the highest temperature that the heating system for the sample holder could achieve was ~550 °C, an attempt was made to create a better ordered TaO_x film by depositing the Ta in an oxygen atmosphere onto the Ag(100) crystal held at an elevated temperature. At ~300 °C, this resulted in only a slight increase in the diffuse background of the LEED with almost no attenuation of the Ag(100) LEED spots. The UPS spectrum for this overlayer taken at a photon energy of 80 eV is shown in Fig 3. As compared to the clean Ag(100) spectrum, there is only a slight attenuation of the Ag 4d emission. This indicates that the mobility of the Ta has increased sufficiently to cause the Ta to island at this temperature, even with the presence of oxygen at the surface. In comparison, deposition of ~15 Å of Ta without oxygen onto the Ag(100) crystal held at -130 °C is shown in Fig. 3. The Ta 5d valence emissions are observed just below the Fermi level, and the attenuated emissions from the Ag 4d band are observed between 4 eV and 7 eV. This indicates that the Ta is mobile enough at this low temperature to form islands.

CONCLUSIONS

The high mobility of the Ta atoms on the Ag(100) substrate indicates that there is a very low adsorbate-substrate interaction. The presence of oxygen in the system reduces the mobility of the Ta sufficiently to result in the formation of a uniform TaO_x overlayer at RT. However, the interaction of the TaO_x and the Ag(100) is rather weak and does not play much of a role in determining the temperature at which the TaO_x begins to order. Since the TaO_x films grown on silver should be free of contamination from the substrate, 500 °C can be taken as an estimate of the lower limit for the onset of crystallization of TaO_x films. For instance, TaO_x films grown by CVD processes on Si(100) are contaminated to some degree with carbon and silicon and begin to crystallize at a temperatures around 700 °C [2,3]. On the other hand, a previous study of the growth of TaO_x on Si(100) using MBE techniques showed that the overlayer began to order at

520 °C [6]. Secondary ion mass spectroscopy (SIMS) measurements of these films give evidence for an abrupt TaO_x/SiO_2 interface with no diffusion of Si into the tantalum oxide film.

For the deposition rate and oxygen pressure used in this study, the TaO_x films were fully oxidized. The UPS spectra show a clear band gap, and a core level shift of ~3.5 eV is observed for the Ta emissions with no broadening after oxidation. These results give evidence that the Ta_2O_5 phase of tantalum oxide is being formed. The onset of valence emission observed in the UPS spectra of the TaO_x films is 3.5 eV below the Fermi level. However, the position of the Fermi level may be sensitive to the defect density of the TaO_x films. Since the position of the Fermi level as a function of the order of the TaO_x films was not measured in this study, a direct comparison of the position of the Fermi level for films grown on Ag and Si has not been attempted.

ACKNOWLEDGMENTS

This work was supported in part by Louisiana Board of Regents Support Fund grant number LEQSF(1999-2002)-RD-A-46. The research was also supported by U. S. DOE contract number DE-FG02-98ER45712 and NSF contract number DMR-9705406. In addition, M. M. Howard would like to thank the Louisiana Board of Regents Support Fund for his support through the superior graduate student fellowship program (LEQSF(1997-2000)-GF-20).

REFERENCES

[1] X. M. Wu, P. K. Wu, T.-M. Lu, E. J. Rymaszewski, Appl. Phys. Lett. **62**, 3264 (1993).

[2] D. Laviale, J. C. Oberlin, and R. A. B. Devine, Appl. Phys. Lett. **65**, 2021 (1994).

[3] H. Kimura, J. Mizuki, S. Kamiyama, and H. Suzuki, Appl. Phys. Lett. **66**, 2209 (1995).

[4] K. Chen, G. R. Yang, M. Nielsen, T.-M. Lu, and E. J. Rymasewski, Appl. Phys. Lett. **70**, 399 (1997).

[5] E. Atanassova and D. Spassov, Appl. Surf. Sci. **135**, 71 (1998).

[6] J. V. Grahn, P.-E. Hellberg, and E. Olsson, J. of Appl. Phys. **84**, 1632 (1998).

[7] G. S. Oehlein, F. M. d'Heurle, and A. Reisman, J. Appl. Phys. **55**, 3715 (1984).

[8] C. A. Ventrice, Jr. and H. Geisler, in *Thin Films: Heteroepitaxial Systems*, edited by W. K. Liu and M. B. Santos (World Scientific Publishers, Singapore, 1999), pp. 167-210.

[9] G. V. Samsonov, *The Oxide Handbook*, IFI/Plenum, New York, 1973.

[10] R. W. G. Wyckoff, *Crystal Structures*, 2nd edition, Interscience, New York, 1969.

[11] M. Hansen, *Constitution of Binary Alloys*, McGraw-Hill, New York, 1958.

[12] S. C. Khanin and A. L. Ivanovskii, Phys. Stat. Sol. B **174**, 449 (1992).

[13] P. A. Murawala *et al.*, Jpn. J. Appl. Phys. **32**, 368 (1993).

MICRO PATTERN OF TiO₂ THIN FILM FORMATION BY DIRECT SYNTHESIS FROM AQUEOUS SOLUTION AND TRANSCRIPTION OF RESIST PATTERN

Takeshi YAO , Yoshiharu UCHIMOTO and Hiroki YABE
Department of Fundamental Energy Science,
Graduate School of Energy Science, Kyoto University,
Yoshida, Sakyo-ku, Kyoto-shi, 606-8501, JAPAN, yao@scl.kyoto-u.ac.jp

ABSTRACT

Direct synthesis from aqueous solution (DSAS) is the method for synthesizing ceramic oxides in crystalline state directly from aqueous solution at ordinary temperature and ordinary pressure. DSAS is advantageous because of the applicability to making films with wide areas and/or complicated shapes with no requirement of vacuum or high temperature. Micro patterning of materials having high dielectric constant is important for manufacturing electronic devices such as dynamic random access memory, ferroelectric random access memory and so on. By using DSAS, TiO₂ thin film was formed on the glass plate surface where micro pattern had been printed by commercial organic photoresist material. Then the resist material was dissolved off by acetone or ethanol with the TiO₂ thin film just on. From SEM observation, it is indicated that micro pattern of TiO₂ thin film transcribing the resist pattern with minimum line width of 1 μm was obtained. This method will be applicable to the construction of highly integrated dielectric devices.

INTRODUCTION

The downsizing of the electronic devises are actively developed in order to increase the accumulation. Because the electric capacity of silicon is going to be short with the downsizing, the usage of ceramics having high dielectric constants for electric devises, such as dynamic random access memory (DRAM), ferroelectric random access memory (FeRAM) and so on, is investigated eagerly. Micro patterning of the ceramics is important for the production of the electronic devices. Plasma etching is used for forming micro pattern of the ceramics, however, expensive vacuum equipment is required and the production efficiency is restricted.

Direct synthesis from aqueous solution (DSAS) is the method for synthesizing ceramic oxides in crystalline state directly from aqueous solution at ordinary temperature and ordinary pressure[1,2]. DSAS is advantageous because of the applicability to making films with wide areas and/or complicated shapes with no requirement of vacuum or high temperature, and because of lower cost. In this study, we have formed micro pattern of TiO₂ thin film by using DSAS and transcription of the resist pattern.

EXPERIMENT

We have considered that the chemical equilibrium between hexafluorotitanate ion and titanium oxide will hold as in reaction(1),

$$TiF_6^{2-} + 2H_2O \rightleftarrows TiO_2 + 6F^- + 4H^+ \qquad (1)$$

and that, when boric acid is added, the fluoride ion is consumed by reaction (2)

$$BO_3^{3-} + 4F^- + 6H^+ \rightarrow BF_4^- + 3H_2O \qquad (2)$$

then the chemical equilibrium in reaction (1) is shifted from left to right in order to increase the amount of fluoride ion, resulting in the formation of titanium oxide.

We prepared a glass plate on which surface micro pattern with minimum line width of 1 μm had been printed by commercial organic photoresist material (AZ1500, Hekist Corp.). Ammo-

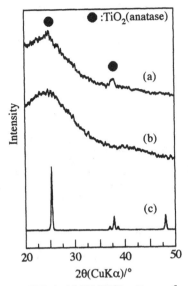

FIG. 1 (a) TF-XRD patterns of the glass plate after the first soak. (b) TF-XRD patterns of the glass plate not soaked. (c) Powder XRD pattern of reagent TiO$_2$ (anatase).

nium hexafluorotitanate : $(NH_4)_2TiF_6$ was dissolved in distilled water. The glass plate was soaked in the solution as the substrate. We added boric acid to the solution. After the soaking, the substrate was washed with distilled water and dried at room temperature. No heat treatment was conducted. We name this soaking operation as the first soak. Then the glass plate was soaked in acetone or ethanol for 30 to 60 min. with ultrasonic vibration, which we name as the second soak. The surface of the glass plate was analyzed by a thin-film X-ray diffraction [TF-XRD] (RINT-2500, Rigaku Co., Tokyo, Japan). Gold film was coated on the surface of the substrate, and scanning electron microscopic images were observed with a scanning electron microscope [SEM] (ESEM-2700, Nikon Co., Tokyo, Japan).

RESULTS and DISCUSSION

Characteristic peaks for anatase TiO$_2$ were observed in TF-XRD patterns for the glass plate after the first soak (Fig. 1). The intensity of the 004 reflection was the strongest. The orientation of (001) plane is indicated. In Fig. 2, SEM micrograph of the surface of the glass plate is given. The glass plate was coated with dense and homogeneous TiO$_2$ thin film with around 0.2 μm thickness. This film was as hard as 6H to 7H pencils and attached to the glass surface strongly. In Fig. 3, SEM micrograph of the glass plate after the second soak is given. The resist material was dissolved off with TiO$_2$ film just on and the micro pattern of TiO$_2$ thin film transcribing the resist pattern was obtained. Various micro pattern of TiO$_2$ thin film are shown in Fig. 4. The minimum line width of the pattern was 1 μm. In Fig. 5, 60^d tilted SEM view of TiO$_2$ thin film was given. It is confirmed that the boundaries are clearly outlined.

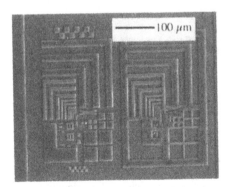

FIG. 2 SEM micrograph of the surface of the glass plate after the first soak.

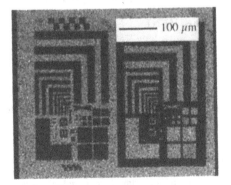

FIG. 3 SEM micrograph of the surface of the glass plate after the second soak.

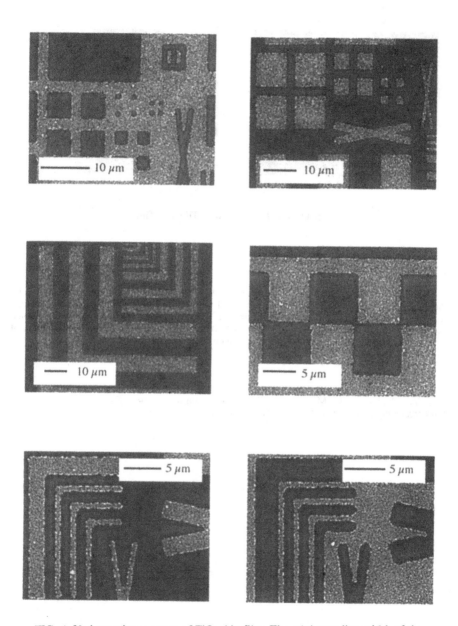

FIG. 4 Various micro pattern of TiO$_2$ thin film. The minimum line width of the pattern was 1 μm.

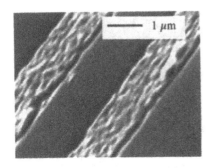

FIG. 5 60° tilted SEM view of TiO$_2$ thin film.

CONCLUSION

TiO$_2$ thin film was formed on the glass plate with micro pattern printed by resist material by using DSAS. The resist material was dissolved off with TiO$_2$ film just on by soaking in acetone or ethanol and the micro pattern of TiO$_2$ thin film transcribing the resist pattern was obtained with the minimum line width of 1 μm. This method is promising for producing micro pattern of ceramics having high dielectric constants for electric devises.

REFERENCES

1. T.Yao, J. Mater. Res., **13**, 1091 (1998)
2. T. Yao, Y. Uchimoto, M. Inobe, K. Satoh, and S. Omi, *The 1999 Electrochemical Society Joint International Meeting Abstracts* No.1399 (1999)

AUTHOR INDEX

428

SUBJECT INDEX

ab initio calculations, 65
Ag(100), 417
AlGaAs oxidation, 407
α-Fe$_2$O$_3$-FeTiO$_3$, 191
amorphous, 291
anatase TiO$_2$, 43
aqueous, 149
a-site, 25
atomic layer epitaxy, 329

Ba$_2$Bi$_4$Ti$_5$O$_{18}$, 173
barium hexaferrite, 137
(Ba,Sr)TiO$_3$, 119, 161
Ba(Ti,Zr)O$_3$, 179
Bi2212 superconductor, 411
bias-dependent tunneling, 51
bismuth
    layer structure, 377
    tungstate, 377

cadmium, 317
carbonate fuel cell cathode, 395
carrier induced ferromagnetism, 65
channel, 25
charge transport measurements, 199
co-doping, 65, 223
complex sol-gel process, 395
conductivity, 291
copper aluminum oxide, 271
crystal growth of Li$_x$Ni$_{1-x}$O, 57
crystallization, 245
Cu
    -free precursors, 411
    tape, 411
cuprate, 3

deep levels, 97
delafossite, 235
devices, 383
dielectric, 383
diffusitives, 289
dipping-pyrolysis process, 401
direct synthesis from aqueous solution, 423

effective mass, 259
electrical
    properties, 383
    tunability, 109
electrochemistry, 289
electrode, 131
electron energy loss spectroscopy, 271
electronic
    device, 423

    structures, 223
electrostatic shutter, 365
epitaxial, 25, 131, 143, 347
    film, 89, 191
    growth, 253
    thin film, 377
exciton, 353

Faraday rotation, 89
ferrimagnetic, 137
ferroelectric(s), 125, 131, 143, 149, 341, 377
    films, 31, 109, 155
    memory, 155
    thin film, 167, 173
flatband potential, 43

grain
    boundaries, 173
    size, 119

half-metallic oxide, 191
heteroepitaxy, 137
heterostructure, 143, 341
high-resolution electron microscopy, 271
high-T$_c$ superconductor, 329
hydrothermal, 179
hyperthermal surface ionization, 37

indium
    oxide, 245
    tin oxide, 253, 277
infrared spectroscopy, 271
In$_2$O$_3$ based ternary compound, 211
interface stability, 51
intrinsic low permittivity layer, 97
I:O$_2$, 31

junction magneto-resistance, 51

Kelvin probe, 37

(100)LaAlO$_3$, 347
lanthanum nickel oxide, 389
laser
    ablation, 277, 341
    MBE, 329
lattice
    defects, 173
    engineering, 329
    parameters, 161
layered oxide, 167
Li-Ni mixed layers in Li$_x$Ni$_{1-x}$O, 57
liquid reaction, 417

431

valance control. 65
volatile, 317

wide band-gap semiconductors, 223
work function, 37

Yb/23 film, 401

YSZ, 253

zinc, 317
   oxide, 259, 329
ZnO, 223, 283, 353, 359

Printed in the United States
By Bookmasters